Chi-Huey Wong (Ed.)

**Carbohydrate-based Drug Discovery
Volume 1**

Further Titles of Interest

K. C. Nicolaou, R. Hanko, W. Hartwig (Eds.)

Handbook of Combinatorial Chemistry

2002
ISBN 3-527-30509-2

M. Demeunynck, C. Bailly, W. D. Wilson (Eds.)

DNA and RNA Binders – From Small Molecules to Drugs

2002
ISBN 3-527-30595-5

B. Ernst, G. W. Hart, P. Sinaÿ (Eds.)

Carbohydrates in Chemistry and Biology

2000
ISBN 3-527-29511-9

T. K. Lindhorst

Essentials of Carbohydrate Chemistry and Biochemistry (2nd Ed.)

2002
ISBN 3-527-30664-1

N. Sewald, H.-D. Jakubke

Peptides: Chemistry and Biology

2002
ISBN 3-527-30405-3

Chi-Huey Wong (Ed.)

Carbohydrate-based Drug Discovery
Volume 1

WILEY-VCH GmbH & Co. KGaA

Prof. Dr. Chi-Huey Wong
Ernest W. Hahn Chair in Chemistry
The Scripps Research Institute
10550 N. Torrey Pines Road
La Jolla, CA 92037
USA
wong@scripps.edu

■ This book was carefully produced. Nevertheless, authors, editor and publisher do not warrant the information contained therein to be free of errors. Readers are advised to keep in mind that statements, data, illustrations, procedural details or other items may inadvertently be inaccurate.

Library of Congress Card No.: applied for

British Library Cataloguing-in-Publication Data
A catalogue record for this book is available from the British Library.

Bibliographic information published by Die Deutsche Bibliothek
Die Deutsche Bibliothek lists this publication in the Deutsche Nationalbibliografie; detailed bibliographic data is available in the Internet at <http://dnb.ddb.de>

© 2003 WILEY-VCH Verlag GmbH & Co. KGaA, Weinheim

All rights reserved (including those of translation in other languages). No part of this book may be reproduced in any form – by photoprinting, microfilm, or any other means – nor transmitted or translated into machine language without written permission from the publishers. Registered names, trademarks, etc. used in this book, even when not specifically marked as such, are not to be considered unprotected by law.

Printed in the Federal Republic of Germany
Printed on acid-free paper

Typesetting K+V Fotosatz GmbH, Beerfelden
Printing Strauss Offsetdruck GmbH, Mörlenbach
Bookbinding J. Schäffer GmbH & Co. KG, Grünstadt

ISBN 3-527-30632-3

Contents

Volume 1

Preface XXV

List of Contributors XXVII

1	**Synthetic Methodologies** 1	
	Chikako Saotome and Osamu Kanie	
1.1	Introduction 1	
1.2	Tactical Analysis for Overall Synthetic Efficiency 1	
1.3	Methodological Improvements 2	
1.3.1	Chemistry 3	
1.3.2	Protecting Group Manipulations 4	
1.3.3	Modulation of the Reactivity of Glycosyl Donors 6	
1.3.4	Block Synthesis 8	
1.4	Accessibility 11	
1.4.1	Solution-based Chemistry 11	
1.4.2	One-Pot Glycosylation 13	
1.4.3	Solid-Phase Chemistry 16	
1.4.3.1	Fundamentals of Solid-Phase Oligosaccharide Synthesis 16	
1.4.3.2	The Support 16	
1.4.3.3	Linkers to the Support 20	
1.4.3.4	Protecting Groups used in Solid-Phase Oligosaccharide Synthesis 20	
1.4.3.5	Solid-Phase Oligosaccharide Synthesis 20	
1.4.3.6	Monitoring of Reaction Progress 26	
1.4.4	Automation 29	
1.5	Concluding Remarks 32	
1.6	References 33	
2	**Complex Carbohydrate Synthesis** 37	
	Makoto Kiso, Hideharu Ishida, and Hiromune Ando	
2.1	Introduction 37	
2.2	Synthetic Gangliosides 38	
2.2.1	Gangliosides GM4 and GM3, and their Analogues and Derivatives 38	

2.2.2	Sialylparagloboside (SPG) Analogues and Derivatives	40
2.2.3	Selectin Ligands	43
2.2.3.1	Sialyl Lewis x	44
2.2.3.2	Novel 6-Sulfo sLe^x Variants	45
2.2.4	Siglec ligands	46
2.2.4.1	Chol-1 (a-Series) Gangliosides	47
2.2.4.2	Novel Sulfated Gangliosides	50
2.3	Toxin Receptor	50
2.4	Summary and Perspectives	52
2.5	References	52

3 The Chemistry of Sialic Acid 55
Geert-Jan Boons and Alexei V. Demchenko

3.1	Introduction	55
3.2	Chemical and Enzymatic Synthesis of Sialic Acids	56
3.3	Chemical Glycosidation of Sialic Acids	59
3.3.1	Direct Chemical Sialylations	60
3.3.1.1	2-Chloro Derivatives as Glycosyl Donors	61
3.3.1.2	2-Thio Derivatives as Glycosyl Donors	62
3.3.1.3	2-Xanthates as Glycosyl Donors	69
3.3.1.3	2-Phosphites as Glycosyl Donors	71
3.3.1.4	Miscellaneous Direct Chemical Methods	71
3.3.2	Indirect Chemical Methods with the Use of a Participating Auxiliary at C-3	73
3.3.2.1	3-Bromo- and other 3-O-Auxiliaries	73
3.3.2.2	3-Thio and 3-Seleno Auxiliaries	74
3.3.3	Synthesis of (2→8)-Linked Sialosides	77
3.4	Enzymatic Glycosidations of Sialic Acids	83
3.4.1	Sialyltransferases	84
3.4.1.1	Metabolic Engineering of the Sialic Acid Biosynthetic Pathway	90
3.4.2	Sialidases	90
3.5	Synthesis of C- and S-Glycosides of Sialic Acid	91
3.6	Modifications at N-5	94
3.7	References	95

4 Solid-Phase Oligosaccharide Synthesis 103
Peter H. Seeberger

4.1	Introduction	103
4.2	Pioneering Efforts in Solid-Phase Oligosaccharide Synthesis	104
4.3	Synthetic Strategies	105
4.3.1	Immobilization of the Glycosyl Acceptor	106
4.3.2	Immobilization of the Glycosyl Donor	106
4.3.3	Bi-directional Strategy	107
4.4	Support Materials	107
4.4.1	Insoluble Supports	107

4.4.2	Soluble Supports 108	
4.5	Linkers 108	
4.5.1	Silyl Ethers 108	
4.5.2	Acid- and Base-Labile Linkers 109	
4.5.3	Thioglycoside Linkers 110	
4.5.4	Linkers Cleaved by Oxidation 110	
4.5.5	Photocleavable Linkers 111	
4.5.6	Linkers Cleaved by Olefin Metathesis 111	
4.6	Synthesis of Oligosaccharides on Solid Support by Use of Different Glycosylating Agents 112	
4.6.1	1,2-Anhydrosugars – The Glycal Assembly Approach 112	
4.6.2	Glycosyl Sulfoxides 113	
4.6.3	Glycosyl Trichloroacetimidates 114	
4.6.4	Thioglycosides 115	
4.6.5	Glycosyl Fluorides 118	
4.6.6	n-Pentenyl Glycosides 118	
4.6.7	Glycosyl Phosphates 118	
4.7	Automated Solid-Phase Oligosaccharide Synthesis 118	
4.7.1	Fundamental Considerations 119	
4.7.2	Automated Synthesis with Glycosyl Trichloroacetimidates 121	
4.7.3	Automated Synthesis with Glycosyl Phosphates 121	
4.7.4	Automated Oligosaccharide Synthesis by Use of Different Glycosylating Agents 121	
4.7.5	"Cap-Tags" to Suppress Deletion Sequences 123	
4.7.6	Current State of the Art of Automated Synthesis 123	
4.8	Conclusion and Outlook 124	
4.9	References 125	

5   Solution and Polymer-Supported Synthesis of Carbohydrates   129
     Shin-Ichiro Nishimura
5.1   Introduction   129
5.2   Mimicking Glycoprotein Biosynthetic Systems   130
5.3   References   136

6   Enzymatic Synthesis of Oligosaccharides
     Jianbo Zhang, Jun Shao, Prezemk Kowal, and Peng George Wang
6.1   Introduction   137
6.2   Sugar Nucleotide Biosynthetic Pathways   140
6.2.1   Basic Principle   140
6.2.2   Regeneration Systems for nine Common Sugar Nucleotides   142
6.2.2.1   Regeneration Systems for UDP-Gal, UDP-Glc, UDP-GlcA and UDP-Xyl   142
6.2.2.2   Regeneration Systems for UDP-GlcNAc and UDP-GalNAc   144
6.2.2.3   Regeneration Systems for GDP-Man and GDP-Fuc   147
6.2.2.4   CMP-Neu5Ac Regeneration   149

6.2.3	Novel Energy Source in Sugar Nucleotide Regeneration	*150*
6.3	Enzymatic Oligosaccharide Synthesis Processes	*151*
6.3.1	Cell-Free Oligosaccharide Synthesis	*151*
6.3.1.1	Immobilized Glycosyltransferases and Water-Soluble Glycopolymer	*152*
6.3.1.2	"Superbeads"	*154*
6.3.2	Large-Scale Syntheses of Oligosaccharides with Whole Cells	*156*
6.3.2.1	Kyowa Hakko's Technology	*157*
6.3.2.2	Wang's "Superbug"	*157*
6.3.2.3	Other Whole Cell-Based Technologies	*161*
6.4	Future Directions	*162*
6.5	References	*162*

7	**Glycopeptides and Glycoproteins: Synthetic Chemistry and Biology**	**169**
	Oliver Seitz	
7.1	Introduction	*169*
7.2	The Glycosidic Linkage	*169*
7.3	The Challenges of Glycopeptide Synthesis	*171*
7.4	Synthesis of Preformed Glycosyl Amino Acids	*173*
7.4.1	*N*-Glycosides	*173*
7.4.2	*O*-Glycosides	*176*
7.4.2.1	*O*-Glycosyl Amino Acids bearing Mono- or Disaccharides	*176*
7.4.2.2	*O*-Glycosyl Amino Acids bearing Complex Carbohydrates	*179*
7.5	Synthesis of Glycopeptides	*181*
7.5.1	*N*-Glycopeptide Synthesis in Solution	*181*
7.5.2	*O*-Glycopeptide Synthesis in Solution	*185*
7.5.3	Solid-Phase Synthesis of *N*-Glycopeptides	*188*
7.5.4	Solid-Phase Synthesis of *O*-Glycopeptides	*192*
7.6	Biological and Biophysical Studies	*200*
7.6.1	Conformations of Glycopeptides	*200*
7.6.2	Glycopeptides as Substrates of Enzymes and Receptors	*203*
7.6.3	Glycopeptides and Cancer Immunotherapy	*204*
7.6.4	Glycopeptides and T Cell Recognition	*206*
7.7	Summary and Outlook	*208*
7.8	References	*209*

8	**Synthesis of Complex Carbohydrates: Everninomicin 13,384-1**	**215**
	K. C. Nicolaou, Helen J. Mitchell, and Scott A. Snyder	
8.1	Introduction	*215*
8.2	Retrosynthetic Analysis and Strategy	*218*
8.2.1	Overview of Synthetic Strategies and Methodologies	*218*
8.2.2	Retrosynthetic Analysis: Overall Approach	*222*
8.3	Total Synthesis of Everninomicin 13,384-1 (1)	*223*
8.3.1	Approaches Towards the $A_1B(A)C$ Fragment	*223*
8.3.1.1	Initial Model Studies	*223*
8.3.1.2	Construction of the Building Blocks	*225*

8.3.1.3	Assembly and Completion of the $A_1B(A)C$ Fragment 229
8.3.2	Construction of the $FGHA_2$ Fragment 231
8.3.2.1	First Generation Approach to the $FGHA_2$ Fragment 231
8.3.2.2	Second Generation Strategy Towards the $FGHA_2$ Fragment 232
8.3.2.3	Assembly of the $FGHA_2$ Fragment 235
8.3.3	Construction of the DE Disaccharide 241
8.3.3.1	Retrosynthetic Analysis and Construction of Building Blocks for the DE Fragment 241
8.3.3.2	Assembly of the DE Fragment 243
8.3.3.3	Test of Strategies 244
8.3.4	Assembly of the $DEFGHA_2$ Fragment 245
8.3.5	Completion of the Total Synthesis of Everninomicin 13,384-1 247
8.4	Conclusion 249
8.5	References 250

9 Chemical Synthesis of Asparagine-Linked Glycoprotein Oligosaccharides: Recent Examples 253
Yukishige Ito and Ichiro Matsuo

9.1	Introduction 253
9.2	Synthesis of Asn-Linked Oligosaccharides: Basic Principles 257
9.3	Chemical Synthesis of Complex Oligosaccharides 261
9.3.1	Classical Examples 261
9.3.2	Trichloroacetimidate Approach to Complex-Type Glycan Chains 265
9.3.3	n-Pentenyl Glycosides as Glycosyl Donors 265
9.3.4	Glycal Approach to Complex Oligosaccharides 267
9.3.5	Intramolecular Aglycon Delivery Approach 269
9.3.6	New Protecting Group Strategy 273
9.3.7	Linear Synthesis of Branched Oligosaccharide 274
9.3.8	Chemoenzymatic Approach to Complex-type Glycans 275
9.4	References 278

10 Chemistry and Biochemistry of Asparagine-Linked Protein Glycosylation 281
Barbara Imperiali and Vincent W.-F. Tai

10.1	Protein Glycosylation 281
10.1.1	Introduction 281
10.1.2	Asparagine-Linked Glycosylation and Oligosaccharyl Transferase 281
10.2	Small-Molecule Probes of the Biochemistry of Oligosaccharyl Transferase 283
10.2.1	Photoaffinity and Affinity Labeling of Oligosaccharyl Transferase 284
10.2.2	Investigation of Peptide-Based Substrate Analogues as Inhibitors of Oligosaccharyl Transferase 287
10.2.2.1	Inhibitors of N-Linked Glycosylation and Glycoprotein Processing 287
10.2.2.2	Peptide-Based Analogues and Inhibitors 288
10.2.2.3	Interim Summary 292

10.2.3	Investigation of Carbohydrate-Based Substrate Analogues as Probes of Oligosaccharyl Transferase Function *292*
10.2.3.1	Possible Mechanisms for Glycosyl Transfer *294*
10.2.3.2	Probing of the Mechanism of Oligosaccharyl Transferase with Potential Inhibitors *296*
10.2.3.3	Interim Summary *300*
10.3	Conclusions *301*
10.4	References *301*

11	**Conformational Analysis of *C*-Glycosides and Related Compounds: Programming Conformational Profiles of *C*- and *O*-Glycosides** *305*
	Peter G. Goekjian, Alexander Wei, and Yoshito Kishi
11.1	Introduction *305*
11.2	Stereoelectronic Effects and the *exo*-Anomeric Conformation *306*
11.3	Conformational Analysis of *C*-Glycosides: *C*-Monoglycosides *309*
11.4	1,4-Linked *C*-Disaccharides: the Importance of *syn*-Pentane Interactions *314*
11.5	Prediction of Conformational Preference and Experimental Validation *318*
11.6	Programming Oligosaccharide Conformation *322*
11.7	Conformational Design of *C*-Trisaccharides based on a Human Blood Group Antigen *323*
11.8	Conformational Design: Relationship to Biological Activity *330*
11.8.1	*C*-Lactose vs. *O*-Lactose *331*
11.8.2	Human Blood Group Trisaccharides *334*
11.9	Concluding Remarks *336*
11.10	Acknowledgements *337*
11.11	References *337*

12	**Synthetic Lipid A Antagonists for Sepsis Treatment** *341*
	William J. Christ, Lynn D. Hawkins, Michael D. Lewis, and Yoshito Kishi
12.1	Background *341*
12.2	Hypothesis and Approach *342*
12.2.1	Monosaccharide Antagonists: Lipid X Analogues *343*
12.2.2	Disaccharide Antagonist of Lipid A: First Generation *344*
12.2.3	Disaccharide Antagonist of Lipid A: Second Generation *348*
12.3	Conclusion *351*
12.4	Acknowledgement *353*
12.5	References *353*

13	**Polysialic Acid Vaccines** *357*
	Harold J. Jennings
13.1	Introduction *357*
13.2	Group C Meningococcal Vaccines *358*

13.2.1	Structure and Immunology of GCMP 358
13.2.2	Group C Conjugate Vaccines 360
13.3	Group B Meningococcal Vaccines 362
13.3.1	Structure of GBMP 362
13.3.2	Immunology of GBMP 362
13.3.3	B Polysaccharide-Protein Conjugates 363
13.3.4	Extended Helical Epitope of PSA 364
13.4	Chemically Modified Group B Meningococcal Vaccines 366
13.4.1	N-Propionylated PSA Conjugate Vaccine 366
13.4.2	Immunology of NPr PSA 368
13.4.3	Protective Epitope mimicked by NPr PSA 370
13.4.4	Safety Concerns 370
13.5	Cancer Vaccines 371
13.5.1	PSA on Human Cells 371
13.5.2	Potential of NPr PSA as a Cancer Vaccine 373
13.6	Acknowledgements 375
13.7	References 375

14   Synthetic Carbohydrate-Based Vaccines 381
Stacy J. Keding and Samuel J. Danishefsky
14.1	Introduction 381
14.2	Cancer Vaccines 382
14.2.1	Carrier Proteins 384
14.2.2	Lipid Carriers 392
14.2.3	T-Cell Epitopes 394
14.2.4	Dendrimers 396
14.3	Bacterial Polysaccharide Vaccines 397
14.4	Synthetic Parasitic Polysaccharide Conjugate Vaccine 402
14.5	Conclusions 403
14.6	References 403

15   Chemistry, Biochemistry, and Pharmaceutical Potentials of Glycosaminoglycans and Related Saccharides 407
Tasneem Islam and Robert J. Linhardt
15.1	Introduction 407
15.1.1	Biological Activities 408
15.1.2	Heparin and Heparan Sulfate 409
15.1.2.1	Structure and Properties 409
15.1.2.2	Biosynthesis and Biological Functions 410
15.1.2.3	Applications of Heparin and Heparan Sulfate 411
15.2	Dermatan and Chondroitin Sulfates 417
15.2.1	Structure and Biological Role 417
15.2.2	Therapeutic Applications 418
15.2.2.1	Dermatan Sulfate 418

15.2.2.2	Chondroitin Sulfates 419
15.3	Hyaluronan 419
15.3.1	Structure and Properties 419
15.3.2	Tissue Distribution and Biosynthesis 420
15.3.3	Functions and Applications 421
15.3.3.1	Medical Applications 422
15.3.3.2	Hyaluronic Acid Biomaterials 423
15.4	Keratan Sulfate 423
15.4.1	Structure and Distribution 423
15.4.2	Chemistry and Biosynthesis of Linkage Regions 424
15.4.2.1	Keratan Sulfate on Cartilage Proteoglycans 424
15.4.2.2	Keratan Sulfate on Corneal Proteoglycans 424
15.4.3	Biological Roles of Keratan Sulfate 425
15.4.3.1	Role of KS in Macular Corneal Dystrophy 425
15.5	Other Acidic Polysaccharides 425
15.5.1	Acharan Sulfate 425
15.5.2	Fucoidins 426
15.5.3	Carrageenans 427
15.5.4	Sulfated Chitins 427
15.5.5	Dextran Sulfate 427
15.5.6	Alginates 428
15.5.7	Fully Synthetic Sulfated Molecules 428
15.5.7.1	Polymers 428
15.5.7.2	Small Sulfonated Molecules 428
15.6	Pharmaceutical Potential and Challenges 430
15.6.1	GAG-Based Agents Are Heterogenous 431
15.6.2	GAG-Based Agents and Sulfonated Analogues Have Low Bioavailability 431
15.6.3	GAGs Have a Myriad of Biological Activities 432
15.6.4	Carbohydrate-Based Drugs Are Expensive and Difficult to Prepare 432
15.7	Conclusion 432
15.8	References 433

16 A New Generation of Antithrombotics Based on Synthetic Oligosaccharides 441
Maurice Petitou and Jean-Marc Herbert

16.1	Introduction 441
16.2	Heparin and Its Mechanism of Action as an Antithrombotic Agent 442
16.2.1	Heparin, a Complex Polysaccharide with Blood Anticoagulant Properties 442
16.2.2	Which Coagulation Factor must be Inhibited? 442
16.2.3	The Structure of Heparin in Relation to Antithrombin Activation 444
16.2.4	The Limitations of Heparin 445
16.3	Synthetic Pentasaccharides, Selective Factor Xa Inhibitors, are Antithrombotic Agents 446

16.3.1	New Synthetic Oligosaccharides Required in Order to Validate A Pharmacological Hypothesis 446
16.3.2	A Strategy for the Synthesis of an Active Pentasaccharide 446
16.3.3	A Strategy for the Synthesis of the First Pentasaccharide 448
16.3.4	Activation of Antithrombin: Structure/Activity Relationship 449
16.3.5	Clinical Trials Results 449
16.3.6	The Second Generation of Antithrombotic Pentasaccharides 451
16.4	Synthetic Thrombin-Inhibiting Oligosaccharides: The Next Generation? 452
16.4.1	First Approach: Oligomerization of a Disaccharide 452
16.4.2	Second Approach: Molecules Containing Two Identified Domains 453
16.4.3	Introduction of a Neutral Domain 454
16.5	The Mechanism of Antithrombin Activation by Synthetic Oligosaccharides 456
16.6	Conclusion and Perspectives 456
16.7	References 457

Volume 2

17	**Sequencing of Oligosaccharides and Glycoproteins** 461
	Stuart M. Haslam, Kay-Hooi Khoo, and Anne Dell
17.1	Mass Spectrometry 462
17.1.1	EI-, FAB-, and MALDI-MS 462
17.1.2	ES, NanoES, and LC-MS 464
17.1.3	MS/MS and Mass Analyzers 465
17.2	MS-Based Sequencing Strategies 466
17.2.1	Chemical Derivatization 467
17.2.2	MS/MS Fragmentation Patterns 467
17.2.3	Permethylation and Sequence Assignment from Fragment Ions 468
17.3	Glycan Sequencing and Structural Determination – A Case Study 470
17.3.1	GC-MS Sugar Analysis 471
17.3.2	Glycan Derivatization 471
17.3.3	FAB-MS of the Deuteroreduced Permethylated HSP Sample 473
17.3.4	ES-MS/MS 473
17.3.5	Linkage Analysis 474
17.3.6	Chemical Hydrolysis 474
17.3.7	Exo-Glycosidase Digestion 474
17.4	Mammalian Glycomics 475
17.5	Some Special Case Strategies 477
17.6	References 481

18		**Preparation of Heterocyclic 2-Deoxystreptamine Aminoglycoside Analogues and Characterization of their Interaction with RNAs by Use of Electrospray Ionization Mass Spectrometry** *483*
		Richard H. Griffey, Steven A. Hofstadler, and Eric E. Swayze
18.1		Introduction *483*
18.1.1		RNA as a Target *483*
18.1.2		Functional RNA Subdomains *483*
18.1.3		Aminoglycosides are a Privileged Class of RNA Ligands *484*
18.2		ESI-MS for Characterization of Aminoglycoside-RNA Interactions *484*
18.2.1		Aminoglycoside-16S and 18S A Site RNA Models *484*
18.2.2		Neomycin and TAR RNA *489*
18.2.3		Interim Summary *490*
18.3		Preparation of Heterocyclic 2-Deoxystreptamines and Binding to a 16S A Site RNA Model *490*
18.3.1		4-Substituted 2-Deoxystreptamine Derivatives *491*
18.3.2		16S rRNA Binding Affinity Study in an ESI-MS Assay *493*
18.3.3		Isolation of Sugar Ring Fragments from Neomycin *494*
18.4		Preparation, Binding, and Biological Activity of Substituted Paromomycin Derivatives *495*
18.4.1		Synthesis of Racemic A Ring-Substituted Paromomycin Analogues *495*
18.4.2		Synthesis of Chiral A Ring-Substituted Paromomycin Analogues *496*
18.5		Future Prospects *498*
18.6		Acknowledgements *498*
18.7		References *498*
19		**Glycosylation Analysis of a Recombinant P-Selectin Antagonist by High-pH Anion-Exchange Chromatography with Pulsed Electrochemical Detection (HPAEC/PED)** *501*
		Mark R. Hardy and Richard J. Cornell
19.1		Introduction *501*
19.2		Use of HPAEC/PED in the Development of Biopharmaceuticals *502*
19.3		Biology of P-Selectin *503*
19.3.1		Structures of PSGL-1 and rPSGL-Ig *503*
19.4		HPAEC/PED as an Adjunct to rPSGL-Ig Process Development *504*
19.4.1		Materials and Methods *504*
19.4.2		HPAEC/PED O-Linked Oligosaccharide Profile Analysis *506*
19.4.3		N-Linked Oligosaccharide Profile Analysis *507*
19.5		Results and Discussion *508*
19.5.1		HPAEC/PED Oligosaccharide Profile Analysis of a Developmental Batch of rPSGL-Ig *508*
19.5.2		Repeatability of the O-Linked Oligosaccharide Profile Method *510*
19.5.3		O-Glycosylation of rPSGL-Ig Expressed by Different Cell Lines *510*
19.5.4		N-Glycosylation of rPSGL-Ig *515*

19.6	Summary 515
19.7	Acknowledgements 516
19.8	References 516

20	**Analytical Techniques for the Characterization and Sequencing of Glycosaminoglycans** 517
	Ram Sasisekharan, Zachary Shriver, Mallik Sundaram, and Ganesh Venkataraman
20.1	Introduction to GAG Linear Complex Polysaccharides 517
20.2	Depolymerization of Nascent GAG Chains 521
20.2.1	Enzymes that Degrade GAGs 521
20.2.2	Chemical Methods for Degrading GAG Oligosaccharides 524
20.3	Detection of GAG Oligosaccharides 525
20.3.1	$\Delta^{4,5}$ Bond Formation and UV Detection 525
20.3.2	Fluorescent Tagging 526
20.3.3	Metabolic Labeling 526
20.4	Analytical Tools Used in the Structural Characterization of GAGs 527
20.4.1	High Pressure Liquid Chromatography 527
20.4.1.1	Amino-Bonded Silica 528
20.4.1.2	High-Performance Gel Permeation 528
20.4.1.3	Weak and Strong Anion Exchange 528
20.4.1.4	Pellicular Anion Exchange 529
20.4.1.5	IP-RPHPLC 529
20.4.1.6	Sequencing GAGs I: HPLC Methods 529
20.4.2	Polyacrylamide Gel Electrophoresis 530
20.4.2.1	Sequencing GAGs II: PAGE Methods 530
20.4.3	Capillary Electrophoresis 530
20.4.4	NMR Spectroscopy 532
20.4.5	Mass Spectrometry 533
20.4.5.1	Sequencing GAGs III: Mass Spectrometric Methodologies 536
20.4.6	Oligosaccharide Array Technologies 536
20.5	Future Directions 536
20.6	Acknowledgements 537
20.7	References 537

21	**Thermodynamic Models of the Multivalency Effect** 541
	Pavel I. Kitov and David R. Bundle
21.1	Introduction 541
21.2	Concept of Distribution Free Energy 542
21.2.1	Binding Isotherm 542
21.2.2	Competitive Inhibition Isotherm 544
21.3	Multivalent Receptor vs. Monovalent Ligand 546
21.3.1	Interim Summary 549
21.4	Multivalent Receptor vs. Multivalent Ligand 551
21.5	Topological Classification of Multivalent Systems 553

21.5.1   Indifferent Presentation 553
21.5.2   Linear Presentation 554
21.5.3   Circular Presentation 554
21.5.4   Radial Presentation 554
21.6   Determination of Microscopic Binding Parameters by Molecular Modeling 555
21.6.1   Optimization of the Tether in Bivalent P^k-Trisaccharide Ligands for Shiga-Like Toxin 557
21.7   Determination of Microscopic Binding Parameters from Binding Data 561
21.8   Thermodynamic Analysis of Multivalent Interaction 562
21.8.1   Radially Arranged Multivalent Ligands for Shiga-Like Toxin 565
21.9   Conclusions 570
21.10   Mathematical Appendix 570
21.10.1   Calculation of Statistical Coefficients 570
21.10.2   Multivalent Receptor and Monovalent Ligand 571
21.10.3   Multivalent Binding with Linear and Circular Topology 571
21.10.4   Multivalent Binding with Radial Topology 572
21.10.5   Derivation of Eq. (24) 572
21.11   References 573

22   Synthetic Multivalent Carbohydrate Ligands as Effectors or Inhibitors of Biological Processes 575
          Laura L. Kiessling, Jason K. Pontrello, and Michael C. Schuster
22.1   Introduction 575
22.1.1   Mechanisms of Binding of Multivalent Ligands 576
22.1.2   Investigating the Structure/Function Relationship of a Series of Ligand Classes 577
22.2   Multivalent Carbohydrate Ligands as Inhibitors 581
22.2.1   Multivalency with AB_5 Toxins 581
22.2.1.1 Bundle's Decavalent Ligand for the *E. coli* Shiga-Like Toxin 582
22.2.1.2 Fan's Pentavalent Ligands for Cholera Toxin and the *E. coli* Heat-Labile Enterotoxin 584
22.2.2   Multivalency in Anti-adhesives 587
22.2.2.1 Low Molecular Weight Multivalent Carbohydrate Inhibitors of Bacterial Adhesion 587
22.2.2.2 Polymeric Multivalent Carbohydrate Inhibitors of Influenza Virus 592
22.2.3   Multivalent Carbohydrate Ligands as Inhibitors of Immune Responses 595
22.3   Multivalent Carbohydrate Ligands as Effectors 596
22.3.1   Low Molecular Weight Multivalent Effectors 597
22.3.2   Multivalency in Targeting Strategies 599
22.3.3   Multivalent Bacterial Chemoattractants 600
22.3.4   Multivalent Ligand-Mediated Cell Aggregation 602
22.3.5   Multivalent Ligands and the Selectins 603

22.4	Conclusions	605
22.5	References	605

23 Glycosyltransferase Inhibitors 609
Karl-Heinz Jung and Richard R. Schmidt
23.1 Introduction 609
23.2 Glycosyltransferases Utilizing NDP-Sugar Donors 610
23.2.1 Inverting Glycosyltransferases 610
23.2.1.1 β-Glucosyltransferases 611
23.2.1.2 β-Galactosyltransferases 616
23.2.1.3 β-N-Acetylglucosaminyltransferases 620
23.2.1.4 α-Fucosyltransferases 625
23.2.1.5 β-Glucuronosyltransferases 632
23.2.2 Retaining Glycosyltransferases 636
23.2.2.1 α-Galactosyltransferases 637
23.2.2.2 α-N-Acetylgalactosaminyltransferases 640
23.3 Glycosyltransferases Utilizing NMP-Sugar Donors 641
23.3.1 α(2–6)Sialyltransferases 641
23.3.2 α(2–3)Sialyltransferases and α(2–8)Sialyltransferases 647
23.4 Bisubstrate Analogues as Inhibitors 648
23.5 Conclusion 653
23.6 References 654

24 RNA-Aminoglycoside Interactions 661
Haim Weizman and Yitzhak Tor
24.1 RNA as an Emerging Therapeutic Target 661
24.2 Aminoglycoside Antibiotics: Past and Present 664
24.3 Aminoglycosides as RNA Binders 666
24.4 Identifying RNA Targets and Developing Binding Assays 670
24.5 Dimeric Aminoglycosides 673
24.6 Aminoglycoside-Intercalator Conjugates 675
24.7 Guanidinoglycosides 677
24.8 Summary and Outlook 679
24.9 Acknowledgements 680
24.10 References 680

25 Glycosylated Natural Products 685
Jon S. Thorson and Thomas Vogt
25.1 Introduction 685
25.2 A Summary of Bioactive Glycosylated Secondary Metabolites 686
25.2.1 Agents that Interact with DNA 686
25.2.1.1 Enediynes 686
25.2.1.2 Bleomycins 688
25.2.1.3 Diazobenzofluorenes 689
25.2.1.4 Anthracyclines 689

25.2.1.5	Pluramycins	689
25.2.1.6	Aureolic Acids	690
25.2.2	Agents that Interact with RNA	692
25.2.2.1	Orthosomycins	692
25.2.2.2	Macrolides	692
25.2.2.3	Aminoglycosides	694
25.2.2.4	Amicetins	695
25.2.3	Agents that Interact with Cell Walls and Cell Membranes	695
25.2.3.1	Non-Ribosomal Peptides	695
25.2.3.2	Polyenes	697
25.2.3.3	Saccharomicins	699
25.2.4	Agents that Interact with Proteins	699
25.2.4.1	Indolocarbazoles	699
25.2.4.2	Coumarins	699
25.2.4.3	Benzoisochromanequinones	701
25.2.4.4	Avermectins	701
25.2.4.5	Angucyclines	701
25.2.4.6	Cardiac Glycosides	702
25.2.4.7	Lignans	703
25.2.4.8	Anthraquinone Glycosides	703
25.2.4.9	Ginsenosides	704
25.2.4.10	Glycoalkaloids	704
25.2.4.11	Glucosinolates	705
25.2.5	Agents that Interact with Other (or Undefined) Targets	706
25.2.5.1	Plant Phenolics	706
25.2.5.2	Mono- and Triterpenoid Glycosides	707
25.2.5.3	Plant Polymeric Natural Glycosides	707
25.3	Conclusions	707
25.4	References	707
26	**Novel Enzymatic Mechanisms in the Biosynthesis of Unusual Sugars**	**713**
	Alexander Wong, Xuemei He, and Hung-Wen Liu	
26.1	Introduction	713
26.2	Biosynthesis of Deoxysugars	714
26.2.1	E_{od}-Catalyzed C-O Bond-Cleavage at the *C*-6 Position in the Biosynthesis of 6-Deoxyhexose	715
26.2.1.1	Catalytic Mechanism of E_{od}	715
26.2.1.2	Stereochemical Course of E_{od}-Catalyzed Reactions	716
26.2.2	E_1- and E_3-Catalyzed C-O Bond-Cleavage at the *C*-3 Position in the Biosynthesis of Ascarylose	717
26.2.2.1	Catalytic Properties of CDP-6-Deoxy-L-Threo-D-Glycero-4-Hexulose 3-Dehydrase (E_1)	718
26.2.2.2	Catalytic Properties of CDP-6-Deoxy-L-Threo-D-Glycero-4-Hexulose 3-Dehydrase Reductase (E_3)	719
26.2.2.3	Formation of Radical Intermediates During E_1 and E_3 Catalysis	719

26.2.3	TylX3- and TylC1-Catalyzed C-O Bond-Cleavage at the C-2 Position in the Biosynthesis of Mycarose	720
26.2.3.1	Biochemical Characterization of Enzymes Involved in C-2 Deoxygenation	721
26.2.3.2	Mechanism of C-2 Deoxygenation	721
26.2.4	DesI- and DesII-Catalyzed C-O Bond-Cleavage at the C-4 Position in the Biosynthesis of Desosamine	722
26.2.4.1	Genetic Disruption of DesI and DesII Genes	723
26.2.4.2	Proposed Mechanisms for C-4 Deoxygenation	723
26.3	Biosynthesis of Aminosugars	725
26.3.1	C-N Bond-Formation by GlmS-Catalyzed Transamidation in the Biosynthesis of Glucosamine-6-Phosphate	727
26.3.1.1	Catalytic Properties of Glucosamine-6-Phosphate Synthetase	727
26.3.1.2	The Glutaminase Activity of Glucosamine-6-Phosphate Synthetase	727
26.3.1.3	The Synthetase Activity of Glucosamine-6-Phosphate Synthetase	728
26.3.2	C-N Bond Formation by TylB-Catalyzed Transamination in the Biosynthesis of Mycaminose	729
26.4	Biosynthesis of Branched-Chain Sugars	730
26.4.1	YerE- and YerF-Catalyzed Two-Carbon Branched-Chain Attachment in the Biosynthesis of Yersiniose A	731
26.4.1.1	Biochemical Properties and Catalytic Mechanism of YerE	731
26.4.1.2	Biochemical Properties of YerF	732
26.4.2	TylC3-Catalyzed One-Carbon Branched-Chain Attachment in the Biosynthesis of Mycarose	732
26.4.2.1	Biochemical Properties and Catalytic Mechanism of TylC3	733
26.5	Epimerization Reactions	734
26.5.1	UDP-N-acetylglucosamine 2-Epimerase-Catalyzed C-2 Epimerization in the Biosynthesis of N-Acetylmannosamine	734
26.5.1.1	Catalytic Properties of UDP-N-Acetylglucosamine 2-Epimerase	734
26.5.1.2	Mechanism of C-2 Epimerization	735
26.5.2	CDP-Tyvelose 2-Epimerase-Catalyzed C-2 Epimerization in the Biosynthesis of Tyvelose	735
26.5.2.1	Biochemical Properties of CDP-Tyvelose 2-Epimerase	736
26.5.2.2	Possible Mechanisms for C-2 Epimerization	737
26.5.2.3	Distinguishing Between Mechanisms Involving C-2 or C-4 Oxidation	737
26.6	Rearrangement of Hexose Skeletons: UDP-Galactopyranose Mutase-Catalyzed Biosynthesis of Galactofuranose	738
26.6.1	Catalytic Properties of UDP-Galactopyranose Mutase	738
26.6.2	Mechanism of Ring Contraction	739
26.7	Summary	740

26.8	Acknowledgements	741
26.9	References	741

27 Neoglycolipids: Identification of Functional Carbohydrate Epitopes 747
Ten Feizi, Alexander M. Lawson, and Wengang Chai

27.1	Rationale for Developing Neoglycolipids as Oligosaccharide Probes	747
27.2	The First and Second Generation Neoglycolipids	749
27.3	Mass Spectrometry of Neoglycolipids	750
27.4	Scope of the Neoglycolipid Technology	752
27.4.1	Novel Sulfated Ligands for the Selectins	752
27.4.2	Novel Class of O-Glycans (O-Mannosyl) in the Brain	754
27.4.3	Unique Tetrasaccharide Sequence on Heparan Sulfate	755
27.5	Oligosaccharide Microarrays	755
27.6	Summary and Perspectives	757
27.7	Acknowledgement	757
27.8	References	757

28 A Preamble to Aglycone Reconstruction for Membrane-Presented Glycolipid Mimics 761
Murugesapillai Mylvaganam and Clifford A. Lingwood

28.1	Introduction	761
28.2	The Role of Ceramide Subtype Composition	762
28.3	Effects of Ceramide Subtype Composition in the Binding of Gb_3Cer to Verotoxins	764
28.4	Hypothesis Regarding Lipid Replacement Structural Motifs (LRSMs)	766
28.5	Effect of Replacement of GSL Fatty Acyl Chains with Rigid, Non-Planar Hydrophobic Groups	768
28.6	Ada-Gb_3Cer, a Functional Mimic of Membrane Presented Gb_3Cer for VT Binding	769
28.7	Ceramide Subtype-Dependent Binding of Heat Shock Protein Hsp70 to Sulfogalactosyl Ceramide	772
28.8	Adamantyl-Acyl Ceramide is a Functional Replacement for a Ceramide-Cholesterol Composition: A Study with HIV Coat Protein gp120	775
28.9	Acknowledgement	777
28.10	References	777

29 Small Molecule Inhibitors of the Sulfotransferases 781
Dawn E. Verdugo, Lars C. Pedersen, and Carolyn R. Bertozzi

29.1	Introduction: Sulfotransferases and the Biology of Sulfation	781
29.2	EST as a Model ST for Inhibitor Design	783
29.2.1	Inhibitors of EST Targeted Toward the PAPS Binding Site	784
29.2.2	A Bisubstrate Analogue Approach to EST Inhibition	788
29.2.3	Discovery of EST Inhibitors from a Library of PAP Analogues	789

29.2.4	Inhibition of EST by Dietary Agents and Environmental Toxins	791
29.3	Inhibition of Representative Golgi-Resident Sulfotransferases: GST-2, GST-3, and TPST-2	792
29.3.1	Heterocyclic Inhibitors of GST-2 and GST-3	792
29.3.2	Tethered Inhibitors of TPST-2	793
29.4	Assays for High-Throughput Screening of STs	794
29.4.1	A Continuous ST Assay	794
29.4.2	Immobilized Enzyme Mass Spectrometry (IEMS) Assay	795
29.4.3	A 96-Well Direct Capture 'Dot-Blot' Assay for Carbohydrate STs	795
29.5	New Directions in Inhibitor Discovery	796
29.6	Conclusions	796
29.7	Acknowledgements	796
29.8	References	797

30 Carbohydrate-Based Treatment of Cancer Metastasis 803
Reiji Kannagi

30.1	Implication of Carbohydrate Determinants in Cancer Metastasis	803
30.1.1	Distant Hematogenous Metastasis of Cancer Cells	803
30.1.2	Multiple Organ Infiltration of Leukemic Cells	806
30.1.3	Lymph Node Infiltration Mediated by L-Selectin	807
30.1.4	Other Carbohydrate Determinants Involved in Distant Metastasis	807
30.2	Tumor Angiogenesis and Cancer-Endothelial Interaction	808
30.2.1	Possible Involvement of Selectin-Mediated Cell Adhesion in Tumor Angiogenesis	808
30.2.2	Roles of Humoral Factors and Cell Adhesion Molecules in Tumor Angiogenesis	809
30.3	Use of Monoclonal Antibodies for Inhibition of Cancer Cell-Endothelial Interaction	809
30.3.1	Diversity of Selectin Ligand Expression on Cancer Cells	809
30.3.2	Internally Fucosylated Ligands for Selectins	810
30.3.3	Sulfated Ligands for Selectins	811
30.3.4	O-Acetylation and Other Sialic Acid Modifications in Carbohydrate Ligands	812
30.4	Inhibitors of Selectin-Mediated Cell Adhesion	812
30.4.1	Use of Carbohydrate Derivatives	812
30.4.2	Use of Peptide Mimetics	813
30.5	Regulation of Selectin Expression on Endothelial Cells	814
30.5.1	Enhanced E-Selectin Expression on Vascular Beds in Cancer Patients	814
30.5.2	Factors Affecting Endothelial E-Selectin Expression in Patients with Cancers	815
30.5.3	Chemoprophylaxis of Cancer Metastasis	815
30.6	Enhanced Expression of Sialyl Le^x and Sialyl Le^a in Malignant Cells and its Modulation	816

30.6.1	Fucosyltransferases Involved in Sialyl Le^a and Sialyl Le^x Synthesis and Antisense Gene Therapy *816*	
30.6.2	Therapy Targeting Transcriptional Regulation of Fucosyltransferases VII and IV in Cancer and Leukemia *817*	
30.6.3	Cancer-Associated Alteration of Sialyltransferase Isoenzymes *818*	
30.6.4	Sialyltransferase and the Concept of Cancer-Associated "Incomplete Synthesis" of Carbohydrate Determinants *820*	
30.6.5	Sulfotransferase and Differentiation Therapy of Cancer with Histone Deacetylase Inhibitors *820*	
30.6.6	Effect of Sialidases and Membrane Recycling on Sialyl Le$^{x/a}$ Expression in Cancer *822*	
30.6.7	Substrate Competition with A- and B-Transferases and DNA Methylation *822*	
30.6.8	Altered Carbohydrate Intermediate Metabolism and Sialyl Le$^{x/a}$ Expression in Cancer – Possible Relation to Warburg Theory *823*	
30.7	References *824*	

31 *N*-Acetylneuraminic Acid Derivatives and Mimetics as Anti-Influenza Agents *831*
Robin Thomson and Mark von Itzstein

31.1	Introduction *831*	
31.1.1	Influenza, the Disease *831*	
31.1.2	The Virus *832*	
31.1.3	Influenza Virus Sialidase *834*	
31.2	Structure-Based Design of Inhibitors of Influenza Virus Sialidase *836*	
31.3	Structure/Activity Relationship Studies of *N*-Acetylneuraminic Acid-Based Influenza Virus Sialidase Inhibitors *840*	
31.3.1	C-4 Modifications *840*	
31.3.2	C-5 Modifications *842*	
31.3.3	C-6 Modifications *843*	
31.3.4	Glycerol Side Chain Modifications *845*	
31.3.5	Glycerol Side Chain Replacement *850*	
31.4	Concluding Remarks *856*	
31.5	Acknowledgements *856*	
31.6	References *857*	

32 Modified and Modifying Sugars as a New Tool for the Development of Therapeutic Agents – The Biochemically Engineered *N*-Acyl Side Chain of Sialic Acid: Biological Implications and Possible Uses in Medicine *863*
Rüdiger Horstkorte, Oliver T. Keppler, and Werner Reutter

32.1	Introduction *863*	
32.2	*N*-Acyl Side Chain-Modified Precursors of Sialic Acid *865*	

32.2.1	Biosynthetic Engineering of Cell Surface Sialic Acid as a Potent Tool for Study of Virus-Receptor Interactions *865*	
32.2.2	Immunotargeting of Tumor Cells Expressing Unnatural Polysialic Acids *868*	
32.2.3	Activation of Human T-Lymphocytes by ManProp *869*	
32.2.4	N-Acyl-Modified Sialic Acids can Stimulate Neural Cells *869*	
32.2.4.1	Stimulation of Glial Cells *869*	
32.2.4.2	Stimulation of Neurons *870*	
32.3	Outlook *871*	
32.4	Acknowledgements *872*	
32.5	Abbreviations *872*	
32.6	References *872*	
33	**Modified and Modifying Sugars as a New Tool for the Development of Therapeutic Agents – Glycosidated Phospholipids as a New Type of Antiproliferative Agents** *875*	
	Kerstin Danker, Annette Fischer, and Werner Reutter	
33.1	Introduction *875*	
33.2	Structures of Synthetic Glycosidated Phospholipid Analogues *876*	
33.3	Antiproliferative Effect and Cytotoxicity of Glycosidated Phospholipid Analogues in Cell Culture Systems *876*	
33.4	Effect of Glycosidated Phospholipid Analogues on Cell Matrix Adhesion *878*	
33.5	Mechanisms of Action *879*	
33.6	Outlook and New Developments *880*	
33.7	Acknowledgements *881*	
33.8	References *881*	
34	**Glycoside Primers and Inhibitors of Glycosylation** *883*	
	Jillian R. Brown, Mark M. Fuster, and Jeffrey D. Esko	
34.1	Introduction *883*	
34.2	Glycoside-Based Substrates *883*	
34.3	Glycoside Primers – Xylosides *884*	
34.4	Other Types of Primers *885*	
34.5	Glycosides as Metabolic Decoys *888*	
34.6	Analogues *890*	
34.7	References *892*	
35	**Carbohydrate-Based Drug Discovery in the Battle Against Bacterial Infections: New Opportunities Arising from Programmable One-Pot Oligosaccharide Synthesis** *899*	
	Thomas K. Ritter and Chi-Huey Wong	
35.1	Introduction *899*	
35.2	Cell-Surface Carbohydrates *900*	
35.3	Peptidoglycan *904*	

35.4	Macrolide Antibiotics	*913*
35.5	Aminoglycosides	*917*
35.6	Programmable One-Pot Oligosaccharide Synthesis	*922*
35.7	Summary	*927*
35.8	References	*928*

Subject Index *933*

Preface

This book is about carbohydrate-based drug discovery, a subject of current interest and challenge. It contains 35 chapters and covers a broad range of topics, including, for example, synthesis of carbohydrates and their mimetics, development of inhibitors targeting enzymes and receptors associated with carbohydrate recognition, design and synthesis of carbohydrate-based vaccines and pharmaceuticals, conformational analysis and sequencing of saccharides, array development, multivalency in sugar receptor interaction, glycoprotein assembly and function, among others.

The contributors are all renowned experts in the field, and their views represent the most recent development and future direction of this subject.

Of the three major classes of biomolecules – proteins, nucleic acids, and carbohydrates – it is carbohydrates that are the least exploited. Despite the important roles that carbohydrates play in numerous biological recognition events (e.g. bacterial and viral infection, cancer metastasis, and inflammatory reactions) the molecular details of these recognition processes are generally not well understood, and consequently the pace of development of carbohydrate-based therapeutics has been relatively slow. This slow pace of development is further hindered by the lack of practical synthetic and analytical methods available for carbohydrate research and the problems associated with undesirable properties of carbohydrates as drug candidates.

Recent advances in the field as described in these chapters, however, have demonstrated that many of these problems can be circumvented with the development of new analytical and synthetic methods and new concepts for carbohydrate research. In addition, new developments in glycobiology research have helped our understanding of numerous carbohydrate-mediated biological processes and provided new targets for therapeutic discovery.

It is hoped that this book will help those who are interested in the field develop useful strategies to tackle the problem of drug discovery associated with carbohydrate recognition.

La Jolla
June 2003

Chi-Huey Wong

List of Contributors

HIROMUNE ANDO
Department of Applied Bioorganic
Chemistry
Gifu University
Gifu 501-1193
Japan

CAROLYN R. BERTOZZI
Department of Chemistry
University of California
Berkeley, CA 94720-1460
USA

GEERT-JAN BOONS
Complex Carbohydrate Research Center
The University of Georgia
220 Riverbend Road
Athens, GA 30602-4712
USA

JILLIAN R. BROWN
Department of Cellular and Molecular
Medicine
University of California, San Diego
9500 Gilman Drive
La Jolla, CA 92090-0687
USA

DAVID R. BUNDLE
Department of Chemistry
University of Alberta
Edmonton, AB T6G 2G2
Canada

WENGANG CHAI
The Glycosciences Laboratory
Faculty of Medicine
Imperial College London
Northwick Park Institute
for Medical Research
Watford Road
Harrow, HA1 3UJ
UK

WILLIAM J. CHRIST
Ancora Pharmaceuticals
700 Huron Av., Suite 18K
Cambridge, MA 02138
USA

RICHARD J. CORNELL
Wyeth BioPharma
One Burtt Road
Andover, MA 01810
USA

KERSTIN DANKER
Institute of Molecular Biology
and Biochemistry
Free University of Berlin
Arnimallee 22
14195 Berlin
Germany

SAMUEL J. DANISHEFSKY
Laboratory of Bioorganic Chemistry
Sloan-Kettering Institute
for Cancer Research
1275 York Avenue
New York, NY 10021
USA

ANNE DELL
Department of Biological Sciences
Imperial College of Science,
Technology and Medicine
South Kensington
London SW7 2AY
UK

ALEXEI V. DEMCHENKO
Department of Chemistry and
Biochemistry
University of Missouri – St. Louis
8001 Natural Bridge Rd.
St. Louis, MO 63121-4499
USA

JEFFREY D. ESKO
Department of Cellular and Molecular
Medicine
University of California, San Diego
9500 Gilman Drive
La Jolla, CA 92090-0687
USA

TEN FEIZI
The Glycosciences Laboratory
Faculty of Medicine, Imperial College
London
Northwick Park Institute for Medical
Research
Watford Road
Harrow, HA1 3UJ
UK

ANNETTE FISCHER
Institute of Molecular Biology and
Biochemistry
Free University of Berlin
Arnimallee 22
D-14195 Berlin
Germany

MARK M. FUSTER
Division of Pulmonary Medicine
University of California, San Diego
200 W. Arbor St.
San Diego, CA 92103-0687
USA

PETER G. GOEKJIAN
Laboratoire Chimie Organique 2
UMR 5622
Université Claude Bernard – Lyon 1
3 rue Victor Grignard
69622 Villeurbanne Cedex
France

RICHARD H. GRIFFEY
Isis Pharmaceuticals
2292 Faraday Avenue
Carlsbad, CA 92008
USA

MARK R. HARDY
Wyeth BioPharma
One Burtt Road
Andover, MA 01810
USA

STUART M. HASLAM
Department of Biological Sciences
Imperial College of Science Technology
and Medicine
South Kensington
London SW7 2AY
UK

LYNN D. HAWKINS
Eisai Research Institute of Boston, Inc.
4 Corporate Drive
Andover, MA 01810
USA

XUEMEI HE
Division of Medicinal Chemistry
College of Pharmacy
University of Texas
Austin, TX 78712
USA

JEAN-MARC HERBERT
Sanofi-Synthélabo Recherche
Cardiovascular/Thrombosis Research
Department
195 Route d'Espagne
31036 Toulouse Cedex
France

STEVEN A. HOFSTADLER
Ibis Therapeutics
2292 Faraday Avenue
Carlsbad, CA 92008
USA

RÜDIGER HORSTKORTE
Institute of Molecular Biology and
Biochemistry
Free University of Berlin
Arnimallee 22
14195 Berlin
Germany

BARBARA IMPERIALI
Department of Chemistry
Massachusetts Institute of Technology
77 Massachusetts Avenue
Cambridge, MA 02139
USA

HIDEHARU ISHIDA
Department of Applied Bioorganic
Chemistry
Gifu University
Gifu 501-1193
Japan

TASNEEM F. ISLAM
College of Pharmacy
University of Iowa
1155 Grand Avenue
Iowa City, Iowa 52242
USA

YUKISHIGE ITO
Synthetic Cellular Chemistry Laboratory
RIKEN
2-1 Hirosawa, Wako-shi
Saitama 351-0198
Japan

HAROLD J. JENNINGS
Institute for Biological Sciences
National Research Council of Canada
100 Sussex Drive
Ottawa, ON K1A 0R6
Canada

KARL-HEINZ JUNG
Department of Chemistry
University of Konstanz
P.O. Box 5560
78457 Konstanz
Germany

OSAMU KANIE
Glycoscience Lab
Mitsubishi Kagaku Institute of Life
Sciences
Minami-Ooya II, Machida-shi
Tokyo 194-8511
Japan

REIJI KANNAGI
Program of Molecular Pathology
Aichi Cancer Center
1-1 Kanokoden, Chikusa-ku
464-8681 Nagoya
Japan

STACY J. KEDING
Laboratory of Bioorganic Chemistry
Sloan-Kettering Institute
for Cancer Research
1275 York Avenue
New York, NY 10021
USA

OLIVER T. KEPPLER
Department of Virology
University of Heidelberg
Im Neuenheimer Feld 324
69120 Heidelberg
Germany

KAY-HOOI KHOO
Institute of Biological Chemistry
Academia Sinica
128 Academia Road, Sec. 2
Nankang, Taipei 115
Taiwan

LAURA L. KIESSLING
Department of Chemistry
University of Wisconsin
1101 University Avenue
Madison, WI 53706-1396
USA

YOSHITO KISHI
Department of Chemistry
and Chemical Biology
Harvard University
12 Oxford Street
Cambridge, MA 02138
USA

MAKOTO KISO
Department of Applied Bioorganic
Chemistry
Gifu University
Gifu 501-1193
Japan

PAVEL I. KITOV
Department of Chemistry
University of Alberta
Edmonton, AB T6G 2G2
Canada

PREZEMK KOWAL
Department of Chemistry
Wayne State University
Detroit, MI 48202
USA

ALEXANDER M. LAWSON
The Glycosciences Laboratory
Faculty of Medicine, Imperial College
London
Northwick Park Institute
for Medical Research
Watford Road
Harrow, HA1 3UJ
UK

MICHAEL D. LEWIS
Eisai Research Institute of Boston, Inc.
4 Corporate Drive
Andover, MA 01810
USA

CLIFFORD A. LINGWOOD
Section of Infection,
Immunity, Injury & Repair
Hospital for Sick Children
555 University Avenue
Toronto, Ontario M5G 1X8
Canada

ROBERT J. LINHARDT
College of Pharmacy
University of Iowa
1155 Grand Avenue
Iowa City, Iowa 52242
USA

HUNG-WEN LIU
Division of Medicinal Chemistry
College of Pharmacy
University of Texas
Austin, TX 78712
USA

ICHIRO MATSUO
The Institute of Physical
and Chemical Research
RIKEN
2-1 Hirosawa, Wako-shi
Saitama 351-0198
Japan

HELEN J. MITCHELL
Merck & Co., Inc.
P.O. Box 4
West Point, PA 19486
USA

MURUGESAPILLAI MYLVAGANAM
Hospital for Sick Children
555 University Avenue
Toronto, Ontario M5G 1X8
Canada

K.C. NICOLAOU
Department of Chemistry
The Scripps Research Institute
10550 N. Torrey Pines Road, BCC 405
La Jolla, CA 92037
USA
and
Department of Chemistry
and Biochemistry
University of California, San Diego
9500 Gilman Drive
La Jolla, CA 92093
USA

SHIN-ICHIRO NISHIMURA
Graduate School of Science
Hokkaido University
Kita-10 Nishi-8
Kitaku, Sapporo 060-0810
Japan

LARS C. PEDERSEN
Laboratory of Reproductive
and Developmental Toxicology
National Institute of Environmental
Health Science National Institutes
of Health Research
Triangle Park, NC 27709
USA

MAURICE PETITOU
Sanofi-Synthélabo Recherche
Cardiovascular/Thrombosis Research
Department
195 Route d'Espagne
31036 Toulouse Cedex
France

JASON K. PONTRELLO
Department of Chemistry
University of Wisconsin
1101 University Avenue
Madison, WI 53706
USA

WERNER REUTTER
Institute of Molecular Biology
and Biochemistry
Free University of Berlin
Arnimallee 22
14195 Berlin
Germany

THOMAS K. RITTER
Department of Chemistry
The Scripps Research Institute
10550 N. Torrey Pines Road
La Jolla, CA 92037
USA

CHIKAKO SAOTOME
Glycoscience Lab
Mitsubishi Kagaku Institute of Life
Sciences
Minami-Ooya II, Machida-shi
Tokyo 194-8511
Japan

RAM SASISEKHARAN
Biological Engineering Division, 16-560
Massachusetts Institute of Technology
77 Massachusetts Avenue
Cambridge, MA 02139
USA

RICHARD R. SCHMIDT
Department of Chemistry
University of Konstanz
P.O. Box 5560
78457 Konstanz
Germany

MICHAEL C. SCHUSTER
Department of Chemistry
University of Wisconsin
1101 University Avenue
Madison, WI 53706
USA

PETER H. SEEBERGER
Department of Chemistry
Massachusetts Institute of Technology
77 Massachusetts Avenue
Cambridge, MA 02139
USA

OLIVER SEITZ
Department of Chemistry
Humboldt University Berlin
Brook-Taylor-Str. 2
12489 Berlin
Germany

JUN SHAO
Department of Chemistry
Wayne State University
Detroit, MI 48202
USA

ZACHARY SHRIVER
Biological Engineering Division, 16-560
Massachusetts Institute of Technology
77 Massachusetts Avenue
Cambridge, MA 02139
USA

SCOTT A. SNYDER
Department of Chemistry and
Biochemistry
University of California, San Diego
9500 Gilman Drive
La Jolla, CA 92093
USA

MALLIK SUNDARAM
Biological Engineering Division, 16-560
Massachusetts Institute of Technology
77 Massachusetts Avenue
Cambridge, MA 02139
USA

List of Contributors

Eric E. Swayze
Isis Pharmaceuticals
2292 Faraday Avenue
Carlsbad, CA 92008
USA

Vincent W.-F. Tai
Department of Chemistry
Massachusetts Institute of Technology
77 Massachusetts Av.
Cambridge, MA 02139
USA

Robin Thomson
Institute for Glycomics
Griffith University
PMB 50 Gold Coast Mail Centre
Queensland 9726
Australia

Jon S. Thorson
Laboratory for Biosynthetic Chemistry
University of Wisconsin School
of Pharmacy
777 Highland Avenue
Madison, WI 53705
USA

Yitzhak Tor
Department of Chemistry and
Biochemistry
University of California, San Diego
9500 Gilman Drive
San Diego, CA 92093-0358
USA

Ganesh Venkataraman
Biological Engineering Division, 16-560
Massachusetts Institute of Technology
77 Massachusetts Avenue
Cambridge, MA 02139
USA

Dawn E. Verdugo
Department of Chemistry
University of California
Berkeley, CA 94720-1460
USA

Thomas Vogt
Department of Plant Secondary
Metabolism
Leibniz Institute for Plant Biochemistry
Weinbergweg 3
06210 Halle/Saale
Germany

Mark von Itzstein
Institute for Glycomics
Griffith University
PMB 50 Gold Coast Mail Centre
Queensland 9726
Australia

Peng George Wang
Department of Chemistry
Wayne State University
Detroit, MI 48202-3489
USA

Alexander Wei
Department of Chemistry
Purdue University
West Lafayette, IN 47907
USA

Haim Weizman
Department of Chemistry and
Biochemistry
University of California, San Diego
9500 Gilman Drive
San Diego, CA 92093-0358
USA

Alexander Wong
Division of Medicinal Chemistry
College of Pharmacy
University of Texas
Austin, TX 78712
USA

Chi-Huey Wong
Department of Chemistry
The Scripps Research Institute
10550 N. Torrey Pines Road
La Jolla, CA 92037
USA

Jianbo Zhang
Department of Chemistry
Wayne State University
Detroit, MI 48202
USA

1
Synthetic Methodologies
CHIKAKO SAOTOME and OSAMU KANIE

1.1
Introduction

Research directed toward revealing the functions of oligosaccharides is currently the subject of great attention, and so the synthesis of oligosaccharides as probes for functional investigation is being widely investigated. After decades of efforts since the first synthesis of the disaccharide sucrose [1], it has now become possible to synthesize a variety of oligosaccharides. For the successful synthesis of oligosaccharides, both chemical reactions and tactics are important concerns. This chapter focuses on the strategic aspect of oligosaccharide synthesis.

1.2
Tactical Analysis for Overall Synthetic Efficiency

For the efficient synthesis of oligosaccharides, both stepwise and convergent methods have to be employed (Fig. 1.1). The former format can be further divided into two subclasses: one in which synthesis starts from the reducing end (A), which has classically been used, and another in which synthesis starts from the non-reducing end (C). Format A has traditionally been used in oligosaccharide synthesis as it was difficult to transform anomeric protecting groups into the leaving groups required for the C format. The recent development of new anomeric protecting groups and some substituent groups that can be directly used as the leaving groups, however, have enabled the alternative format (C) to be used. The concept of stepwise synthesis is especially important for the construction of relatively small oligomers, but the convergent format (B) has to be employed for the synthesis of larger saccharides. This can be easily understood by simple tactical analysis in the cases of the synthesis of large oligosaccharides or oligosaccharides possessing repeating units in their structures.

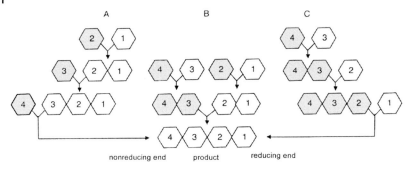

Fig. 1.1 Two stepwise methods and a convergent method in the synthesis of oligosaccharides. A: One of the stepwise methods, in which synthesis starts from the reducing end. B: The convergent method is especially advantageous for the synthesis of oligosaccharide with repeating structure. C: The other stepwise method, in which synthesis starts from the non-reducing end. Open hexagons: acceptors or protected forms. Gray hexagons: donors.

1.3
Methodological Improvements

One of the most important improvements in oligosaccharide synthesis is the discovery of the use of "stable" leaving groups that can function as protecting groups until exposed to certain activation conditions. This type of "potential leaving group" at the anomeric position is an ideal candidate intermediate in flexible synthetic strategies for oligosaccharides [2] (Scheme 1.1). The chemoselective glycosylation strategy that has emerged is based on tactical analysis aiming at efficient oligosaccharide synthesis [3–6, 7–23] (see Section 1.3.3).

The advancement of oligosaccharide synthesis is largely based on the development of good anomeric leaving groups and methods to control stereochemistry [24–26]. Regardless of the method used to control the stereochemistry of a given newly formed glycosidic linkage, one of the key factors is the reactivities and the stabilities of the leaving groups and the conditions used to activate one over another selectively.

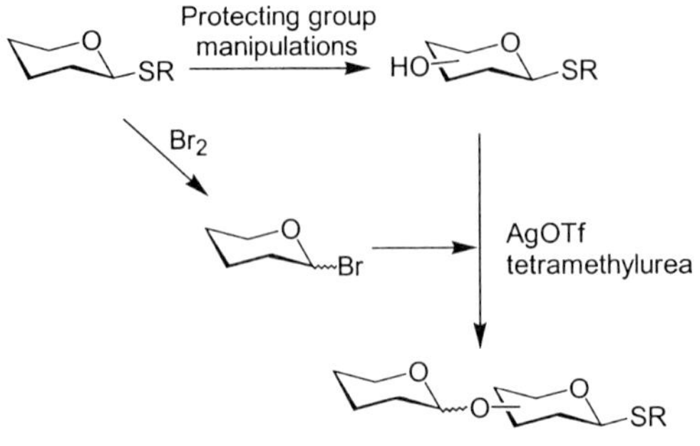

Scheme 1.1

1.3.1
Chemistry

The first species to be recognized as a form of protected carbohydrate synthetic unit were alkyl- or phenylthio glycosides, the use of which allows anomeric centers to be readily converted into halides [3–5, 27] (Scheme 1.1). This so-called two-stage activation (see Section 1.3.3) is possible thanks to the stabilities of these compounds towards the acidic conditions generally used for glycosylation reactions and protecting group manipulations. In addition, they can be activated directly, which allows extremely flexible synthetic schemes for oligosaccharide synthesis, including the "armed and disarmed" concept [28], orthogonal strategy [29], the "active and latent" concept [30], and one-pot glycosylation [6, 21]. Thioglycosides can also be converted into more reactive sulfoxides, which have been shown to be useful both in solution and in solid-phase reactions [6, 31, 32].

One of the most powerful and popular anomeric leaving groups is the trichloroacetimidate group, which has been used for the synthesis of oligosaccharides in solution [33, 34]. One special characteristic of this group is its applicability for transferring large oligosaccharides onto aglycon moieties, such as in the case of azido sphingosine, a commonly used ceramide precursor, to afford a ganglioside precursor [35–37] (Scheme 1.2). Glycosyl trichloroacetimidates have also been shown to react in highly polar solvents such as DMF, which has allowed the glycosylation of unprotected glycosyl acceptors in a random manner [38–40]. Activation of the imidate donor can be achieved by use of $BF_3 \cdot OEt_2$, trimethylsilyl triflate (TMSOTf), triethylsilyl triflate (TESOTf), or silver triflate (AgOTf) [41]. TESOTf was introduced to avoid by-product formation (glycosyl fluoride in the case of $BF_3 \cdot OEt_2$ [42], or TMS ethers of the acceptor in the case of TMSOTf). Recently, the use of dibutylboron triflate (DBBOTf) to address both problems has been reported [43]. Trichloroacetimidate is also used in polymer-supported oligosaccharide synthesis and has been shown to be compatible with a variety of supports, including PEG [43, 44] (see Section 1.4.1), Merrifield-type resin [45–48], and CPG [49] (see Section 1.4.3).

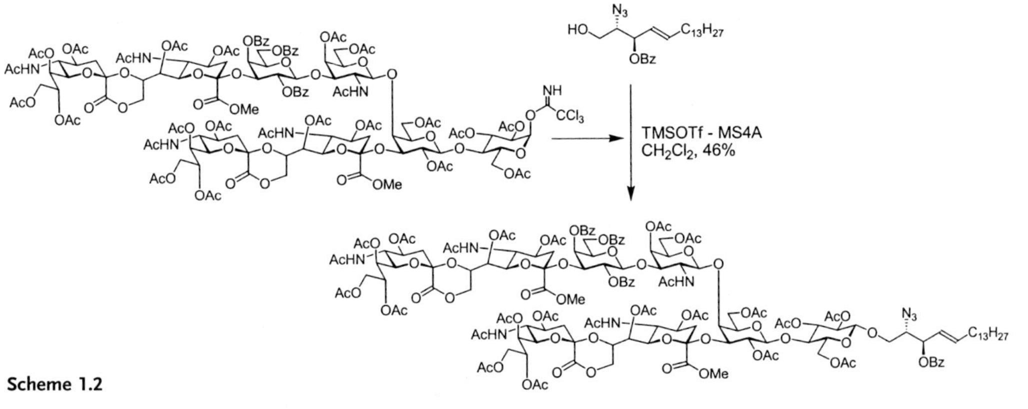

Scheme 1.2

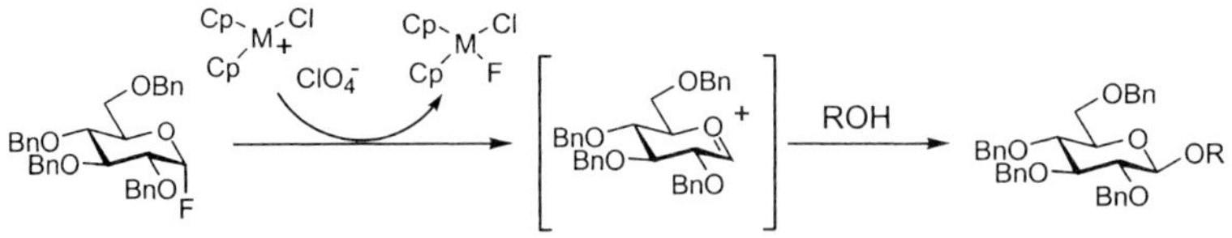

Scheme 1.3

Mukaiyama, in 1981, used SnCl$_2$/AgClO$_4$ as an activation system for glycosyl fluorides in ether to form glycosidic linkages [50]. Generally, however, it was the case that glycosyl fluorides were too stable to act as glycosylating agents in complex oligosaccharide syntheses. The situation changed after Suzuki's discovery of mild conditions with the use of Cp$_2$MCl$_2$/AgClO$_4$, where M is Hf or Zr [51, 52] (Scheme 1.3). Under these conditions, glycosyl fluorides can usually be activated at lower temperatures.

Glycals act as 1,2-protected sugars and are used as glycosylating agents. Traditionally, glycals were used to synthesize glycosides of 2-deoxy sugars by Fisher glycosylation or through 2-halo intermediates. Glycals were also used to produce ordinary glycosides via epoxides as the active agents. The advantage of glycals is their flexibility in the synthetic scheme, as has been shown in Danishefsky's research [53, 54] [see Section 1.4.3.5]. 2-Deoxy halosugars can be transformed into 2-amino-2-deoxy sugars by substitution reactions, which makes the glycals more useful strategically.

Other leaving groups (n-pentenyl, phosphite, phenylselenyl, etc.) have also been used for successful oligosaccharide synthesis [25, 26, 55]. In recent methods of synthesis of complex oligosaccharides, selective activation of a certain leaving group among others has allowed highly efficient syntheses [2, 56].

1.3.2
Protecting Group Manipulations

In oligosaccharide synthesis, particular sets of protecting groups have to be used, due to the multifunctional nature of carbohydrates (Scheme 1.4). The incorporation of protecting groups on functional groups needing to be protected and the order of deprotection have to be considered before the synthesis.

A general term "selectivity" has been used to describe these complex protecting group manipulations, but in 1977 a concept of chemical distinctiveness was introduced. The idea of orthogonal protection was defined by Baranay and Merrifield as "a set of completely independent classes of protection groups, such that each class can be removed in any order and in the presence of all other classes" [57]. Orthogonal protecting group manipulations are widely accepted, not only in peptide chemistry, but also in other fields including carbohydrate chemistry. The concept is summarized in Fig. 1.2. When individual hydroxy groups (two to five OHs) are protected with A, B, C, and D, respectively, and individual protecting groups can be removed in any order under certain conditions, the protecting groups can

1.3 Methodological Improvements

Scheme 1.4

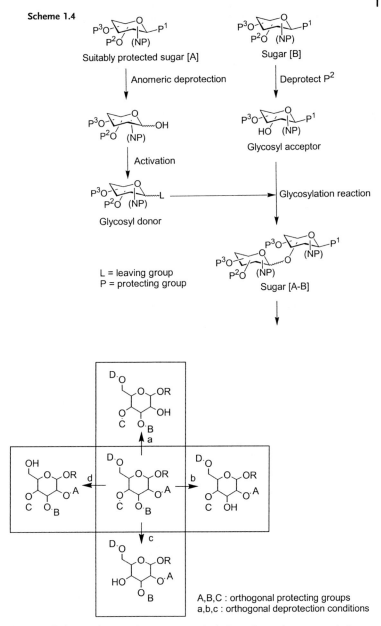

Fig. 1.2 Orthogonal protecting group manipulations. Protecting groups A–D can be removed in any order, eliminating tedious protecting group manipulations during complex oligosaccharide syntheses.

be said to be in an orthogonal relationship. The use of the concept is described well in Wong's work. As a representative set of orthogonal hydroxy protecting groups in carbohydrate chemistry, one combination of protecting groups and corresponding orthogonal deprotection conditions to have been used successfully [58] is A: chloroacetyl (a: $NaHCO_3/MeOH/H_2O$), B: methoxybenzyl (b: TFA/CH_2Cl_2), C: levulinyl (c: $NH_2NH_2/AcOH/THF/MeOH$), and D: TBDPS (d: HF/Pyr/AcOH/THF). Other sets are being investigated [59].

1.3.3
Modulation of the Reactivity of Glycosyl Donors

The reactivity of glycosyl donors can be controlled either through protecting groups or by anomeric leaving groups. Through the use of a set of molecules with suitable reactivities, oligosaccharides can be synthesized in the minimum possible number of operations.

The "armed and disarmed" concept, which employs a single potential leaving group (the n-pentenyloxy group) at the anomeric positions both of the donor and of the acceptor was developed from the observation that the reactivities of glycosyl donors are affected by the protecting groups (i.e., ether or ester) [28] (Scheme 1.5). The utility of this methodology is obvious, since small fragments of oligosaccharides can be systematically synthesized in short steps, in which a "disarmed" unit can be transformed into an "armed" unit by exchanging the protecting groups. Alternatively, the coupling product can be directly used as a donor if exposed to slightly stronger activation conditions. The armed and disarmed concept has also proven to be applicable to glycals [8], thioglycosides [60], selenyl glycosides [61], and glycosyl phosphoroamidates [62]. Furthermore, it has also been shown that the reactivities of these potential glycosyl donors can be controlled by selection of protecting groups at positions other than O-2 [61, 63–65].

A strategically related but conceptionally independent method, orthogonal glycosylation, has been developed. The key feature of the orthogonal coupling concept

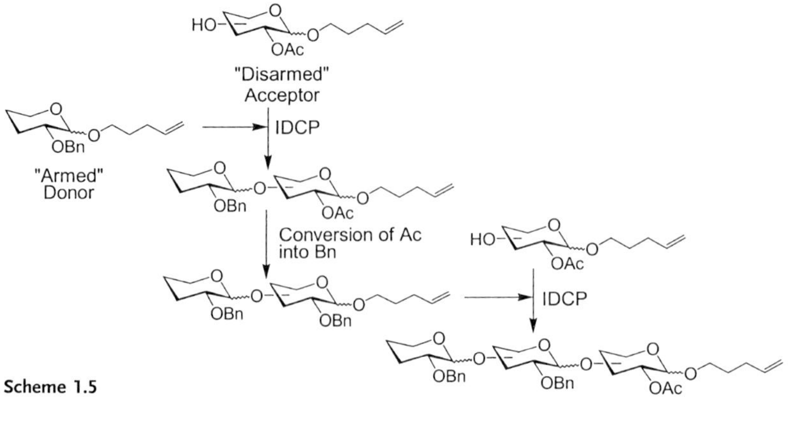

Scheme 1.5

Scheme 1.6

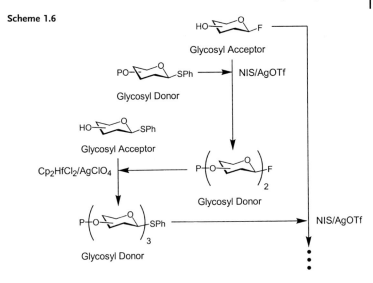

is the combined use of two chemically distinct glycosylation reactions [29, 66, 67] (Scheme 1.6). A set of potential leaving groups and activation conditions for each group – phenylthio group and fluoride, and NIS/AgOTf and $Cp_2HfCl_2/AgClO_4$ – were used. Since the reactions of the set are mutually distinct, there is no need for reactivity control, so this methodology is conceptionally different from reactivity modulation methods. In addition, it has also been shown that the strategy can be applied to a polymer-supported oligosaccharide synthesis [67]. Another set of potential leaving groups with orthogonal reactivities has also been investigated [68].

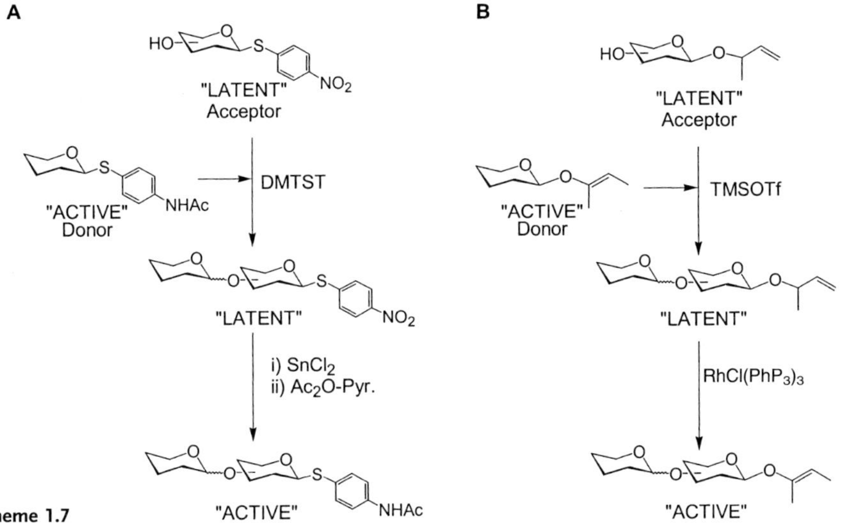

Scheme 1.7

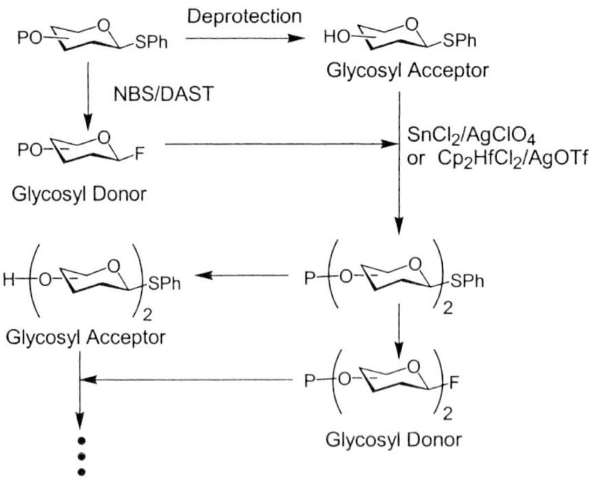

Scheme 1.8

Another tactic in oligosaccharide synthesis is the so called "active and latent" method. This method may be regarded as an extension of the traditional method without cleavage of the group but its transformation into an active species. One of the technique's successes is in the use of an allyloxy group as the protecting group at anomeric centers. It is later converted into a vinyl ether, which is readily activated in the glycosylation reaction [30, 69] (Scheme 1.7). However, this method is more likely related to the two-stage method discussed below.

The phenylthio group has commonly been used as a precursor of glycosyl donors such as glycosyl fluorides. The glycosyl fluorides can be activated chemoselectively without affecting the parent thioglycoside [3–5] (Scheme 1.8). In this way, extremely efficient syntheses of oligosaccharides possessing repeating sequences have been achieved in a convergent manner.

1.3.4
Block Synthesis

The importance of the convergent method (Fig. 1.1) is obvious in the synthesis of larger oligosaccharides (see also Section 1.2). This section covers several examples of oligosaccharide synthesis with special emphasis on the tactics. Because of the structural heterogeneity of the oligosaccharides involved, block synthesis is more suitable term than convergent synthesis to describe the synthesis.

The first example is the synthesis of a heptasaccharide reported by Boons et al., based on profound knowledge of carbohydrate chemistry [70] (Scheme 1.9).

The target heptasaccharide was first retrosynthetically taken into four blocks as shown in Scheme 1.9, the key feature of the synthesis being a reduced number of chemical steps after having four synthetic units. Sequential glycosylation reactions involving a 4,6-di-O-tritylated n-pentenyl glycoside derivative of glucosamine as a key unit were carried out with a methylthio glycoside, a cyanoethylidene, and an n-pentenyl glycoside as glycosyl donors. Another tactical ploy employed in this in-

Scheme 1.9

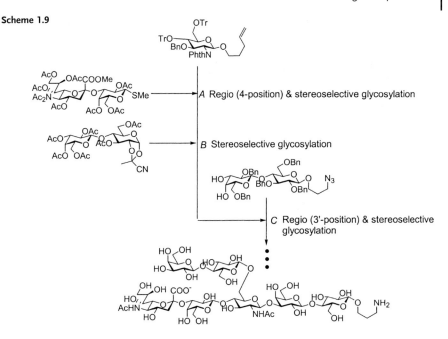

vestigation is the use of least protected acceptors. It is obvious that the use of least protected acceptors in a synthetic strategy involving multifunctional components as in carbohydrate synthesis is advantageous because one can eliminate time-consuming protective group manipulations [71]. In addition, the coupling reaction is free from any influence of nearby bulky protecting groups.

The issue, however, is the regiospecificity. One of the regiospecific glycosylations was carried out in the synthesis of sialyl galactose donor. A *N*-diacetylated methylthio glycoside of sialic acid and 4,6-benzylidene TMS ethyl galactoside were used as the donor and acceptor. The advantage of the diacetylated sialyl donor is the enhanced reactivity of thioglycoside due to the long range electronic effect [72]. It was reported that a higher yield than with *N*-acetyl derivative was obtained in a shorter reaction time, the stereochemistry being controlled through solvent effects.

The coupling product was further transformed into a thioglycoside. The second glycosylation was between the sialyl galactose donor and the 4,6-di-*O*-tritylated monosaccharide. The large steric hindrance of the 4-*O*-trityl group gave rise to polarization of the C–O bond of the secondary trityl ether, which enhanced reactivity and enabled regioselective glycosylation at the 4-*O* position. The neighboring participating effect of the 2-*O*-acetyl group permitted β-stereoselective glycosylation. The coupling product bearing a 6-*O*-trityl group was directly used as an acceptor for the next glycosylation reaction with cyanoethylidene lactosyl donor. Furthermore, since the anomeric position of the GlcN derivative was protected as an *n*-pentenyl glycoside, the formed pentasaccharide could again be directly used as a donor to couple with another lactose derivative. In this glycosylation reaction, regioselectivity toward the equatorial 3-*O* position was achieved. After removal of all protecting groups, the introduced amino functionality at the reducing terminal was used to incorporate the saccharide onto polyacrylamide for biological assays.

When the target is a series of oligosaccharides, a more systematic and unified strategy is required. Common building blocks have to be carefully designed and used in the synthesis of multiple target saccharides. Excellent examples of this kind of research can be found in a course of synthetic work carried out by Hasegawa and Kiso [73] and by Schmidt [74].

As a representative systematic oligosaccharide synthesis, we focus on a synthesis of a ganglioside known as GQ1b*a* [75–77] (Scheme 1.10). In this synthesis, the synthetic plan is carefully designed on the basis of the frequency of the existence of a certain unit in oligosaccharide, and also on its natural abundance generally, which affects on availability of a unit. The lactose unit, which is always found as

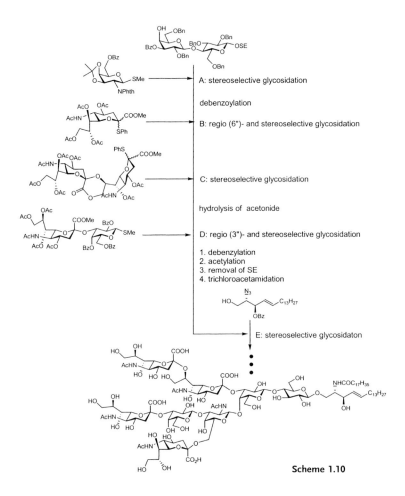

Scheme 1.10

the reducing terminal of mammalian glycolipids, is therefore used. A commercially available α-2-8-linked dimer of sialic acid was utilized, eliminating difficult problems in constructing an α-2-8 sialyl sialic acid linkage. Sialyl α-2-3 Gal was selected as a donor unit for the reason that it is commonly found in a variety of gangliosides. Indeed, the disaccharide was used as the donor in the synthesis of GM1b, GD1a, GD1*a*, GT1a*a* etc. A stepwise method was applied for the introduction of the sialyl 2-6 GalNAc sequence, since it is a special case for the so-called α-series gangliosides. The 2-(trimethylsilyl)-ethyl group was used as a persistent protecting group for the anomeric position of the lactose unit. The stability of the group, together with the mild and selective conditions needed for its removal, enabled multiple glycosylation reactions and other protecting manipulations to be performed. Thioglycosides were used as the glycosyl donors throughout synthesis, except for the coupling of the octasaccharide unit and azidosphingosine. The trichloroacetimidate approach was used for this particular glycosylation, as it has been shown to be very successful for construction of this type of glycosidic linkage.

1.4
Accessibility

When a biologist wants to investigate the functions of oligosaccharides, one of the most important issues will be the accessibility of particular oligosaccharides. The strategic considerations described above are thus very important. To this end, automation in the synthesis of oligosaccharides strongly deserves consideration, as in the cases of functional investigations of oligonucleotides and oligopeptides. One evident approach for automation is based on solid-phase synthesis, through which tedious workup and chromatographic purification after every reaction are eliminated. This, however, can only be achieved once a reliable synthetic method – especially for the glycosylation reaction – has been developed, because there is only one chance for the purification. For this reason, strategic analysis with regard to the overall reaction yield is also required. PEG-based polymer-supported chemistry and also the recently developed fluorous-phase chemistry may be alternatives [78]. One-pot reactions can be considered to be advantageous if smaller numbers of coupling reactions are in mind, although in this case a different approach has to be taken to access larger oligosaccharides [6, 21]. Convergent synthesis is very useful in this instance.

1.4.1
Solution-based Chemistry

Polyethyleneglycol monomethyl ether (MPEG) of molecular weight of approximately 5000 has been used as a support in oligosaccharide synthesis [43, 44, 67, 79–81]. A unique characteristic of this soluble polymer is that it can be precipitated by addition of *tert*-butyl methyl ether, facilitating isolation of polymer-bound substances from the glycosylating agents and reagents used in the coupling reaction. In addition, since MPEG is soluble in various solvents used in solution-phase oligosaccharide synthesis, these reactions are solution-phase reactions, and so reaction conditions used for solution-phase oligosaccharide synthesis can be employed (Scheme 1.11). Alternatively, relatively short-chain MPEG can be used to facilitate rapid chromatographic isolation [82, 83]. Another advantage of the use of MPEG is that reaction progress can be monitored either by NMR or by mass spectrometric methods [83, 84] without cleavage from the support.

The recent target molecule in oligosaccharide synthesis is a heptasaccharide phytoalexin elicitor [80, 85]. A successful approach to the synthesis by use of the MPEG approach was reported in 1993 [79]. The MPEG was attached at the 4-OH group of a glucose unit through an ester linkage [80] (Scheme 1.12). On the basis of retrosynthetic analysis, three synthetic blocks were prepared. All glycosyl donors were synthesized as thioglycosides; protecting groups used were the minimum. After four coupling and deprotection reaction cycles, the heptasaccharide was synthesized in 18% overall yield. The synthetic scheme is very simple, which in turn indicates the strength of the method. In another case, an amino functionality was also used as an anchoring point [86].

Use of the orthogonal strategy [29, 66] described in Section 1.3.3 has also been reported [67] (Scheme 1.13). The synthetic plan also takes advantage of introduced hydrophobicity at the end of polymer supported synthesis, which facilitates the isolation of desired products, in addition to the advantage of the self-correction ef-

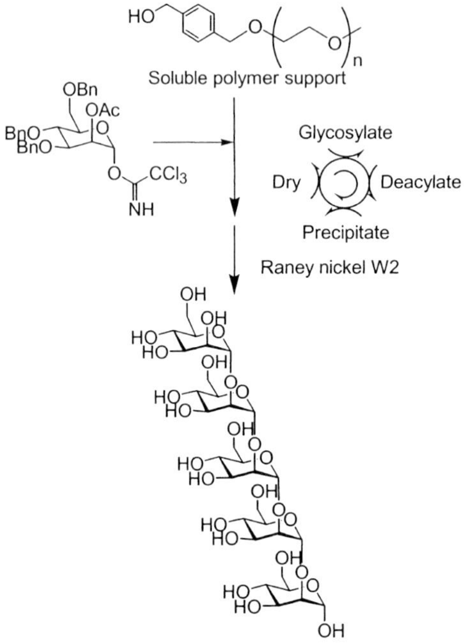

Scheme 1.11

Scheme 1.12

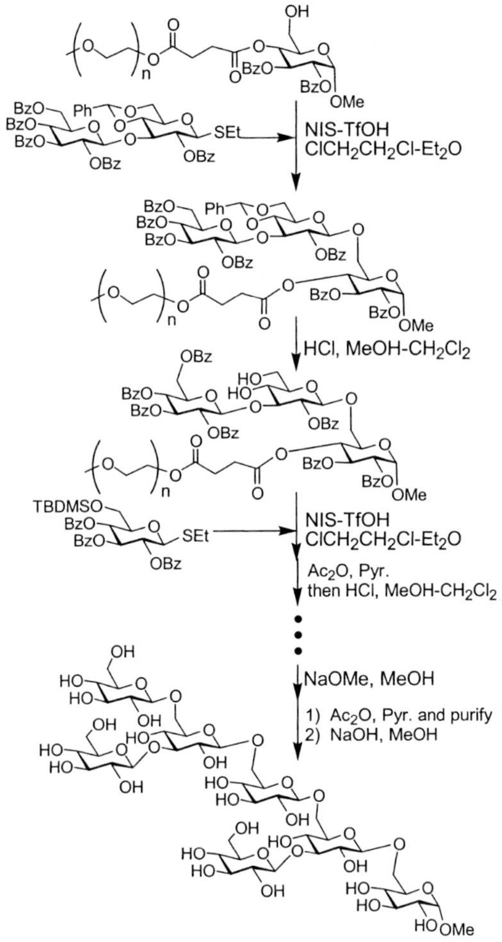

fect described in Section 1.4.3.5. In addition, thanks to the use of orthogonal sets of potential donors such as thioglycoside and glycosyl fluoride with leaving groups already installed, there is no need for activation on the support. It was later shown that the method can be used in combination with intramolecular aglycon delivery system [87–90] for the construction of β-mannopyranosides [91].

1.4.2
One-Pot Glycosylation

One of the important applications of methods based on anomeric reactivity modulation is the one-pot glycosylation method. If the reactivities of leaving groups at the anomeric centers of glycosyl units are differently controlled, a series of glycosylation reactions can be performed either all-in-one (A) [6] or sequentially (B) [20,

Scheme 1.13

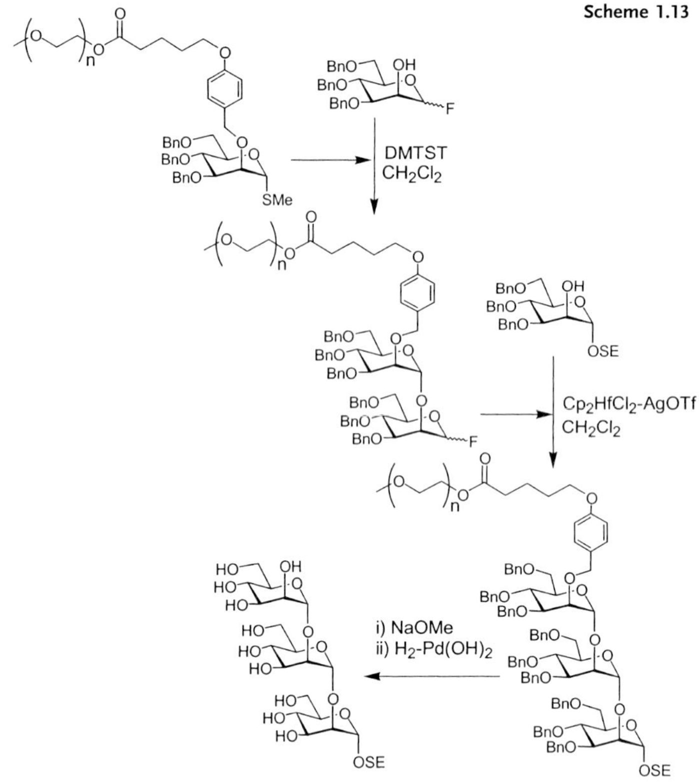

21, 92–101]. Despite its limitations arising from the identification and acquisition of glycosylating units with differently modulated reactivities, the method allows multiple coupling reactions to be performed in one-pot fashion, with obvious advantages over standard methods. Because this method eliminates the need for workup and purification steps during the operations, it should be regarded as equally as important as solid-phase synthesis. Careful purification must be performed after the reaction since there is no way to prevent the formation of deletion compounds. In an example of Format A, a phenylsulfoxide and a methoxyphenylsulfoxide were used as leaving groups. The reactivities of the acceptor molecules were also controlled by silylation of one of the hydroxy groups in the system [6] (Scheme 1.14).

One-pot sequential glycosylation (B) typically uses a series of leaving groups, requiring either that they can be activated under the same conditions or that a promoter used for the first coupling does not affect the other potential glycosyl donors [20, 21] (Scheme 1.15) (see also Section 1.4.6).

Scheme 1.14

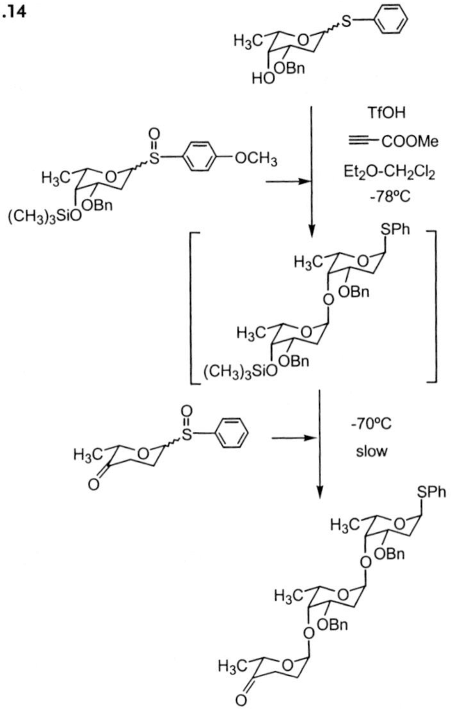

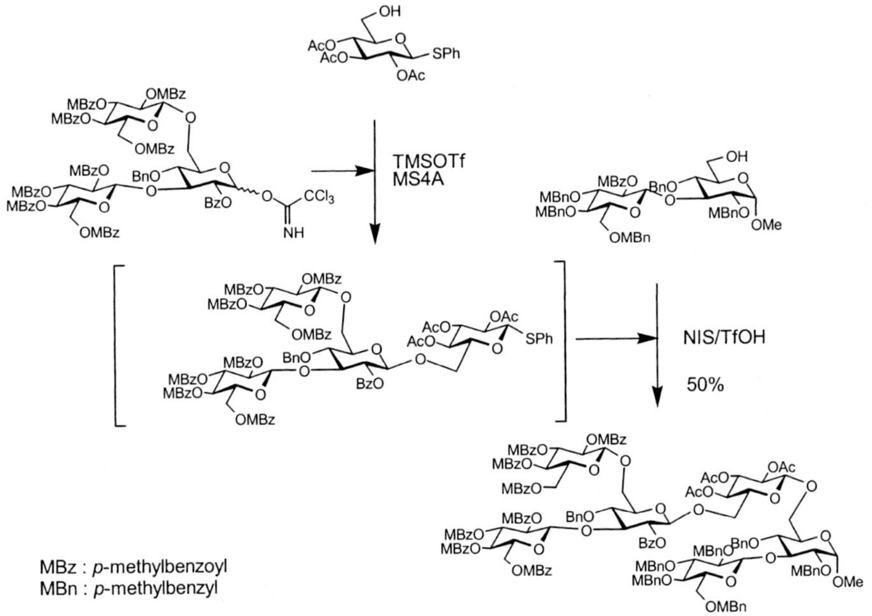

Scheme 1.15

MBz : *p*-methylbenzoyl
MBn : *p*-methylbenzyl

1.4.3
Solid-Phase Chemistry

The advantage of solid-phase reactions is the quick and simple workup process. Because only the growing molecule is attached on the support, other reagents used can be washed away by simple filtration. Higher reaction yields can generally be achieved by use of excess amounts of reagents. Furthermore, because of the simplicity of the process, it can be automated, allowing non-specialists to synthesize oligosaccharides.

There are basically two methods employed for solid-phase oligosaccharide synthesis. They differ in the direction of chain elongation: one starts from the reducing sugar (A) and the other is the opposite (B) (Fig. 1.3). Approach A is generally advantageous when both the polymer-supported glycosyl acceptor and the glycosyl donor are reactive enough to ensure completion of all glycosylations. The application of Approach B, on the other hand, is less straightforward, due to the following considerations. Firstly, every glycosylation inevitably gives rise to side reaction(s) (elimination, hydrolysis, etc.) together with the formation of the desired *O*-glycoside. These products arise largely from the glycosyl donor, so all products are accumulated on the support. In addition, transformation of the reducing-end anomeric position into a particular leaving group is required after each glycosylation. Nevertheless, it is necessary to make the choice of which method is to be used throughout the synthesis, because the synthetic schemes are completely reversed and so the choice of leaving group and protecting groups, including the linker, are different. Polymer-supported oligosaccharide syntheses appearing to date have been categorized in this context (Tab. 1.1).

1.4.3.1 Fundamentals of Solid-Phase Oligosaccharide Synthesis
To facilitate solid-phase oligosaccharide synthesis, several issues have to be addressed. The support may have a major influence on the reactions because of the demanding steric bulk close to the reaction site as well as physical properties. The choice of a suitable linker and protecting groups are important factors for the synthesis of multifunctional molecules such as oligosaccharides. Decisions regarding requirements for the reducing anomeric position after cleavage from the resin (i.e., hydroxy free or with a linker for further conjugation etc.) have to be made before the synthesis.

1.4.3.2 The Support
Polystyrene divinylbenzene cross-linked (PS) resins are mostly used, not only in carbohydrate chemistry but also in the synthesis of peptides and other small organic molecules. The main reason for this is their chemical stability toward a variety of chemical reaction conditions. However, there is room for improvement in areas such as swelling properties, which sometimes restrict the synthetic plan. Because of the multifunctionality of carbohydrates, it is to be expected that, for the

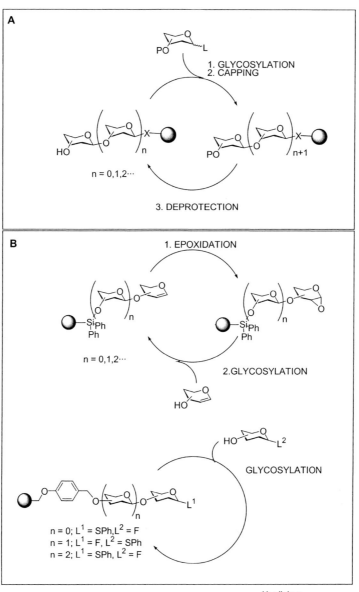

Fig. 1.3 Schematic representation of solid-phase oligosaccharide synthesis. A: A reducing sugar is attached on the support. The method typically operates in three steps for a cycle. B: A non-reducing end sugar is attached on the support, two examples being shown. One involves activation and glycosylation steps as a cycle and the other uses an orthogonal set of leaving groups, enabling a single step per cycle.

1 Synthetic Methodologies

Tab. 1.1 Structures of linkers and the conditions of cleavage.

Category	Structure of linkers	Condition of cleavage	Product	Support	Ref.
A	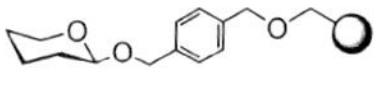 (sugar)-O-CH₂-C₆H₄-CH₂-O-◯	Raney nickel W2; Pd, H₂	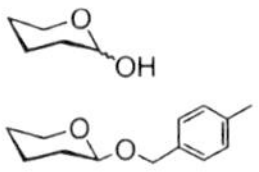 (sugar)-OH + (sugar)-O-CH₂-C₆H₄-CH₃	MPEG	44
A	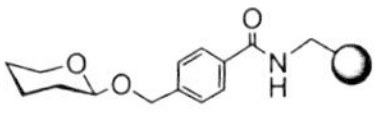 (sugar)-O-CH₂-C₆H₄-C(O)NH-◯	H₂, Pd-C, 40 psi	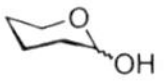 (sugar)-OH	MPEG	139
A	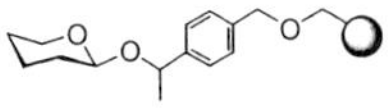 (sugar)-O-CH(CH₃)-C₆H₄-CH₂-O-◯	Sc(OTf)₃, Ac₂O	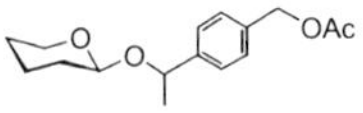 (sugar)-O-CH(CH₃)-C₆H₄-OAc	MPEG, PEG-PS	81
A, C	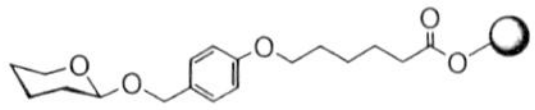 (sugar)-O-CH₂-C₆H₄-O-(CH₂)ₙ-C(O)O-◯	H₂, Pd(OH)₂ or TFA; H₂NNH₂·H₂O	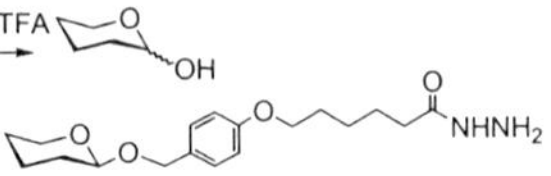 (sugar)-OH + (sugar)-O-CH₂-C₆H₄-O-(CH₂)ₙ-C(O)NHNH₂	MPEG, PS	140
A	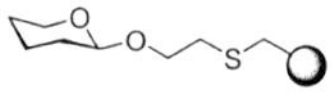 (sugar)-O-CH₂CH₂-S-◯	dioxirane then NaOMe	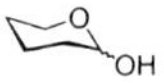 (sugar)-OH	MPEG	139
A	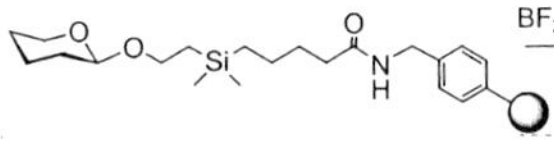 (sugar)-O-CH₂CH₂-Si(Me)₂-(CH₂)₃-C(O)NH-CH₂-C₆H₄-◯	BF₃·OEt₂, (RCO)₂O	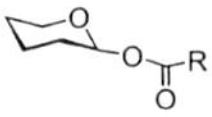 (sugar)-O-C(O)-R	PS	141
A	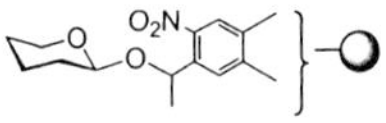 (sugar)-O-CH(CH₃)-C₆H₂(NO₂)(CH₃)-◯	hγ, THF	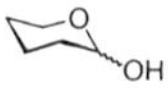 (sugar)-OH	PS	85, 104, 128
A	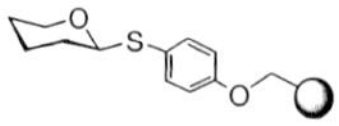 (sugar)-S-C₆H₄-O-CH₂-◯	Hg⁺⁺	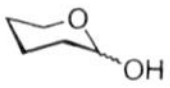 (sugar)-OH	PEG-Ps, PS	31, 32
A	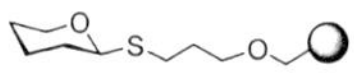 (sugar)-S-(CH₂)₃-O-◯	NBS, DTBP, ROH	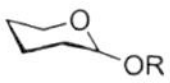 (sugar)-OR	PS, CPG	46, 49
A, B	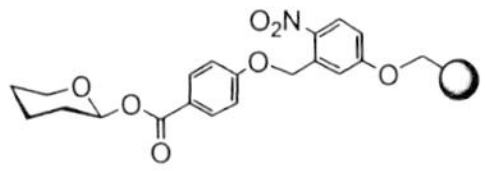 (sugar)-O-C(O)-C₆H₃(NO₂)-O-CH₂-C₆H₄-CH₂-◯	hγ, THF; TMSSPh, ZnI₂	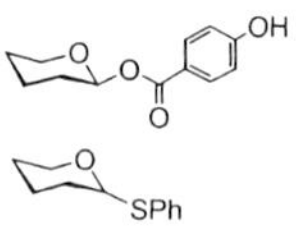 (sugar)-O-C(O)-C₆H₄-OH + (sugar)-SPh	PS	103
B	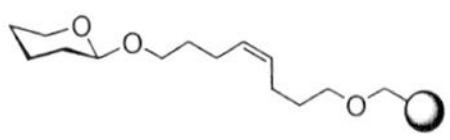 (sugar)-O-CH₂CH₂-CH=CH-CH₂CH₂-O-◯	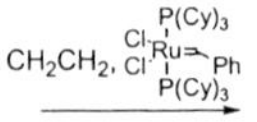 Cl₂(PCy₃)₂Ru=CHPh, CH₂=CH₂	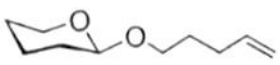 (sugar)-O-CH₂CH₂-CH=CH₂	PS	48
C	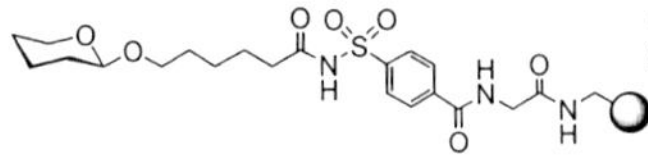 (sugar)-O-(CH₂)₄-C(O)NH-SO₂-C₆H₄-C(O)NH-CH₂-C(O)NH-◯	1) TMSCHN₂, THF; 2) 0.5 N NaOH, THF	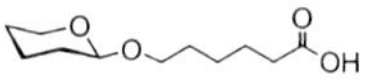 (sugar)-O-(CH₂)₄-C(O)OH	PEG-PS	134

Tab. 1.1 (cont.)

Category	Structure of linkers	Condition of cleavage	Product	Support	Ref.
D	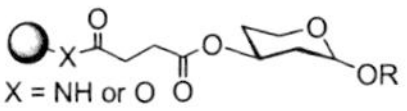 X = NH or O	NH₄OH		MPEG PEG-PS	79, 80, 116, 142
D	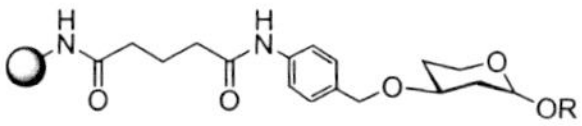	DDQ		PS	127
D	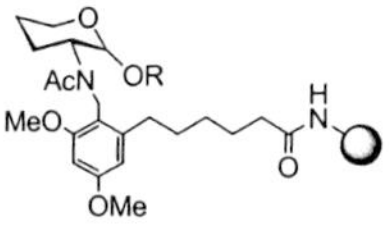	TFA, H₂O	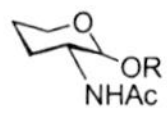	PS	86
D	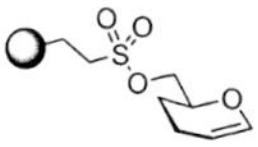	NaI, 2-butanone		PS	143
E	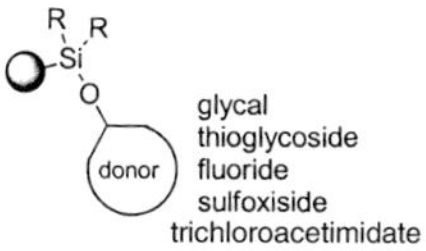 glycal, thioglycoside, fluoride, sulfoxiside, trichloroacetimidate (donor)	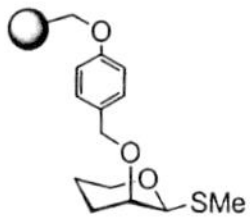		PS MPEG	67, 105–113

synthesis of an oligosaccharide, many reactions will have to be performed on the support. The suitability of the resin under various reaction conditions therefore has to be taken into account [102].

Merrifield-type polystyrene (PS) resins have most often been used in oligosaccharide synthesis [45–48, 85, 103–113]. Their chemical stability and compatibility in organic synthesis are the reasons for this choice. The commonly used PS resin is cross-linked with divinyl benzene ($\sim 2\%$) with a relatively large "active" reaction surface (up to 3 mmol g^{-1}), depending on the type of functional group and linker. Solvent-dependent swelling properties and the steric bulk of the polymer may affect reactivity and stereochemical outcome.

In order to address problems with PS resins, polyethyleneglycol (PEG) has been incorporated into PS resins, allowing reactions to proceed under quasi-solution conditions. In addition, PEG is compatible with reactions that require polar solvents. PEG-PS resins can also directly be used in biological assays, which is important for high-throughput screening [32].

To improve the polymer further, PEG-based polyether-type resins have been developed, eliminating the swelling problem seen in PS-based supports [114, 115]. Large pore sizes facilitate enzymatic reactions, receptor-ligand binding studies, and so on, thus enabling biological assay after the completion of synthesis. Furthermore, some PEG cross-linked polymers do not contain any UV-absorbing components, and so can be used effectively in photometric assays.

Although controlled pore glass (CPG) has been used extensively in oligonucleotide synthesis, it has seldom been used in the field of oligosaccharide synthesis [49, 102, 116]. It has recently been shown, though, that CPG can serve as a solid support in oligosaccharide synthesis, glycosyl trichloroacetimidate and TMSOTf being successfully used as the glycosyl donor and activator, respectively.

1.4.3.3 Linkers to the Support

A variety of linkers have been used to connect protected carbohydrate units to the support. Some of the linkers used are categorized by the position of the functional group on the sugar unit (i.e., anomeric position and others; Tab. 1.1), since a linker at the anomeric position may have to be cleaved when the completely deprotected oligosaccharide is the target (A), while in other cases the aglycon part of the linker may be converted into a potential leaving group or other protecting group to be used in the synthesis to follow (B). A spacer may be left and can be used to connect the formed oligosaccharide with other materials (C). A linker at other hydroxy functions can be regarded as one of the protecting groups (D, E).

1.4.3.4 Protecting Groups used in Solid-Phase Oligosaccharide Synthesis

The basic combination of protecting groups consists of one for the anomeric position, one for the hydroxy group involved in chain-elongation, and "persistent" protecting groups for the others. Selective sequential reactions of these protecting groups are the minimum requirement not only in solid-phase oligosaccharide synthesis but also in solution-phase synthesis. When branching structures have to be constructed on the support, an orthogonal protection scheme has to be employed (see Section 1.3.2).

As long as this requirement is fulfilled, any kind of protecting groups can be used if deprotection is planned after cleavage from the support. If, however, deprotection is completed while the oligosaccharide is still attached to the support, the beads can be used in screening assays directly [31, 117]. For this approach, acid- and base-sensitive protecting groups are frequently used. In addition, substituted benzyl groups have been introduced not only to provide more flexibility in the synthetic scheme but also to compensate the problem of inability of removing the benzyl-protecting group frequently used in solution-phase synthesis [118–123]. Removal of benzyl-protecting groups cannot be achieved by catalytic hydrogenolysis, probably due to steric problems. To address this, it was shown recently that palladium nanoparticles can be used for the catalytic hydrogenolysis of solid-supported compounds [124].

1.4.3.5 Solid-Phase Oligosaccharide Synthesis

There is a choice in the positions of hydroxy groups that can be used to anchor a synthetic unit onto a polymer support. The majority of researchers have selected the anomeric position to be connected to the support (Fig. 1.3A). Attachment of the first sugar through the anomeric position is much more straightforward, since attachment and protection of the anomeric position are achieved at once. Substituted benzyl-, ethyl-, pentenyl-, and ester-type linkers have been used to attach reducing sugars to the support (Tab. 1.1 A–C). However, linking of sugars through the other hydroxy groups offers a number of advantages. One can install temporary anomeric protecting groups to facilitate transformation of the oligosaccharide bound to the support into a glycosyl donor [79, 80, 86, 116, 125, 126]. More options in the synthetic scheme gives flexibility in the synthesis of oligosaccharides, glycoconjugates, and libraries. When, however, the donor is attached to the support (Fig. 1.3B), several improvements have to be employed for this method to be useful or advantageous.

Synthesis Starting from the Reducing Terminus
Glycosyl trichloroacetimidates have been shown to be powerful glycosyl donors in solid-phase oligosaccharide synthesis [45–48, 127]. These donors appear to be unaffected by the polymer support used in the synthesis, so PS resins [45–48], PEG [43, 44], and CPG [49] have all been used as supports. Scheme 1.16 illustrates one approach, in which it was shown that this leaving group could be used on CPG. A very straightforward synthesis of an α-$(1\rightarrow 2)$-linked trimannoside has been reported. In order to push the reaction to consume unreacted acceptor, the glycosylation reaction was carried out twice per cycle, and over 95% yield was achieved for each coupling step. The reducing terminus was coupled to the support through a thioglycosidic linkage, which was cleaved the last stage. The phenoxyacetyl group was used as a temporary protecting group and removed by treatment with guanidine in the presence of benzyl groups as the persistent protecting group.

Glycosylation reactions with a carbohydrate monomer already attached to a peptide sequence constructed on a PEGA resin have been investigated [127] (Scheme 1.17). A solid-phase-bound glycosylated octapeptide, the sequence of which is a part of mucin MUC 2 protein, was used as an acceptor and coupled with a glycosyl trichloroacetimidate. Di- and trisaccharide portions were constructed through the use of mono- and disaccharide donors. In addition, the removal of the benzylidene group in one of the products followed by further glycosylation of the diol was achieved in stereo- and regiospecific manner. The glycosyl acetimidates were activated by TMSOTf throughout the synthesis, but interestingly it was reported that only freshly distilled reagent was effective, which is unlike the observation in solution-phase chemistry. Some influence of the support on the stereochemical outcome was also reported.

A solid-phase synthesis of one of the most complex oligosaccharide structures is depicted in Scheme 1.18 [103]. In this scheme, two cleavage sites were introduced in an extremely flexible and powerful approach. The linker consists of a nitrobenzyl group and an ester function. The reducing ester linkage was cleaved upon activation by a Lewis acid in the presence of a thiol to provide an oligosaccharide glycosyl donor for the convergent synthesis, and the other part of the linker is a

Scheme 1.16

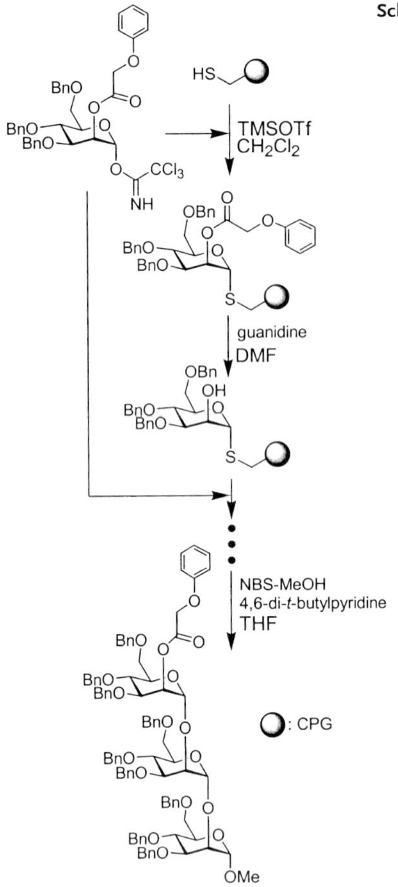

photolabile group [85, 104, 128], so the constructed oligosaccharide can be released at the end of the chain-elongation without affecting other protecting groups. This helps in determining of the structure of the constructed oligosaccharide, since anomerically pure compounds are released. Suitably protected thioglycosides were utilized as glycosylation agents throughout the synthesis. After iterative coupling and deprotection reactions, the trisaccharide was released as a thioglycoside, which was used in the following convergent synthesis (see Fig. 1.1 B). Thus, a dodecasaccharide was synthesized on PS resin.

Glycosyl sulfoxides have also been used in the solid-phase synthesis of oligosaccharides (Scheme 1.19). A β-(1 → 6)-galacto-trioside was synthesized [32]. A combination of an acid-labile trityl group as a temporary protecting group, pivaloyl groups as persistent protecting groups and also as auxiliaries for β-selectivity, and a phenylsulfonyl group were used for the synthesis on PS resin. Anchoring was achieved through a thioglycoside, which was cleaved by the action of $Hg(OCOCF_3)_2$ at the end of the synthesis.

Scheme 1.17

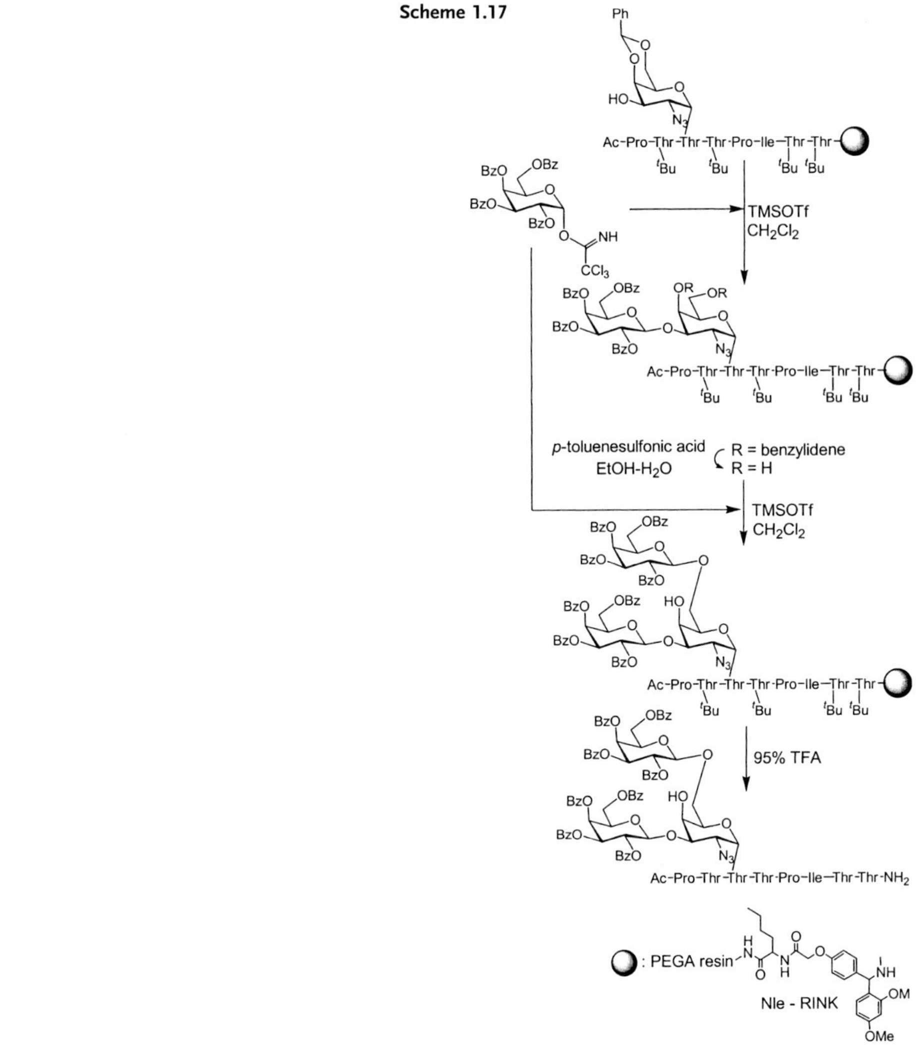

One challenge in oligosaccharide synthesis is the sialylation reaction, and this has been addressed through solid-phase synthesis. The general strategy is the use of benzyl groups as the persistent protecting group, the acetyl group as a short-term protecting group, and thioglycosides as general glycosyl donors. Sialyl LeX tetrasaccharide was successfully synthesized, together with all possible anomers. The tetrasaccharides have carboxylic acid functionalities at the reducing terminus, and these can be used to form conjugates with various materials. In addition, the reaction process was monitored nondestructively by a gated decoupling ^{13}C NMR technique (Scheme 1.20; see Section 1.4.3.6).

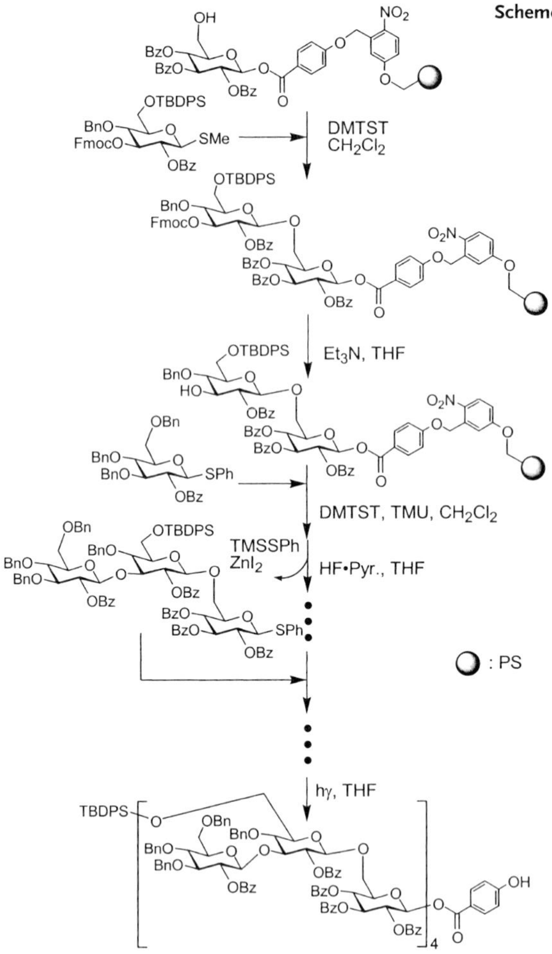

Scheme 1.18

Pentenyl glycosides have also been used successfully as glycosyl donors in combination with PS resin as the support. The synthesis of a branched trimannoside was achieved through a chemoselective deprotection scheme [104]. An overall yield of 42% was achieved, an average yield of 87% per step.

Synthesis Starting from the Non-Reducing Terminus
The assembly of oligosaccharides through the use of glycals as precursors of glycosyl donors has been investigated intensively [112, 129]. A characteristic of this method is the reversed synthetic direction (Fig. 1.1C), the glycosyl donor being attached to the support (Fig. 1.3B). As stated in the literature, a major advantage of the reversed method is the self-suppressing effect of formation of sequence deletion compounds. When the activation step to obtain a highly active epoxy or

Scheme 1.19

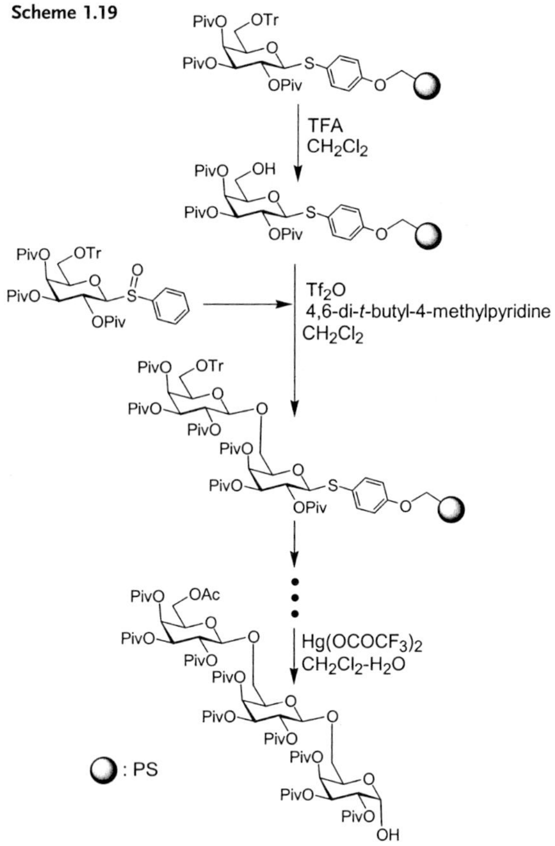

equivalent donor is complete, the donor undergoes a glycosylation reaction, producing only the coupling product and the hydrolyzed donor. Since the formed byproduct can be neither a substrate for the next activation reaction nor an acceptor for the next coupling, because of steric factors, these by-products do not affect further reactions. Therefore, regardless of the coupling yields, no capping step is required. Scheme 1.21 illustrates the glycal method [105, 112], by which a homotetrasaccharide was synthesized. The first carbohydrate was attached to a PS resin by silylation, and the double bond was oxidized with 2,2-dimethyldioxirane. The 1,2-epoxide produced was coupled with the primary OH group of a galactal acceptor to afford a β-galactosyl-linked disaccharide. Iterative reactions gave the tetrasaccharide in 32% overall yield. The β-stereocontrol of glycosylation with *galacto*-glycals notwithstanding, difficulty was reported in the case of *gluco*-type species, resulting in α-glycoside formation. To overcome this problem, the epoxide was converted into a thioglycoside with subsequent acylation, which acts as auxiliary to enhance β-stereoselectivity [110].

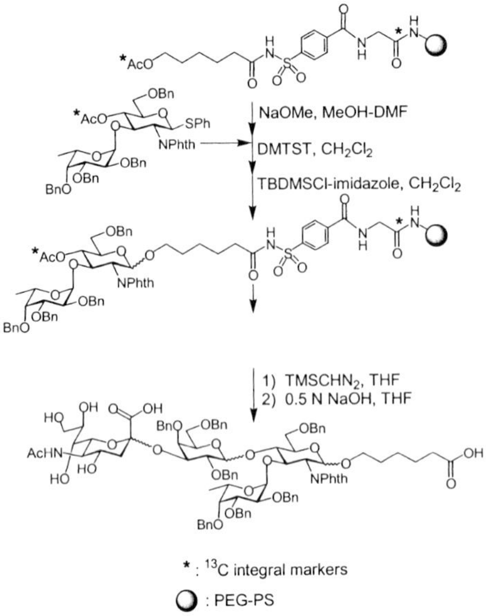

Scheme 1.20

*: ^{13}C integral markers
◯ : PEG-PS

A route to 2-amino sugars was addressed by transformation of the glycal into a 2-phenylsulfonamido-thioglycoside by rearrangement-displacement of 2-iodo-1-phenylsulfonamide [111]. A structure found in the reducing terminus of N-linked oligosaccharide was also synthesized, the oligopeptide part being incorporated on the support at the final stage of the coupling process. This is one of the advantages of the strategy, since diversity other than carbohydrate in nature can be added after completion of the oligosaccharide synthesis [108].

1.4.3.6 Monitoring of Reaction Progress

Monitoring of reaction progress in solid-phase synthesis is very important, especially for the optimization of reaction conditions, because cleavage of the product from a support to analyze the reaction by TLC negates the advantage of solid-phase synthesis. Gravimetric analysis has classically been used, but resin breakdown and difficulties associated with incomplete "dryness" for analysis prevents quantitative measurements. For qualitative analysis, IR [45], gel-phase ^{13}C NMR [130], and MALDI TOF MS [49] have been used.

Methods based on nuclear magnetic resonance (NMR) are nondestructive and reliable techniques widely used in oligosaccharide synthesis. ^1H NMR spectroscopy is one of the most informative analytical methods, frequently used in organic chemistry. However, the spectra typically obtained for solid-bound compounds are broadened and it is difficult to obtain quantitative or even qualitative information, due to the short relaxation time of the macromolecule. An exception is found, however, in the case of the soluble polymer MPEG (see Section 1.4.1). Reaction progress in this case can be monitored easily by standard experimental techniques

Scheme 1.21

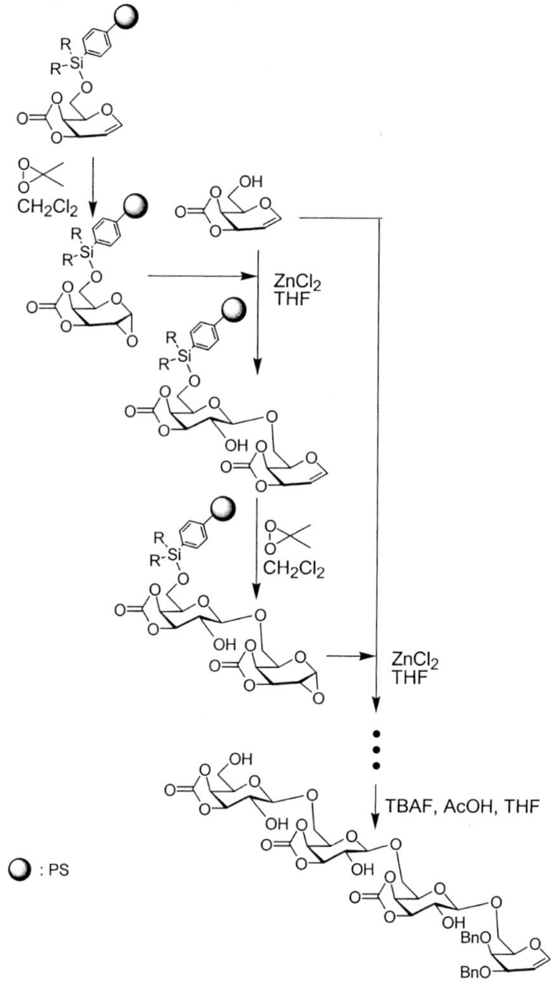

[67, 78]. Monitoring of the reaction progress of MPEG-based synthesis can also be performed by use of MALDI TOF MS [83, 131] (Fig. 1.4). Alternatively, high-resolution magic angle spinning NMR (HR-MAS) has also been used [106], and it has been shown that TOCSY is useful for obtaining coupling constants [132, 133] (Fig. 1.5).

To analyze the molecular structure of a product attached on the support, HR-MAS is probably the only method. Although the "high resolution" required for the coupling constants was not achieved by ^{1}H NMR, a trisaccharide bound to PS resin was analyzed on the basis of the chemical shifts of the anomeric protons and carbons after ^{1}H, ^{13}C, and HMQC experiments.

A conceptionally important and useful monitoring approach making use of ^{13}C NMR to monitor reaction progress focusing at a ^{13}C-enriched carbon center has

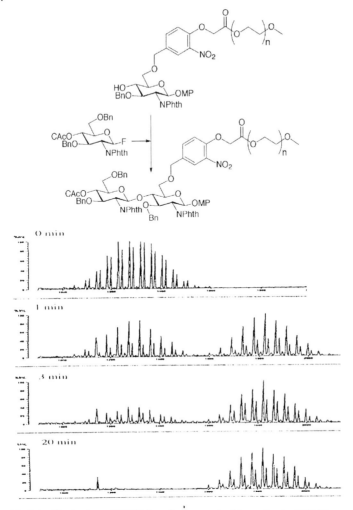

Fig. 1.4 Monitoring of MPEG-based oligosaccharide synthesis. MALDI-TOF mass spectrometry is used to monitor reaction progress in MPEG-supported oligosaccharide synthesis.

been reported [135] (Fig. 1.6; see Sections 1.4.3.5 and 1.4.4). In order to allow quantitative analysis of the solid-phase reaction by the so-called gel-phase ^{13}C NMR [130], the gated decoupling technique was used [136, 137]. ^{13}C-enriched temporary protecting groups and an internal ^{13}C marker have also been used, as well as a relaxation agent to obtain a short T_1 value. The method is particularly useful when only a small quantity of material has been synthesized and quantitative information is required. Through the use of gated decoupling techniques, the reaction progress can be monitored quantitatively without cleavage of the molecule from the resin. Since the yields are always given relative to an internal integral

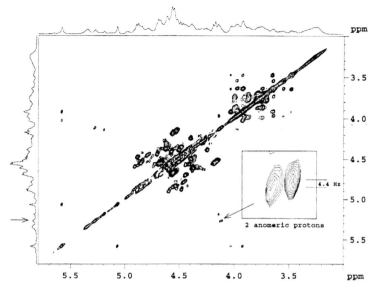

Fig. 1.5 Monitoring of solid-phase synthesis with HR-MAS. A TOCSY spectrum obtained by HR-MAS even gives anomeric coupling constants while the oligosaccharide is attached on the Merrifield resin.

marker, the chemical yields are determined regardless of the isolated yield, which is important for optimization of solid-phase reactions.

A more practical method, comparable to a Kaiser test in peptide synthesis, has been investigated [83]. In this method, a chloroacetyl group was used as a short-term protecting group for a hydroxy function for the next coupling reaction. Sequential treatment at the chloroacetyl stage with 4-(4-nitrobenzyl)pyridine and piperidine formed a zwitterion possessing a red color. Reaction progress could thus be monitored colorimetrically.

1.4.4
Automation

The emerging area of automation of oligosaccharide synthesis should contribute greatly not only to glycobiology but also to cell biology in general [59, 92, 93]. Access to structurally defined complex oligosaccharides has been very laborious, contrary to the needs for biological investigations. The final stage of a chemical synthetic method is the development of an automated system, and this has been addressed recently.

Seeberger used a modified peptide synthesizer equipped with a temperature-controlled reactor [135] (see Sections 1.4.3.5 and 1.4.4). The method was demonstrated in the cases of trichloroacetimidate and phosphate as the leaving groups; the trichloroacetimidate method is depicted in Fig. 1.7, which shows the sequential process program. In this way, non-specialists may soon be able obtain particu-

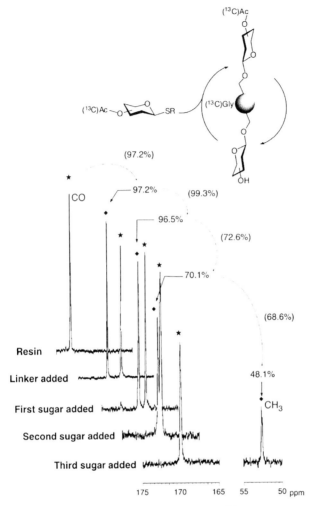

Fig. 1.6 Monitoring of solid-phase synthesis by ^{13}C NMR. Inverse gated decoupling ^{13}C NMR spectra with conventional NMR and use of a ^{13}C-enriched integral marker provide non-destructive monitoring of resin-bound (TentaGel) oligosaccharide synthesis.

lar oligosaccharides, although one issue pertinent for this type of system is the carbohydrate synthetic units. A variety of units have to be prepared one by one in a laboratory, and this matter still has to be resolved.

A different approach for automation by sequential one-pot glycosylation has also been developed (Fig. 1.8; see Section 1.4.2). A key issue in this approach is the use of a program to determine the glycosyl units needed in the synthesis of oligosaccharide. The basis of the method lies in analysis of the relative reactivities of carefully chosen suitably protected synthetic blocks in association with the armed and dis-

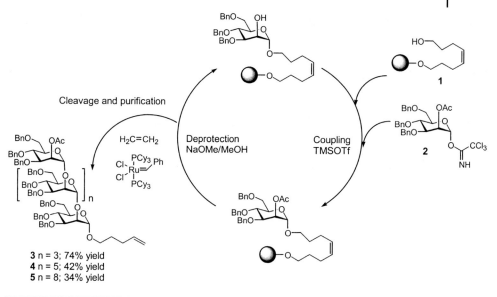

Step	Function	Reagent	Time (min)
1	Couple	10 equiv, donor and 0.5 equiv. TMSOTf	30
2	Wash	CH_2Cl_2	6
3	Couple	10 equiv, donor and 0.5 equiv. TMSOTf	30
4	Wash	CH_2Cl_2	6
5	Wash	1:9 MeOH:CH_2Cl_2	6
6	Deprotection	2×10 equiv. NaOMe (1:9 MeOH:CH_2Cl_2)	60
7	Wash	1:9 MeOH:CH_2Cl_2	4
8	Wash	0.2 M AcOH-THF	4
9	Wash	THF	4
10	Wash	CH_2Cl_2	6

Fig. 1.7 Automation of oligosaccharide synthesis based on solid-phase operations. Solid-phase oligosaccharide synthesis was automated for the first time with trichloroacetimidate and phosphate being successfully used as leaving groups.

armed concept [92, 93] (see Section 3.3). The choice of building blocks is stored in a database, from which researchers can select "suitable combinations" of glycosylating agents to be used in the one-pot sequential glycosylation. Expansion of the database is crucial for the success of this method, since estimation of the anomeric reactivities of differently protected carbohydrates is difficult. Another related problem can be seen when alkylthio and arylthio groups are used as glycosyl donor and acceptor. In some cases the former is preferentially activated, while in other cases the reactivity is reversed [137]. However, as long as a single substituent group at the anomeric position is used, the "programmable" oligosaccharide synthesis is considered effective since it has been shown that the relative reactivity number correlates with the chemical shift of the anomeric proton in the ^{1}H NMR [93].

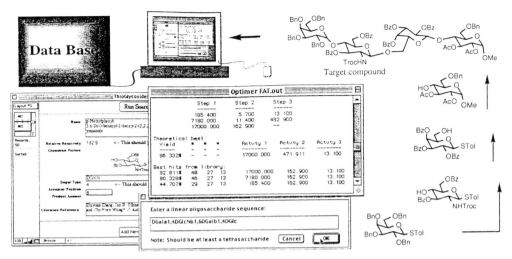

Fig. 1.8 A computer-assisted approach to the sequential one-pot synthesis of oligosaccharides. A database containing relative reactivities of synthetic units assists chemists in synthesizing target oligosaccharides.

1.5
Concluding Remarks

A variety of probes are needed for biological investigation of oligosaccharide functions. Methodological investigation is necessary for this purpose. Organic synthesis, enzymatic synthesis, isolation from natural sources, and/or combinations of each method can be utilized. This review summarizes the status of the organic synthesis of oligosaccharides, focusing on tactical aspects. The method best to be relied upon is an often debated matter, but it is important to obtain oligosaccharides by taking advantages of individual methods. One of the advantages of the synthetic method is that it is possible to access non-natural structures. Combinatorial oligosaccharide synthesis represents challenging but very important research in connection with approaches addressing infectious diseases. One has to take account of every aspect of current methods in oligosaccharide synthesis and to develop synthetic and engineering methods further in order for oligosaccharide probes to be available to all researchers.

1.6 References

1. LEMIEUX, R. U. *J. Am. Chem. Soc.* **1953**, *75*, 4118.
2. BOONS, G.-J.; HUBER, G. *Tetrahedron* **1996**, *52*, 1095–1121.
3. KOTO, S.; UCHIDA, T.; ZEN, S. *Bull. Chem. Soc. Jpn.* **1973**, *46*, 2520–2523.
4. NICOLAOU, K. C.; DOLLE, R. E.; PAPAHATJIS, D. P.; RANDALL, J. L. *J. Am. Chem. Soc.* **1984**, *106*, 4189–4192.
5. LÖNN, H. *Carbohydr. Res.* **1985**, *139*, 105–113.
6. RAGHAVAN, S.; KAHNE, D. *J. Am. Chem. Soc.* **1993**, *115*, 1580–1581.
7. PAULSEN, H.; TIETZ, H. *Carbohydr. Res.* **1985**, *144*, 205–229.
8. FRIESEN, R. W.; DANISHEFSKY, S. J. *J. Am. Chem. Soc.* **1989**, *111*, 6656–6660.
9. TRUMTEL, M.; VEYRIÈRES, A.; SINAŸ, P. *Tetrahedron Lett.* **1989**, *30*, 2529–2532.
10. VEENEMAN, G. H.; van BOOM, J. H. *Tetrahedron Lett.* **1990**, *31*, 275–278.
11. VEENEMAN, G. H.; van LEEUWEN, S. H.; ZUURMOND, H.; van BOOM, J. H. *J. Carbohydr. Chem.* **1990**, *9*, 783–796.
12. MORI, M.; ITO, Y.; UZAWA, J.; OGAWA, T. *Tetrahedron Lett.* **1990**, *31*, 3191–3194.
13. MEHTA, S.; PINTO, B. M. *Tetrahedron Lett.* **1991**, *32*, 4435–4438.
14. MARRA, A.; GAUFFENY, F.; SINAŸ, P. *Tetrahedron* **1991**, *47*, 5149–5160.
15. ZEGELAAR-JAARSVELD, K.; van der MAREL, G. A.; van BOOM, J. H. *Tetrahedron* **1992**, *48*, 10133–10148.
16. JAIN, R. K.; MATTA, K. L. *Carbohydr. Res.* **1992**, *226*, 91–100.
17. MARRA, A.; ESNAULT, J.; VEYRIÈRES, A.; SINAŸ, P. *J. Am. Chem. Soc.* **1992**, *114*, 6354–6360.
18. MEHTA, S.; PINTO, B. M. *J. Org. Chem.* **1993**, *58*, 3269–3276.
19. SLIEDREGT, L. A. J. M.; ZEGELAAR-JAARSVELD, K.; van der MAREL, G. A.; van BOOM, J. H. *Synlett* **1993**, 335–337.
20. YAMADA, H.; HARADA, T.; MIYAZAKI, H.; TAKAHASHI, T. *Tetrahedron Lett.* **1994**, *35*, 3979–3982.
21. YAMADA, H.; HARADA, T.; TAKAHASHI, T. *J. Am. Chem. Soc.* **1994**, *116*, 7919–7920.
22. CHENAULT, H. K.; CASTRO, A. *Tetrahedron Lett.* **1994**, *35*, 9145–9148.
23. GEURTSEN, R.; CÔTÉ, F.; HAHN, M. G.; BOONS, G.-J. *J. Org. Chem.* **1999**, *64*, 7828–7835.
24. PAULSEN, H. *Angew. Chem. Int. Ed. Engl.* **1982**, *21*, 155–173.
25. TOSHIMA, K.; TATSUTA, H. *Chem. Rev.* **1993**, *93*, 1503–1531.
26. SHIMIZU, M.; TOGO, H.; YOKOYAMA, M. *Synthesis* **1998**, 799–822.
27. GAREGG, P. J. *Adv. Carbohydr. Chem. Biochem.* **1997**, *52*, 179–205.
28. MOOTOO, D. R.; KONRADSSON, P.; UDODONG, U.; FRASER-REID, B. *J. Am. Chem. Soc.* **1988**, *110*, 5583–5584.
29. KANIE, O.; ITO, Y.; OGAWA, T. *J. Am. Chem. Soc.* **1994**, *116*, 12,073–12,074.
30. ROY, R.; ANDERSSON, F. O.; LETELLIER, M. *Tetrahedron Lett.* **1992**, *33*, 6053–6056.
31. LIANG, R.; YAN, L.; LOEBACH, J.; GE, M.; UOZUMI, Y.; SEKANINA, K.; HORAN, N.; GILDERSLEEVE, J.; THOMPSON, C.; SMITH, A.; BISWAS, K.; STILL, W. C.; KAHNE, D. *Science* **1996**, *274*, 1520–1522.
32. YAN, L.; TAYLOR, C. M.; GOODNOW, JR. R.; Kahne, D. *J. Am. Chem. Soc.* **1994**, *116*, 6953–6954.
33. SCHMIDT, R. R.; MICHEL, J. *Angew. Chem. Int. Ed. Engl.* **1980**, *19*, 731–732.
34. SCHMIDT, R. R. *Angew. Chem. Int. Ed. Engl.* **1986**, *25*, 212–235.
35. LASSALETTA, J. M.; CARLSSON, K.; GAREGG, P. J.; SCHMIDT, R. R. *J. Org. Chem.* **1996**, *61*, 6873–6880.
36. ISHIDA, H. K.; ISHIDA, H.; KISO, M.; HASEGAWA, A. *Carbohydr. Res.* **1994**, *260*, c1–c6.
37. MATSUZAKI, Y.; ITO, Y.; OGAWA, T. *Tetrahedron Lett.* **1992**, *33*, 6343–6346.
38. KANIE, O.; BARRESI, F.; DING, Y.; LABBE, J.; OTTER, A.; FORSBERG, L. S.; ERNST, B.; HINDSGAUL, O. *Angew. Chem. Int. Ed. Engl.* **1995**, *34*, 2720–2722.
39. DING, Y.; KANIE, O.; LABBE, J.; PALCIC, M. M.; ERNST, B.; HINDSGAUL, O. in *Glycoimmunology*, ALAVI, A.; AXFORD, J. S. Eds, Plenum Press, New York, **1995**, pp 261–269.
40. DING, Y.; LABBE, J.; KANIE, O.; HINDSGAUL, O. *Bioorg. Med. Chem.* **1996**, *4*, 683–692.

41 Douglas, S. P.; Whitfield, D. M.; Krepinsky, J. J. *J. Carbohydr. Chem.* **1993**, *12*, 131–136.
42 Nakahara, Y.; Nakahara, Y.; Ito, Y.; Ogawa, T. *Carbohydr. Res.* **1998**, *309*, 287–296.
43 Wang, Z.-G.; Douglas, S. P.; Krepinsky, J. J. *Tetrahedron Lett.* **1996**, *37*, 6985–6988.
44 Douglas, S. P.; Whitfield, D. M.; Krepinsky, J. J. *J. Am. Chem. Soc.* **1995**, *117*, 2116–2117.
45 Rademann, J.; Schmidt, R. R. *Tetrahedron Lett.* **1996**, *37*, 3989–3990.
46 Redemann, J.; Geyer, A.; Schmidt, R. R. *Angew. Chem. Int. Ed. Engl.* **1998**, *37*, 1241–1245.
47 Shimizu, H.; Ito, Y.; Kanie, O.; Ogawa, T. *Bioorg. Med. Chem. Lett.* **1996**, *6*, 2841–2846.
48 Andrade, R. B.; Plante, O. J.; Melean, L. G.; Seeberger, P. H. *Organic Lett.* **1999**, *1*, 1811–1814.
49 Heckel, A.; Mross, E.; Jung, K.-H.; Rademann, J.; Schmidt, R. R. *Synlett* **1998**, 171–173.
50 Mukaiyama, T.; Murai, Y.; Shoda, S.; *Chem. Lett.*, **1981**, 431–432.
51 Suzuki, K.; Maeta, H.; Matsumoto, T.; late Tsuchihashi, G. *Tetrahedron Lett.*, **1988**, *29*, 3571–3574.
52 Suzuki, K.; Maeta, H.; Matsumoto, T. *Tetrahedron Lett.* **1989**, *30*, 4853–4856.
53 Danishefsky, S. J.; Bilodeau, M. T. *Angew. Chem. Int. Ed. Engl.* **1996**, *35*, 1380–1419.
54 Danishefsky, S. J.; Allen, J. R. *Angew. Chem. Int. Ed. Engl.* **2000**, *39*, 836–863.
55 Fraser-Reid, B.; Udodong, U. E.; Wu, Z.; Ottosson, H.; Merritt, J. R.; Rao, C. S.; Roberts, C.; Madsen, R. *Synlett* **1992**, 927–942.
56 Schelhaas, M.; Waldmann, H. *Angew. Chem. Int. Ed. Engl.* **1996**, *35*, 2056–2083.
57 Barany, G.; Merrifield, R. B., *J. Am. Chem. Soc.* **1977**, *116*, 7363–7365.
58 Wong, C.-H.; Ye, X.-S.; Zhang, Z. *J. Am. Chem. Soc.* **1998**, *120*, 7137–7138.
59 Zhu, T.; Boons, G. J. *Tetrahedron: Asymmetry* **2000**, *11*, 199–205.
60 Veeneman, G. H.; van Leeuwen, S. H.; van Boom, J. H. *Tetrahedron Lett.* **1990**, *31*, 1331–1334.
61 Grice, P.; Ley, S. V.; Pietruszka, J.; Priepke, H. W. M.; Walther, E. P. E. *Synlett* **1995**, 781–784.
62 Hashimoto, S.; Sakamoto, H.; Honda, T.; Abe, H.; Nakamura, S.; Ikegami, S. *Tetrahedron Lett.* **1997**, *38*, 8969–8972.
63 Wilson, B. G.; Fraser-Reid, B. *J. Org. Chem.* **1995**, *60*, 317–320.
64 Douglas, N. L.; Ley, S. V.; Lücking, U.; Warriner, S. L. *J. Chem. Soc., Perkin Trans. 1* **1998**, 51–66.
65 Cheung, M.-K.; Douglas, N. L.; Hinzen, B.; Ley, S. V.; Pannecoucke, X. *Synlett* **1997**, 257–260.
66 Kanie, O.; Ito, Y.; Ogawa, T. *Tetrahedron Lett.* **1996**, *37*, 4551–4554.
67 Ito, Y.; Kanie, O.; Ogawa, T. *Angew. Chem. Int. Ed. Engl.* **1996**, *35*, 2510–2512.
68 Chang, G. X.; Lowary, T. L. *Org. Lett.* **2000**, *2*, 1505–1508.
69 Boons, G.-J.; Isles, S. *Tetrahedron Lett.* **1994**, *35*, 3593–3596.
70 Demchenko, A. V.; Boons, G.-J. *J. Org. Chem.* **2001**, *66*, 2547–2554.
71 Hindsgaul, O.; Kanie, O. *Curr. Opin. Struct. Biol.* **1992**, *2*, 674–681.
72 Demchenko, A. V.; Boons, G.-J. *Chem. Europ. J.* **1999**, *5*, 1278–1283.
73 Ishida, H. *Carbohydrates in Chemistry and Biology*, Ernst, B.; Hart, G. W.; Sinaÿ, P.; Eds, Vol. 1, Chapt. 12, pp 305–317, **2000**.
74 Schmidt, R. R.; Jung, K.-H. *Carbohydrates in Chemistry and Biology*, Ernst, B.; Hart, G. W.; Sinaÿ, P. Eds, Vol. 1, Chap. 2, pp 5–59, **2000**.
75 Hotta, K.; Ishida, H.; Kiso, M.; Hasegawa, A. *J. Carbohydr. Chem.* **1995**, *14*, 491–506.
76 Kameyama, A.; Ishida, H.; Kiso, M.; Hasegawa, A. *Carbohydr. Res.* **1990**, *200*, 269–285.
77 Ishida, H.; Kiso, M. *J. Synth. Org. Chem. Jpn.* **2000**, *58*, 1108–1113.
78 Miura, T.; Hirose, Y.; Ohmae, M.; Inazu, T. *Org. Lett.* **2001**, *3*, 3947–3950.
79 Douglas, S. P.; Whitfield, D. M.; Krepinsky, J. J. *J. Am. Chem. Soc.* **1991**, *113*, 5095–5097.
80 Verduyn, R.; van der Klein, P. A. M.; Douwes, M.; van der Marel, G. A.; van Boom, J. H. *Recl. Trav. Chim. Pays-Bas* **1993**, *112*, 464–466.

81 Mehta, S.; Whitfield, D. *Tetrahedron Lett.* **1998**, *39*, 5907–5910.
82 Jiang, L.; Hartly, C.; Chan, T.-H. *Chem. Commun.* **1996**, 2193.
83 Ando, H.; Manabe, S.; Nakahara, Y.; Ito, Y. *J. Am. Chem. Soc.* **2001**, *123*, 3848–3849.
84 Thürmer, R.; Meisenbach, M.; Echner, H.; Weiler, A.; Al-Qawasmeh, R. A.; Voelter, W.; Korff, U.; Schmitt-Sody, W. *Rapid Commun. Mass Spectrometry* **1998**, *12*, 398–402.
85 Nicolaou, K. C.; Winssinger, N.; Pastor, J.; DeRoose, F. *J. Am. Chem. Soc.* **1997**, *119*, 449–450.
86 Tolborg, J. F.; Jensen, K. J. *Chem. Commun.* **2000**, 147–148.
87 Barresi, F.; Hindsgaul, O. *J. Am. Chem. Soc.* **1991**, *113*, 9376–9377.
88 Stork, G.; Kim, G. *J. Am. Chem. Soc.* **1992**, *114*, 1087–1088.
89 Bols, M. *Tetrahedron*, **1993** *49*, 10049–10060.
90 Ito, Y.; Ogawa, T. *Angew. Chem. Int. Ed. Engl.* **1994**, *33*, 1765–1767.
91 Ito, Y.; Ogawa, T. *J. Am. Chem. Soc.* **1997**, *119*, 5562–5566.
92 Zhang, Z.; Ollmann, I. R.; Ye, X.-S.; Wischnat, R.; Baasov, T.; Wong, C.-H. *J. Am. Chem. Soc.* **1999**, *121*, 734–753.
93 Ye, X.-S.; Wong, C.-H. *J. Org. Chem.* **2000**, *65*, 2410–2431.
94 Manabe, S.; Ito, Y.; Ogawa, T. *Molecules Online* **1998**, *2*, 40–45.
95 Ley, S. V.; Priepke, H. W. M. *Angew. Chem. Int. Ed. Engl.* **1994**, *33*, 2292–2294.
96 Tsukida, T.; Yoshida, M.; Kurokawa, K.; Nakai, Y.; Achiha, T.; Kiyoi, T.; Kondo, H. *J. Org. Chem.* **1997**, *62*, 6876–6881.
97 Yamada, H.; Kato, T.; Takahashi, T. *Tetrahedron Lett.* **1999**, *40*, 4581–4584.
98 Yu, B.; Yu, H.; Hui, Y.; Han, X. *Tetrahedron Lett.* **1999**, *40*, 8591–8594.
99 Takeuchi, K.; Tamura, T.; Mukaiyama, T. *Chem. Lett.* **2000**, 122–123.
100 Takeuchi, K.; Tamura, T.; Mukaiyama, T. *Chem. Lett.* **2000**, 124–125.
101 Yoshida, M.; Kiyoi, T.; Tsukida, T.; Kondo, H. *J. Carbohydr. Chem.* **1998**, *17*, 673–682.
102 Adinolfi, M.; Barone, G.; Napoli, L. D.; Iadonisi, A.; Piccialli, G. *Tetrahedron Lett.* **1998**, *39*, 1953–1956.
103 Nicolaou, K. C.; Watanabe, N.; Li, J.; Pastor, J.; Wissinger, N. *Angew. Chem. Int. Ed. Engl.* **1998**, *37*, 1559–1561.
104 Rodebaugh, R.; Fraser-Reid, B.; Geysen, H. M. *Tetrahedron Lett.* **1997**, *38*, 7653–7656.
105 Danishefsky, S. J.; McClure, K. F.; Randolph, J. T.; Ruggeri, R. B. *Science* **1993**, *260*, 1307–1309.
106 Seeberger, P. H.; Beebe, X.; Sukenick, G. D.; Pochapsky, S.; Danishefsky, S. J. *Angew. Chem. Int. Ed. Engl.* **1997**, *36*, 491–493.
107 Randolph, J. T.; Danishefsky, S. J. *Angew. Chem. Int. Ed. Engl.* **1994**, *33*, 1470–1473.
108 Roberge, J. Y.; Beebe, X.; Danishefsky, S. J. *Science* **1995**, *269*, 202–204.
109 Randolph, J. T.; McClure, K. F.; Danishefsky, S. J. *J. Am. Chem. Soc.* **1995**, *117*, 5712–5719.
110 Zheng, C.; Seeberger, P. H.; Danishefsky, S. J. *J. Org. Chem.*, **1998**, *63*, 1126–1130.
111 Zheng, C.; Seeberger, P. H.; Danishefsky, S. J. *Angew. Chem. Int. Ed. Engl.* **1998**, *37*, 786–787.
112 Danishefsky, S. J.; Bilodeau, M. T. *Angew. Chem. Int. Ed. Engl.* **1996**, *35*, 1380–1419.
113 Doi, T.; Sugiki, M.; Yamada, H.; Takahashi, T. *Tetrahedron Lett.* **1999**, *40*, 2141–2144.
114 Renil, M.; Meldal, M. *Tetrahedron Lett.* **1996**, *37*, 6185–6188.
115 Rademann, J.; Grøtli, M.; Meldal, M.; Bock, K. *J. Am. Chem. Soc.* **1999**, *121*, 5459–5466.
116 Adinolfi, M.; Barone, G.; De Napoli, L.; Iadonisi, A.; Piccialli, G. *Tetrahedron Lett.* **1996**, *37*, 5007–5010.
117 Liang, R.; Loebach, J.; Horan, N.; Ge, M.; Thompson, C.; Yan, L.; Kahne, D. *Proc. Natl. Acad. Sci. USA* **1997**, *94*, 10554–10559.
118 Yan, L.; Kahne, D. *Synlett* **1995**, 523–524.
119 Oikawa, Y.; Yoshioka, T.; Yonemitsu, O. *Tetrahedron Lett.* **1982**, *23*, 885–888.
120 Jobron, L.; Hindsgaul, O. *J. Am. Chem. Soc.* **1999**, *121*, 5835–5836.
121 Bouzide, A.; Sauvé, G. *Synlett* **1997**, 1153–1154.

122 Fukase, K.; Hashida, M.; Kusumoto, S. *Tetrahedron Lett.* **1991**, *32*, 3557–3558.
123 Fukase, K.; Yoshimura, T.; Hashida, M.; Kusumoto, S. *Tetrahedron Lett.* **1991**, *32*, 4019–4022.
124 Kanie, O.; Grotenberg, G.; Wong, C.-H. *Angew. Chem. Int. Ed. Engl.* **2000**, *39*, 4545–4547.
125 Zhu, T.; Boons, G.-J. *Angew. Chem. Int. Ed. Engl.* **1998**, *37*, 1898–1900.
126 Fukase, K.; Nakai, Y.; Egusa, K.; Porco, J. A. Jr.; Kusumoto, S. *Synlett* **1999**, 1074–1078.
127 Paulsen, H.; Schleyer, A.; Mathieux, N.; Meldal, M.; Bock, K. *J. Chem. Soc., Perkin Trans. 1* **1997**, 281–294.
128 Kantchev, A. B.; Parquette, J. R. *Tetrahedron Lett.* **1999**, *40*, 8049–8053.
129 Danishefsky, S. J.; Allen, J. R. *Angew. Chem. Int. Ed. Engl.* **2000**, *39*, 836–863.
130 Rodebaugh, R.; Joshi, S.; Fraser-Reid, B.; Geysen, H. M. *J. Org. Chem.* **1997**, *62*, 5660–5661.
131 Thürmer, R.; Meisenbach, M.; Echner, H.; Weiler, A.; Al-Qawasmeh, R. A.; Voelter, W.; Korff, U.; Schmitt-Sody, W. *Rapid Commun. in Mass Spectrometry* **1988**, *12*, 398–402.
132 Plante, O. J.; Palmacci, E. R.; Seeberger, P. H. *Science* **2001**, *291*, 1523–1527.
133 Haase, W.-C.; Seeberger, P. H.; Pochapsky, S. S. *Solid Support Oligosaccharide Synthesis and Combinatorial Carbohydrate Libraries*, Seeberger, P. H. Ed, Chapt. 8, pp 165–174.
134 Choudhury, A. K.; Mukherjee, I.; Mukhopadhyay, B.; Roy, N. *J. Carbohydr. Chem.* **1999**, *18*, 361–367.
135 Look, G. C.; Holmes, C. P.; Chinn, J. P.; Gallop, M. A. *J. Org. Chem.* **1994**, *59*, 7588–7590.
136 Kanemitsu, T.; Kanie, O.; Wong, C.-H. *Angew. Chem. Int. Ed. Engl.* **1998**, *37*, 3415–3418.
137 Kanemitsu, T.; Wong, C.-H.; Kanie, O. *J. Am. Chem. Soc.* **2002**, *124*, 3591–3599.
138 Jiang, L.; Hartley, R. C.; Chan, T.-H. *Chem. Commun.* **1996**, 2193–2194.
139 Manabe, S.; Ito, Y.; Ogawa, T. *Synlett* **1998**, 628–630.
140 Weigelt, D.; Magnusson, G. *Tetrahedron Lett.* **1998**, *39*, 2839–2842.
141 Kononov, L. O.; Ito, Y.; Ogawa, T. *Tetrahedron Lett.* **1997**, *38*, 1599–1602.
142 Hunt, J. A.; Roush, W. R. *J. Am. Chem. Soc.* **1996**, *118*, 9998–9999.

2
Complex Carbohydrate Synthesis
Makoto Kiso, Hideharu Ishida, and Hiromune Ando

2.1
Introduction

Complex carbohydrates on cell surfaces (Fig. 2.1) have been recognized to play crucial roles in various biological processes such as infection, reception of toxins and hormones, fertilization, cell adhesion, cell differentiation and proliferation, tumor progression, aging, immune responses, brain-neural functions, and so on. The oligosaccharide components, exposed to the external environment as ligands or receptors, code a variety of biological information to regulate the physiological functions.

Gangliosides, glycosphingolipids containing sialic acid (Sia), are a class of structurally diverse molecules common in vertebrate plasma membranes and especially abundant in nerve tissues. These molecules not only express many prime physiological activities by themselves, but are also highly useful as versatile tools for elucidation of carbohydrate functions (Fig. 2.2).

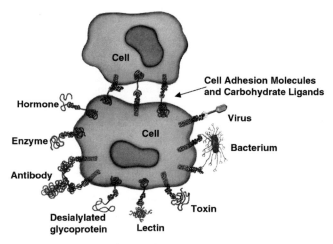

Fig. 2.1 Interactions of cell surface complex carbohydrates with various receptor proteins.

38 | 2 Complex Carbohydrate Synthesis

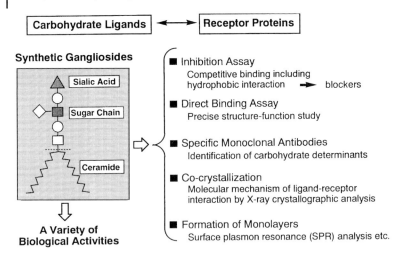

Fig. 2.2 Synthetic ganglioside probes: versatile tools for elucidation of carbohydrate functions.

2.2
Synthetic Gangliosides

The systematic synthesis of gangliosides, their analogues, and derivatives over 500 species has been achieved by the development of a facile regioselective and α-predominant sialidation in acetonitrile [1–3]. Fig. 2.3 shows some typical synthetic gangliosides in which Sia-$\alpha(2 \to 3)$-Gal and Sia-$\alpha(2 \to 6)$-Gal structures are common glyco-codons involved in both immune and neural systems.

Additional recognition specificities are given by the elongation of glycan chains containing GlcNAc (Lacto series) and GalNAc (Ganglio series). These synthetic gangliosides have successfully been utilized to elucidate not only direct biological activities, but also the molecular mechanisms of interactions between carbohydrate ligands and receptor proteins [4].

2.2.1
Gangliosides GM4 and GM3, and their Analogues and Derivatives

Gangliosides GM4 and GM3, and their analogues and derivatives have been systematically synthesized [5–7] by starting from the Sia-$\alpha(2 \to 3)/\alpha(2 \to 6)$-Gal donors (Fig. 2.4) and the corresponding Sia-$\alpha(2 \to 3)/(2 \to 6)$-Lac donors. It has been shown that chemically synthesized gangliosides GM3 and GM4 had the same high degree of immunosuppressive activity as the natural GM3 and GM4 gangliosides [8a].

GSC-53, a GM3 analogue containing a 2-(tetradecyl)hexadecyl (B30) group in place of ceramide, has been found to be a potent inhibitor of cellular immune response, comparable to cyclosporin A [8b] (Fig. 2.5). The corresponding synthetic GM4 analogue GSC-188, again with a B30 group, has also been found to be highly immunosuppressive.

2.2 Synthetic Gangliosides

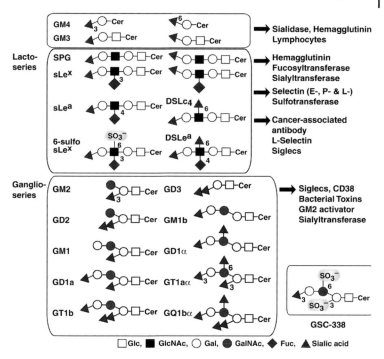

Fig. 2.3 Synthetic gangliosides and receptor proteins.

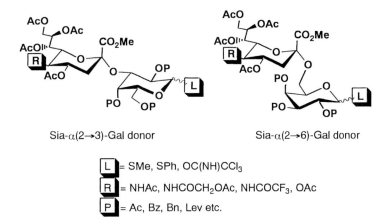

Fig. 2.4 The common building blocks of Sia-$\alpha(2 \rightarrow 3)/\alpha(2 \rightarrow 6)$-Gal donors.

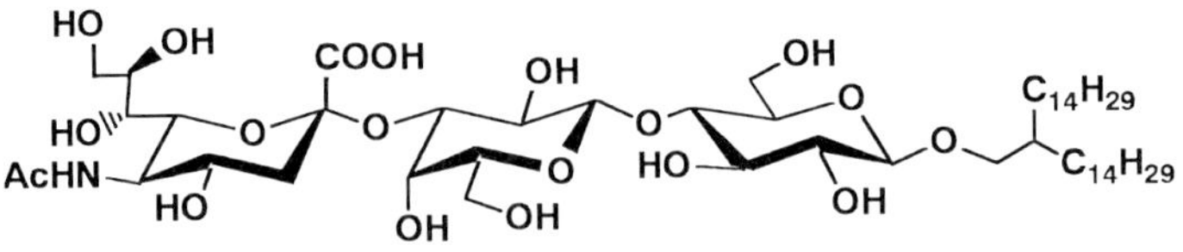

GSC-53 (GM3 analog)

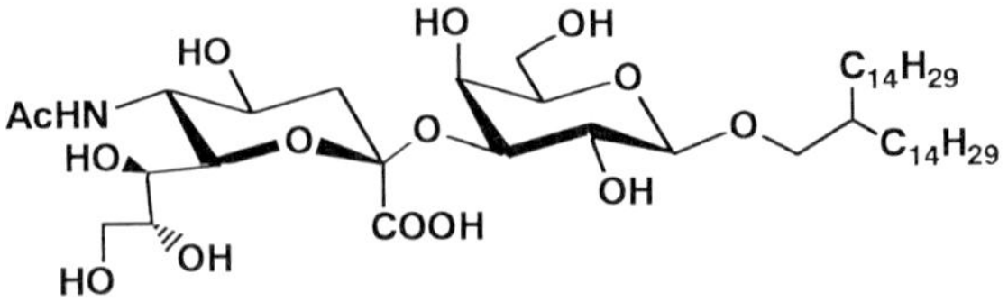

GSC-188 (GM4 analog)

Fig. 2.5 Potent immunosuppressive ganglioside analogues.

In a search for new ligand structures for human L-selectin involved in lymphocyte homing and leukocyte recruitment to sites of inflammation, novel ganglioside GM4 analogues containing N-deacetylated (2) or lactamized (3) sialic acid in place of the usual N-acetylneuraminic acid have been synthesized in a highly efficient manner [9] (Scheme 2.1).

2.2.2
Sialylparagloboside (SPG) Analogues and Derivatives

Sialyl lacto series type I [Siaα(2→3)(6)-Galβ1–3GlcNAcβ-] and type II [Siaα(2→3)(6)-Galβ1–4GlcNAcβ-] chains are highly useful for elucidation of the recognition specificities of influenza virus hemagglutinin (HA) and sialidase, as well as fucosyl- and sialyltransferases involved in the biosynthesis of sialyl Lewis x (sLe^x), sialyl Lewis a (sLe^a), and related cancer-associated antigens. As shown in Fig. 2.6, four kinds of Sia-α(2→3)/α(2→6)-linked sialylparagloboside (SPG) analogues containing N-acetyl- and N-glycolylneuraminic acids have been synthesized by coupling of Sia-α(2→3)/α(2→6)-Gal donors (Fig. 2.4) with suitably protected lactotriose derivatives [10].

Our recent studies employing these synthetic SPG analogues have demonstrated that: (1) substitution of the amino acid residue in influenza A virus HA affects recognition of sialooligosaccharides containing NeuGc [11], (2) the recognition of NeuGc linked to Gal residue through the α2–3 linkage is associated with intestinal replication of influenza A virus in ducks [12], and (3) the sialic acid species (NeuAc or NeuGc) and the difference in sialyl-Gal linkages (α2–3 or α2–6) determine the host ranges of influenza A viruses [13]. As shown in Fig. 2.7, human influenza viruses preferentially bind Siaα2–6Gal containing NeuAc, while bird and horse influ-

Scheme 2.1 a) TMSOTf/CH$_2$Cl$_2$ (72%), b) NaOMe/MeOH, then H$_2$O, c) WSC·HCl/DMSO, 60 °C (71%).

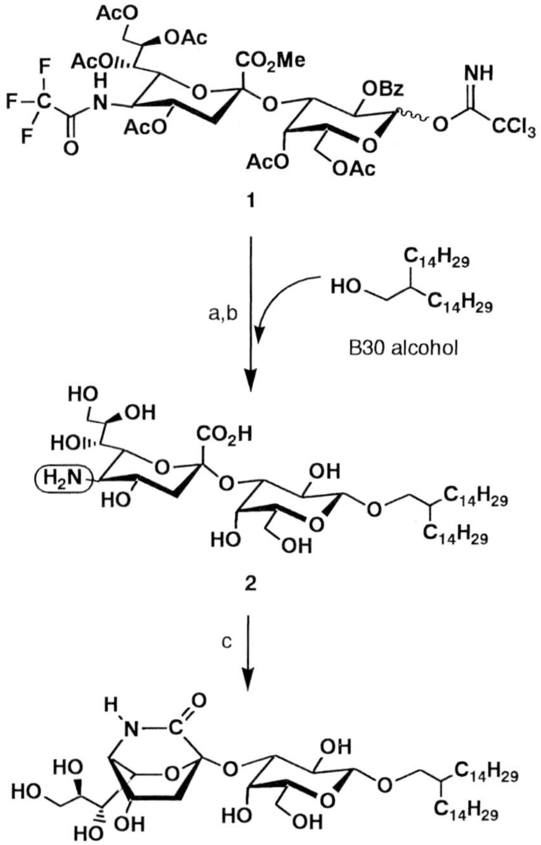

enza viruses preferentially bind Sia$a$2–3Gal containing NeuAc/NeuGc (duck), NeuAc (chicken), or NeuGc (horse), respectively. On the other hand, pig influenza viruses could bind both Sia$a$2–6Gal and Sia$a$2–3Gal containing NeuAc and/or NeuGc. These binding specificities correlate well with the sialooligosaccharide structures that express on the tracheal or intestinal epithelial cells in host animals.

The systematic synthesis of sialyl-a(2 → 3)-neolactotetraose derivatives modified at C-5 of Sia [14] and at C-2 of GlcNAc [15], and sulfated at C-6 of GlcNAc and/or Gal [16] has been achieved (Fig. 2.8), and their acceptor specificity for an a(1 → 3)-fucosyltransferase (Fuc-TVII) involved in the biosynthesis of L-selectin ligand was examined by inhibition assay. The 6-sulfo derivative **5**, in which the hydroxyl group at C-6 of GlcNAc is sulfated, showed much higher binding affinity than the nonsulfated SPG oligosaccharide **4**, while 6′-sulfo and 6,6′-disulfo derivatives (**6** and **7**) did not exhibit any binding affinity up to 100 µM (Fig. 2.9). It therefore seems more plausible that the regiospecific 6-sulfation of GlcNAc catalyzed by 6-sulfotransferase occurs prior to Fuc-TVII-catalyzed fucosylation, ultimately to form the 6-sulfo sLe^x determinant, an endogenous L-selectin ligand found on high endothelial venules (HEVs) of lymph nodes [17] (Fig. 2.10).

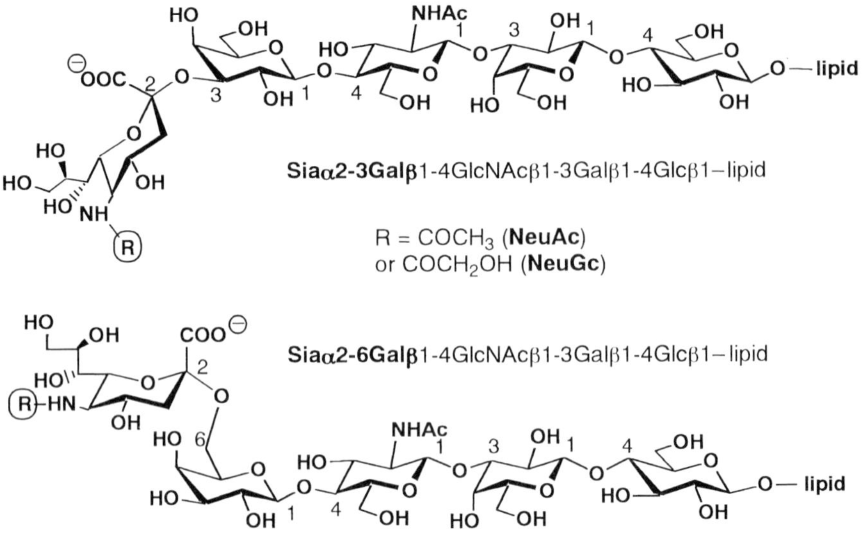

2 Complex Carbohydrate Synthesis

Fig. 2.6 Synthetic sialylparagloboside analogues.

Siaα2-3Galβ1-4GlcNAcβ1-3Galβ1-4Glcβ1–lipid

R = COCH$_3$ (**NeuAc**)
or COCH$_2$OH (**NeuGc**)

Siaα2-6Galβ1-4GlcNAcβ1-3Galβ1-4Glcβ1–lipid

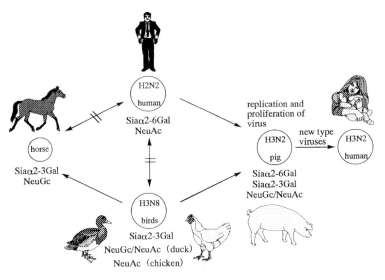

Fig. 2.7 The host ranges of influenza A viruses are determined by a combination of the sialic acid species (NeuAc/NeuGc) and differences in sialyl-Gal linkages (α2–3/α2–6).

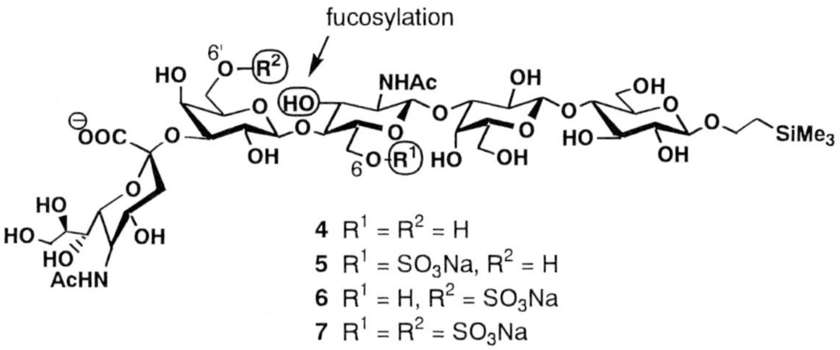

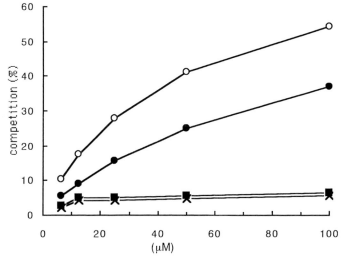

Fig. 2.8 Sulfated sialyl-α(2 → 3)-neolactotetraose derivatives.

Fig. 2.9 Competition between non-sulfated (**4**) and sulfated (**5–7**) sialyl-α(2 → 3)-neolactotetraose derivatives against the pyridylaminated derivative for Fuc-TVII. Non-sulfated **4** (●), 6-sulfo **5** (○), 6'-sulfo **6** (■), and 6,6'-bissulfo **7** (×).

2.2.3
Selectin Ligands

E-, P-, and L-selectin are a family of cell adhesion receptors belonging to C-type lectin, each possessing an *N*-terminal carbohydrate recognition domain [18–20]. E- and P-selectin are induced on the endothelial surface and platelet, whereas L-selectin is constitutively expressed on all classes of circulating leukocytes. Interaction between selectins and specific carbohydrate ligands, which is dependent on calcium ion concentration, results in cell adhesion. This is intimately involved in the behavior of leukocytes in sites of inflammation, the homing of lymphocytes,

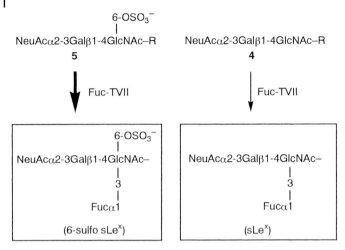

Fig. 2.10 Regiospecific 6-sulfation of GlcNAc catalyzed by 6-sulfotransferase probably occurs prior to Fuc-TVII-catalyzed fucosylation.

the permeation and the metastasis of malignant cells, and the development of blood vessel disease such as arteriosclerosis.

2.2.3.1 Sialyl Lewis x

Since the first total synthesis [21] of sialyl Lewis x (sLe^x) ganglioside (**8**, Fig. 2.11), it has been demonstrated that: (1) the three members of the selectin receptor family (E-, P-, and L-) recognize a common carbohydrate epitope, the sLe^x oligosaccharide [22], and (2) the five hydroxyl groups (2-, 3-, and 4-OH of Fuc and 4- and 6-OH of Gal) and the carboxyl group of sialic acid are essential for selectin binding [23], which has resulted in the selectin blocker GSC-150 [23, 24] (Fig. 2.12) and related selectin binding inhibitors [25–27].

This structure-activity mapping has recently been strongly supported by an X-ray crystallographic study of P- and E-selectins in complexation with sLe^x and PSGL-1 [28].

For L-selectin ligands, a novel O-glycan analogue of GlyCAM-1 has been synthesized [29] in an efficient manner.

A series of sulfated sLe^x gangliosides (**9–11**; Fig. 2.11) has been systematically synthesized [30], and a binding assay with L-selectin revealed that 6-sulfo sLe^x (**9**) possessed the highest binding activity. The hierarchy of binding strengths was 6-sulfo sLe^x (**9**) > sLe^x (**8**) = 6,6′-disulfo sLe^x (**11**) ≫ 6′-sulfo sLe^x (**10**) [31]. Moreover, the newly generated monoclonal antibody (G152), which was obtained by immunization against 6-sulfo sLe^x ganglioside (**9**), clearly reacted with the human high endothelial venule (HEV), and strongly inhibited the L-selectin-dependent leukocyte adhesion [17]. This is the first report that established the endogenous ligand structure for L-selectin on HEV through the use of the specific monoclonal anti-

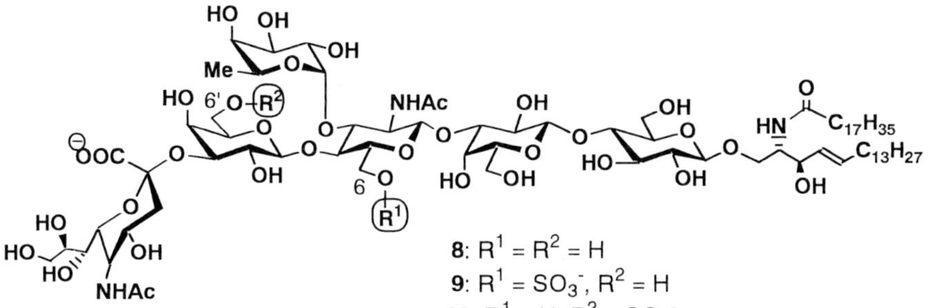

Fig. 2.11 The molecular structures of sialyl Lewis x (sLe^x) ganglioside (**8**) and 6/6' mono- or disulfated sLe^x gangliosides **9–11**.

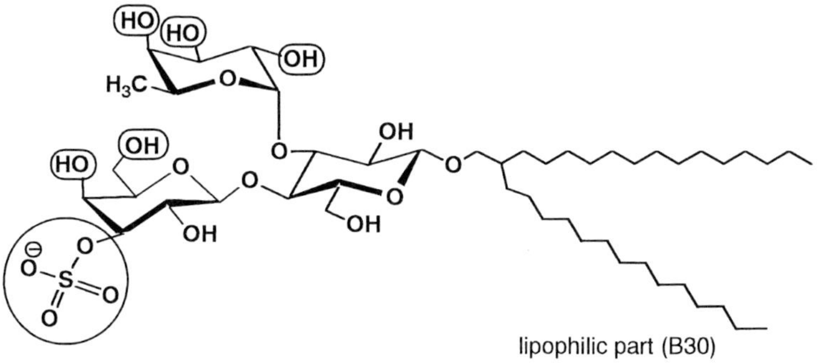

8: R^1 = R^2 = H
9: R^1 = SO$_3^-$, R^2 = H
10: R^1 = H, R^2 = SO$_3^-$
11: R^1 = R^2 = SO$_3^-$

Fig. 2.12 GCS-150.

body (MoAb). The result was further supported by investigations of the acceptor specificity of the synthetic 6/6'-O-sulfated sialyl-α-(2–3)-neolactotetraose probes for a human α-(1–3)-fucosyltransferase (Fuc-TVII) (see Section 2.2.2).

2.2.3.2 Novel 6-Sulfo sLe^x Variants

In 1996 and 1997, two novel 6-sulfo sLe^x gangliosides (Fig. 2.13) – 6-sulfo de-*N*-acetylsialyl Lewis x (**12**), obtained by de-*N*-acetylation of the acetamide group, and 6-sulfo cyclic sialyl Lewis x (**13**), resulting from intramolecular amide bond formation – were accidentally discovered in the course of the chemical synthesis of **9** [30, 31]. The total synthesis of **12** has been achieved [32] by coupling of the *N*-trifluoroacetylneuraminyl-α-(2 → 3)-galactopyranosyl imidate donor (**1** in Scheme 2.1) with a suitably protected lactotriose acceptor (Scheme 2.2). Fucosylation, introduction of the ceramide part into the reducing end, sulfation, and complete removal of the *O*-acyl and *N*-trifluoroacetyl groups, followed by saponification of the methyl ester group of sialic acid, afforded the desired 6-sulfo de-*N*-acetyl sLe^x ganglioside **12**. This synthetic method was further applied to the systematic synthesis

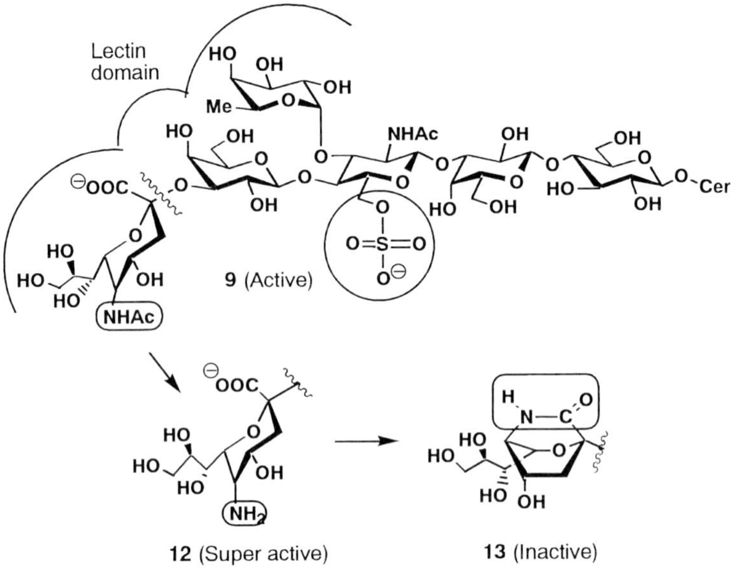

Fig. 2.13 Hypothetical metabolic pathway for 6-sulfo sLe^x (**9**).

of a series of de-N-acetylated and lactamized sLe^x glycolipids [33]. The non-sulfated de-N-acetyl sLe^x ganglioside was also synthesized [34] and the strengths of L-selectin binding for a series of sLe^x variants were compared. Surprisingly, the 6-sulfo de-N-acetyl sLe^x (**12**; Fig. 2.13) was a better L-selectin ligand than the N-acetyl form (**9**), while the cyclic form (**13**) was inactive in binding (Fig. 2.13). The hierarchy of the binding strengths was 6-sulfo de-N-acetyl sLe^x (**12**) > 6-sulfo sLe^x (**9**) = de-N-acetyl sLe^x > sLe^x (**8**) ≫ 6-sulfo cyclic sLe^x (**13**). It is therefore plausible that an endogenous L-selectin ligand **9** (N-acetyl form) may be further activated by de-N-acetylation to give **12** as a super-active ligand, which is then converted to the inactive cyclic form **13** [35]. Such dramatic changes in physiological activities through the modification of the sialic acid residue have not so far been recognized. The total synthesis of **13** has also very recently been achieved [36] to clarify the hypothetical metabolic pathway of 6-sulfo sLe^x (**9**; Fig. 2.13) at the molecular level.

2.2.4
Siglec Ligands

Siglecs (Sialic acid binding Ig-like lectins) are another group of carbohydrate-binding proteins belonging to the immunoglobulin (Ig) superfamily. Eleven members of the family have so far been reported in humans [37], including sialoadhesin (Siglec-1, expressed on macrophages), CD22 (Siglec-2, on B lymphocytes), CD33 (Siglec-3, on myeloid precursors and monocytes), etc. [38]. Myelin-associated glycoprotein (MAG, Siglec-4a) and Schwann cell myelin protein (SMP, Siglec-4b) are

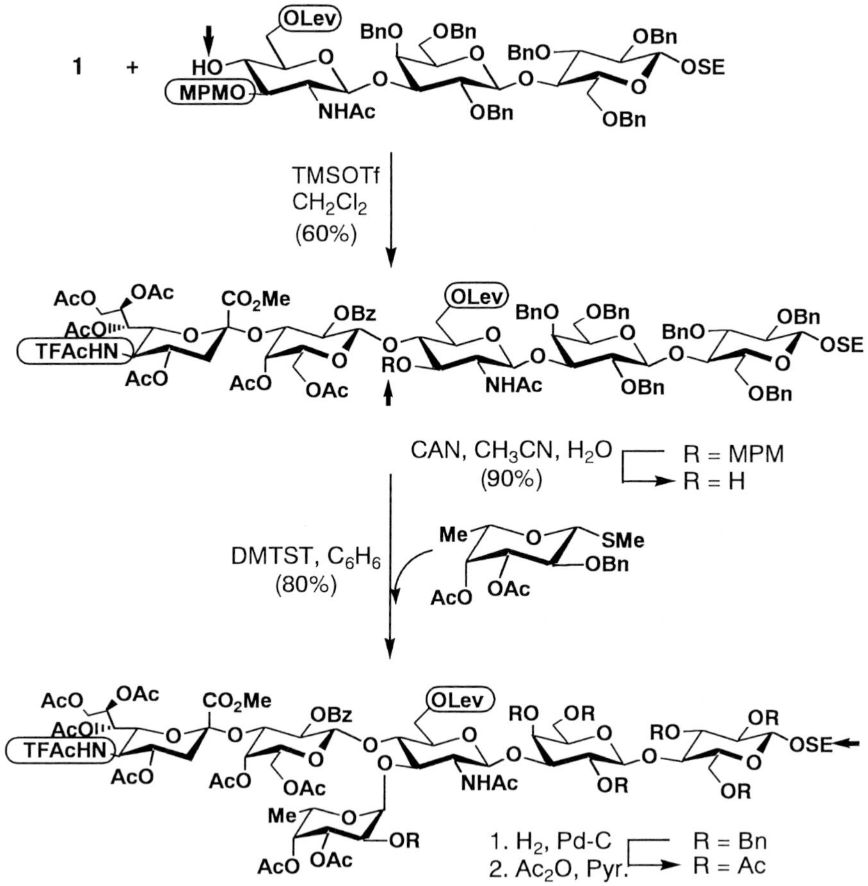

Scheme 2.2 Total synthesis of 6-sulfo de-N-acetylsialyl Le^x ganglioside.

representatives of nervous system Siglecs involved in myelin maintenance and in myelin-axon interactions [39]. Although it is hypothesized that gangliosides are functional MAG ligands, the details of the structure required for the binding are still obscure.

2.2.4.1 Chol-1 (a-Series) Gangliosides

Chol-1 (a-series) gangliosides, such as GD1a, GT1aa, and GQ1ba (Fig. 2.14) are defined as a new series of gangliosides containing NeuAc linked to the C-6 of Gal-NAc of the gangliotetraosyl backbone (Fig. 2.3). They are thought to be only minor components, and little is known about their physiological functions. Because the expression of these gangliosides is restricted to a particular region and a particular population in brain tissues, it has been suggested that these gangliosides

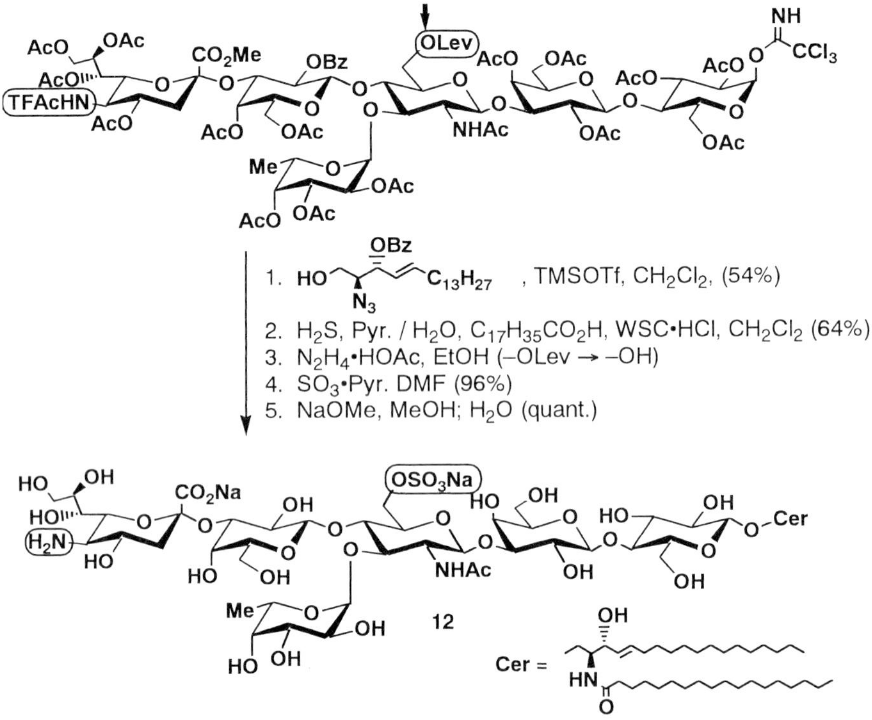

1. HO−CH(N₃)−CH=CH−C₁₃H₂₇ (OBz), TMSOTf, CH₂Cl₂, (54%)
2. H₂S, Pyr. / H₂O, C₁₇H₃₅CO₂H, WSC·HCl, CH₂Cl₂ (64%)
3. N₂H₄·HOAc, EtOH (−OLev → −OH)
4. SO₃·Pyr. DMF (96%)
5. NaOMe, MeOH; H₂O (quant.)

Scheme 2.2 (cont.)

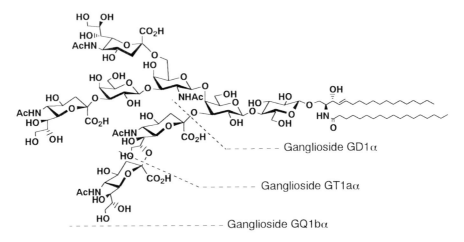

Fig. 2.14 Structures of Chol-1 (α-series) gangliosides.

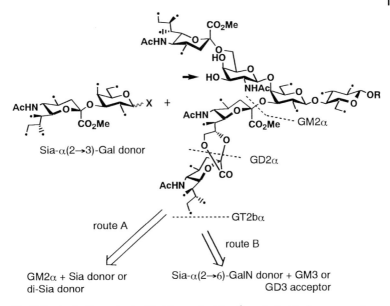

Fig. 2.15 Synthetic routes to Chol-1 gangliosides (•: protected hydroxy groups).

Tab. 2.1 Siglec-mediated cell adhesion to gangliosides [39].

Gangliosides	Gangliosides concentration (pmol/well) supporting half-maximal cell adhesion by:		
	MAG	SMP	Sialoadhesin
GSC-338	1.5	3.0	9.5
GQ1bα	6.0	8.5	21
GT1aα	17	30	22
GD1α	19	22	23
GT1b	50	>100	47
GD1a	50	>100	24
GM1b	80	150	87

may serve as ligands for some neural proteins. The total syntheses of GD1α [40], GT1aα [41], and GQ1bα gangliosides have been achieved [42] by coupling of Sia-$a(2 \to 3)$-Gal donor (Fig. 2.4) with the suitably protected GM2α, GD2α, and GT2bα oligosaccharides, respectively, in the final stage (Fig. 2.15).

All these synthetic Chol-1 gangliosides (GD1α, GT1aα, GQ1bα) showed binding activities to MAG, SMP, and sialoadhesin higher than those of the corresponding GM1b, GD1a, and GT1b gangliosides (Tab. 2.1 and Fig. 2.3), indicating that the sialic acid $a(2 \to 6)$-linked to C-6 of the GalNAc residue enhances the binding activity [39].

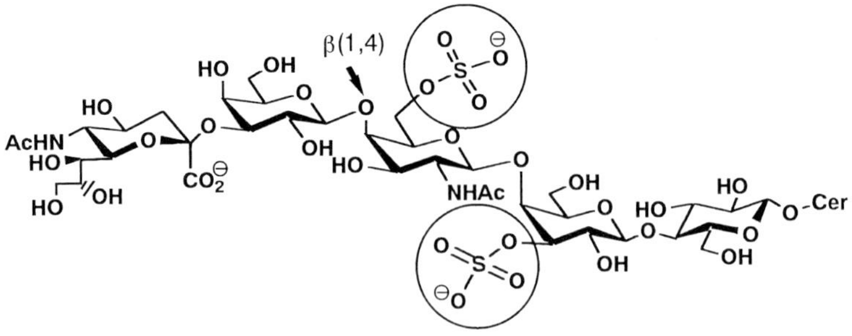

Fig. 2.16 GSC-338.

In addition, a series of GD1a analogues bearing 7-deoxy-, 8-deoxy-, and 9-deoxy-N-acetylneuraminic acids at C-6 of the GalNAc residue were able to support adhesion as strong as the parent GD1a, suggesting that the internal sialic acid may be replaced by other anionic substituents [43].

2.2.4.2 Novel Sulfated Gangliosides

Several novel sulfated gangliosides have been designed and synthesized on the basis of the results of structure/binding activity studies of Chol-1 (a-series) gangliosides and neural siglecs. Among those, GSC-338 (Fig. 2.16) [44], with two sulfate groups at C-6 of GalNAc and C-3 of Gal, showed the highest binding activity to MAG, SMP, and sialoadhesin [39] (Tab. 1). It was ten times more potent than GT1aa in supporting MAG and SMP binding, and four times more potent than GQ1ba, previously the highest affinity ligand for MAG.

2.3
Toxin Receptor

Tetanus toxin (TeNT) and the botulinum toxins (BoNTs) are extremely potent neurotoxins produced by the anaerobic Gram-positive bacteria *Clostridium tetani* and *Clostridium botulinum*, respectively [45–47]. The two toxins are structurally and functionally related, each being synthesized as a 150 kDa single polypeptide consisting of two disulfide-bonded components: a 50 kDa amino-terminal light chain (L-chain) with protease (toxin) activity, and a 100 kDa carboxyl-terminal heavy chain (H-chain) (Fig. 2.17). The H-chain can be further divided into two fragments, H_N and H_C, believed to have translocation function and recognition binding activity to neural cells, respectively [47]. Ganglioside GM1 – Galβ(1–3)Gal-NAcβ(1–4)[NeuAcα(2–3)]Galβ(1–4) Glcβ(1–1)Cer – was first identified as the receptor of BoNT and TeNT. Later studies demonstrated that the gangliosides of series b, especially GT1b and GD1b (Fig. 2.18a), have the highest affinity to these toxins. In an X-ray crystal structure investigation on the binding mechanism between sialooligosaccharides and the H_C fragment, initial attempts to crystallize a

2.3 Toxin Receptor | 51

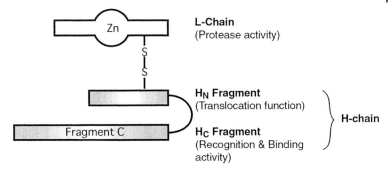

Fig. 2.17 Matured structure of tetanus toxin (TeNT).

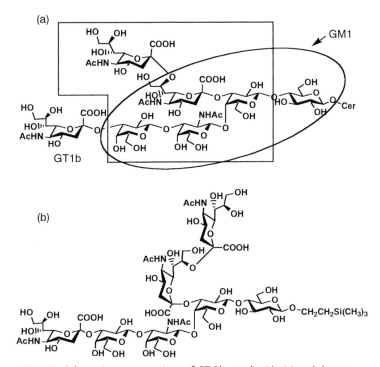

Fig. 2.18 Schematic representations of GT 1b ganglioside (a) and the synthetic GT 1b ganglioside analogue (b). The elliptical area corresponds to the GM 1 ganglioside, and the boxed outline encloses the portion of the oligosaccharide of GM 1 essential for binding of TeNT.

complex between TeNT H_C and native GT1b ganglioside were unsuccessful. The crystals were obtained for the first time from a complex between TeNT H_C and a synthetic GT1b analogue (Fig. 2.18b) [48]. This analogue has a 2-(trimethylsilyl)ethyl group instead of ceramide as an aglycon and the disialic acid attached to the central galactose residue through a β(2–3) linkage. The X-ray crystal structure analysis of this complex clearly showed the cross-linking between the sialooligosaccharide and the toxin, suggesting that two binding sites on the GT1b molecules, Gal β(1–4) GalNAc and NeuAcα(2–8)NeuAc, provide the key interactions for the TeNT-ganglioside recognition. The terminal sialic acid linked to the outer Gal has no significant interaction to the H_C fragment. Although the internal NeuAc-Gal linkage of the synthetic ganglioside probe is β, the binding activity is expected to be stronger than that of the natural ganglioside.

This binding mechanism is closely related to that of BoNT, so these results provide extremely important information for the design of anti-tetanus and anti-botulinum therapeutic agents.

2.4
Summary and Perspectives

Our recent achievements in sialoglycochemistry and glycobiology research through the use of a variety of synthetic ganglioside probes have been described, focussing on the specific interactions between carbohydrate ligands and receptor proteins including hemagglutinin, selectins, siglecs, and tetanus toxin. As cancer-associated antigens, sialyl lacto-series type I gangliosides such as sialyl Lewis a (sLe^a) [49] and disialyl Lewis a (DSLe^a) (Fig. 2.3) [50] have been synthesized. In this series of gangliosides, disialyl lactotetraosyl ceramide (DSLc$_4$) (Fig. 2.3) has recently been identified as a ligand of siglec-7 [51]. On the other hand, carbohydrate-carbohydrate interactions have also been postulated to be important in cellular recognition. Gg3-GM3 interactions, for example, have been investigated through the use of surface pressure-area (π-A) isotherms of glycolipid monolayers in contact with glycoconjugate polystyrenes at the airwater interface [52, 53]. Elucidation of the specific sialyl glyco-codons recognized by the receptor proteins or carbohydrate epitopes should lead to the design of various drugs in the near future.

2.5
References

1 A. Hasegawa, M. Kiso in *Carbohydrates – Synthetic Methods and Applications in Medicinal Chemistry* (Eds.: H. Ogura, A. Hasegawa, T. Suami), Kodansha/VCH, Tokyo/Weinheim, **1992**, pp 243–266.

2 A. Hasegawa, M. Kiso in *Preparative Carbohydrate Chemistry* (Ed.: S. Hanessian), Marcel Dekker, Inc., **1997**, pp 357–379.

3 a) M. Kiso, H. Ishida, H. Ito in *Carbohydrates in Chemistry and Biology*, Vol. 1 (Eds.: B. Ernst, G.W. Hart, P. Sinäy), Wiley-VCH, Weinheim, **2000**, pp 345–365; b) H. Ishida in *Carbohydrates in Chemistry and Biology*, Vol. 1 (Eds.: B. Ernst, G.W. Hart, P. Sinäy), Wiley-VCH, Weinheim, **2000**, pp 305–317.

4 a) H. Ishida, M. Kiso, *Trends Glycosci. Glycotechnol. (TIGG)* **2001**, *13*, 57–64; b) T. Ando, H. Ando, M. Kiso, *Trends Glycosci. Glycotechnol. (TIGG)* **2001**, *13*, 573–586.

5 a) M. Kiso, A. Hasegawa, *Methods Enzymol.* **1994**, *242*, 173–183; b) H. Ishida,

M. Kiso, A. Hasegawa, *Methods Enzymol.* **1994**, *242*, 183–197.

6 a) A. Hasegawa, N. Suzuki, H. Ishida, M. Kiso, *J. Carbohydr. Chem.* **1996**, *15(5)*, 623–637; b) A. Hasegawa, N. Suzuki, F. Kozawa, H. Ishida, M. Kiso, *J. Carbohydr. Chem.* **1996**, *15*, 639–648.

7 X. Zhang, T. Kamiya, N. Otsubo, H. Ishida, M. Kiso, *J. Carbohydr. Chem.* **1999**, *18*, 225–239.

8 a) S. Ladisch, A. Hasegawa, R. Li, M. Kiso, *Biochemistry* **1995**, *34*, 1197–1202; b) S. Ladisch, A. Hasegawa, R. Li, M. Kiso, *Biochem. Biophys. Res. Commun.* **1994**, *203*, 1102–1109.

9 a) N. Otsubo, H. Ishida, M. Kiso, *Carbohydr. Res.* **2001**, *330*, 1–5; b) T. Hamada, H. Hirota, S. Yokoyama, N. Otsubo, H. Ishida, M. Kiso, A. Kanamori, R. Kannagi, *Magn. Reson. Chem.* **2002**, *40*, 517–523.

10 K. Fukunaga, T. Toyoda, H. Ishida, M. Kiso, *J. Carbohydr. Chem.* **2003**, in press.

11 H. Masuda, T. Suzuki, Y. Sugiyama, G. Horiike, K. Murakami, D. Miyamoto, K. Hidari, T. Ito, H. Kida, M. Kiso, K. Fukunaga, M. Ohuchi, T. Toyoda, A. Ishihara, Y. Kawaoka, Y. Suzuki, *FEBS Lett.* **1999**, *464*, 71–74.

12 T. Ito, Y. Suzuki, T. Suzuki, A. Takada, T. Horimoto, K. Wells, H. Kida, K. Otsuki, M. Kiso, H. Ishida, Y. Kawaoka, *J. Virol.* **2000**, *74*, 9300–9305.

13 Y. Suzuki, T. Ito, T. Suzuki, R. E. Holland Jr., T. M. Chambers, M. Kiso, H. Ishida, Y. Kawaoka, *J. Virol.* **2000**, *74*, 11825–11831.

14 E. Tanahashi, K. Fukunaga, Y. Ozawa, T. Toyoda, H. Ishida, M. Kiso, *J. Carbohydr. Chem.* **2000**, *19*, 747–768.

15 K. Fukunaga, N. Ikami, H. Ishida, M. Kiso, *J. Carbohydr. Chem.* **2002**, *21*, 385–409.

16 K. Fukunaga, K. Shinoda, H. Ishida, M. Kiso, *Carbohydr. Res.* **2000**, *328*, 85–94.

17 C. Mitsuoka, M. Sawada-Kasugai, K. Ando-Furui, M. Izawa, H. Nakanishi, S. Nakamura, H. Ishida, M. Kiso, R. Kannagi, *J. Biol. Chem.* **1998**, *273*, 11,225–11,233.

18 R. D. Cummings, D. F. Smith, *BioEssays.* **1992**, *14*, 849–856.

19 L. A. Lasky, *Science* **1992**, *258*, 964–969.

20 R. P. McEver, *Glycoconjugate J.* **1997**, *14*, 585–591.

21 a) A. Kameyama, H. Ishida, M. Kiso, A. Hasegawa, *Carbohydr. Res.* **1991**, *209*, c1–c4; b) *ibid.*, *J. Carbohydr. Chem.* **1991**, *10*, 549–560; c) A. Hasegawa, M. Kiso, *Methods Enzymol.* **1994**, *242*, 158–173.

22 C. Foxall, S. R. Watson, D. Dowbenko, C. Fennie, L. A. Lasky, M. Kiso, A. Hasegawa, D. Asa, B. K. Brandley, *J. Cell Biol.* **1992**, *117*, 895–902.

23 B. K. Brandley, M. Kiso, S. Abbas, P. Nikrad, O. Srivasatva, C. Foxall, Y. Oda, A. Hasegawa, *Glycobiology* **1993**, *3*, 633–641.

24 H. Tsujishita, Y. Hiramatsu, N. Kondo, H. Ohmoto, H. Kondo, M. Kiso, A. Hasegawa, *J. Med. Chem.* **1997**, *40*, 362–369.

25 A. Hasegawa, M. Kiso in *Carbohydrates in Drug Design* (Eds.: Z. J. Witczak, K. A. Nieforth), Marcel Dekker, New York, **1997**, pp 137–155.

26 T. Ikami, H. Ishida, M. Kiso, *Methods Enzymol.* **1999**, *311*, 547–568.

27 H. Furui, K. Ando-Furui, H. Inagaki, T. Ando, H. Ishida, M. Kiso, *J. Carbohydr. Chem.* **2001**, *20*, 789–812.

28 W. S. Somers, J. Tang, G. D. Shaw, R. T. Camphausen, *Cell* **2000**, *103*, 467–479.

29 a) N. Otsubo, H. Ishida, M. Kiso, *Aust. J. Chem.* **2002**, *55*, 105–112; b) *ibid. Tetrahedron Lett.* **2000**, *41*, 3879–3882.

30 S. Komba, H. Ishida, M. Kiso, A. Hasegawa, *Bioorg. Med. Chem.* **1996**, *4*, 1833–1847.

31 C. Galustian, A. M. Lawson, S. Komba, H. Ishida, M. Kiso, T. Feizi, *Biochem. Biophys. Res. Commun.* **1997**, *240*, 748–751.

32 a) S. Komba, C. Galustian, H. Ishida, T. Feizi, R. Kannagi, M. Kiso, *Angew. Chem. Int. Ed.* **1999**, *38*, 1131–1133; b) S. Komba, M. Yamaguchi, H. Ishida, M. Kiso, *Biol. Chem.* **2001**, *382*, 233–240.

33 N. Otsubo, M. Yamaguchi, H. Ishida, M. Kiso, *J. Carbohydr. Chem.* **2001**, *20*, 329–334.

34 M. Yamaguchi, H. Ishida, C. Galustian, T. Feizi, *Carbohydr. Res.* **2002**, *337*, 2111–2117.

35 C. Mitsuoka, K. Ohmori, N. Kimura, A. Kanamori, S. Komba, H. Ishida, M.

Kiso, R. Kannagi, *Proc. Natl. Acad. Sci. U.S.A.* **1999**, *96*, 1597–1602.

36 M. Yamaguchi, H. Ishida, M. Kiso, unpublished result.

37 T. Angata, S. C. Kerr, D. R. Greaves, N. M. Varki, P. R. Crocker, A. Varki, *J. Biol. Chem.* **2002**, *277*, 24466–24474.

38 P. R. Crocker, *Immunology* **2001**, *103*, 137–145.

39 B. E. Collins, H. Ito, N. Sawada, H. Ishida, M. Kiso, R. L. Schnaar, *J. Biol. Chem.* **1999**, *274*, 37637–37643.

40 H. Ito, H. Ishida, M. Kiso, *J. Carbohydr. Chem.* **2001**, *20*, 207–225.

41 H. Ito, H. Ishida, H. Waki, S. Ando, M. Kiso, *Glycoconjugate J.* **1999**, *16*, 585–598.

42 H. Ishida, M. Kiso, *J. Synth. Org. Chem., Jpn.* **2000**, *58*, 1108–1113.

43 N. Sawada, H. Ishida, B. E. Collins, R. L. Schnaar, M. Kiso, *Carbohydr. Res.* **1999**, *316*, 1–5.

44 H. Ito, H. Ishida, B. E. Collins, S. E. Fromholt, R. L. Schnaar, M. Kiso, *Carbohydr. Res.* **2003**, in press.

45 G. Schiavo, F. Benfenati, B. Poulain, O. Rossetto, P. P. de Laureto, B. R. Das-Gupta, C. Montecucco, *Nature* **1992**, *359*, 832–839.

46 C. Montecucco, G. Schiavo, *Mol. Microbiol.* **1994**, *13*, 1–8.

47 J. Herreros, G. Lalli, C. Montecucco, G. Schiavo in *The Comprehensive Sourcebook of Bacterial Protein Toxins* (Eds.: J. E. Alouf and J. H. Freer), Academic Press, **1999**, pp 202–228.

48 C. Fotinou, P. Emsley, I. Black, H. Ando, H. Ishida, M. Kiso, K. A. Sinha, N. F. Fairweather, N. W. Isaacs, *J. Biol. Chem.* **2001**, *276*, 32,274–32,281.

49 Y. Makimura, H. Ishida, M. Kiso, A. Hasegawa, *J. Carbohydr. Chem.* **1996**, *15*, 1097–1118.

50 T. Ando, H. Ishida, M. Kiso, *J. Carbohydr. Chem.* **2001**, *20*, 425–430.

51 A. Ito, K. Handa, D. A. Withers, M. Satoh, S. Hakomori, *FEBS Lett.* **2001**, *498*, 116–120.

52 K. Matsuura, H. Kitakouji, N. Sawada, H. Ishida, M. Kiso, K. Kitajima, K. Kobayashi, *J. Am. Chem. Soc.* **2000**, *122*, 7406–7407.

53 K. Matsuura, H. Kitakouji, R. Oda, Y. Morimoto, A. H. H. Ishida, M. Kiso, K. Kitajima, K. Kobayashi, *Langmuir* **2002**, *18*, 6940–6945.

3
The Chemistry of Sialic Acid

Geert-Jan Boons and Alexei V. Demchenko

3.1
Introduction

Sialic acids are a family of naturally occurring 2-keto-3-deoxy-nonionic acids involved in a wide range of biological processes, 43 different derivatives thus far having been reported [1, 2]. The C-5-amino derivative is the long-known neuraminic acid, and its amino function can be either acetylated (Neu5Ac) or glycolylated (Neu5Gc). The hydroxy groups of these derivatives can be further acetylated, most commonly at C-9, but di- and tri-O-acetylated derivatives are also known. Lactoylation or phosphorylation may also occur at C-9, while the C-8 hydroxy group can be methylated or sulfated. The most abundant sialic acid, however, is N-acetylneuraminic acid (5-acetamido-3,5-dideoxy-D-*glycero*-D-*galacto*-non-2-ulopyranosonic acid). 3-Deoxy-D-*glycero*-D-*galacto*-non-2-ulopyranosonic acid (KDN) is an important form of sialic acid that does not possess an amino functionality [2]. The sugar ring in Neu5Ac has a $^2C_5$ conformation in which the bulky side chain and the C-5 acetamido moiety adopt equatorial orientations (Fig. 3.1).

Sialic acids seem to have appeared late in evolution and are mainly found in animals of the *deuterostoma* lineage, which comprises the vertebrates and some higher invertebrates. However, notable exceptions include sialic acid in *Drosophila* embryos, while certain strains of bacteria contain large quantities of sialic acids in their capsular polysaccharides.

Most sialic acids occur as glycosides of oligosaccharides and are typically found at the outermost ends of N-glycans, O-glycans, and glycosphingolipids. The natural equatorial glycosides are classified as α anomers, whereas the unnatural axial ones are termed β. In N-linked glycoproteins, sialic acids appear essentially as terminal sugars α(2→3)- or α(2→6)-linked to galactosides or α(2→6)-linked to N-acetyl-galactosaminides (such as in Neu5Acα(2→3)Gal, Neu5Acα(2→6)Gal, and Neu5Acα(2→6)GalNAc), whereas terminal Neu5Acα(2→6)GalNAc moieties can often be found in O-linked glycoproteins. The disialosyl structures Neu5Acα(2→8) Neu5Ac and Neu5Acα(2→9)Neu5Ac have also been found as constituents of glycoproteins and lipids.

Neu5Ac or Neu5Gc also occur in linear homopolymers, in which they are usually linked internally by α(2→8), α(2→9), or alternating α(2→8)/α(2→9) glycosidic

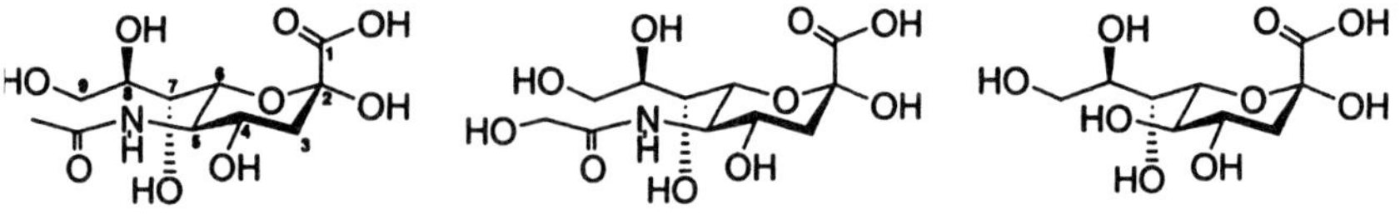

Fig. 3.1 Major naturally occurring sialic acids.

linkages [1, 3–6]. These polysialic acids are found in glycoproteins of embryonic neural membranes, in which they play a role as neural cell adhesion molecules. They are also found in fish eggs and in the capsules of certain bacteria such as *Neisseria meningitidis* group B. An $a(2 \rightarrow 4)$-linked homopolymer of 5,7-diacetamido-8-O-acetyl-3,5,7,9-tetradeoxy-D-*glycero*-D-*galacto*-nonulosonic acid [7], $a(2 \rightarrow 5)$-linked derivatives [8, 9], and galactose-substituted subterminal Neu5Ac [10] have also been described. Another unusual polysialic acid is composed of Neu5Gc moieties, which are glycosidically linked to the hydroxyl of the glycolyl moiety [11].

Striking differences in the sialylation patterns of cells during development, activation, aging, and oncogenesis have been found [12–19]. As terminal substituents of cell surface glycoproteins and glycolipids, sialic acids are ideally positioned to participate in carbohydrate-protein interactions mediating recognition phenomena. Indeed, they serve as ligands for microbial toxins, for microbial adhesion, mediation of attachment to a host cell, and for animal lectins important for cell-cell adhesion [20–25].

Sialic acids also play important roles as masks to prevent biological recognition [5]. Acetylation of the C-9 hydroxy group, for example, prevents this monosaccharide from acting as a receptor for the attachment of influenza A and B viruses [26]. Modifications of sialic acids can interfere with the mode of cell interaction, O-acetylation or N-acetyl hydroxylation, for example, hindering the action of sialidases and resulting in longer lifetimes of rat erythrocytes [27].

3.2
Chemical and Enzymatic Synthesis of Sialic Acids

The approach most commonly used for the synthesis of Neu5Ac involves condensation between N-acetyl mannosamine (ManNAc) and pyruvate, with catalysis by the enzyme Neu5Ac aldolase (Scheme 3.1) [28–31]. This enzyme has been isolated from bacteria and from eukaryotes and has also been overexpressed in *E. coli* [32–35]. The mechanism of the enzyme-catalyzed reaction involves the formation of a Schiff base between a lysine moiety in the active site of the enzyme and pyruvate. This then condenses with ManNAc in a general acid/base-catalyzed reaction (Scheme 3.1). The biological function of Neu5Ac aldolase is the degradation of Neu5Ac, and as a consequence the reaction is shifted towards the retro-aldol side with an equilibrium constant of 12.7 M^{-1}. For synthetic purposes, however, the equilibrium can be shifted towards Neu5Ac by employing an excess of pyruvate [36–39]. ManNAc

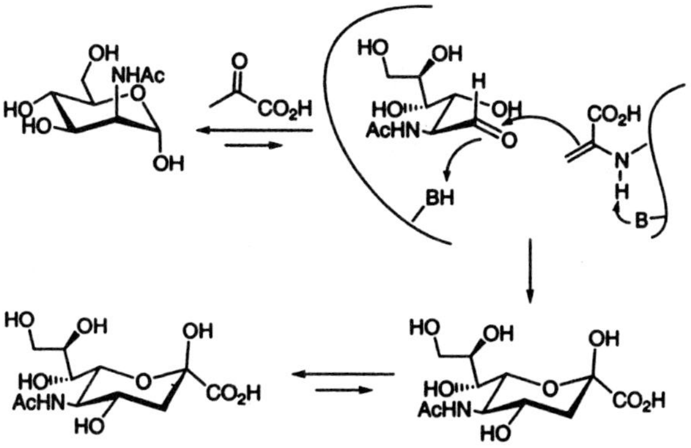

Scheme 3.1 Mechanism of Neu5Ac-catalyzed reaction.

can either be obtained via epimerization of *N*-acetylglucosamine using moderately strong basic conditions [40] or the enzyme *N*-acetylglucosamine-2-epimerise [41].

Neu5Ac aldolase has an absolute substrate requirement for pyruvate, but accepts modifications of ManNAc, thus opening a route to other sialic acid derivatives [28, 31, 42, 43]. The use of mannose as a substrate, for example, resulted in the facile synthesis of KDN. An elegant and high-yielding synthesis of 9-*O*-Ac-Neu5Ac was achieved by regioselective acylation of ManNAc with the serine protease subtilisin, by use of vinyl acetate as the acylating reagent, to give 6-*O*-acetyl-ManNAc, which was used as a substrate in a Neu5Ac-catalyzed reaction [44]. It has also been shown that derivatives of Neu5Ac can be selectively acylated at C-9 by use of subtilisin and vinyl acetate [45].

Several elegant approaches for the chemical synthesis of sialic acids have also been reported [46]. A short route involves condensation between ManNAc and ethyl α-bromoethyl acrylate in the presence of indium in dilute aqueous hydrochloric acid (Scheme 3.2) to give the corresponding enoate as a mixture of diastereoisomers (*threo*:*erythro* 4:1) [47]. Ozonolysis of the vinyl group followed by oxidative workup gave protected Neu5Ac in a yield of 51%. The indium-mediated allylation has also been used for the synthesis of KDN [48], KDO, and a six-carbon truncated sialic acid derivative [49]. The last derivative could not be obtained by an enzymatic approach because aldehyde precursors with fewer than five carbons are generally poor substrates for Neu5Ac aldolase.

Dondoni and co-workers have employed their thiazole methodology for the synthesis of a sialic acid derivative possessing an amine at C-4 [50, 51]. The carbon framework was obtained by means of a Wittig reaction between a selectively protected mannose derivative and a thiazole ylide (Scheme 3.3). Conjugative addition of benzylamine to the resulting enone predominantly gave a *syn* adduct, which was then converted into a 4-acetamido-4-deoxy-KDN derivative.

The first *de novo* synthesis of Neu5Ac was reported by Danishefsky and co-workers, the key steps in their approach being a hetero Diels-Alder reaction between an aldehyde and a diene, a stereoselective osmolation, and the use of a furan as a

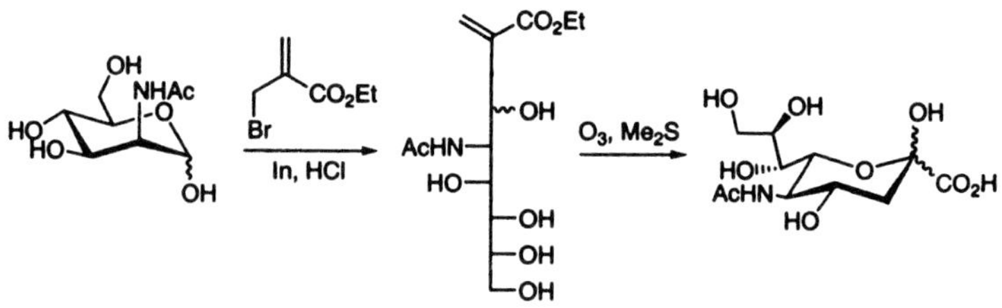

Scheme 3.2 Indium-mediated allylation strategy for Neu5Ac synthesis.

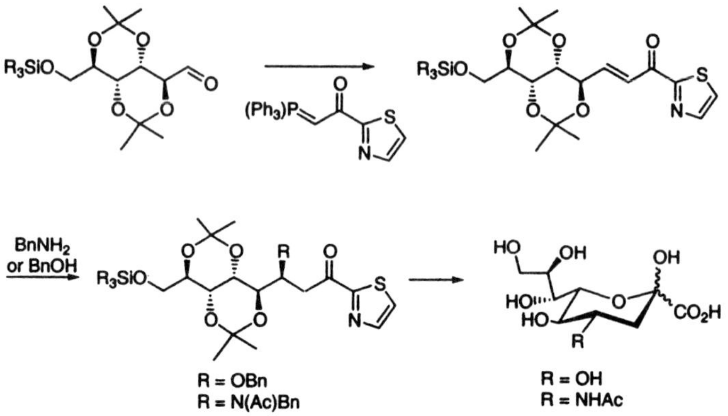

Scheme 3.3 Thiazole methodology for the synthesis of 4-acetamido-4-deoxy-KDN.

mask for the carboxylic acid [46]. More recently, Neu5Ac has also been obtained through a salen Co(II) hetero Diels-Alder reaction and oxidative azidation of a silylenol ether as key steps [52].

An overall high-yielding formal total synthesis of KDN was achieved through a novel ketalization/ring-closing metathesis reaction (Scheme 3.4) [53, 54]. Thus, ring-closing metathesis of a C-2 symmetric triene in the presence of Grubbs' catalyst gave a substituted dioxabicyclo[3.2.1]oct-2-ene ring system. A double Sharpless asymmetric dihydroxylation of the two double bonds, followed by selective tosylation through the use of intermediate tin acetals, gave a di-O-tosylate, which was then peracetylated. Displacement of the tosyl groups of this derivative with CsOAc in the presence of 18-crown-6 provided a peracetylated derivative with the appropriate stereochemistry. Finally, the electron-rich 3,4-dimethoxybenzene moiety was unmasked by oxidation with RuO_4 produced in situ by $NaIO_4$-mediated oxidation of a catalytic amount of RuO_3. KDN has also been prepared from cis-1,2-dihydroctechol, which was obtained by enantioselective microbial oxidation of chlorobenzene. This compound was converted into a suitably protected C-6 aldehyde, which was extended by the indium methodology to give KDN [48].

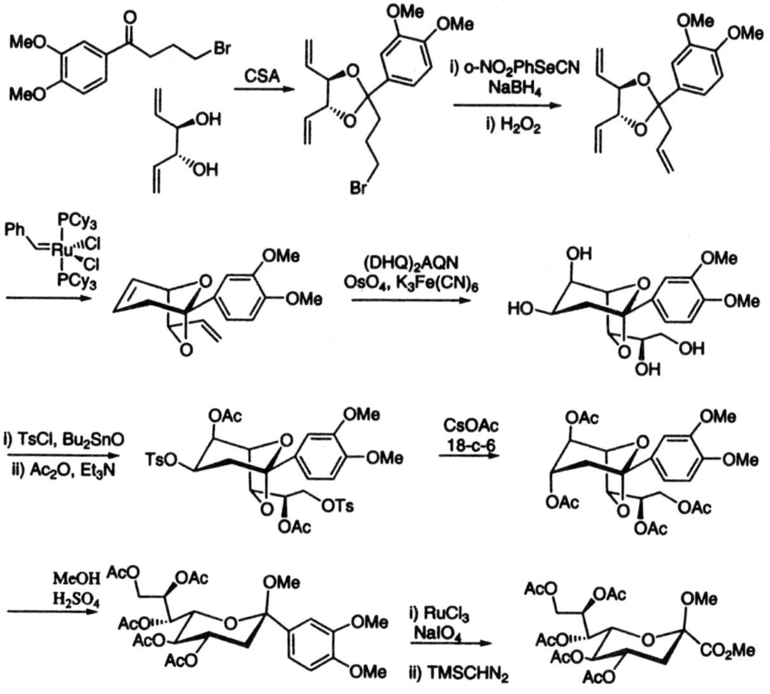

Scheme 3.4 Total synthesis of KDN.

3.3
Chemical Glycosidation of Sialic Acids

Advances in both chemical and enzymatic syntheses have provided reliable routes for the production of many complex sialosides [31, 55–59]. Synthetic compounds have proven to be of key importance for determination of the biological roles of these glycoconjugates. The next two sections survey recent progress in chemical and enzymatic sialylation.

Inter-glycosidic bond formation is generally achieved through condensation between a fully protected glycosyl donor, bearing a potential leaving group at its anomeric center, and a suitably protected glycosyl acceptor often containing only one free hydroxy group [60–66]. The glycosylation can result in a and/or a β anomer, and stereocontrol of this condensation reaction is one of the most challenging topics in oligosaccharide chemistry. The nature of the protecting group at C-2 of the glycosyl donor is a major determinant of the anomeric selectivity. A C-2 protecting group capable of entering into neighboring group participation during glycosylation should favor the formation of a 1,2-*trans*-glycosidic linkage. On the other hand, the reaction conditions (e.g., solvent, temperature, and promoter) will determine the anomeric selectivity when a non-assisting functionality is present at C-2. In addition, the constitutions of the glycosyl donor and the acceptor (e.g., type of saccharide, leaving group at the anomeric center, protection and substitution pattern) have major effects on the α/β selectivity.

Glycosides of N-acetylneuraminic acid can be introduced by similar glycosylation approaches, and these coupling procedures are classified as direct methods [30, 55, 67–69]. The use of glycosyl donors of Neu5Ac is complicated, however, by the fact that there is no C-3 functionality present to direct the stereochemical outcome of glycosylations. In addition, the deoxy moiety, in combination with the electron-withdrawing carboxylic acid at the anomeric center, makes these derivatives prone to elimination (formation of glycals) and, moreover, glycosylations of Neu5Ac take place at a sterically hindered tertiary oxo-carbenium ion intermediate. The most successful glycosyl donors of Neu5Ac use rather "unusual" anomeric leaving groups. For example, anomeric fluorides and trichloroacetimidates are most widely applied for glycosidation of common glycosyl donors, whereas sialyl donors rarely or never possess these leaving groups. On the other hand, less common phosphites and xanthates are very attractive leaving groups for sialyl donors.

3.3.1
Direct Chemical Sialylations

Direct O-sialylations include those methods that result in the formation of O-sialosides in one synthetic step, involving coupling between a glycosyl acceptor possessing a free hydroxy group and a glycosyl donor with an appropriate leaving group at C-2 (Scheme 3.5). These methods have found broad application and nowadays offer a reliable and efficient means for the synthesis of the natural and unnatural sialosides. Conversely, *indirect* methods afford O-sialosides in two or

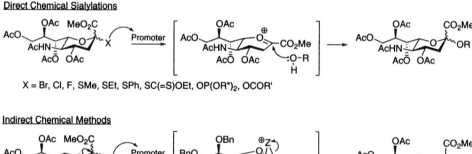

Scheme 3.5

3.3 Chemical Glycosidation of Sialic Acids | 61

more synthetic steps, one of which may be a glycosylation (Scheme 3.5). Most of these approaches involve modified sialyl donors possessing a participating functionality at C-3, in order to achieve better stereocontrol of sialylations.

3.3.1.1 2-Chloro Derivatives as Glycosyl Donors

The synthesis of sialosides from 2-chloro derivatives was the main tool for the synthesis of compounds containing N-acetylneuraminic acid from the 1960s until the 1980s [30, 55, 67]. Nowadays, however, the use of these derivatives is limited to the glycosylation of simple alcohols. 2-Chloro derivatives can be prepared from the corresponding C-2 acetates by treatment with HCl/AcCl [70], $TiCl_4$ [71], or AcCl/MeOH in $CHCl_3$ [72]. The first synthesis of a disaccharide containing Neu5Ac was accomplished in 1971, by Khorlin and co-workers, from a fully protected 2-chloro derivative in the presence of Ag_2CO_3 [73]. Further studies by Paulsen and co-workers [74], however, reached the conclusion that silver salt-catalyzed glycosylations give satisfactory results only in reaction with highly reactive acceptors such as 1,2:3,4-di-O-isopropylidene-galactose (Scheme 3.6). Glycal (2,3-dehydro derivative) formation by competing elimination was mainly observed when these conditions were applied to less reactive secondary or sterically hindered primary hydroxy groups (e.g., tetra-O-benzyl galactose); in the latter case only trace amounts of the desired disaccharide were isolated. In this context, glycal formation can be suppressed by employment of $Hg(CN)_2/HgBr_2$ as catalyst [74]. For example, a Neu5Ac(2→6)Gal derivative was obtained in an excellent yield of 84% as a 3/4 mixture of α/β anomers. Conversely, Neu5Ac(2→3)Gal derivatives were obtained with rather low efficiency (15%, α/β 2/3), presumably due to the lower reactivity of secondary hydroxy groups resulting mainly in the elimination of the Neu5Ac donor [75].

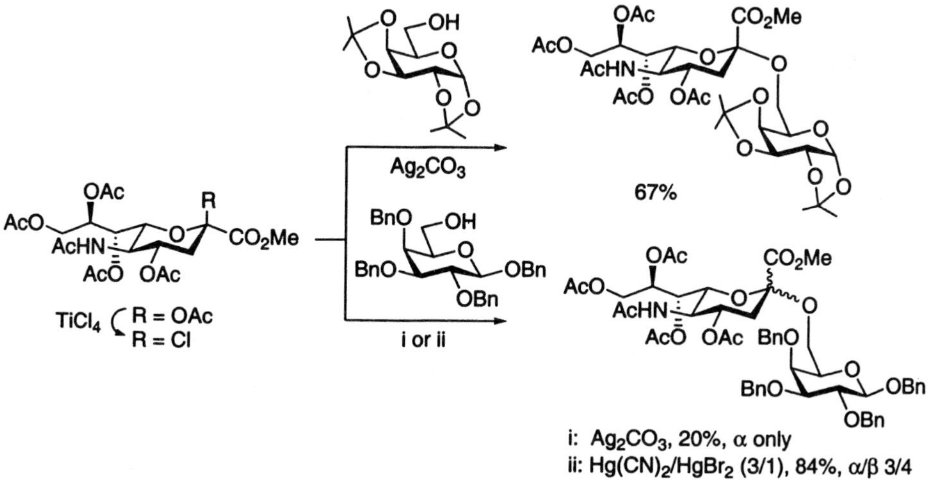

Scheme 3.6

Another improvement emerged with the use of polymer-based silver salts, such as silver polymaleate, polymethacrylate [76], or salicylate [77]. Initially, insoluble silver salts were introduced for glycosylations with inversion of configuration for the preparation of β-mannosides [78, 79]. When the silver salts were applied for coupling between acetochloroneuraminic acid and simple alcohols, the corresponding α-sialosides were isolated in reasonable yields (51–64%). In most cases these glycosylations were complete within few minutes, and so found application for the synthesis of several simple glycosides of Neu5Ac [80–84].

Other promoters, including $ZnBr_2$ [85], silver triflate [86], silver zeolite [87], I_2/2,3-dichloro-5,6-dicyano-1,4-benzoquinone (DDQ) [88], diisopropylethylamine [89], and the phase-transfer catalyst $BnNEt_3Cl$ [90], have also been applied for the sialylation of 2-chloro derivatives of Neu5Ac. It has also been shown that glycosylation of aliphatic alcohols can be effectively performed in the absence of added promoter both for free acid and for the corresponding methyl ester [91].

In summary, the application of 2-chloro derivatives of Neu5Ac typically gives good yields when applied for the sialylation of simple or primary sugar alcohols, whereas complex glycosides require modern glycosylation methods, discussed in the subsequent sections. It should be noted that 2-bromo [71, 72, 92] and 2-fluoro derivatives [93–95] of Neu5Ac have also been used as sialyl donors but the results were disappointing. The 2-bromo derivative of KDN has found wider application for the synthesis of KDN-glycosides, presumably due to its higher stability relative to the analogous Neu5Ac derivative [96–99].

3.3.1.2 2-Thio Derivatives as Glycosyl Donors

Alkyl(aryl) thioglycosides have emerged as versatile building blocks for oligosaccharide synthesis [100]. Thanks to their excellent chemical stabilities, anomeric alkyl(aryl) thio groups offer efficient protection of anomeric centers and are compatible with many reaction conditions often employed in carbohydrate chemistry. However, in the presence of soft electrophiles, thioglycosides can be activated and used in direct glycosylations. The most commonly used activating reagents for O-sialylations include dimethyl(methylthio)sulfonium trifluoromethanesulfonate (DMTST), N-iodosuccinimide (NIS)/trifluoromethanesulfonic acid (TfOH), and phenylselenyl triflate (PhSeOTf). Another attractive feature of thioglycosides is that they can be transformed into a range of other glycosyl donors [55].

The 2-thiomethyl sialosides are obtained from the corresponding 2-acetates by treatment with trimethyl(methylthio)silane (TMSSMe) in the presence of trimethylsilyl triflate (TMSOTf) [101]. As a result, a fully acetylated thioglycoside is obtained as a 1/1 mixture of α/β anomers. It should be noted that, in sialylations, the α and β anomers have very similar glycosyl donor properties and therefore do not need to be separated [102]. The synthesis of 2-thiomethyl glycosides of Neu5Ac and their use as sialyl donors has been developed by Hasegawa and coworkers [103–106].

Good yields and high α anomer selectivities were achieved when the synthesis of Neu5Ac α(2–3)Gal derivatives was carried out in the presence of DMTST at low temperature in the participating solvent acetonitrile (MeCN) [103]. It should be pointed out that the best results were achieved when the glycosyl acceptors were protected only at the anomeric and the primary (C-6) positions (3-O-sialylation of a 2,3,4-triol) [107–109]. The regioselectivity of this sialylation is attributed to the greater reactivity of the equatorial alcohol in relation to the axial C-4 hydroxy group. Furthermore, the C-2 hydroxy group has a lower nucleophilicity, due to the electron-withdrawing effect of the adjacent anomeric center. When a similar reaction was performed with a galactosyl acceptor possessing only a free 3,4-diol, the yield and anomeric stereoselectivity were significantly reduced. Further improvement came with the application of the highly reactive NIS/catalytic TfOH promoter system, which has proven to be especially valuable when applied for glycosylations of sterically more hindered hydroxy groups [110]. As an example, NIS/TfOH-mediated glycosylation of a 3′,4′-diol of a lactoside gave a much higher yield of the (2 → 3)-linked product (69%) than the DMTST-promoted sialylation, together with improved anomeric stereoselectivity (α/β 6/1) (Scheme 3.7).

A reaction mechanism for the activation of thiosialoside in MeCN has been proposed [110]. This reaction starts with the in situ generation of the electrophilic species – such as $^+$SMe or I$^+$ from DMTST or NIS/TfOH, respectively – and this reacts with the lone pair of sulfur, resulting in the formation of a sulfonium intermediate (Scheme 3.7). The sulfonium moiety is an excellent leaving group and can be displaced by a hydroxy group of a glycosyl acceptor or, alternatively, by the nitrogen of acetonitrile to give a nitrilium ion. The nitrilium ion preferentially adopts an axial (β) configuration, so subsequent nucleophilic substitution with an alcohol predominantly gives an equatorial α-sialoside. It has been observed that less reactive (secondary) alcohols provide better α-selectivities than primary alcohols.

A significantly more reactive 2-thioglycosyl donor of Neu5Ac, bearing a di-N-acetyl (N-acetylacetamido) functionality at C-5, has recently been introduced [111]. It was observed that the presence of the additional N-acetyl moiety in the glycosyl donor dramatically increases its reactivity, resulting in improved yields of glycosylation products (Scheme 3.8). For example, an NIS/TfOH-promoted coupling between 2-(trimethylsilyl)ethyl-6-O-benzoyl-β-D-galactopyranoside and the di-N-acetylated sialyl donor proceeded with complete regioselectivity to give, after a reaction time of less than 5 min, an $\alpha(2 \to 3)$-linked disaccharide in a yield of 72% [111]. These results compare very favorably with the classic sialylation approach with the corresponding mono-N-acetylated donor [110]. Moreover, only a small excess of donor was required to achieve the high yield (1.1 mol equiv. of the donor instead of the conventional 1.7–2.0 mol equiv.). An excellent yield (81%) was also obtained when a 4,6-O-benzylidene-protected galactoside was used as glycosyl acceptor [112]. The resulting glycosylation products could easily be converted into glycosyl donors by conversion of the temporary anomeric OTE functionality into a thiomethyl leaving group. This approach was successfully applied for the synthesis of a branched heptasaccharide derived from group B type III *Streptococcus* [112]. The additional N-acetyl function could be easily cleaved under Zémplen deacetylation conditions with concomitant O-acetyl group removal.

3 The Chemistry of Sialic Acid

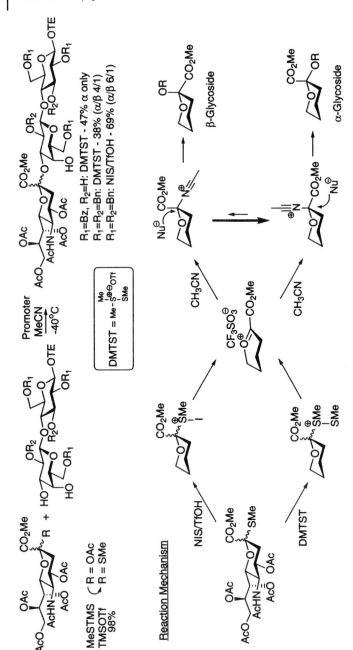

Scheme 3.7

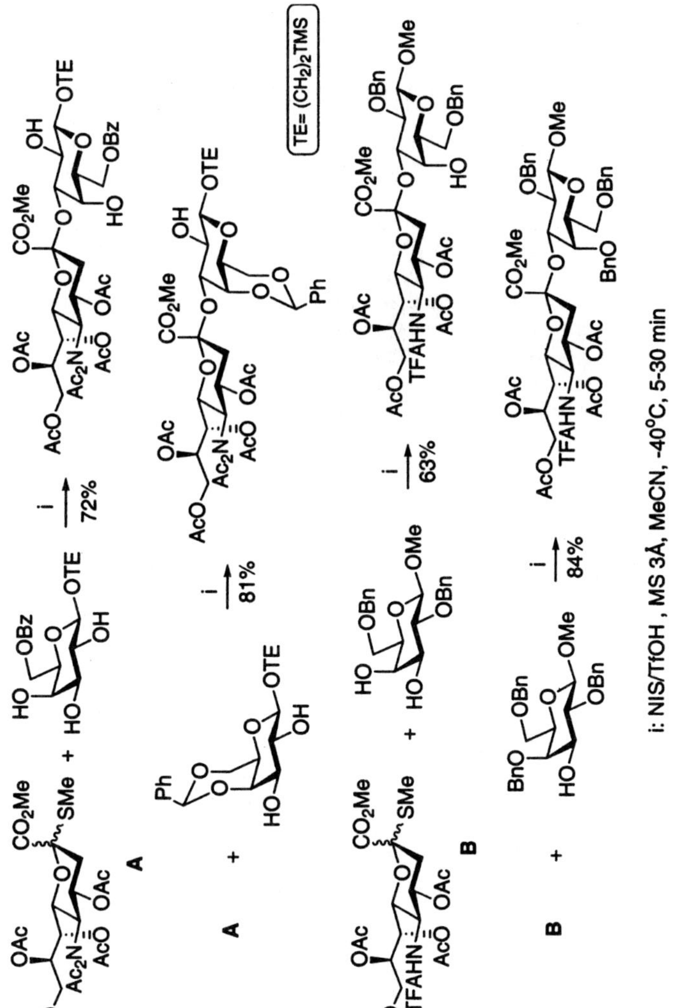

Scheme 3.8

66 | 3 The Chemistry of Sialic Acid

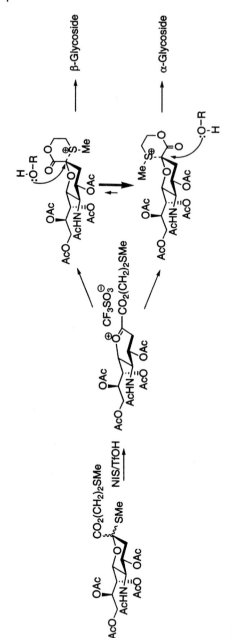

Scheme 3.9

Further improvements came with the introduction of an N-trifluoroacetylated derivative as the sialyl donor [113–115]. Interestingly, the best results were obtained when a sterically hindered acceptor was glycosylated (Scheme 3.8) [115]. The N-trifluoroacetamido moiety can be removed by treatment with 1 M aq. NaOH in MeOH, and the liberated amino function can then be acetylated or glycosylated. Both N-acetylacetamido- and trifluoroacetamido-substituted methylthiosialosides have also been applied for the direct synthesis of (2→8)-linked dimers of N-acetylneuraminic acid (see below) [114, 116].

An interesting technique for control of the stereoselectivity of sialylation through long-range participation was recently reported [117]. In this approach, the carboxylic acid at C-1 of 2-thiomethyl Neu5Ac was protected as a 2-thioethyl ester. It was envisaged that activation of the anomeric center of the Neu5Ac donor would give an oxonium ion intermediate, which would be stabilized through long-range participation by the thiomethyl moiety, resulting in the formation of a sulfonium intermediate (Scheme 3.9). Glycosylation of the thermodynamically more stable β-sulfonium intermediate should give mainly formation of α glycosides. Reasonable yields and selectivities for α anomers were achieved when this approach was applied to both primary and secondary glycosyl acceptors. The proposed mechanism was supported by the finding that anomeric selectivities were not affected by the reaction solvents, α-linked disaccharides mainly being formed, in moderate yields (20–50%), even when the glycosylation was performed in ethylene glycol dimethyl ether. The method was applied for the preparation of a taxol-sialyl conjugate [118]. A similar approach was reported by Gin and co-workers, who employed an N,N-dimethylglycolamido (OCH_2CONMe_2) ester functionality as an auxiliary at C-1. This strategy allowed the synthesis of a number of sialosides with high α stereoselectivity [119].

Thiomethyl sialosides have also been applied in the polymer-supported synthesis of sialyl Lewis[x] [120] and in the orthogonal glycosylation strategy [121, 122]. The latter technique exploits the finding that a thioglycoside can be activated in the presence of PhSeOTf [123] without affecting an anomeric fluoride and vice versa. This methodology allowed the convergent synthesis of complex oligosaccharides without the need for extensive manipulations at the anomeric center.

Thioethyl glycosides of Neu5Ac, readily available from the corresponding 2-O-acetyl derivative [124, 125], have been demonstrated to possess sialyl donor properties similar to those of the corresponding S-methyl glycosides [126–128]. 2-Thiophenyl glycosides of Neu5Ac can be synthesized by treatment of 2-O-acetyl, 2-chloro or 2-fluoro derivatives of Neu5Ac with thiophenol [124, 129]. In general, 2-thioaryl glycosyl donors can be activated by promoter systems similar to those used for S-alkyl derivatives. In some cases, 2-thiophenyl glycosides proved to be more efficient than the corresponding 2-thiomethyl derivatives [130], and in particular higher yields were obtained when a benzoylated glycosyl donor was used [131]. 2-Thiophenyl glycosides were also employed for the introduction of KDN glycosides, NIS/TMSOTf being used as the promoter in this case [132, 133]. The most valuable application of thiophenyl sialosides is for the formation of (2→8)-linked oligomeric sialyl units.

Roy and co-workers have reported the synthesis and glycosidation of p-substituted 2-thioaryl glycosides [125, 134]. It was found that a p-methoxyphenyl 2-thio glycoside has glycosyl donor properties similar to those of an S-phenyl derivative. On the other hand, an anomeric p-nitrophenyl substituent was found to be *latent* toward activation with NIS/TfOH. This moiety could, however, easily be converted into the *active* p-acetamidophenyl species by treatment with $SnCl_2$/EtOH, followed by N-acetylation [134]. Such a derivative could be activated with NIS/TfOH as the promoter system. Tuning of the anomeric reactivity of glycosides in this fashion offers an attractive way to synthesize complex oligosaccharides with a minimal number of protecting group manipulations.

It was recently shown that 2-thiocresol sialyl donors in which the 5-acetamido function is masked as an azido moiety give significantly improved α stereoselectivities in sialylation, especially when applied to the synthesis of 2–9-linked dimers [135]. The high α stereoselectivity was explained in terms of both steric and electronic effects, and it has been proposed that a 5-azido group can stabilize an acetonitrilium intermediate more effectively than a 5-acetamido derivative can. Furthermore, the bottom face of the molecule is more accessible, due to the linear structure of the azido function. However, the stereoselectivity was significantly compromised when this approach was applied for the glycosylation of the primary or secondary hydroxy groups of galactosyl acceptors [135, 136]. It was noted that these sialyl donors are three to five times less reactive than the corresponding 5-acetamido derivatives [135].

An example of a highly convergent synthesis of an oligosaccharide containing four sialyl moieties is shown in Scheme 3.10 [137]. NIS/TfOH-promoted sialylation of a trisaccharide possessing a 3′,6″-diol with a 2-thiophenyl Neu5Ac glycosyl donor gave the expected α(2→6)-linked tetrasaccharide in 45% yield. Subsequent glycosylation of the resulting tetrasaccharide with a dimeric sialyl donor under similar reaction conditions afforded a hexasaccharide (42%). Removal of the 3,4-O-isopropylidene acetal and subsequent DMTST-promoted glycosylation with a Neu5Acα(2→3)GalSMe donor furnished the desired octasaccharide in excellent overall yield. This example clearly illustrates three different ways of introducing sialic acid fragments into complex oligosaccharides. The first glycosylation entails an NIS/TfOH-mediated regioselective glycosylation of a monomeric 2-thiophenyl sialyl donor. A 2-thiophenyl donor is also employed in the second glycosylation, but in this case the donor is a more complex dimeric Neu5Acα(2→8)Neu5Ac derivative. The last glycosylation step cannot be classified as an O-sialylation, because it uses a galactosyl donor with a Neu5Ac moiety at C-3. Nevertheless, this synthetic block is important for the synthesis of complex sialylated oligosaccharides, because the sialic acid unit is introduced into a simpler structure prior to complex oligosaccharide assembly.

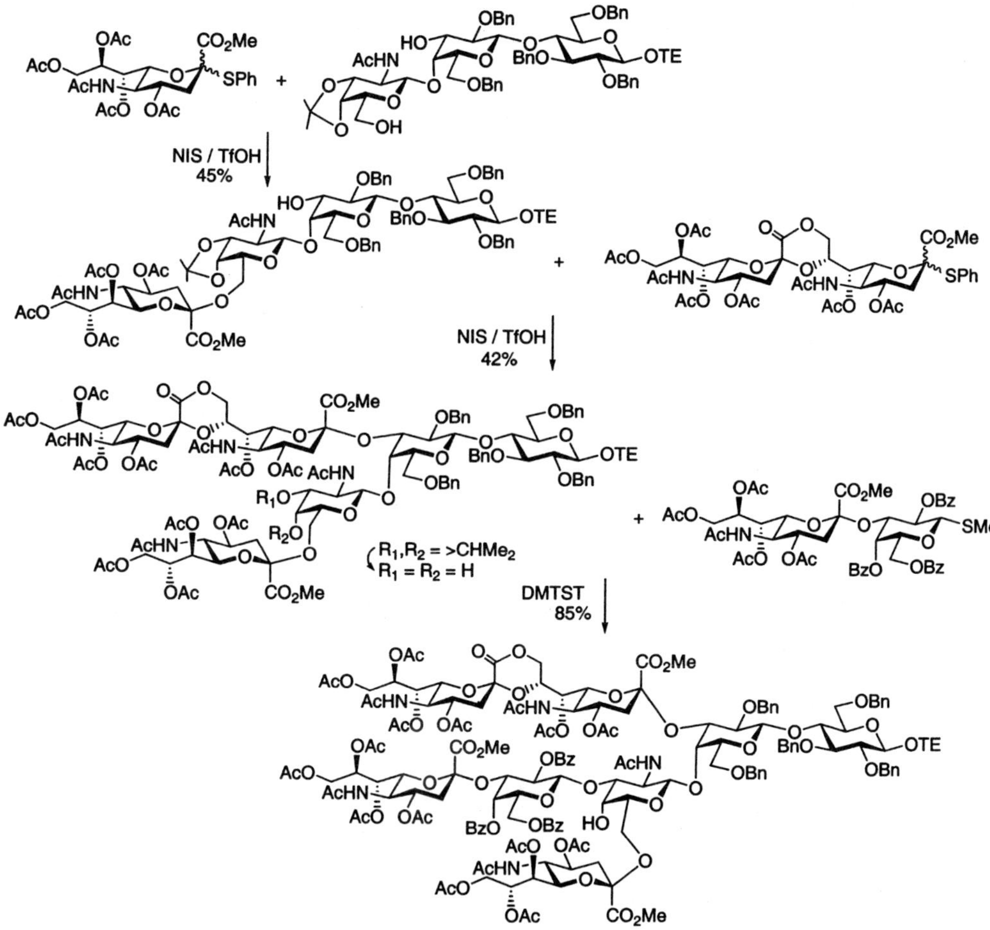

Scheme 3.10

3.3.1.3 2-Xanthates as Glycosyl Donors

2-(Ethoxy)dithiocarbonates, or 2-xanthates, of Neu5Ac have been synthesized from 2-chlorides by treatment with potassium ethoxydithiocarbonate in EtOH [124]. Activation of xanthates with DMTST [138] or NIS/TfOH [126, 139, 140] has been successfully employed in the synthesis of sialylated oligosaccharides. Xanthates can also readily be activated with methylsulfenyl triflate (MeSOTf), a highly reactive thiophilic reagent, generated in situ by treatment of methylsulfenyl bromide (MeSBr) with AgOTf. An important feature of this glycosylation procedure is that sialyl xanthates can be selectively activated in the presence of the disarmed thioglycosides (Scheme 3.11). Phenylsulfenyl trifluoromethanesulfonate (PhSOTf), obtained in a similar fashion from PhSCl and AgOTf, has proven to be an even more efficient promoter, especially when applied in combination with the hin-

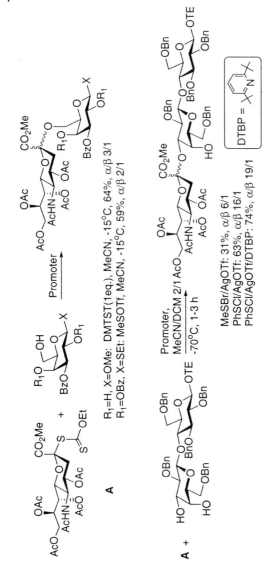

Scheme 3.11

dered base 2,6-di(tert-butyl)pyridine (DTBP) at low temperatures (−70 °C) [141]. This procedure has been used for the synthesis of a trisaccharide, isolated in an excellent yield of 74% and mainly as the $α$ anomer ($α/β$ 19/1, Scheme 3.11).

3.3.1.3 2-Phosphites as Glycosyl Donors

Sialyl phosphites, introduced independently by Wong [142] and by Schmidt [143], have found wide application in chemical O-sialylation [66, 139, 144, 145]. Sialyl phosphites are highly reactive glycosyl donors, and only a catalytic amount of TMSOTf (typically 10–20 mol %) is required for their activation.

Diethyl $β$-sialyl phosphites can be prepared in high yields (97%) by treatment of a suitably protected 2-hydroxy derivative of Neu5Ac with ClP(OEt)$_2$ in the presence of the hindered base i-Pr$_2$NEt [143, 146]. Glycosylation of a 6-hydroxy derivative of a glucoside in the presence of TMSOTf (0.1 equiv.) in MeCN at −40 °C afforded a (2 → 6)-linked disaccharide as a 4/1 mixture of $α/β$ anomers (70%). The use of a di-O-benzyl phosphite derivative in the same model reaction gave a somewhat higher yield and improved anomeric selectivity (80%, $α/β$ 5/1–6/1, Scheme 3.12) [142, 147, 148]. Despite the latter result, the di-O-benzyl phosphite methodology has not found wide application, probably due to the fact that the dibenzyl-N,N-diethylphosphoroamidite (DDP) reagent required for its introduction is not commercially available. Moreover, treatment of a 2-hydroxysialyl derivative with DDP in the presence of tetrazole provided the desired phosphite in a yield of 70%, whereas a yield of 97% was obtained for the diethyl analogue.

Several other examples of the use of 2-(diethyl)phosphite [140, 149–153] and dibenzyl phosphite [153–155] derivatives of Neu5Ac have been reported. Other sialyl phosphites such as -O(nBu)-, -OCH$_2$CH$_2$Cl, -OCH$_2$CCl$_3$, -O(CH$_2$)$_3$O-, O-CH$_2$CH$_2$Me$_2$CH$_2$O- [146], and 1,2-O-cyclopentyl [156] were found to be less efficient sialyl donors than their OEt or OBn counterparts. However, dimethyl phosphite proved to be rather reactive [157], especially when promoted with ZnCl$_2$/AgClO$_4$ in CH$_2$Cl$_2$ at room temperature, providing sialosides in high yields (83–85%) and with good $β$ stereoselectivities ($α/β$ 1/5–1/6) when treated with simple acceptors.

3.3.1.4 Miscellaneous Direct Chemical Methods

Protic acid promoted sialidations (Fischer-type glycosylations) of 2-hydroxy derivatives of Neu5Ac have found some use in the synthesis of simple alkyl sialosides [158]. Although the method provides the shortest route towards O-sialosides, it only gives access to unnatural $β$-sialosides. Similar results were obtained for the synthesis of methyl glycosides of KDN by the use either of cation-exchange resin (Dowex-50, H$^+$) [159, 160] or of HCl produced in situ from AcCl and MeOH [161].

2-O-Acyl derivatives of Neu5Ac have been applied as glycosyl donors, but their use is rather limited, due variously to low-yielding reactions or to poor anomeric selectivity. Anomeric acetates, for example, gave mainly $β$-sialosides when activated with a Lewis acid in either MeCN or CH$_2$Cl$_2$ [85, 162, 163].

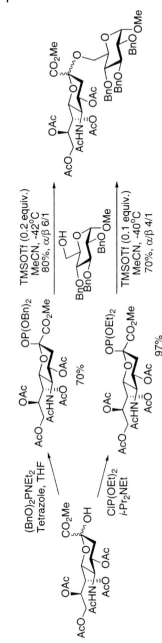

Scheme 3.12

3.3.2
Indirect Chemical Methods with the Use of a Participating Auxiliary at C-3

Several glycosyl donors derived from Neu5Ac and possessing an auxiliary at C-3 have been prepared. This auxiliary should control the anomeric selectivity of a glycosylation through neighboring group participation, resulting in the formation of 2,3-*trans*-glycosides [55]. Thus, α-glycosides are favored in the equatorial auxiliary case (Scheme 3.5), whereas β-glycosides are preferentially formed when the auxiliary is axial. Auxiliaries should also prevent 2,3-elimination, which is often a major side reaction in direct O-sialylations. Apart from these features, an auxiliary should be easily installed prior to a glycosylation and easily removed afterwards. Auxiliaries are usually introduced by chemical modification of a 2,3-dehydro derivative of Neu5Ac, either via a 2,3-oxirane derivative or by an addition reaction to the double bond. It should be noted that the 2,3-dehydro derivatives are easily accessible from the methyl ester of acetochloroneuraminic acid [164]. Other methods such as the application of glycosyl donors with a protected C-1 hydroxymethyl group have been reported [165].

3.3.2.1 3-Bromo- and other 3-O-Auxiliaries

Addition of bromine to the double bond of the glycal of Neu5Ac gave a diaxially substituted 2,3-dibromo derivative, which was immediately glycosidated in the presence of AgOTf/NaHPO$_4$ [164, 166, 167]. Because of the axial orientation of the bromide at C-3, only the unnatural β-linked disaccharides were formed. Alternatively, addition of NBS to the glycal at low temperatures (–20 °C) in aqueous dimethyl sulfoxide (DMSO) predominantly afforded a diequatorial substitution product (84%) with a bromo substituent at C-3 and a hydroxy group at the anomeric center [164]. This compound could be converted into a C-2 diethyl phosphite derivative, which gave good anomeric selectivities and yields when applied in glycosylations with reactive hydroxy groups [168]. Conversely, predominantly β anomer was isolated when the synthesis of a (2→8)-linked sialyl dimer was attempted. The bromo auxiliary could be removed by reduction with Bu$_3$SnH to furnish β-O-sialosides [168]. A chemoenzymatic approach was applied for the synthesis of a 3(ax)-bromo donor with 2-dibenzylphosphite as leaving group, also resulting in the formation of β-glycosides [148].

Several approaches for the synthesis of 3-O derivatives of Neu5Ac have been reported. For example, synthesis of a 3-hydroxy-substituted compound with a halide leaving group at C-2 (Cl, Br, F) was accomplished by a three-step reaction sequence starting from the glycal [164, 169]. Although this donor gave poor anomeric selectivity, due to the lack of neighboring group participation by the unprotected hydroxy group, it was efficiently used for the first synthesis of both (2→8)-linked dimeric structures and (2→9)-linked derivatives [170, 171]. This auxiliary could be removed by a two-step reaction involving phenoxythiocarbonylation in the presence of DMAP (95%), followed by reduction with Bu$_3$SnH in the presence of 2,2′-azobisisobutyronitrile (AIBN, 97%).

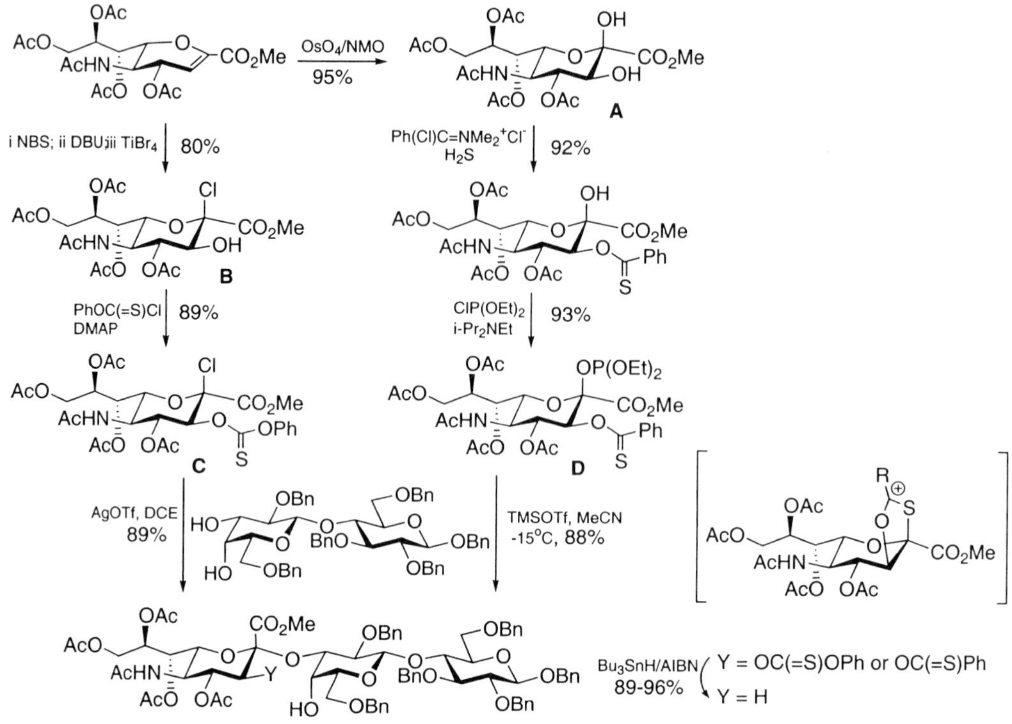

Scheme 3.13

More efficient anomeric control could be achieved by use of a 3-O-thiocarbamate functionality, introduced into the 3-hydroxy-2-chloro derivative of Neu5Ac (**B** in Scheme 3.13) prior to glycosylation to provide a 2-chloro glycosyl donor (**C**). Alternatively, 2,3-diol **A**, obtained by a stereoselective dihydroxylation, could be treated with Ph(Cl)C=NMe$_2^+$Cl$^-$ in the presence of H$_2$S to give a 3-O-thiocarbonyl derivative. The anomeric hydroxy group was subsequently converted into a phosphite leaving group by treatment with (EtO)$_2$PCl to afford the glycosyl donor **D**. It was anticipated that these auxiliaries would provide neighboring group participation through a five-membered cyclic intermediate (Scheme 3.13) [152, 168] and indeed, excellent α stereoselectivities and high yields were achieved when the donors **C** and **D** were used.

3.3.2.2 3-Thio and 3-Seleno Auxiliaries

Neu5Ac derivatives containing a 3-thio or 3-seleno auxiliary in combination with an appropriate anomeric leaving group constitute the most reliable group of glycosyl donors [55]. It has been proposed that these glycosylations proceed through an episulfonium (or episelenium) intermediate, which induces excellent α stereoselectivity [172].

These glycosyl donors can be obtained by addition of PhSCl to the double bond of a 2,3-dehydro derivative of Neu5Ac, giving predominantly the 3-equatorial diastereoisomer (77%) together with 15% of the 3-axial derivative when the addition was performed in CH_2Cl_2 (Scheme 3.14) [173]. After separation of this mixture of diastereomers, the equatorially substituted compound could be used as a glycosyl donor, although it was shown that better results were obtained if the anomeric chloride was first converted into a thioglycoside [174, 175]. It was demonstrated that the thioglycosyl donor could be efficiently activated with MeSBr/AgOTf, and glycosidation of a sterically hindered 3′-hydroxy group of a lactoside at −40 °C afforded a required a-linked trisaccharide in a good yield of 67%. This method gives results that compare very favorably with the use of other popular glycosyl donors, especially when applied to glycosidations of sterically hindered alcohols. Indeed, direct sialidations either of 2-xanthate or of 2-methylthio derivatives of N-acetylneuraminic with the lactosyl acceptor gave trisaccharides as mixtures of anomers in lower yields [174, 175].

Additions of "PhSX"-type electrophiles afford mixtures of diastereomers, requiring separation or epimerization at C-3 prior to glycosylation and so making the synthesis and application of these compounds inconvenient. It has been reported that treatment of sialyl glycal with 2,4-dimethylbenzenesulfenyl chloride in CH_2Cl_2 affords the corresponding 2-chloro-3-thioaryl derivative with good diastereoselectivity and in high yield (ax/eq 1/21, 89%) [176]. The glycosyl donor properties of this compound were investigated by glycosylation with a 3′,4′-dihydroxy lactoside. The desired O-sialoside was isolated in an excellent yield of 71% after O-acetylation and reductive removal of the S-aryl auxiliary.

2-Chloro and 2-fluoro derivatives possessing 3-S-phenyl moieties have been employed for the preparation of a tetrasaccharide by an elegant glycosylation strategy (Scheme 3.15) [177, 178]. It was established that the glycosyl fluoride is stable towards the conditions required for activation of the anomeric chloride. Therefore, coupling of a sialyl glycosyl fluoride acceptor bearing an 8-hydroxy group with a fully protected 2-chloro sialyl donor in the presence of AgOTf gave a (2 → 8)-linked dimer in a 49% yield. As in the previous cases, high a selectivity was achieved thanks to neighboring group participation by the 3-thiophenyl moiety. The obtained dimer was used in the subsequent glycosylation without the need for any chemical modifications. Indeed, glycosylation with a thio-lactosyl acceptor in the presence of AgOTf/$SnCl_2$ afforded a tetrasaccharide in 39% yield. This reaction exploits the finding that glycosyl fluorides can be activated in the presence of thioglycosyl acceptors. The thioglycoside can subsequently be activated by use of a thiophilic reagent, and in this case the promoter DMTST was used in a coupling with an azidosphingosine acceptor to give a glycosyl azidosphingosine.

A 2,3-dithio-substituted derivative was also used in a polymer-supported synthesis of sialyl glycosides, in which the glycosyl donor was attached to the soluble polymer polyethylene glycol (PEG) through a succinyl linker (Scheme 3.16) [179]. Glycosylation of a 6-hydroxy derivative of a galactoside in the presence of DMTST, followed by cleavage from the polymer support by base treatment, gave the target disaccharide in 70% yield. The 3-thiophenyl moiety was conveniently removed to

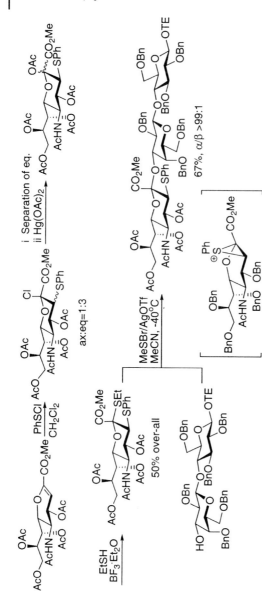

Scheme 3.14

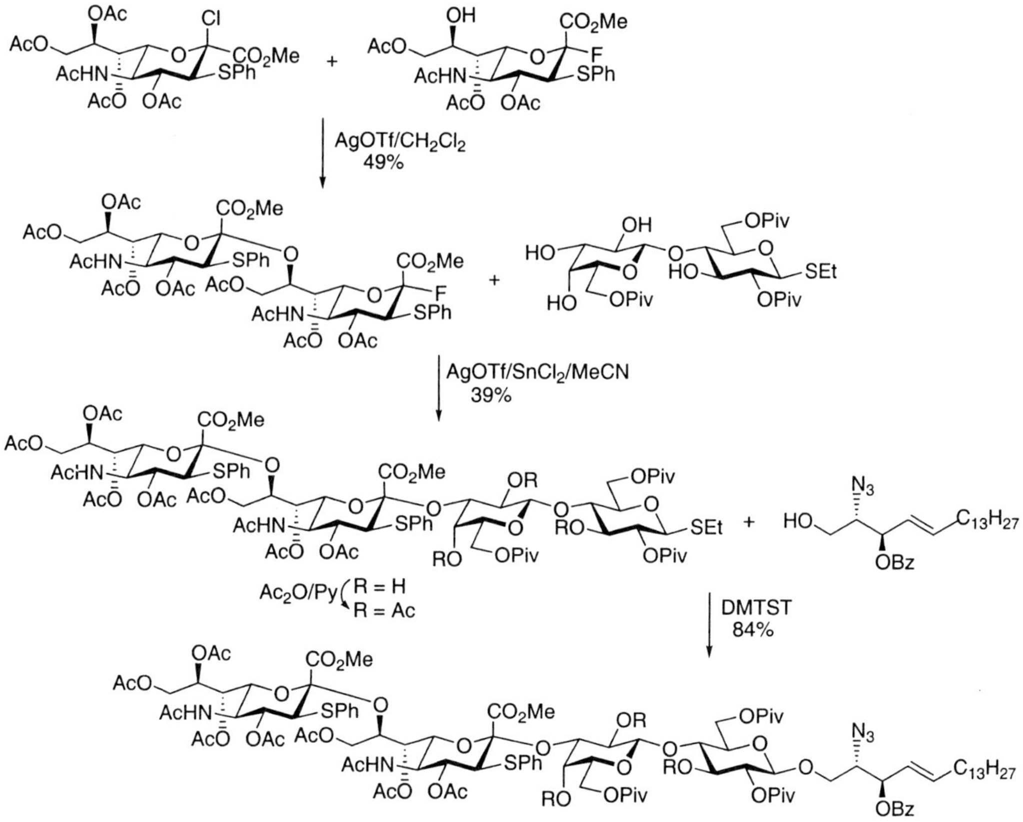

Scheme 3.15

furnish a (2 → 6)-linked O-sialoside along with a trace amount of the β anomer. The applicability of this approach for the synthesis of (2 → 3)-linked derivatives has also been documented.

3.3.3
Synthesis of (2 → 8)-Linked Sialosides

Neu5Ac or Neu5Gc occur as linear homopolymers, in which they are usually linked internally through $a(2 \to 8)$, $a(2 \to 9)$, or alternating $a(2 \to 8)/a(2 \to 9)$ glycosidic linkages. These polysialic acids – together with some other unusual structures – are found in glycoproteins of embryonic neural membranes, where they play roles as neural cell adhesion molecules [1, 3–9, 11]. The dimer Neu5Acα(2 → 8)Neu5Ac is a principal terminal constituent of a number of glycoconjugates and plays important roles in numerous biological phenomena, being a tumor-associated antigen and a receptor for bacterial toxins and viruses. The dimeric structure is important

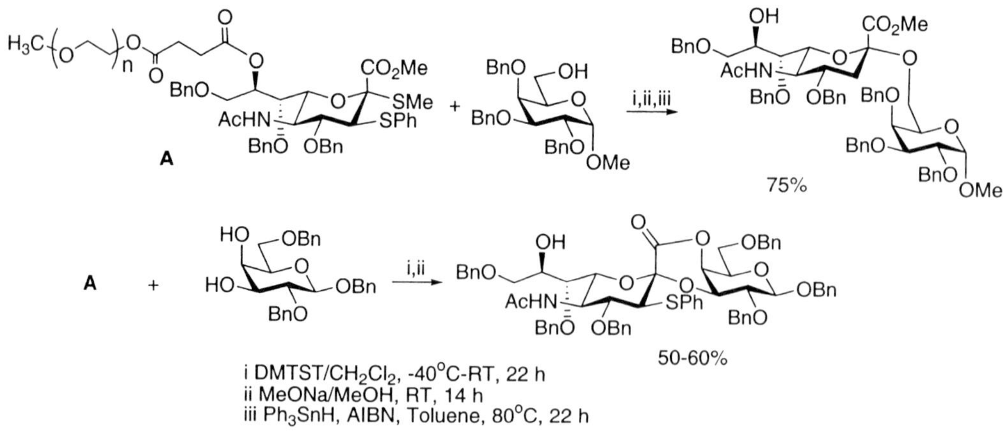

Scheme 3.16

i DMTST/CH$_2$Cl$_2$, -40°C-RT, 22 h
ii MeONa/MeOH, RT, 14 h
iii Ph$_3$SnH, AIBN, Toluene, 80°C, 22 h

for these biological properties, as removal of one of the two Neu5Ac residues often results in dramatic or total loss of activity.

The synthesis of oligosaccharides containing (2 → 8)-linked fragments is complicated by the low reactivity of the C-8 hydroxy group of Neu5Ac. Despite this complication, several successful syntheses of this dimer have been reported, most approaches having been based on the use of indirect methods. For example, a 2-bromo-3-hydroxy-substituted glycosyl donor afforded a (2 → 8)-linked dimer in 34% yield and as a mixture of anomers (α/β3/1, Scheme 3.17) [170, 171]. A more favorable outcome was obtained on application of the 2-bromo-3$_{eq}$-phenylthio-substituted glycosyl donor [180, 181]. In this case, the dimer was obtained exclusively as the α anomer and in a good yield of 64%. Further improvements came with the application of glycosyl donors possessing a 2-phosphite leaving group in combination with a 3-thiophenyl [182] or 3-O-phenylthiocarbonyl [168] participating functionality. On employment of the latter derivative, the α anomer was formed in an excellent yield of 83%. Use of a 2-thioethyl leaving group in combination with a 3-thiophenyl participating auxiliary afforded α(2 → 8)-linked product in a disappointing yield of 28% [175, 183], but this yield was significantly improved (44%) [184] when the C-5 acetamido group was protected as an N-acetylacetamido group [111, 116]. As discussed above (Scheme 3.15), the selective glycosylation approach has also been applied for the efficient preparation of this class of compounds [177, 178].

Early attempts to obtain (2 → 8)-linked dimers by direct methods gave very disappointing results. As an example, direct glycosylation of an 8-hydroxy sialyl acceptor with a 2-thiomethyl neuraminyl donor in the presence of DMTST in MeCN at −40 °C afforded only trace amounts (5%) of the desired α-linked dimer (Scheme 3.18) [103]. Glycosyl phosphites were shown to be more effective for the direct sialylation; in this case (2 → 8)-linked derivatives were obtained in good yields (22–68%) [149, 152], but unfortunately the unnatural β anomers were formed predominantly.

The 8-hydroxy group of Neu5Ac is of low nucleophilicity, due to a combination of obvious steric effects, interactions with the acetamido group at C-5, and/or the

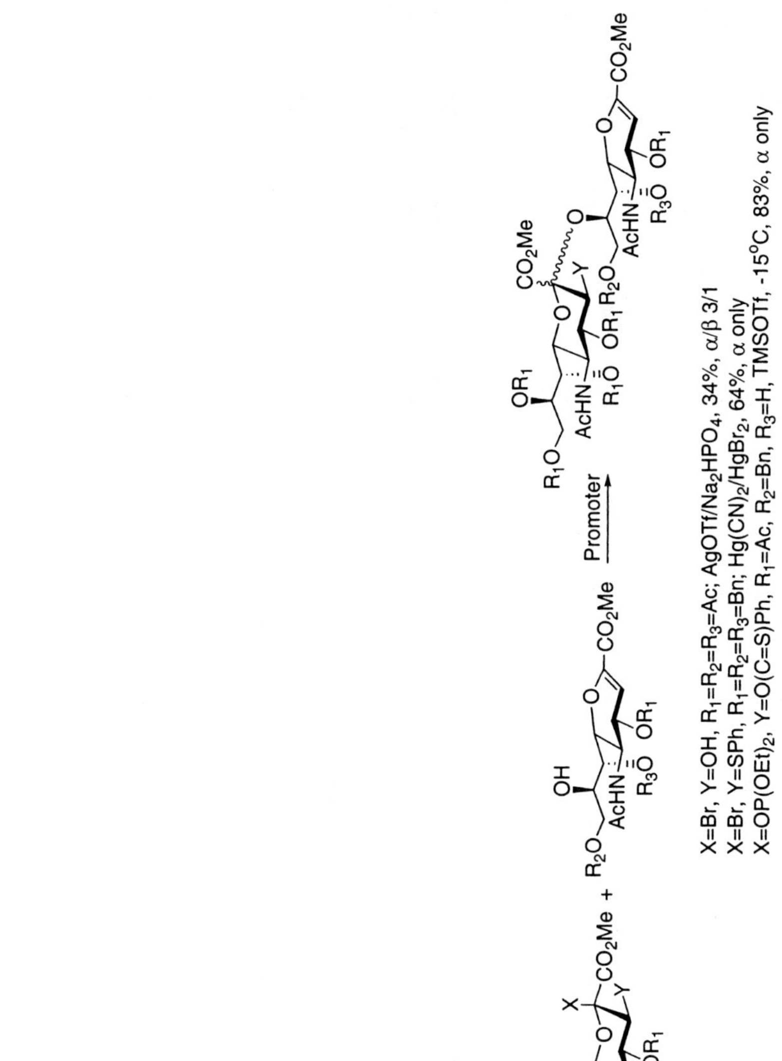

3.3 Chemical Glycosidation of Sialic Acids

Scheme 3.17

X=Br, Y=OH, $R_1=R_2=R_3$=Ac; AgOTf/Na_2HPO_4, 34%, α/β 3/1
X=Br, Y=SPh, $R_1=R_2=R_3$=Bn; $Hg(CN)_2$/$HgBr_2$, 64%, α only
X=OP(OEt)$_2$, Y=O(C=S)Ph, R_1=Ac, R_2=Bn, R_3=H, TMSOTf, -15°C, 83%, α only

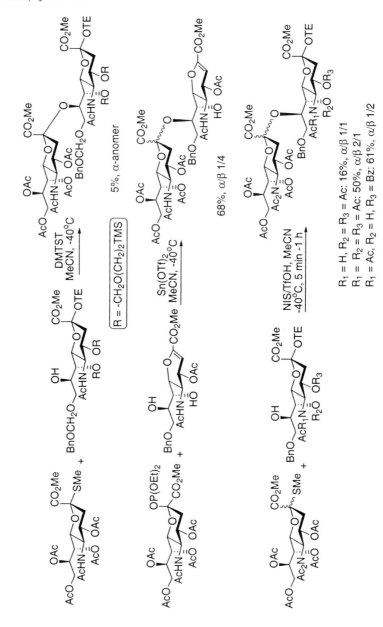

Scheme 3.18

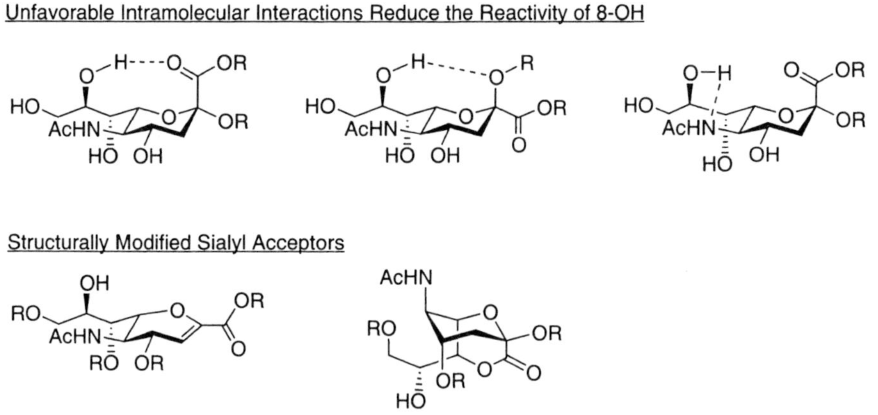

3.3 Chemical Glycosidation of Sialic Acids | 81

Unfavorable Intramolecular Interactions Reduce the Reactivity of 8-OH

Structurally Modified Sialyl Acceptors

Fig. 3.2 The low nucleophilicity of the 8-hydroxy group in Neu5Ac.

presence of an internal hydrogen bond between the C-8 hydroxy group and C-1 carboxyl or 2-OR (R = Me, Ac, H, etc.) moieties (Fig. 3.2). Formation of the internal 1,7-lactone results in a change of the ring conformation to favor $^5C_2$, in which the unfavorable intramolecular interactions are removed. Some of these interactions may also be eliminated through introduction of the 2,3-dehydro moiety in the acceptor. In this case, the C-1 carbonyl moiety is positioned farther from C-8 hydroxy group, so no hydrogen bonding can take place. The improvement in the sialyl acceptor properties resulted in good yields of coupling products, but unfortunately in only modest a anomer selectivity.

An alternative approach to addressing the difficulties of synthesizing Neu5Aca (2→8)Neu5Ac dimers by direct sialylation was recently reported [116]. As mentioned above, the presence of an additional N-acetyl moiety at C-5 of a glycosyl donor dramatically increases its reactivity and gives improved yields in sialylations [111]. In this context, coupling between a 2-thiomethyl-5-acetylacetamido glycosyl donor and the C-8 hydroxy group of a mono-N-acetylated acceptor in the presence of NIS/TfOH in MeCN at –40 °C gave a dimer as a mixture of anomers in modest yield (16%, a/β 1/1, Scheme 3.18). This result, however, was an improvement on previously reported attempts with a similar acceptor [103]. The application of a di-N-acetylated glycosyl acceptor under identical reaction conditions gave the dimer in a much improved yield and with better anomeric stereoselectivity (50%, a/β 2/1). These results illustrate that high yields of coupling products can be obtained when the 5-acetamido moieties both of the neuraminyl glycosyl donor and of the acceptor are derivatized into N-acetylacetamido functionalities. An attempt to reduce the steric hindrance around O-8 by use of a 7,8-diol as an acceptor resulted in an excellent yield (61%) of coupling product, but in this case mainly β anomer was formed (a/β 1/2). Schmidt and co-workers made a similar observation [152], and these combined findings are remarkable since reduction of steric hindrance around C-3 of a galactosyl acceptor results in higher a-stereoselectivities.

82 | 3 The Chemistry of Sialic Acid

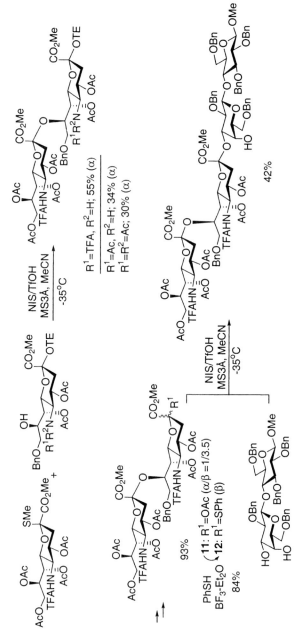

Scheme 3.19

Further screening of suitable substituents at C-5 revealed that 2-thiomethyl or 2-thiophenyl sialyl donors bearing a trifluoroacetamido functionality at C-5 possess glycosyl donor properties even better than those of the acetamido and N-acetylacetamido derivatives [114]. Remarkably, complete stereoselectivity was achieved when this approach was applied for the synthesis of (2–8)-linked dimers. Surprisingly high yields were achieved when both sialyl donor and sialyl acceptor were protected at N-5 with trifluoroacetamido functionalities. The advantageous features of this technique were further exploited for the synthesis of sialoconjugates; as an example, a tetrasaccharide bearing a dimeric neuraminic acid sequence was efficiently synthesized (Scheme 3.19).

A totally different strategy for obtaining oligosaccharides containing the Neu5Acα(2→8)Neu5Ac moiety, avoiding the difficulties associated with the chemical formation of the α(2→8) linkage, has also been developed [185]. In this case, mild acid treatment of colominic acid (a homopolymer made up of Neu5Acα(2→8)Neu5Ac moieties) results in glycosidic bond cleavage to give the dimer, together with higher oligomers and the monomer. After separation, the dimer could be converted into a glycosyl donor (most commonly 2-SPh) by chemical methods, but such a donor unfortunately typically gives lower yields of coupling products than the corresponding monomeric donor, especially when applied for sialylation of secondary sugar alcohols [186–188]. In this context, a trimeric and a tetrameric glycosyl donor of Neu5Ac have also been coupled with a lactosyl acceptor to give oligosaccharides in reasonable yields [189, 190].

It should also be noted that both direct [116, 135, 191–193] and indirect [170, 171] chemical syntheses of Neu5Acα(2–9)Neu5Ac and Neu5Acα(2–5)Neu5Gc [194] have been reported.

3.4
Enzymatic Glycosidations of Sialic Acids

The need for increasingly efficient methods for oligosaccharide synthesis has stimulated the development of enzymatic approaches [6, 28, 29, 68, 69, 144, 154, 195–206]. These enzymatic methods bypass the need for protecting groups since they govern both the regioselectivity and the stereoselectivity of glycosylations. Two fundamentally different approaches for enzymatic oligosaccharide synthesis have been developed; (i) the use of glycosyltransferases, and (ii) the application of glycosyl hydrolases.

Glycosyltransferases are essential enzymes for oligosaccharide biosynthesis. These enzymes can be classified into enzymes of the Leloir pathway and those of the non-Leloir pathway. Glycosyltransferases of the Leloir pathway are involved in the biosynthesis of most N- and O-linked glycoproteins in mammalians and utilize sugar nucleotide mono- or diphosphates as glycosyl donors. In contrast, glycosyltransferases of the non-Leloir pathway use sugar phosphates as substrates.

Glycosyltransferases are highly regio- and stereoselective enzymes and have been successfully applied for enzymatic synthesis of oligosaccharides. These enzymes can be isolated from milk, serum, and organ tissues and are most commonly purified by affinity column chromatography using immobilized sugar nucleotide diphosphates. Several glycosyltransferases have been cloned and over-expressed and are now readily available in reasonable quantities. However, the number of easily available glycosyltransferases is still very limited. Furthermore, these enzymes are highly substrate-specific and the possibilities of preparing analogues are therefore limited.

Nature employs glycosylhydrolases for the degradation of oligosaccharides. However, the reversed hydrolytic activity of these enzymes can also be exploited in glycosidic bond formation. This method has allowed the preparation of several di- and trisaccharides. Glycosylhydrolases are much more readily available than glycosyltransferases but are generally less regioselective and the transformations are lower-yielding.

3.4.1
Sialyltransferases

Several $\alpha(2 \rightarrow 6)$- and $\alpha(2 \rightarrow 3)$-sialyltransferases have been used for oligosaccharide synthesis [28, 29, 207–216]. These enzymes are used to transfer the neuraminic acid moiety of activated CMP-Neu5Ac to the *C*-6 or *C*-3 hydroxy groups of terminal galactosides and *N*-acetyl-galactosaminides (Scheme 3.20). $\alpha(2 \rightarrow 8)$-Sialyltransferase is involved in the synthesis of $\alpha(2 \rightarrow 8)$-linked polysialic acids.

CMP-Neu5Ac has been obtained both by chemical and by enzymatic approaches. The Wong group employed a glycosylphosphoamidate, which was coupled with an appropriately protected cytosine [217]. Halcomb and co-workers also employed the phosphite coupling methodology, but in this case a cytosine

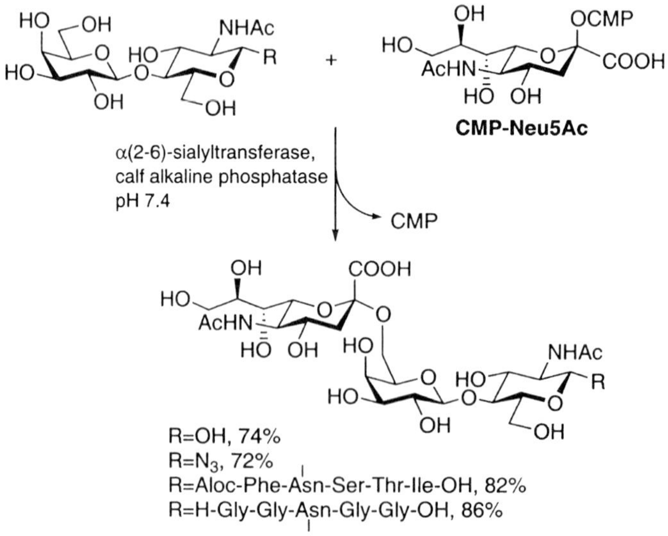
CMP-Neu5Ac
α(2-6)-sialyltransferase,
calf alkaline phosphatase
pH 7.4
CMP
R=OH, 74%
R=N3, 72%
R=Aloc-Phe-Asn-Ser-Thr-Ile-OH, 82%
R=H-Gly-Gly-Asn-Gly-Gly-OH, 86%

Scheme 3.20

phosphoamidate was coupled with the C-2 hydroxy group of Neu5Ac [218]. Oxidation of the resulting intermediate phosphite diesters with tBuOOH gave phosphotriesters, which were converted into phosphodiesters by removal of an allyl or cyanoethyl moiety. Schmidt and co-workers used an anomeric diethyl phosphite of Neu5Ac, which was coupled with the 5-phosphate of cytosine [218].

The most attractive procedure for obtaining CMP-Neu5Ac involves a procedure similar to the biosynthetic pathway, in which Neu5Ac and CTP are coupled with the aid of the enzyme CMP-Neu5Ac synthase [36, 40]. CTP can be obtained from cheap CMP, with phosphoenol pyruvate as the phosphate source, by means of several phosphokinase-catalyzed phosphate-transfer reactions. Neu5Ac has been prepared by means of an aldolase-mediated condensation between pyruvate and N-acetyl-mannosamine, obtained in turn by epimerization of GlcNAc (see above). The CTP and Neu5Ac were used as crude preparations and the only purification step in the synthesis was the final separation of CMP-Neu5Ac from the reaction mixture by ion-exchange chromatography (Scheme 3.21). Multi-gram quantities of pure CMP-Neu5Ac have been obtained by use of a synthetase cloned and over-expressed in *E. coli* [219] and the procedure could be further improved by use of a deacetylase for selective removal of any non-epimerized GlcNAc, which inhibits the ensuing aldolase reaction [220]. The sialic acid synthetase has also been employed for the preparation of several modified CMP-Neu5Ac analogues [42, 221], a 9-amino Neu5Ac derivative having been used, for example, for the preparation of a photolabeled CMP-Neu5Ac [43].

Another attractive way to synthesize sialylated oligosaccharides enzymatically involves the in situ regeneration of CMP-Neu5Ac [222, 223]. This procedure re-

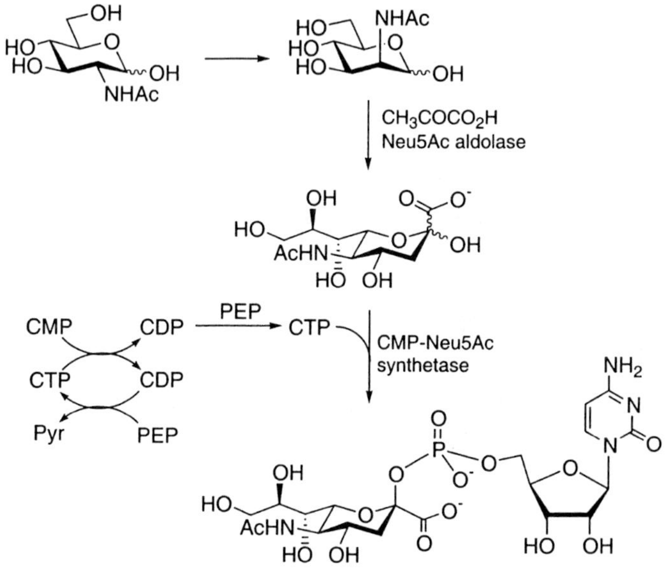

Scheme 3.21

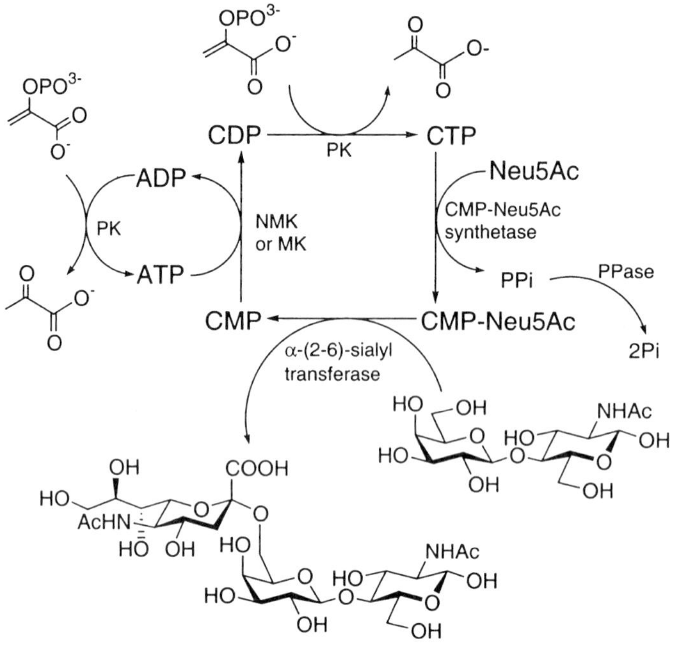

Scheme 3.22

quires only a catalytic amount of CMP-Neu5Ac, the CMP formed after the sialyltransferase being regenerated in situ by a series of enzymatic transformations (Scheme 3.22). An even more sophisticated procedure involves the in situ enzymatic synthesis of Neu5Ac by an aldolase-mediated reaction between ManNAc and pyruvate, coupled with the regeneration of CMP-Neu5Ac. This procedure was employed for the synthesis of Neu5Acα(2 → 3)Galβ(1 → 4)GlcNAcβ1-R and Neu5Acα(2 → 6)Galβ(1 → 4)GlcNAcβ1-R derivatives.

α(2 → 6)-Sialyltransferases have been isolated from porcine liver, bovine colostrum, and *Photobacterium damsela* [224, 225], while α(2 → 3)-sialyltransferases have been obtained from porcine liver and porcine submaxillary glands. The latter enzyme has also been cloned and over-expressed [226].

Sialyltransferases accept particular modifications in their acceptor substrates [224, 227–235] and the use of modified acceptors has resulted in the synthesis of modified sialylated oligosaccharides. Several sialyltransferases with different substrate specificities have been characterized and purified [236–239]. The most striking example is that of the α(2 → 6)-sialyltransferases of *Photobacterium damsela* [224, 225, 240], which can transfer Neu5Ac to 2′-fucosyllactose and 3′-sialyllactose and can use both O- and N-linked glycoproteins as substrates [225]. The α(2–3)-sialyltransferase from *Neisseria gonnorrheae* is another enzyme with a relatively broad acceptor substrate specificity and was shown [216] to transfer sialic acid to galactose and N-acetylgalactosamine, sulfated N-acetyllactosamine, 6-O-sulfochitotriose and -tetrose, and to tyrosine-sulfated glycopeptides and lactosides with a short aliphatic or aromatic aglycon. In terms of its donor substrate specificity, this enzyme accepts Neu5Ac as an excellent substrate and Neu5Gc, Neu5Boc, and KDN as poor substrates, whereas Neu5N$_3$ and Neu5Acα(2–9)Neu5Ac are not incorporated in acceptors.

Sialyltransferases accept some modifications of CMP-Neu5Ac and so can be used for the incorporation of modified sialic acids. It has been shown that an $\alpha(2 \rightarrow 3)$-sialyltransferase was able to accept CMP-sialic acid analogues with N-glycolyl, benzyloxycarbamate, or hydroxy groups at C-5 [241]. A 5-acetamido-3,5,9-trideoxy-β-D-*glycero*-D-*galacto*-nonulopyranosidonic acid analogue was also employable [242], incorporation of that unit into glycoproteins giving glycoproteins that aided structure determination by X-ray crystallography. The modified donors for these enzymatic glycosylations were prepared by chemical methods.

CMP 3-fluoroneuraminic acid is an inhibitor of $\alpha(2 \rightarrow 6)$ sialyltransferase. The enzyme recognizes this compound, but the 3-fluoro moiety destabilizes the oxycarbenium ion formed at the transition state [243]. A CMP-9-fluoro-[3-^{13}C]-Neu5Ac derivative has been synthesized by a combined chemical and enzymatic approach, and this substrate could be used for the sialylation of a glycoprotein through the use of an $\alpha(2 \rightarrow 6)$-sialyltransferase [244]. Only 60% of the galactosides of the glycoprotein were sialylated and those that had not reacted were removed. The ^{13}C and fluoride labels enabled NMR experiments to be performed on the neoglycoprotein.

Sialyltransferases have also been used for the synthesis of neolacto series gangliosides [245]. It was observed that a purified $\alpha(2 \rightarrow 6)$-sialyltransferase would not accept a neutral glycosphingolipid paragloboside as an acceptor (Scheme 3.23, R=CO(CH$_2$)$_{18}$CH$_3$). This result was rather surprising, since the oligosaccharide of the glycolipid is known to be an excellent substrate for the enzyme, so the ceramide must have a very profound effect on the enzymatic transformation. Modified ceramides with no substitution or with an acetyl (COCH$_3$) or chloroacetyl (COCH$_2$Cl) moiety at the amino function proved to be good substrates for the enzyme.

Clustered acceptors proved to be appropriate substrates for sialyltransferases [246–248]. A trimeric β-lactosyl cluster based on 2-(hydroxymethyl)-2-nitropropane-1,3-diol is an effective acceptor for rat liver $\alpha(2 \rightarrow 3)$-sialyltransferase. Its K_m was comparable to those of monomeric lactosyl and N-acetyllactosamine acceptors, whereas its V_{max} was only 1% of that measured for the LacNAc acceptor. While the V_{max} was relatively low, a preparative scale sialylation afforded a trimeric cluster of the GM$_3$ oligosaccharide in good yield. Di-, tetra-, and octavalent sialyl Lewis^x (SLe^x) ligands were also enzymatically prepared, starting from hypervalent dendritic L-lysine cores with a 2-acetamido-2-deoxy-D-glucose attachment.

Sialyltransferases have been employed in the chemoenzymatic synthesis of several oligosaccharides. Most of the procedures involve the chemical synthesis of a complex oligosaccharide and the introduction – after removal of the protecting groups – of a sialyl moiety with the aid of a sialyltransferase. This approach has been used for the synthesis of the core sialyl-containing hexasaccharide found on O-linked glycoproteins [249], a complex-type biantennary N-glycan [250], GM$_3$ [251, 252], ^{13}C-enriched GM$_3$ and sialyl Lewis^x [253], a spacer-modified sialyl

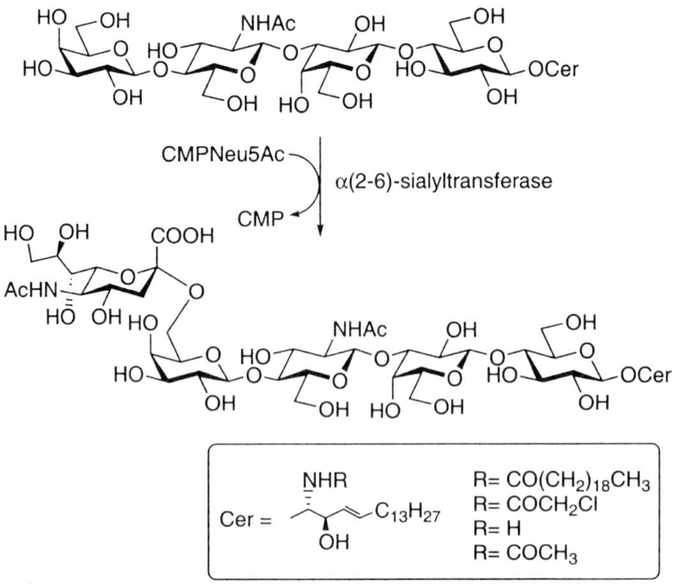

Scheme 3.23

Lewis^x [248, 254], and a *Streptococcus* group B capsular oligosaccharide [255–257]. A dimethyloctylsilylethyl lactoside proved to be an appropriate substrate for $\alpha(2 \rightarrow 3)$-sialyltransferase, but the turnover rate was only 2% of that reported for β-D-Gal-$(1 \rightarrow 3)$-3OEt$_2$.

Several glycosyltransferases, including sialyltransferases, have been used for solid-phase syntheses of oligosaccharides [259–262]. In this procedure, a glycosyl acceptor is immobilized to a resin through a selective, cleavable linker. Subsequently, glycoside bond synthesis is accomplished by the addition of glycosyltransferases and appropriate substrates. This technology has several advantages over traditional solution-based methods, including: (i) ease of product purification by filtration and washing, (ii) convenient use of large excesses of reagents to drive reactions to completion, and (iii) the possibility of automation. The resin has to meet several requirements, such as proper swelling in aqueous buffers and that all immobilized glycosyl acceptors are assessable to the enzymes.

Spacer-containing sialyl Lewis^x has been prepared on a thiol-derivatized Sepharose matrix by immobilization of a derivative of *N*-acetylglucosamine through a disulfide linkage [263]. The resin was incubated successively with $\beta(1–4)$galactosyltransferase, recombinant $\alpha(2–3)$sialyltransferase, and human milk (1,3/4)fucosyltransferase, in the presence of the corresponding sugar nucleotides and alkaline phosphate. Treatment with dithiothreitol gave the product in an overall yield of 57%. The selection of an appropriately long linker was an important requirement for achieving this high yield. Wong and co-workers synthesized several sialyl oligosaccharides on controlled-pore glass by a chemoenzymatic approach [264]. A disaccharide primer was first attached to the support through a spacer group containing an ester linkage, followed by enzymatic incorporation of galactose and Neu5Ac. Two to three equivalents of the sugar nucleotides were used in the enzymatic glycosylation and the conversion in each step was found to be >98%. The compound could easily be cleaved from the resin by ester cleavage with hydrazine. The same group also synthesized a glycopeptide containing an O-linked sialyl Lewis^x moiety by a combination of chemical and enzymatic steps [265, 266]. Firstly, an O-GlcNAc octapeptide was synthesized chemically on solid support by use of Fmoc-Thrβ(GlcNAc)-OH as a glycosylated amino acid-building block. The protecting groups were removed, extension with monosaccharides was carried out with the aid of glycosyltransferases, and the product was finally removed from the solid support. The use of the acid- and base-stable Hycron linker enabled complete removal of all protecting groups without cleavage of the compound from the solid support. The allylic linker, however, can be cleaved under mild conditions by treatment with Pd(0) in the presence of the allyl scavenger morpholine. Poly-(ethyleneglycol)-polyacrylamide copolymer (PEGA) and controlled-pore glass were also considered as polymeric supports, and although controlled-pore glass was found not to be the optimal support for chemical peptide synthesis, it was most suitable for performing enzymatic reactions.

Cummings and co-workers employed a similar strategy for their impressive synthesis of a glycosulfopeptide that binds to the P-selectin glycoprotein ligand-1 (PSGL-1) [267–270]. This group synthesized a glycopeptide by a conventional solid-phase approach, using a tri-O-acetyl-GalNAcFmoc threonine derivative. The glycopeptide was cleaved from the solid support and the crude product was de-O-acetylated with methanolic sodium methoxide. The product was purified and then used in a range of recombinant glycosyl transfer-catalyzed glycosylations, subsequently employing β(1,3)GalT, β(1,6)GlcNAcT, β(1–4)GalT, α(2,3)SialylT, and α(1,3)FucT-VI. Finally, the tyrosine residues were sulfated with the aid of recombinant tyrosyl-protein sulfotransferase-1 (TPST-1) and adenosine 3′-phosphate 5′-phosphosulfate as sulfate source. In an alternative strategy, Fmoc-Tyr(SO$_3$H-H) sodium salt was employed for incorporation of sulfated tyrosines prior to extension of the glycopeptide with glycosyltransferases. Wong and co-workers, however, noted that the presence of sulfated tyrosines may slow the action of the glycosyltransferases [266].

Water-soluble polymers have also been applied for enzymatic sialoside synthesis. These polymers have the advantages that the acceptors are more readily accessible by enzymes and that the macromolecular properties of the support permit product purification by dialysis or nanofiltration. Amino-substituted poly(vinyl alcohol)polyacrylamide-poly(N-acryloxysuccinimide) (PAN) [271] and polyethylene glycol [272] have been employed as soluble polymer supports.

3.4.1.1 Metabolic Engineering of the Sialic Acid Biosynthetic Pathway

Bertozzi and co-workers have exploited the ability of the sialic acid biosynthetic pathway to process modified substrates to incorporate modified sialic acids into cell-surface glycoconjugates [273–277]. The use of analogues such as N-levulinoyl-mannosamine resulted in the incorporation of ketone-modified sialic acids, which have a unique reactivity and could therefore be selectively condensed with probes such as biotin hydrazine. The use of these probes allowed the development of cell selection schemes that exploit the ability of ManLev to transit the entire biosynthetic pathway. Consequently, mutations that either enhance or diminish the ability of the unnatural substrate to produce cell-surface expression are detectable. Metabolic engineering has been used to construct non-natural glycans for lectin binding studies and to target magnetic resonance imaging contrast reagents to cells over-expressing sialic acid residues, a hallmark of many tumors. Unnatural sialic acids bearing ketone groups have also been used to facilitate virally mediated gene delivery.

3.4.2
Sialidases

Sialidases are glycosylhydrolases that catalyze the release of terminal sialic acids α-glycosidically linked to glycoproteins, glycolipids, and polysaccharides. These enzymes have also been used for the synthesis of sialosides by transglycosylation [272]. Thiem and co-workers reported the first example of such a transformation [278], employing immobilized sialidase from *Vibrio cholerae* as the catalyst, p-nitrophenyl Neu5Ac as the glycosyl donor, and several galactosides as acceptors. Both $\alpha(2 \rightarrow 3)$ and $\alpha(2 \rightarrow 6)$ glycosides were formed, in yields ranging from 14–24% (Scheme 3.24). The ratios of the two regioisomers depended upon the acceptor substrate used, but a preference for the formation of $(2 \rightarrow 6)$-linked derivatives was observed in each case.

The origin of the sialidase influences the regioselectivity of the transformations. Sialidases from *Clostridium perfringens*, *Arthrobacter ureafaciens*, and *Vibrio cholerae*

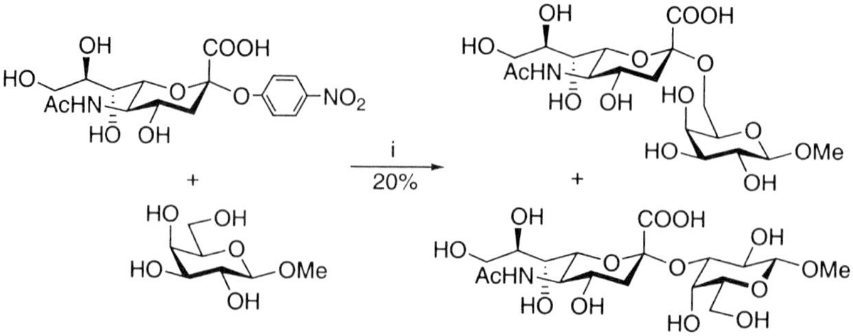

Scheme 3.24 *Conditions*: (i) immobilized enzyme, pH 5.5 buffer (NaOAc, CaCl$_3$, NaN$_3$), in the presence or absence of DMSO.

gave mainly $\alpha(2 \rightarrow 6)$-linked sialyl N-acetyl lactosamine derivatives, whereas the sialidase from Newcastle disease virus gave preferential formation of the $\alpha(2 \rightarrow 3)$ linked regioisomer [279]. In these transformations, p-nitrophenyl Neu5Ac or α(2-8)-linked sialic acid dimer were used as glycosyl donors. It has been suggested that immobilization of these enzymes results in lower regioselectivities [279].

Sialidase-catalyzed transglycosylations have also been applied for regioselective $\alpha(2 \rightarrow 3)$ sialylations of Lewis^x and Lewis^a [280]. Sialidase from *Salmonella typhimurium* LT2 was used for these transformations, with p-nitrophenyl Neu5Ac as the acceptor. Sialyl Lewis^x and allyl Lewis^a were isolated in yields of 9.3% and 12.0%, respectively. These results are significant because mammalian sialyltransferases recognize Le^a or Le^b as acceptors.

The trimer Neu5Ac$\alpha(2 \rightarrow 3)$Gal$\beta(1 \rightarrow 4)$GlcNAc was synthesized by two regioselective transglycosylations with β-galactosidase from *Bacillus circulans* and trans-sialidase from *Trypanosoma cruzi* [281]. A yield of 60% was achieved when 4-methylumbelliferyl Neu5Ac was used as a donor. Colominic acid (a polymer composed of $\alpha(2 \rightarrow 8)$-linked Neu5Ac) has been used as a donor substrate for trans-sialylations [282].

Neu5Ac$\alpha(2 \rightarrow 3)\beta$GalOR derivatives have been prepared by combined use of trans-sialidase and sialyltransferase [283]. This method takes advantage of the trans-sialidase enzyme of *Trypanosoma cruzi*, which has the unique property of catalyzing the reversible transfer of Neu5Ac from a donor substrate of the sequence Neu5Ac$\alpha(2 \rightarrow 3)$GalβOR^1 to virtually any galactoside acceptor β-D-Gal-OR^2 to yield the new product Neu5Ac$\alpha(2 \rightarrow 3)$GalβOR^2. The substrate of this enzyme was prepared in situ by use of $\alpha(2 \rightarrow 3)$-sialyltransferase to transfer free Neu5Ac to its precursor galactoside acceptor with catalytic in situ regeneration of CMP-Neu5Ac. This method circumvents the narrow substrate specificity of sialyltransferases.

The combined use of glycosidases and sialyltransferases has also been reported [284, 285]. In this approach, the synthesis of Gal$\beta(1 \rightarrow 3)$GalNAc was catalyzed by a β-galactosidase and used p-nitrophenyl β-galactoside and N-acetylglucosamine as the substrates. The lactosamine was sialylated by an $\alpha(2 \rightarrow 3)$-sialyltransferase coupled with in situ regeneration of CMP-Neu5Ac. A similar approach was used for the preparation of Neu5Ac$\alpha(2 \rightarrow 3)$Gal$\beta(1 \rightarrow 4)$GlcNAc.

3.5
Synthesis of C- and S-Glycosides of Sialic Acid

Glycosidic linkages of Neu5Ac are cleaved by sialidases. The design of non-hydrolyzable analogues of Neu5Ac glycosides has received attention, efforts having been focused on synthesis of compounds in which the exocyclic oxygen is replaced by a methylene group to give a C-glycoside or by sulfur to give a thioglycoside. These derivatives have also found use as inhibitors of sialidases.

Many methods for the synthesis of C-glycosides have been reported, but most are not applicable to sialic acid, the major problem being that the formation of the new C–C bond results in a quaternary C atom. Lindhardt and co-workers employed a general diastereoselective approach based on the reduction of an anomeric 2-pyridyl sulfone with SmI_2 to generate an N-acetylneuraminyl samarium species in situ. This could then be coupled with carbonyl derivatives under Barbier conditions [286, 287]. The coupling reaction almost exclusively gave α-linked C-glycosides, which was explained by assuming that a bulky anomeric $Sm(III)I_2$ moiety in an intermediate species adopts the thermodynamically more stable equatorial configuration. Furthermore, the diastereochemical outcome of the reaction could be explained by the Felkin-Ahn model, in which the bulkiest ligand to the carbonyl is the C-2 atom bearing an axial α-OMe glycoside. This substituent has a perpendicular relationship to the plane of the carbonyl group and is anticlinal to the Burgi-Dunitz trajectory of the incoming nucleophile. The methodology was applied to the synthesis of five C-glycosides of Neu5Ac, ranging from simple hydroxyalkyl derivatives to a C-disaccharide [288]. These compounds were tested for their ability to inhibit the neuriminidase of C. perfringens and it was found that compounds with a hydrophobic substituent at C-2 displayed the highest level of inhibitory activity. The SmI_2-mediated coupling was also employed for the preparation of C-glycosides of KDO [289]. In this case an anomeric chloride was reduced with SmI_2 to give a reactive intermediate.

Schmidt and co-workers developed a general strategy for the synthesis of C-glycosides of Neu5Ac and applied it to the preparation of a methylene-bridged Neu5Ac-α-(2,3)-Gal C-disaccharide (Scheme 3.26) [290]. The key step was a stereoselective 6-exo-trig electrophilic cyclization of an open-chain precursor. This open-chain precursor was obtained by addition of a lithiated iodide to an open chain aldehyde obtained from D-glucono-δ-lactone by chain elongation. Oxidation of the resulting alcohol and subsequent C-1 incorporation with Tebbe reagent, followed by protecting group manipulations, gave an alkene. This was cyclized with phenylselenyl triflate (PhSeOTf) to give a 7:1 mixture of diastereomers, the major one being that with the required phenylselenyl methylene moiety in an axial position. A selena-Pummerer rearrangement gave an aldehyde, which was oxidized to a car-

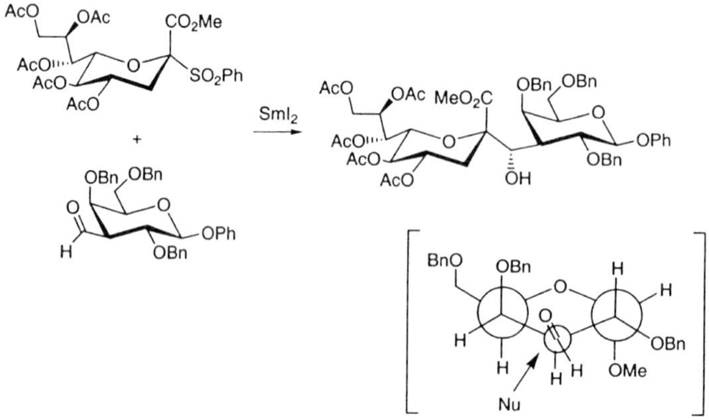

Scheme 3.25 Synthesis of a C-glycoside of Neu5Ac.

boxylic acid with Dess–Martin periodinane and protected as a methyl ester. Subsequent steps involved the introduction of an acetamido function by selective triflation of the axial hydroxy group, displacement with sodium azide, reduction of the resulting azido derivative, and N-acetylation.

Sialic acid thioglycosides or thiosialosides are another class of hydrolytically stable analogues of glycosides of Neu5Ac. The most commonly employed approach for the synthesis of these compounds entails the S-alkylation of an acceptor containing a good leaving group with a 2-thio-Neu5Ac derivative (Scheme 3.27, approach a) [291–295]. Von Itzstein and co-workers employed 2-thioacetyl Neu5Ac, which was deacetylated in situ with diethylamine and coupled with a sugar triflate to give the corresponding thioglycosides in good yields [296–298]. This method has provided access to a range of α-linked thiosialosides. The same group showed by NMR experiments that Neu5Ac-2-S-α-(2–6)Galβ-OMe and Neu5Ac-2-S-α-(2–3)Galβ-OMe are indeed stable towards the sialidase of *Vibrio cholerae* [299]. An S-alkylation strategy was employed to synthesize a 2-S-5′-aminopentenyl derivative, which was coupled to CNBr-activated Sepharose 4B through the terminal amino group [298, 300]. This matrix was employed to purify *Vibrio cholerae* sia-

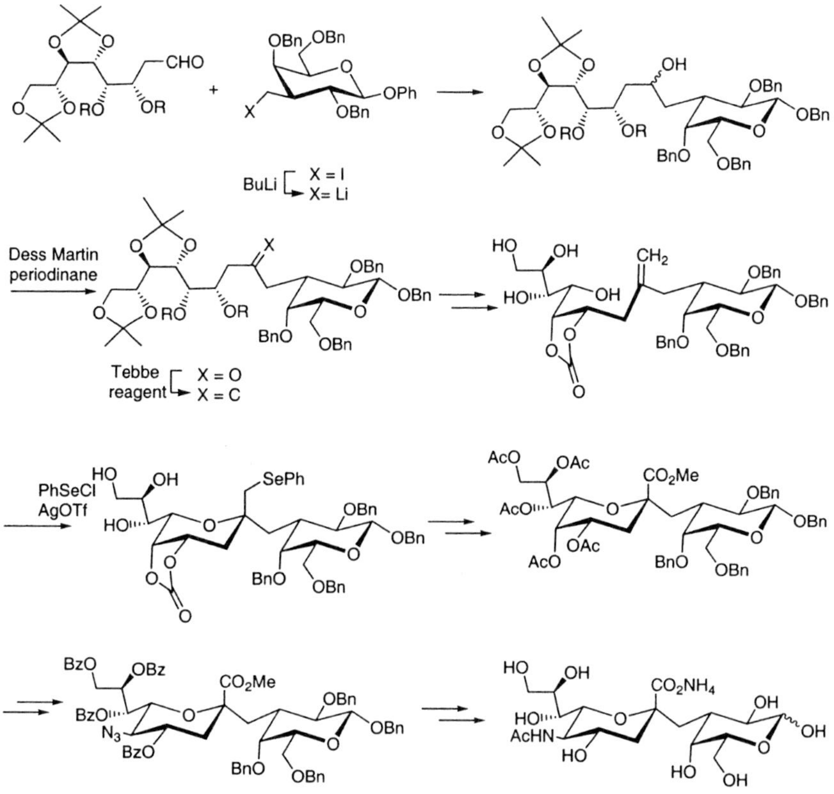

Scheme 3.26 Synthesis of a C-glycoside of Neu5Ac by PhSeOTf-mediated cyclization.

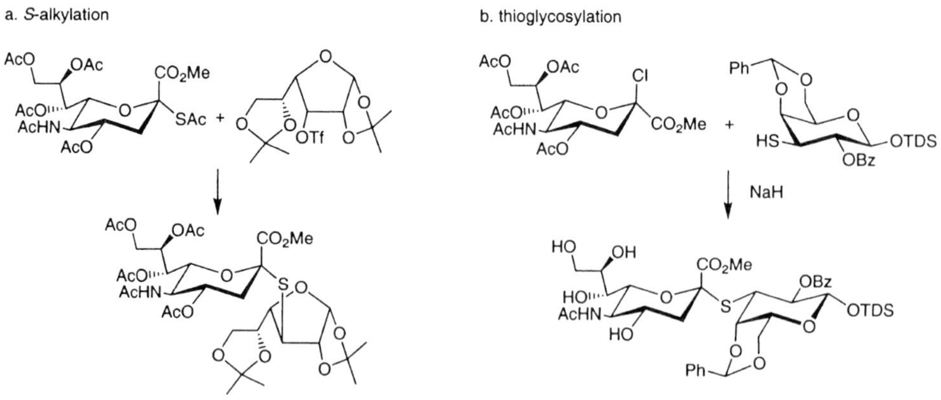

Scheme 3.27 Synthesis of thioglycosides of Neu5Ac.

lidase, sialidase-L from leech, trans-sialidase from *Trypanosoma cruzi*, and sialyltransferases from rat liver in high yield. In an attempt to synthesize Neu5Ac-2-S-α-(2–3)Galβ-OMe, Field and co-workers observed mainly elimination of a sugar triflate [301]. This unwanted reaction could, however, be avoided by use of a 3-O-triflate-gulofuranose as the alkylation reagent (Scheme 3.27, approach a).

An alternative route to thioglycosides of Neu5Ac is based on thioglycosidation strategy in which a sulfide is incorporated into the glycosyl acceptor (Scheme 3.27, approach b), which is then coupled with a conventional glycosyl donor. Glycosidations of peracetylated 2-chloro-Neu5Ac under phase-transfer conditions have proven to be the best conditions and this coupling approach has been used in an impressive synthesis of a sialyl Lewis[x] analogue that has all its exocyclic oxygens replaced by sulfur.

3.6
Modifications at N-5

N-Glycolylneuraminic acid (Neu5Gc) is an important naturally occurring neuraminic acid derivative. Several attempts to synthesize oligosaccharides containing this derivative have been reported, and glycosyl donors containing O-acetyl- [97, 302–304] or O-benzyl-protected [305, 306] N-glycolyl moieties {-C(=O)CH$_2$OAc or -C(=O)CH$_2$OBn} have been prepared for this purpose. Other functional groups at C-5 were introduced for biological studies [307].

Protection of the amino group at C-5 as *tert*-butyloxycarbonyl (Boc) [308], benzyloxycarbonyl (CBz, Z) [83], azido [83, 309, 310], phthalimido (Phth) [311], trifluoroacetamido [114, 115], or *tert*-butyloxycarbonylacetamido [312] groups allowed access after deprotection to a free C-5 amine, which can be then derivatized with glycolyl or other moieties.

3.7 References

1. Schauer, R., *Sialic Acids: Chemistry, Metabolism and Function*; Springer, Wien New York, **1982**, Cell Biology Monographs, Vol. 10.
2. Rosenberg, A. Biology of sialic acids; Plenum Press, New York, London, **1995**.
3. Schauer, R., *Angew. Chem., Int. Ed. Engl.* **1973**, *12*, 127–138.
4. Reglero, A.; Rodriguez-Aparicio, L. B.; Luengo, J. M., *Int. J. Biochem.* **1993**, *25*, 1517–1527.
5. Varki, A., *Glycobiology* **1993**, *3*, 97–130.
6. Inoue, Y.; Inoue, S., *Pure Appl. Chem.* **1999**, *71*, 789–800.
7. Yamada, K.; Harada, Y.; Nagaregawa, Y.; Miyamoto, T.; Isobe, R.; Higuchi, R., *Eur. J. Org. Chem.* **1998**, 2519–2525.
8. Kitazume, S.; Kitajima, K.; Inoue, S.; Haslam, S. M.; Morris, H. R.; Dell, A.; Lennarz, W. J.; Inoue, Y., *J. Biol. Chem.* **1996**, *271*, 6694–6701.
9. Inagaki, M.; Isobe, R.; Higuchi, R., *Eur. J. Org. Chem.* **1999**, 771–774.
10. Smirnova, G. P.; Chekareva, N. V., *Bioorg. Khim.* **1997**, *23*, 586–590.
11. Schauer, R., *Trends Glycosci. Glycotech.* **1997**, *9*, 315–330.
12. Paulson, J. C.; Colley, K. J., *J. Biol. Chem.* **1989**, *264*, 17615–17618.
13. van den Eijnden, D. H.; Joziasse, D. H., *Curr. Opin. Struct. Biol.* **1993**, *3*, 711–721.
14. Powell, L. D.; Varki, A., *J. Biol. Chem.* **1994**, *269*, 10628–10636.
15. Rossen, S. D.; Bertozzi, C. R., *Curr. Opin. Cell Biol.* **1994**, *6*, 663.
16. Springer, T. A., *Annu. Rev. Physiol.* **1995**, *75*, 827–872.
17. Kelm, S.; Pelz, A.; Schauer, R.; Filbin, M. T.; Tang, S.; Bellard, M.-E.; Schnaar, R. L.; Mahoney, J. A.; Hartnell, A.; Bradfield, P.; Crocker, P. R., *Curr. Biol.* **1994**, *4*, 965–972.
18. Hanasaki, K.; Varki, A.; Powell, L. D., *J. Biol. Chem.* **1995**, *270*, 7533–7542.
19. Hennett, T.; Chui, D.; Paulson, J. C.; Marth, J. D., *Proc. Natl. Acad. Sci. USA* **1998**, *95*, 4504–4509.
20. Watowich, S. J.; Skehel, J. J.; Wiley, D. C., *Structure* **1994**, *2*, 719–731.
21. Stehle, T.; Yan, Y.; Benjamin, T. L.; Harrison, S. C., *Nature* **1994**, *369*, 160–163.
22. Roulston, A.; Beauparlant, P.; Rice, N.; Hiscott, J., *J. Virol.* **1993**, *67*, 5235–5246.
23. Dallocchio, F.; Tomasi, M.; Bellini, T., *Biochem. Biophys. Res. Commun.* **1995**, *208*, 36–41.
24. Stein, P. E.; Boodhoo, A.; Armstrong, G. D.; Heerze, L. D.; Cockle, S. A.; Klein, M. H.; Read, R. J., *Nature Struct. Biol.* **1994**, *1*.
25. Sim, B. K.; Chitnis, C. E.; Wasniowska, K.; Hadley, T. J.; Miller, L. H., *Science* **1994**, *264*, 1941–1944.
26. Herrler, G.; Hausmann, J.; Klenk, H. D., Sialic acid as receptor determinant of ortho- and paramyxoviruses; Rosenberg, A. Ed.; Plenum Press, New York, London **1995**. *Biology of the sialic acids*; pp. 315–336.
27. Reutter, W.; Kottgen, E.; Bauer, C.; Gerok, W. Biological significance of sialic acids; Schauer, R. Ed.; Springer, Wien, New York, **1982**. Sialic acids. *Chemistry, Metabolism and function*, pp. 263–305.
28. Wong, C. H.; Halcomb, R. L.; Ichikawa, Y.; Kajimoto, T., *Angew. Chem., Int. Ed. Engl.* **1995**, *34*, 412–432.
29. Wong, C. H.; Halcomb, R. L.; Ichikawa, Y.; Kajimoto, T., *Angew. Chem., Int. Ed. Engl.* **1995**, *34*, 521–546.
30. DeNinno, M. P., *Synthesis* **1991**, 583–593.
31. Kiefel, M. J.; von Itzstein, M., *Chem. Rev.* **2002**, *102*, 471–490.
32. Uchida, Y.; Tsukada, Y.; Sugimori, T., *Agric. Biol. Chem.* **1985**, *49*, 181–187.
33. Deijl, C. M.; Vliegenthart, J. F. G., *Biochem. Biophys. Res. Comm.* **1983**, *111*, 668–674.
34. Aisaka, K.; Tamura, S.; Arai, Y.; Uwajima, T., *Biotechnology Lett.* **1987**, *9*, 633–637.
35. Liu, J. L. C.; Shen, G. J.; Ichikawa, Y.; Rutan, J. F.; Zapata, G.; Vann, W. F.; Wong, C. H., *J. Am. Chem. Soc.* **1992**, *114*, 3901–3910.
36. Auge, A.; Gautheron, C., *Tetrahedron Lett.* **1988**, *29*, 789–790.

37 Auge, C.; David, S.; Gautheron, C., *Tetrahedron Lett.* **1984**, *25*, 4663–4664.
38 Auge, C.; David, S.; Gautheron, C.; Veyrieres, A., *Tetrahedron Lett.* **1985**, *26*, 2439–2440.
39 Kim, M.J.; Hennen, W.J.; Sweers, H.M.; Wong, C.H., *J. Am. Chem. Soc.* **1988**, *110*, 6481–6486.
40 Simon, E.S.; Bernadski, M.D.; Whitesides, G.M., *J. Am. Chem. Soc.* **1988**, *110*, 7159–7163.
41 Kragl, U.; Gygax, D.; Ghisalba, O.; Wandrey, C., *Angew. Chem. Int. Ed. Engl.* **1991**, *30*, 827–828.
42 Gross, H.J.; Brossmer, R., *Glycoconjugate J.* **1995**, *12*, 739–746.
43 Brossmer, R.; Gross, H.J., *Neoglycoconjugates, Pt. B.* **1994**, *247*, 177–193.
44 Hasegawa, A.; Nagamara, T.; Ohki, H.; Kiso, M., *J. Carbohydr. Chem.* **1992**, *11*, 699–714.
45 Takayama, S.; Livingston, P.O.; Wong, C.H., *Tetrahedron Lett.* **1996**, *37*, 9271–9274.
46 Deninno, M.P.; Danishefsky, S.J.; Schulte, G., *J. Am. Chem. Soc.* **1988**, *110*, 3925–3929.
47 Chan, T.H.; Lee, M.C., *J. Org. Chem.* **1995**, *60*, 4228–4232.
48 Banwell, M.; De Savi, C.; Watson, K., *Chem. Comm.* **1998**, 1189–1190.
49 Chappell, M.D.; Halcomb, R.L., *Org. Lett.* **2000**, *2*, 2003–2005.
50 Dondoni, A.; Marra, A.; Boscarato, A., *Chem. Eur. J.* **1999**, *5*, 3562–3572.
51 Dondoni, A.; Marra, A.; Merino, P., *J. Am. Chem. Soc.* **1994**, *116*, 3324–3336.
52 Li, X.L.; Ohtake, H.; Takahashi, H.; Ikegami, S., *Tetrahedron* **2001**, *57*, 4297–4309.
53 Burke, S.D.; Sametz, G.M., *Org. Lett.* **1999**, *1*, 71–74.
54 Burke, S.D.; Voight, E.A., *Org. Lett.* **2001**, *3*, 237–240.
55 Boons, G.J.; Demchenko, A.V., *Chem. Rev.* **2000**, *100*, 4539–4565.
56 Kiso, M.; Ishida, H.; Ito, H., Special problems in glycosylation reactions: *Sialidations*; Ernst, B.; Hart, G.W.; Sinay, P. Eds.; Wiley-VCH, **2000**. Carbohydrates in Chemistry and Biology, Vol. 1; pp. 345–366.
57 Ando, T.; Ando, H.; Kiso, M., *Trends Glycosci. Glycotechnol.* **2001**, *13*, 573–586.
58 Izumi, M.; Wong, C.H., *Trends Glycosci. Glycotechnol.* **2001**, *13*, 345–360.
59 Lin, C.H.; Lin, C.C., *Adv. Exp. Med. Biol.* **2001**, *491*, 215–230.
60 Sinay, P., *Pure Appl. Chem.* **1991**, *63*, 519–528.
61 Toshima, K.; Tatsuta, K., *Chem. Rev.* **1993**, *93*, 1503–1531.
62 Boons, G.J., *Contemp. Org. Synth.* **1996**, *3*, 173–200.
63 Boons, G.J., *Tetrahedron.* **1996**, *52*, 1095–1121.
64 Whitfield, D.M.; Douglas, S.P., *Glycoconjugate J.* **1996**, *13*, 5–17.
65 Veeneman, G.H., Chemical Synthesis of O-glycosides; Boons, G. Ed.; Blackie Academic & Professional **1998**, Carbohydrate Chemistry, pp. 98–174.
66 Schmidt, R.R.; Castro-Palomino, J.C.; Retz, O., *Pure Appl. Chem.* **1999**, *71*, 729–744.
67 Okamoto, K.; Goto, T., *Tetrahedron.* **1990**, *46*, 5835–5857.
68 Ito, Y.; Gaudino, J.J.; Paulson, J.C., *Pure Appl. Chem.* **1993**, *65*, 753–762.
69 von Itzstein, M.; Thomson, R.J., *Curr. Med. Chem.* **1997**, *4*, 185–210.
70 Kuhn, R.; Lutz, P.; MacDonald, D.L., *Chem. Ber.* **1966**, *99*, 611–617.
71 Paulsen, H.; Tietz, H., *Carbohydr. Res.* **1984**, *125*, 47–64.
72 Byramova, N.E.; Tuzikov, A.B.; Bovin, N.V., *Carbohydr. Res.* **1992**, *237*, 161–175.
73 Khorlin, A.Y.; Privalova, I.M.; Bystrova, I.B., *Carbohydr. Res.* **1971**, *19*, 272–275.
74 Paulsen, H.; Tietz, H., *Angew. Chem., Int. Ed. Engl.* **1982**, *21*, 927–928.
75 Ogawa, T.; Sugimoto, M., *Carbohydr. Res.* **1985**, *135*, c5–c9.
76 Eschenfelder, V.; Brossmer, R., *Carbohydr. Res.* **1980**, *78*, 190–194.
77 van der Vleugel, D.J.M.; van Heeswijk, W.A.R.; Vliegenthart, J.F.G., *Carbohydr. Res.* **1982**, *102*, 121–130.
78 Paulsen, H.; Lockhoff, O., *Chem. Ber.* **1981**, *114*, 3102–3114.
79 van Boeckel, C.A.A.; Beetz, T.; van Aelst, S.F., *Tetrahedron* **1984**, *40*, 4097–4107.
80 Roy, R.; Laferriere, C.A., *Can. J. Chem.* **1990**, *68*, 2045–2054.
81 Spaltenstein, A.; Whitesides, G.M., *J. Am. Chem. Soc.* **1991**, *113*, 686–687.

82 Allanson, N. M.; Davidson, A. H.; Martin, F. M., Tetrahedron Lett. **1993**, *34*, 3945–3948.
83 Sparks, M. A.; Williams, K. W.; Lukacs, C.; Schrell, A.; Priebe, G.; Spaltenstein, A.; Whitesides, G. M., Tetrahedron **1993**, *49*, 1–12.
84 Dekany, G.; Wright, K.; Ward, P.; Toth, I., J. Carbohydr. Chem. **1997**, *16*, 11–24.
85 Higashi, K.; Miyoshi, S.; Nakabayashi, S.; Yamada, H.; Ito, Y., Chem. Pharm. Bull. **1992**, *40*, 2300–2303.
86 Sato, S.; Furuhata, K.; Itoh, M.; Shitori, Y.; Ogura, H., Chem. Pharm. Bull. **1988**, *36*, 914–919.
87 Thomas, R. L.; Sarkar, A. K.; Kohata, K.; Abbas, S. A.; Mata, K. L., Tetrahedron Lett. **1990**, *31*, 2825–2828.
88 Kartha, K. P. R.; Aloui, M.; Field, R. A., Tetrahedron Lett. **1996**, *37*, 8807–8810.
89 Kuboki, A.; Sekiguchi, T.; Sugai, T.; Ohta, H., Synlett. **1998**, 479–482.
90 Rothermel, J.; Faillard, H., Carbohydr. Res. **1990**, *196*, 29–40.
91 Kononov, L. O.; Magnusson, G., Acta Chem. Scand. **1998**, *52*, 141–144.
92 Paulsen, H.; von Dessen, U., Carbohydr. Res. **1986**, *146*, 147–153.
93 Sharma, M. N.; Eby, R., Carbohydr. Res. **1984**, *127*, 201–210.
94 Kunz, H.; Waldmann, H., J. Chem. Soc., Chem. Comm. **1985**, 638–640.
95 Kunz, H.; Waldmann, H.; Klinkhammer, U., Helv. Chim. Acta. **1988**, *71*, 1868–1874.
96 Nakamura, M.; Furuhata, K.; Ogura, H., Chem. Pharm. Bull. **1988**, *36*, 4807–4813.
97 Ikeda, K.; Kawai, K.; Achiwa, K., Chem. Pharm. Bull. **1991**, *39*, 1305–1309.
98 Nakamura, M.; Fujita, S.; Ogura, H., Chem. Pharm. Bull. **1993**, *41*, 21–25.
99 Sun, X. L.; Kai, T.; Fujita, S.; Takayanagi, H.; Furuhata, K., Chem. Pharm. Bull. **1997**, *45*, 795–798.
100 Garegg, P. J., Adv. Carbohydr. Chem. Biochem. **1997**, *52*, 179–205.
101 Hasegawa, A.; Ohki, H.; Nagahama, T.; Ishida, H., Carbohydr. Res. **1991**, *212*, 277–281.
102 Ito, Y.; Ogawa, T.; Numata, M.; Sugimoto, M., Carbohydr. Res. **1990**, *202*, 165–175.
103 Kanie, O.; Kiso, M.; Hasegawa, A., J. Carbohydr. Chem. **1988**, *7*, 501–506.
104 Hasegawa, A.; Kiso, M., Systematic Synthesis of Gangliosides Toward the Elucidation and Biomedical Application of Their Biological Functions; Ogura, H.; Hasegawa, A.; Suami, T. Eds.; Nodansha, VCH, **1992**, Carbohydrates – synthetic methods and applications in medicinal chemistry, pp. 243–266.
105 Hasegawa, A. Synthesis of Sialoglycoconjugates; Khan, S.; O'Neill, R. Eds.; Harwood academic publishers, **1996**, Modern methods in carbohydrate synthesis, Vol. 1, pp. 277–300.
106 Hasegawa, A.; Kiso, M., Chemical Synthesis of Sialyl Glycosides; Hanessian, S. Ed.; Marcel Dekker, Inc., **1997**, Preparative carbohydrate chemistry, pp. 357–379.
107 Murase, T.; Ishida, H.; Kiso, M.; Hasegawa, A., Carbohydr. Res. **1988**, *184*, c1–c4.
108 Murase, T.; Ishida, H.; Kiso, M.; Hasegawa, A., Carbohydr. Res. **1989**, *188*, 71–80.
109 Murase, T.; Kameyama, A.; Kartha, K. P. R.; Ishida, H.; Kiso, M.; Hasegawa, A., J. Carbohydr. Chem. **1989**, *8*, 265–283.
110 Hasegawa, A.; Nagahama, T.; Ohki, H.; Hotta, K.; Ishida, H.; Kiso, M., J., Carbohydr. Chem. **1991**, *10*, 493–498.
111 Demchenko, A. V.; Boons, G. J., Tetrahedron Lett. **1998**, *39*, 3065–3068.
112 Demchenko, A. V.; Boons, G. J., J. Org. Chem. **2001**, *66*, 2547–2554.
113 Komba, S.; Glalustian, C.; Ishida, H.; Feizi, T.; Kannagi, R.; Kiso, M., Angew. Chem., Int. Ed. **1999**, *38*, 1131–1133.
114 De Meo, C.; Demchenko, A. V.; Boons, G. J., J. Org. Chem. **2001**, *66*, 5490–5497.
115 De Meo, C.; Demchenko, A. V.; Boons, G. J., Aust. J. Chem. **2002**, *55*, 131–134.
116 Demchenko, A. V.; Boons, G. J., Chem. Eur. J. **1999**, *5*, 1278–1283.
117 Takahashi, T.; Tsukamoto, H.; Yamada, H., Tetrahedron Lett. **1997**, *38*, 8223–8226.
118 Takahashi, T.; Tsukamoto, H.; Yamada, H., Bioorg. Med. Chem. Lett. **1998**, *8*, 113–116.
119 Haberman, J. M.; Gin, D. Y. Org. Lett. **2001**, *3*, 1665–1668.
120 Kanemitsu, T.; Kanie, O.; Wong, C. H., Angew. Chem. Int. Ed. **1998**, *37*, 3415–3418.

121 Iida, M.; Endo, A.; Fujita, S.; Numata, M.; Suzuki, K.; Nunomura, S.; Ogawa, T., *Glycoconjugate J.* **1996**, *13*, 203–211.
122 Iida, M.; Endo, A.; Fujita, S.; Numata, M.; Sugimoto, M.; Nunomura, S.; Ogawa, T., *J. Carbohydr. Chem.* **1998**, *17*, 647–672.
123 Ito, Y.; Ogawa, T., *Tetrahedron Lett.* **1988**, *29*, 1061–1604.
124 Marra, A.; Sinay, P., *Carbohydr. Res.* **1989**, *187*, 35–42.
125 Cao, S.; Meunier, S.J.; Andersson, F.O.; Letellier, M.; Roy, R., *Tetrahedron: Asymmetry* **1994**, *5*, 2303–2312.
126 Nifantiev, N.E.; Tsvetkov, Y.E.; Shashkov, A.S.; Kononov, L.O.; Menshov, V.M.; Tuzikov, A.B.; Bovin, N., *J. Carbohydr. Chem.* **1996**, *15*, 939–953.
127 Tsvetkov, Y.E.; Schmidt, R.R., *Carbohydr. Lett.* **1996**, *2*, 149–156.
128 Simeoni, L.A.; Byramova, N.E.; Bovin, N.V., *Bioorg. Khim.* **1999**, *25*, 62–69.
129 Kirchner, E.; Thiem, F.; Dernick, R.; Heukeshoven, J.; Thiem, J., *J. Carbohydr. Chem.* **1988**, *7*, 453–486.
130 Ito, H.; Ishida, H.; Kiso, M.; Hasegawa, A., *Carbohydr. Res.* **1998**, *306*, 581–585.
131 Ehara, T.; Kameyama, A.; Yamada, Y.; Ishida, H.; Kiso, M.; Hasegawa, A., *Carbohydr. Res.* **1996**, *281*, 237–252.
132 Terada, T.; Toyoda, T.; Ishida, H.; Kiso, M.; Hasegawa, A., *J. Carbohydr. Chem.* **1995**, *14*, 769–790.
133 Kiso, M.; Hasegawa, A., Synthesis of Ganglioside G_{M3} and Analogs Containing Modified Sialic Acids and Ceramides; Lee, Y.; Lee, R. Eds.; Academic Press **1994**. Neoglycoconjugates,; Vol. 242; pp. 173–183.
134 Roy, R.; Andersson, F.O.; Letellier, M., *Tetrahedron Lett.* **1992**, *33*, 6053–6056.
135 Yu, C.S.; Niikura, K.; Lin, C.C.; Wong, C.H., *Angew. Chem., Int. Ed Engl.* **2001**, *40*, 2900–2903.
136 Lu, K.C.; Tseng, S.Y.; Lin, C.C., *Carbohydr. Res.* **2002**, *337*, 755–760.
137 Hotta, K.; Ishida, H.; Kiso, M.; Hasegawa, A., *J. Carbohydr. Chem.* **1995**, *14*, 491–506.
138 Marra, A.; Sinay, P., *Carbohydr. Res.* **1990**, *195*, 303–308.
139 Schmidt, R.R., Chemical Synthesis of Sialylated Glycoconjugates; Kovac, P. Ed., **1994**; *Synthetic Oligosaccharides*; Vol. 560; pp. 276–296.
140 Greilich, U.; Brescello, R.; Jung, K.H.; Schmidt, R.R., *Liebigs Ann.* **1996**, 663–672.
141 Martichonok, V.; Whitesides, G.M., *J. Org. Chem.* **1996**, *61*, 1702–1706.
142 Kondo, H.; Ichikawa, Y.; Wong, C.H., *J. Am. Chem. Soc.* **1992**, *114*, 8748–8750.
143 Martin, T.J.; Schmidt, R.R., *Tetrahedron Lett.* **1992**, *33*, 6123–6126.
144 Wong, C.H., Practical Synthesis of Oligosaccharides Based on Glycosyltransferases and Glycosylphosphites; Khan, S.; O'Neill, R. Eds.; Harwood academic publishers, **1996**; *Modern methods in carbohydrate synthesis,*; Vol. 1; pp. 467–491.
145 Nicolaou, K.C.; Bockovich, N.J. Chemical Synthesis of Complex Carbohydrates; Hecht, S. Ed.; Oxford University Press, New York, Oxford, **1999**; *Bioorganic Chemistry: Carbohydrates*; pp. 134–173.
146 Martin, T.J.; Brescello, R.; Toepfer, A.; Schmidt, R.R., *Glycoconjugate J.* **1993**, *10*, 16–25.
147 Sim, M.M.; Kondo, H.; Wong, C.H., *J. Am. Chem. Soc.* **1993**, *115*, 2260–2267.
148 Kondo, H.; Aoki, S.; Ichikawa, Y.; Halcomb, R.L.; Ritzen, H.; Wong, C.H., *J. Org. Chem.* **1994**, *59*, 864–877.
149 Tsvetkov, Y.E.; Schmidt, R.R., *Tetrahedron Lett.* **1994**, *35*, 8583–8586.
150 Lassaletta, J.M.; Schmidt, R.R., *Tetrahedron Lett.* **1995**, *36*, 4209–4212.
151 Lassaletta, J.M.; Carlsson, K.; Garegg, P.J.; Schmidt, R.R., *J. Org. Chem.* **1996**, *61*, 6873–6880.
152 Castro-Palomino, J.C.; Tsvetkov, Y.E.; Schneider, R.; Schmidt, R.R., *Tetrahedron Lett.* **1997**, *38*, 6837–6840.
153 Schwartz, J.B.; Kuduk, S.D.; Chen, X.T.; Sames, D.; Glunz, P.W.; Danishefsky, S.J., *J. Am. Chem. Soc.* **1999**, *121*, 2662–2673.
154 Halcomb, R.L.; Wong, C.H., *Curr. Opin. Struct. Biol.* **1993**, *3*, 694–700.
155 Aoki, S.; Kondo, H.; Wong, C.H., Glycosyl Phosphites as Glycosylation Reagents; Lee, Y.; Lee, R. Eds.; Academic Press **1994**; *Neoglycoconjugates*, Vol. 247; pp., 193–211.
156 Veeneman, G.H.; van der Hulst, R.G.A.; van Boeckel, C.A.A.; Philipsen,

R. L. A.; Ruigt, G. S. F.; Tonnaer, J. A. D. M.; van Delft, T. M. L.; Konings, P. N. M., *Bioorg. Med. Chem. Lett.* **1995**, *5*, 9–14.
157 Watanabe, Y.; Nakamoto, C.; Yamamoto, T.; Ozaki, S., *Tetrahedron* **1994**, *50*, 6523–6536.
158 Meindl, P.; Tuppy, H., *Monatsh. Chem.* **1965**, *96*, 816–827.
159 Nakamura, M.; Takayanagi, H.; Furuhata, K.; Ogura, H., *Chem. Pharm. Bull.* **1992**, *40*, 879–885.
160 Ogura, H., Sialic Acid Derivatives as Glycolipoids; Ogura, H.; Hasegawa, A.; Suami, T. Eds.; Nodansha, VCH, **1992**; *Carbohydrates – synthetic methods and applications in medicinal chemistry*, pp. 282–303.
161 Kai, T.; Sun, X. L.; Tanaka, M.; Takayanagi, H.; Furuhata, K., *Chem. Pharm. Bull.* **1996**, *44*, 208–211.
162 Mukaiyama, T.; Sasaki, T.; Iwashita, E.; Matsubara, K., *Chem. Lett.* **1995**, 455–456.
163 Kaneko, H.; Murahashi, N.; Sasaki, A.; Yamada, H.; Sakagami, M.; Ikeda, M., *Chem. Pharm. Bull.* **1997**, *45*, 951–956.
164 Okamoto, K.; Kondo, T.; Goto, T., *Bull. Chem. Soc. Jpn.* **1987**, *60*, 631–636.
165 Ye, X. S.; Huang, X. F.; Wong, C. H., *Chem. Commun.* **2001**, 974–975.
166 Okamoto, K.; Kondo, T.; Goto, T., *Chem. Lett.* **1986**, 1449–1452.
167 Okamoto, K.; Kondo, T.; Goto, T., *Tetrahedron* **1987**, *43*, 5909–5918.
168 Castro-Palomino, J. C.; Tsvetkov, Y. E.; Schmidt, R. R., *J. Am. Chem. Soc.* **1998**, *120*, 5434–5440.
169 Okamoto, K.; Kondo, T.; Goto, T., *Tetrahedron Lett.* **1986**, *27*, 5233–5236.
170 Okamoto, K.; Kondo, T.; Goto, T., *Tetrahedron Lett.* **1986**, *27*, 5229–5232.
171 Okamoto, K.; Kondo, T.; Goto, T., *Tetrahedron*; **1988**, *44*, 1291–1298.
172 Ito, Y.; Ogawa, T., *Tetrahedron*; **1990**, *46*, 89–102.
173 Kondo, T.; Abe, H.; Goto, T., *Chem. Lett.* **1988**, 1657–1660.
174 Ercegovic, T.; Magnusson, G., *J. Chem. Soc., Chem. Comm.* **1994**, 831–832.
175 Ercegovic, T.; Magnusson, G., *J. Org. Chem.* **1995**, *60*, 2278–2284.
176 Martichonok, V.; Whitesides, G., *J. Am. Chem. Soc.* **1996**, *118*, 8187–8191.

177 Kondo, T.; Tomoo, T.; Abe, H.; Isobe, M.; Goto, T., *Chem. Lett.* **1996**, 337–338.
178 Kondo, T.; Tomoo, T.; Abe, H.; Isobe, M.; Goto, T., *J. Carbohydr. Chem.* **1996**, *15*, 857–878.
179 Kononov, L. O.; Ito, Y.; Ogawa, T., *Tetrahedron Lett.* **1997**, *38*, 1599–1602.
180 Ito, Y.; Numata, M.; Sugimoto, M.; Ogawa, T., *J. Am. Chem. Soc.* **1989**, *111*, 8508–8510.
181 Ito, Y.; Nunomura, S.; Shibayama, S.; Ogawa, T., *J. Org. Chem.* **1992**, *57*, 1821–1831.
182 Castro-Palomino, J. C.; Simon, B.; Speer, O.; Leist, M.; Schmidt, R. R., *Chem. Eur. J.* **2001**, *7*, 2178–2184.
183 Ercegovic, T.; Magnusson, G., *J. Org. Chem.* **1996**, *61*, 179–184.
184 Hossain, N.; Magnusson, G., *Tetrahedron Lett.* **1999**, *40*, 2217–2220.
185 Roy, R.; Pon, R. A., *Glycoconjugate J.* **1990**, *7*, 3–12.
186 Hasegawa, A.; Ishida, H.; Kiso, M., *J. Carbohydr. Chem.* **1993**, *12*, 371–376.
187 Ishida, H.; Ohta, Y.; Tsukada, Y.; Kiso, M.; Hasegawa, A., *Carbohydr. Res.* **1993**. *246*, 75–88.
188 Ishida, H. K.; Ishida, H.; Kiso, M.; Hasegawa, A., *Tetrahedron: Asymmetry*, **1994**, *5*, 2493–2512.
189 Ishida, H. K.; Ishida, H.; Kiso, M.; Hasegawa, A., *J. Carbohydr. Chem.* **1994**, *13*, 655–664.
190 Ando, H.; Ishida, H.; Kiso, M., *J. Carbohydr. Chem.* **1999**, *18*, 603–607.
191 Ogawa, T.; Sugimoto, M., *Carbohydr. Res.* **1984**, *128*, c1–c4.
192 Shimizu, C.; Achiwa, K., *Carbohydr. Res.* **1987**, *166*, 314–316.
193 Hasegawa, A.; Ogawa, M.; Ishida, H.; Kiso, M., *J. Carbohydr. Chem.* **1990**, *9*, 393–414.
194 Ren, C. T.; Chen, C. S.; Wu, S. H., *J. Org. Chem.* **2002**, *67*, 1376–1379.
195 Heidlas, J. E.; Williams, K. W.; Whitesides, G. M., *Acc. Chem. Res.* **1992**, *25*, 307–314.
196 Ichikawa, Y.; Look, G. C.; Wong, C. H.; Kajimoto, T., *J. Synth. Org. Chem. Jpn.* **1992**, *50*, 441–450.
197 Thiem, J., *Biofutur.* **1993**, *N 125*, 53–59.
198 Thiem, J., *FEMS Microbiol. Rev.* **1995**, *16*, 193–211.

199 Gijsen, H. J. M.; Qiao, L.; Fitz, W.; Wong, C. H., *Chem. Rev.* **1996**, *96*, 443–473.
200 Roy, R., *Trends Glycosci. Glycotech.* **1996**, *8*, 79–99.
201 Gambert, U.; Thiem, J., *Topics Curr. Chem.* **1997**, *186*, 21–43.
202 Kren, V.; Thiem, J., *Chem. Soc. Rev.* **1997**, *26*, 463–473.
203 von Itzstein, M.; Thomson, R. T., *Glycoscience Synthesis of Oligosaccharides and Glycoconjugates*; **1997**, *186*, 119–170.
204 Sears, P.; Wong, C. H., *Chem. Comm.* **1998**, 1161–1170.
205 Simanek, E. E.; McGarvey, G. J.; Jablonowski, J. A.; Wong, C. H., *Chem. Rev.* **1998**, *98*, 833–862.
206 Hendrix, M.; Wong, C. H., Enzymatic Synthesis of Carbohydrates; Hecht, S. Ed.; Oxford University Press, New York, Oxford **1999**; *Bioorganic Chemistry: Carbohydrates*; pp. 198–243.
207 Sabesan, S.; Paulson, J. C., *J. Am. Chem. Soc.* **1986**, *108*, 2068–2080.
208 Thiem, J.; Treder, W., *Angew. Chem. Int. Ed. Engl.* **1986**, *25*, 1096–1097.
209 Nilsson, K. G. I., *Carbohydr. Res.* **1989**, *188*, 9–17.
210 Auge, C.; Gautheron, C.; Pora, H., *Carbohydr. Res.* **1989**, *193*, 288–293.
211 Palcic, M. M.; Venot, A. P.; Ratcliffe, R. M.; Hindsgaul, O., *Carbohydr. Res.* **1989**, *190*, 1–11.
212 Auge, C.; Fernandez-Fernandez, R.; Gautheron, C., *Carbohydr. Res.* **1990**, *200*, 257–268.
213 Thiem, J.; Wiemann, T., *Angew. Chem. Int. Ed. Engl.* **1990**, *29*, 80–82.
214 Unverzagt, C.; Kunz, H.; Paulson, J. C., *J. Am. Chem. Soc.* **1990**, *112*, 9308–9309.
215 Lubineau, A.; Auge, C.; Francois, P., *Carbohydr. Res.* **1992**, *228*, 137–144.
216 Izumi, M.; Shen, G. J.; Wacowich-Sgarbi, S.; Nakatani, T.; Plettenburg, O.; Wong, C. H., *J. Am. Chem. Soc.* **2001**, *123*, 10909–10918.
217 Kondo, H.; Ichikawa, Y.; Wong, C. H., *J. Am. Chem. Soc.* **1992**, *114*, 8748–8750.
218 Chappell, M. D.; Halcomb, R. L., *Tetrahedron* **1997**, *53*, 11109–11120.
219 Kittelmann, M.; Klein, T.; Kragl, U.; Wandrey, C.; Ghisalba, O., *Applied Microbiology and Biotechnology* **1995**, *44*, 59–67.
220 Kuboki, A.; Okazaki, H.; Sugai, T.; Ohta, H., *Tetrahedron* **1997**, *53*, 2387–2400.
221 Hanessian, S.; Prabhanjan, H., *Synlett.* **1994**, 868–870.
222 Ichikawa, Y.; Shen, G. J.; Wong, C. H., *J. Am. Chem. Soc.* **1991**, *113*, 4698–4700.
223 Ichikawa, Y.; Liu, J. L. C.; Shen, G. J.; Wong, C. H., *J. Am. Chem. Soc.* **1991**, *113*, 6300–6302.
224 Kajihara, Y.; Yamamoto, T.; Nagae, H.; Nakashizuka, M.; Sakakibara, T.; Terada, I., *J. Org. Chem.* **1996**, *61*, 8632–8635.
225 Yamamoto, T.; Nagae, H.; Kajihara, Y.; Terada, I., *Biosci. Biotech. Biochem.* **1998**, *62*, 210–214.
226 Wen, D. X.; Livingston, B. D.; Medzihradszky, K. F.; Kelm, S.; Burlingame, A. L.; Paulson, J. C., *J. Biol. Chem.* **1992**, *267*, 21011–21019.
227 Sabesan, S.; Bock, K.; Paulson, J. C., *Carbohydr. Res.* **1991**, *218*, 27–54.
228 DeFrees, S. A.; Gaeta, F. C. A., *J. Am. Chem. Soc.* **1993**, *115*, 7549–7550.
229 Wlasichuk, K. B.; Kashem, M. A.; Nikrad, P. V.; Bird, P.; Jiang, C.; Venot, A. P., *J. Biol. Chem.* **1993**, *268*, 13971–13977.
230 van Dorst, J. A. L. M.; Tikkanen, J. M.; Krezdorn, C. H.; Streiff, M. B.; Berger, E. G.; van Kuik, J. A.; Kamerling, J. P.; Vliegenthart, J. F. G., *Eur. J. Biochem.* **1996**, *242*, 674–681.
231 Hayashi, M.; Tanaka, M.; Itoh, M.; Miyauchi, H., *J. Org. Chem.* **1996**, *61*, 2938–2945.
232 Lubineau, A.; Basset-Carpentier, K.; Auge, C., *Carbohydr. Res.* **1997**, *300*, 161–167.
233 Baisch, G.; Ohrlein, R., *Bioorg. Med. Chem.* **1998**, *6*, 1673–1682.
234 Rodrigues, E. C.; Marcaurelle, L. A.; Bertozzi, C. R., *J. Org. Chem.* **1998**, *63*, 7134–7135.
235 Baisch, G.; Ohrlein, R.; Streiff, M., *Bioorg. Med. Chem. Lett.* **1998**, *8*, 157–160.
236 Hashimoto, H.; Sato, K.; Wakabayashi, T.; Kodama, H.; Kajihara, Y., *Carbohydr. Res.* **1993**, *247*, 179–193.
237 Harduin-Lepers, A.; Recchi, M. A.; Delannoy, P., *Glycobiology* **1995**, *5*, 741–758.

238 Kleineidam, R. G.; Schmelter, T.; Schwartz, R. T.; Schomer, R., *Glycoconjugate J.* **1997**, *14*, 57–66.
239 Schaub, C.; Muller, B.; Schmidt, R. R., *Glycoconjugate J.* **1998**, *15*, 345–354.
240 Yamamoto, T.; Nakashizuka, M.; Kodama, H.; Kajihara, Y.; Terada, I., *J. Biochem.* **1996**, *120*, 104–110.
241 Chappell, M. D.; Halcomb, R. L., *J. Am. Chem. Soc.* **1997**, *119*, 3393–3394.
242 Martin, R.; Witte, K. L.; Wong, C. H., *Bioorg. Med. Chem.* **1998**, *6*, 1283–1292.
243 Burkart, M. D.; Vincent, S. P.; Wong, C. H., *Chem. Comm.* **1999**, 1525–1526.
244 Miyazaki, T.; Sakakibara, T.; Sato, H.; Kajihara, Y., *J. Am. Chem. Soc.* **1999**, *121*, 1411–1412.
245 Gaudino, J. J.; Paulson, J. C., *J. Am. Chem. Soc.* **1994**, *116*, 1149–1150.
246 Earle, M. A.; Manku, S.; Hultin, P. G.; Li, H.; Palcic, M. M., *Carbohydr. Res.* **1997**, *301*, 1–4.
247 Palcic, M. M.; Li, H.; Zanini, D.; Bhella, R. S.; Roy, R., *Carbohydr. Res.* **1998**, *305*, 433–442.
248 Wittmann, V.; Takayama, S.; Gong, K. W.; Weitz-Schmidt, G.; Wong, C. H., *J. Org. Chem.* **1998**, *63*, 5137–5143.
249 Oehrlein, R.; Hindsgaul, O.; Palcic, M. M., *Carbohydr. Res.* **1993**, *244*, 149–159.
250 Unverzagt, C., *Tetrahedron Lett.* **1997**, *38*, 5627–5630.
251 Liu, K. K. C.; Danishefsky, S. J., *J. Am. Chem. Soc.* **1993**, *115*, 4933–4934.
252 Liu, K. K. C.; Danishefsky, S. J., *Chem. Eur. J.* **1996**, *2*, 1359–1362.
253 Probert, M. A.; Milton, M. J.; Harris, R.; Schenkman, S.; Brown, J. M.; Homans, S. W.; Field, R. A., *Tetrahedron Lett.* **1997**, *38*, 5861–5864.
254 Depre, D.; Duffels, A.; Green, L. G.; Lenz, R.; Ley, S. V.; Wong, C. H., *Chem. Eur. J.* **1999**, *5*, 3326–3340.
255 Pozsgay, V.; Brisson, J. R.; Jennings, H. J., *J. Org. Chem.* **1991**, *56*, 3377–3385.
256 Zou, W.; Brisson, J. R.; Yang, Q. L.; van der Zwan, M.; Jennings, H. J., *Carbohydr. Res.* **1996**, *295*, 209–228.
257 Zou, W.; Laferriere, C. A.; Jennings, H. J., *Carbohydr. Res.* **1998**, *309*, 297–301.
258 Stangier, P.; Palcic, M. M.; Bundle, D. R., *Carbohydr. Res.* **1995**, *267*, 153–159.
259 Yamada, K.; Nishimura, S. I., *Tetrahedron Lett.* **1995**, *36*, 9493–9496.
260 Zehavi, U.; Tuchinsky, A., *Glycoconjugate J.* **1998**, *15*, 657–662.
261 Osborn, H. M. I.; Khan, T. H., *Tetrahedron* **1999**, *55*, 1807–1850.
262 Wu, W. G.; Pasternack, L.; Huang, D. H.; Koeller, K. M.; Lin, C. C.; Seitz, O.; Wong, C. H., *J. Am. Chem. Soc.* **1999**, *121*, 2409–2417.
263 Blixt, O.; Norberg, T., *J. Org. Chem.* **1998**, *63*, 2705–2710.
264 Halcomb, R. L.; Huang, H. M.; Wong, C. H., *J. Am. Chem. Soc.* **1994**, *116*, 11315–11322.
265 Schuster, M.; Wang, P.; Paulson, J. C.; Wong, C. H., *J. Am. Chem. Soc.* **1994**, *116*, 1135–1136.
266 Seitz, O.; Wong, C. H., *J. Am. Chem. Soc.* **1997**, *119*, 8766–8776.
267 Leppanen, A.; Penttila, L.; Renkonen, O.; McEver, R. P.; Cummings, R. D., *J. Biol. Chem.* **2002**, *275*, 7839–7853.
268 Leppanen, A.; White, S. P.; Helin, J.; McEver, R. P.; Cummings, R. D., *J. Biol. Chem.* **2000**, *275*, 39569–39578.
269 Leppanen, A.; Mehta, P.; Ouyang, Y. B.; Ju, T. Z.; Helin, J.; Moore, K. L.; van Die, I.; Canfield, W. M.; McEver, R. P.; Cummings, R. D., *J. Biol. Chem.* **1999**, *274*, 24838–24848.
270 Liu, W.; Ramachandran, V.; Kang, J.; Kishimoto, T. K.; Cummings, R. D.; McEver, R. P., *J. Biol. Chem* **1998**, *273*, 7078–7087.
271 Yamada, K.; Fujita, E.; Nishimura, S. I., *Carbohydr. Res.* **1997**, *305*, 443–461.
272 Mehta, S.; Gilbert, M.; Wakarchuk, W. W.; Whitfield, D. M., *Org. Lett.* **2000**, *2*, 751–753.
273 Keppler, O. T.; Stehling, P.; Herrmann, M.; Kayser, H.; Grunow, D.; Reutter, W.; Pawlita, M., *J. Biol. Chem* **1995**, *270*, 1308–1314.
274 Kayser, H.; Zeitler, R.; Kannicht, C.; Grunow, D.; Nuck, R.; Reutter, W., *J. Biol. Chem.* **1992**, *267*, 16934–16938.
275 Mahal, L. K.; Yarema, K. J.; Bertozzi, C. R., *Science* **1997**, *276*, 1125–1128.
276 Yarema, K. J.; Bertozzi, C. R., *Gen. Biol.* **2001**, *2*, 1–10.

277 Lemieux, G. A.; Yarema, K. J.; Jacobs, C. L.; Bertozzi, C. R., *J. Amer. Chem. Soc.* **1999**, *121*, 4278–4279.

278 Thiem, J.; Sauerbrei, B., *Angew. Chem. Int. Ed. Engl.* **1991**, *30*, 1503–1505.

279 Ajisaka, K.; Fujimoto, H.; Isomura, M., *Carbohydr. Res.* **1994**, *259*, 103–115.

280 Makimura, Y.; Ishida, H.; Kondo, A.; Hasegawa, A.; Kiso, M., *J. Carbohydr. Chem.* **1998**, *17*, 975–979.

281 Vetere, A.; Paoletti, S., *FEBS Lett.* **1996**, *399*, 203–206.

282 Tanaka, H.; Ito, F.; Iwasaki, T., *Biosci. Biotech. Biochem.* **1995**, *59*, 638–643.

283 Ito, Y.; Paulson, J. C., *J. Am. Chem. Soc.* **1993**, *115*, 7862–7863.

284 Herrmann, G. F.; Ichikawa, Y.; Wandrey, C.; Gaeta, F. C. A.; Paulson, J. C.; Wong, C. H., *Tetrahedron Lett.* **1993**, *34*, 3091–3094.

285 Kren, V.; Thiem, J., *Angew. Chem., Int. Ed. Engl.* **1995**, *34*, 893–895.

286 Vlahov, I. R.; Vlahova, P. I.; Linhardt, R. J., *J. Amer. Chem. Soc.* **1997**, *119*, 1480–1481.

287 Bazin, H. G.; Du, Y.; Polat, T.; Linhardt, R. J., *J. Org. Chem.* **1999**, *64*, 7254–7259.

288 Wang, Z. G.; Zhang, X. F.; Ito, Y.; Nakahara, Y.; Ogawa, T., *Bioorg. Med. Chem.* **1996**, *4*, 1901–1908.

289 Koketsu, M.; Balagurunathan, K.; Linhardt, R. J., *Organic Lett.* **2000**, *2*, 3361–3363.

290 Notz, W.; Hartel, C.; Waldscheck, B.; Schmidt, R. R. *J. Org. Chem.* **2001**, *66*, 4250–4260.

291 Suzuki, Y.; Sato, K.; Kiso, M.; Hasegawa, A., *Glycoconjugate J.* **1990**, *7*, 349–356.

292 Warner, T. G.; Lee, L. A., *Carbohydr. Res.* **1988**, *176*, 211–218.

293 Zanini, D.; Roy, R., *J. Org. Chem.* **1998**, *63*, 3486–3491.

294 Park, W. K. C.; Meunier, S. J.; Zanini, D.; Roy, R., *Carbohydr. Lett.* **1995**, *1*, 179–184.

295 Eisele, T.; Toepfer, A.; Kretzschmar, G.; Schmidt, R. R., *Tetrahedron Lett.* **1996**, *37*, 1389–1392.

296 Bennett, S.; von Itzstein, M.; Kiefel, M. J., *Carbohydr. Res.* **1994**, *259*, 293–299.

297 Kiefel, M. J.; Beisner, B.; Bennett, S.; Holmes, I. D.; von Itzstein, M., *J. Med. Chem.* **1996**, *39*, 1314–1320.

298 Angus, D. I.; von Itzstein, M., *Carbohydr. Res.* **1995**, *274*, 279–283.

299 Wilson, J. C.; Kiefel, M. J.; Angus, D. I.; von Itzstein, M., *Org. Lett.* **1999**, *1*, 443–446.

300 Abo, S.; Ciccotosto, S.; Alafaci, A.; von Itzstein, M., *Carbohydr. Res.* **1999**, *322*, 201–208.

301 Turnbull, W. B.; Field, R. A., *J. Chem. Soc. Perkin 1* **2000**, 1859–1866.

302 Hasegawa, A.; Uchimura, A.; Ishida, H.; Kiso, M., *Biosci. Biotech. Biochem.* **1995**, *59*, 1091–1094.

303 Simeoni, L. A.; Tuzikov, A. B.; Byramova, N. E.; Bovin, N. V., *Bioorg. Khim.* **1997**, *23*, 139–146.

304 Sugata, T.; Kan, Y.; Nagaregawa, Y.; Miyamoto, T.; Higuchi, R., *J. Carbohydr. Chem.* **1997**, *16*, 917–925.

305 Sugata, T.; Higuchi, R., *Tetrahedron Lett.* **1996**, *37*, 2613–2614.

306 Higuchi, R.; Mori, T.; Sugata, T.; Yamada, K.; Miyamoto, T., *Eur. J. Org. Chem.* **1999**, 145–147.

307 Kelm, S.; Brossmer, R.; Isecke, R.; Gross, H. J.; Strenge, K.; Schauer, R., *Eur. J. Biochem.* **1998**, *255*, 663–672.

308 Fujita, S.; Numata, M.; Sugimoto, M.; Tomita, K.; Ogawa, T., *Carbohydr. Res.* **1992**, *228*, 347–370.

309 Kuznik, G.; Horsch, R.; Kretzschmar, G.; Unverzagt, C., *Bioorg. Med. Chem. Lett.* **1997**, *7*, 577–580.

310 Schneider, R.; Freyhardt, C. C.; Schmidt, R. R., *Eur. J. Org. Chem.* **2001**, 1655–1661.

311 Fujita, S.; Numata, M.; Sugimoto, M.; Tomita, K.; Ogawa, T., *Carbohydr. Res.* **1994**, *263*, 181–196.

312 Sherman, A. A.; Yudina, O. N.; Shashkov, A. S.; Menshov, V. M.; Nifant'ev, N. E., *Carbohydr. Res.* **2001**, *330*, 445–458.

4
Solid-Phase Oligosaccharide Synthesis
PETER H. SEEBERGER

4.1
Introduction

The three major classes of biopolymers – proteins, nucleic acids, and glycoconjugates – are of major importance to biology. Carbohydrate drug development and the field of glycobiology in general still lack rapid access to pure carbohydrates and glycoconjugates. Detailed structure-activity studies and medicinal chemistry efforts require sufficient quantities of defined oligosaccharides, while the procurement of synthetic material presents a formidable challenge to the synthetic chemist [1]. While the need for chemically defined oligosaccharides has steadily increased in recent years, the synthesis of these complex molecules remained time-consuming until very recently and is still carried out by just a few specialized laboratories. Oligonucleotides – [2] and oligopeptides [3], on the other hand, are now routinely prepared on automated synthesizers, providing pure substances in a rapid and efficient manner by solid-phase approaches. The effect of an automated oligosaccharide synthesizer on the field of glycobiology would be expected to be comparable to the impact of automated solid-phase peptide and oligonucleotide synthesis on drug development and the biotechnology industry in these areas. Solid-phase synthesis lends itself particularly well to automation and is the focus of this review. In addition to solid-phase approaches, the programmable one-pot synthesis of oligosaccharides allows for parts of the synthetic process to be automated and is discussed in Chapters 1 and 35 [4].

The solid-phase synthesis approach lends itself particularly well to automation of the synthetic process. Solid-phase synthesis allows for removal of excess reagents used to drive the reaction to completion simply by washing the resin, while purification of the reaction products at the end of the synthesis minimizes the number of chromatographic steps required.

The development of an automated solid-phase oligosaccharide synthesizer requires solutions for several challenges: (a) the design of an overall synthetic strategy with either the "reducing" or the "non-reducing" end of the growing carbohydrate chain attached to the support, (b) selection of a polymer and linker that have to be inert to all reaction conditions during the synthesis but provide smooth and effective cleavage when desired, (c) a protecting group strategy consistent with the complexity of the desired oligosaccharide, (d) stereospecific and high-yielding glycosylation reactions, (e) a capping protocol that minimizes the accumulation of internal deletion sequences, and (f) automation of the synthetic process.

Different aspects of solid-phase oligosaccharide synthesis have been reviewed previously [5]. Here, a summary of chemical solid-phase carbohydrate synthesis relevant to automated synthesis and carbohydrate drug discovery is presented.

4.2
Pioneering Efforts in Solid-Phase Oligosaccharide Synthesis

Inspired by the success of Merrifield's solid-phase peptide synthesis [6], the first studies toward solid-phase oligosaccharide synthesis were carried out in the 1970s [7]. Fréchet and Schuerch reported the first synthesis of di- and trisaccharides on a solid support by utilizing a monosaccharide connected through the anomeric position to allyl alcohol-functionalized Merrifield resin [8]. Cleavage from the resin was accomplished by ozonolysis. Linkers, temporary protecting groups, and glycosylating agents were explored in this context.

A photolabile linkage to the polymer did not meet expectations [9]. A thioglycosidic linkage – to the solid support that would release the reducing end of the oligosaccharide in the form of the lactol was described by Anderson [10]. An ester linker was employed in connecting the C6-hydroxy group of glucosamine with polystyrene. This linkage was stable to glycosylations but was readily cleaved with sodium methoxide [11]. In addition to studies utilizing functionalized polystyrene (Merrifield's resin), controlled pore glass (CPG) was applied to solid-phase oligosaccharide synthesis as a non-swelling inorganic support by Schuerch [12].

Fréchet described an unconventional approach for attaching the first monosaccharide to the solid support. A resin-bound cyclic boronic acid ester was selectively introduced to connect *cis*-1,2- and *cis*-1,3-diols, leaving one hydroxy for further chain elongation. Simple hydrolysis of the cyclic esters resulted in liberation from the polymer [13]. Unfortunately, couplings involving such a support-bound monosaccharide as acceptor proceeded in poor yields.

Guthrie immobilized the first monosaccharide on polymer support by co-polymerization of styrene with a sugar monomer bearing a polymerizable *O*-protecting group [14]. This non-crosslinked, soluble polymer allowed for glycosylation reactions in homogeneous solutions, but could also readily be precipitated to facilitate purification. This work is also significant as the first time the glycosyl donor was attached to the support and allowed to react with an excess of solution-based acceptor [15].

These early attempts explored many of the fundamental issues associated with solid-phase oligosaccharide synthesis, including different strategies (donor- vs. acceptor-bound synthesis), various solid supports (soluble and insoluble), different linkers, and a variety of glycosylating agents. Most of the recent advances in the field have been based on these concepts, developed almost 30 years ago.

Following this initial burst of activity in solid-phase oligosaccharide synthesis, which brought little in the way of encouraging results, interest in the area subsided. Linear β-(1 → 5)-linked galactofuranosyl homopolymers were chosen as targets for an iterative oligosaccharide synthesis by van Boom et al. in the 1980s [16]. These synthetic structures were the basis for studies correlating oligosaccharide length and immunogenicity.

Major advances in solution-phase oligosaccharide synthesis – including the development of more powerful glycosylating agents of improved selectivity [17], greater diversity of available protecting groups [18], and new analytical techniques – provided the impetus for a thorough reexamination of the area in the early 1990s.

4.3
Synthetic Strategies

At the heart of every oligosaccharide synthesis is a glycosylation reaction involving a reactive species (the glycosyl donor) and a nucleophile (the glycosyl acceptor). Two general synthetic strategies are available. Attachment of the glycosyl donor is achieved by connecting the 'non-reducing' end of the first carbohydrate moiety to the solid support (donor-bound strategy, Scheme 4.1). Otherwise, the anomeric position of the glycosyl acceptor is fixed to the support (acceptor-bound strategy). Both alternatives, as well as the bi-directional synthesis, a variation of these two main strategies, have been utilized in syntheses.

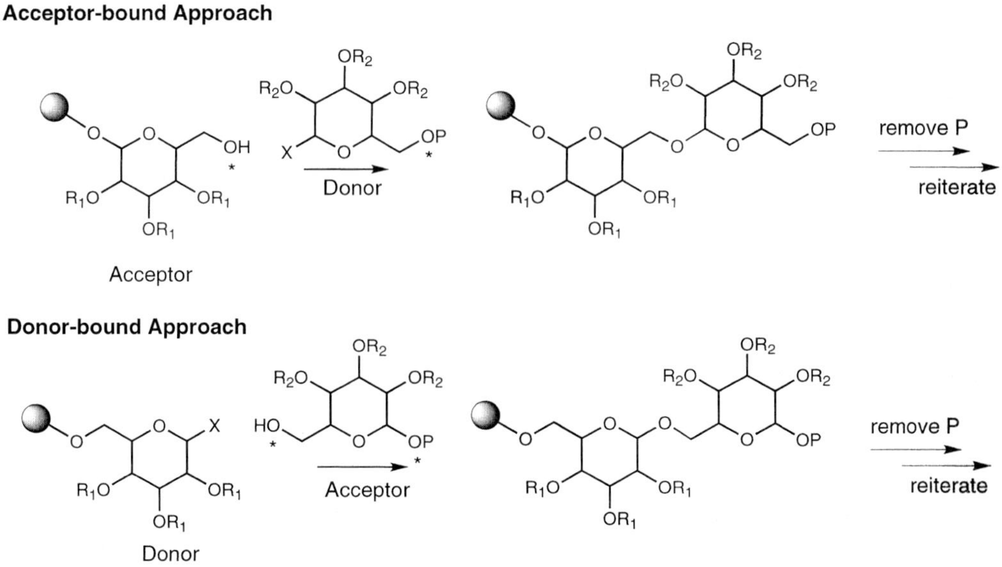

Acceptor-bound Approach

Donor-bound Approach

Scheme 4.1 Glycosyl acceptor (case 1) and donor (case 2) bound to the solid support. ●: Polymer support, P: unique protecting group, X: activating group, *: uniquely differentiated hydroxy group.

4.3.1
Immobilization of the Glycosyl Acceptor

Immobilization of the acceptor on the solid support allows for an excess of side reaction-prone glycosyl donor to be added in order to maximize coupling yields, nonproductive side products being washed away after each coupling. It has been this reasoning that has generated immense interest in the acceptor-bound approach to solid-phase oligosaccharide synthesis. Our group has used this approach in the development of the first automated oligosaccharide synthesizer (see Section 4.7).

4.3.2
Immobilization of the Glycosyl Donor

The donor-bound approach was the basis of the solid-phase glycal assembly method for oligosaccharide synthesis introduced by Danishefsky and co-workers [19]. A 6-O-diphenyl arylsilane linker serves to connect the first glycal monosaccharide to the polymeric support. Treatment with dimethyldioxirane (DMDO) results in formation of the corresponding 1,2-anhydrosugar, which in turn serves as a support-bound glycosyl donor upon activation with zinc chloride. This strategy is illustrated for the preparation of a β-(1 → 6)-linked tetrasaccharide in 32% overall yield (Scheme 4.2).

A principle drawback of the donor-bound strategy has to be taken into account. Most side reactions during glycosylations involve the glycosyl donor and thus result in termination of chain elongation, the consequence being a reduction in the overall yield in the donor-bound strategy. This, though, did not prevent Danishefsky and co-workers from completing the solid-phase synthesis of an impressive array of oligosaccharide structures [20].

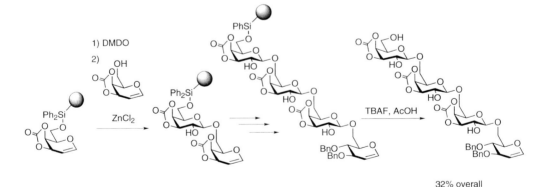

Scheme 4.2 Synthesis of a linear tetrasaccharide by the glycal assembly approach. ●: Merrifield's resin.

4.3.3
Bi-directional Strategy

Elongation of the growing oligosaccharide in both directions ("bi-directional" approach) would be ideal. For such a hybrid between the donor- and the acceptor-bound strategies, the glycosyl donor is attached to the polymer in the "non-reducing" region, thus producing an acceptor site and an anomeric donor function for chain elongation in two directions. The latent donor moiety, present on the reducing end of the saccharide that serves initially as a glycosyl acceptor, has to be completely inert toward the coupling conditions used for elongation of the acceptor-bound branch, as illustrated by the synthesis of a trisaccharide (Scheme 4.3) [21].

4.4
Support Materials

4.4.1
Insoluble Supports

The nature of the support is an important determinant of the synthetic strategy and the type of reaction conditions that may be used during oligosaccharide assembly. Merrifield's resin (polystyrene cross-linked with 1% divinylbenzene) has been most popular, due to its high loading capacity, compatibility with a broad range of reaction conditions, durability, and low price. The major limitation of polystyrene resins is the limited capabilities of the solvents dichloromethane, THF, DMF, and dioxane to affect their swelling and thus access to all reactive sites. Grafting of polyethylene glycol (PEG) chains onto the polystyrene backbone has resulted in resins such as TentaGel [22] and ArgoGel that exhibit more desirable swelling properties (even in water) at the expense of lower loading capacities (0.2–0.6 mmol g^{-1}) and higher price.

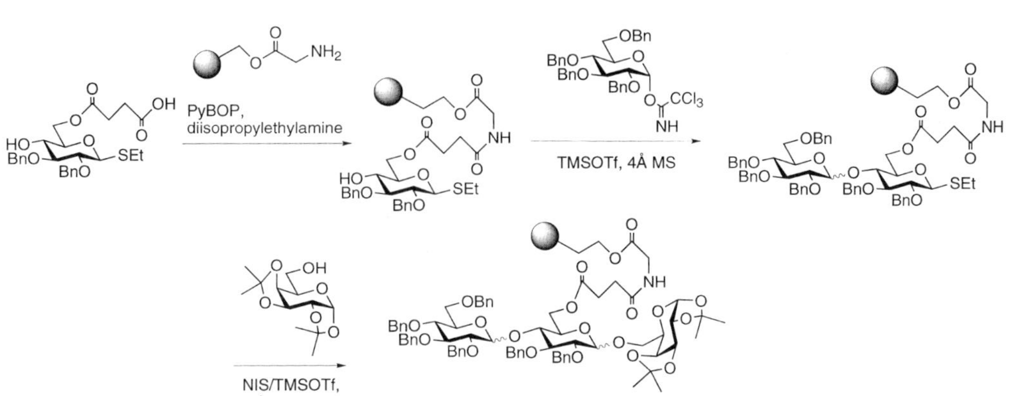

Scheme 4.3 Synthesis of a trisaccharide by the bi-directional approach. ●: TentaGel resin.

Only the surface of non-swelling controlled pore glass (CPG) supports, commonly used for automated DNA synthesis, is functionalized, thus resulting in lower loading (0.03–0.3 mmol g^{-1}) but also possibly in easier reagent access. While glass beads do not require swelling and may be used in a variety of solvents, their fragility complicates their handling. These CPG supports have been evaluated for their performance in oligosaccharide synthesis with trichloroacetimidate donors [23]. A serious shortcoming in the context of oligosaccharide assembly is the incompatibility of CPG with the silyl ether protecting groups commonly employed for temporary hydroxy protection.

4.4.2
Soluble Supports

Procedures developed for solution-phase synthesis often require adjustments to make them amenable to solid-phase oligosaccharide assembly. In addition, insoluble supports preclude the use of many reagents such as metal salts commonly employed in solution-phase synthesis. Soluble polymer supports have become increasingly popular, as they combine the advantages of the solution-phase regime with the easy workups of solid-phase synthesis. The polymer remains in homogeneous solution during all transformations but is precipitated out after each step to remove excess reagents by filtration. The overall yield for the assembly of large structures may be decreased by the loss of material during each precipitation. Still, soluble polymers have been very successfully utilized by Boons et al. in the block synthesis of complex oligosaccharides (Scheme 4.4) [24].

4.5
Linkers

The chemical nature of the entity connecting the solid support with the first sugar determines all the reaction conditions that may be applied during the oligosaccharide assembly process [25]. In principle, the anchor is just a support-bound protecting group and so any type of protective group can be utilized as a linker, although compatibility with other protecting groups needs to be ensured.

4.5.1
Silyl Ethers

Silyl ethers are convenient protective groups for hydroxy groups as they may be removed selectively under mild conditions. While diisopropyl arylsilane linkers preclude the use of further silyl ethers as temporary means of protection they have been used very successfully in Danishefsky's donor-bound strategy employing glycal-derived donors (Scheme 4.5) [20]. The compatibility of the silane linkers with different glycosylation agents such as thioglycosides, anomeric fluorides, trichloroacetimidates, and sulfoxides has been demonstrated [26].

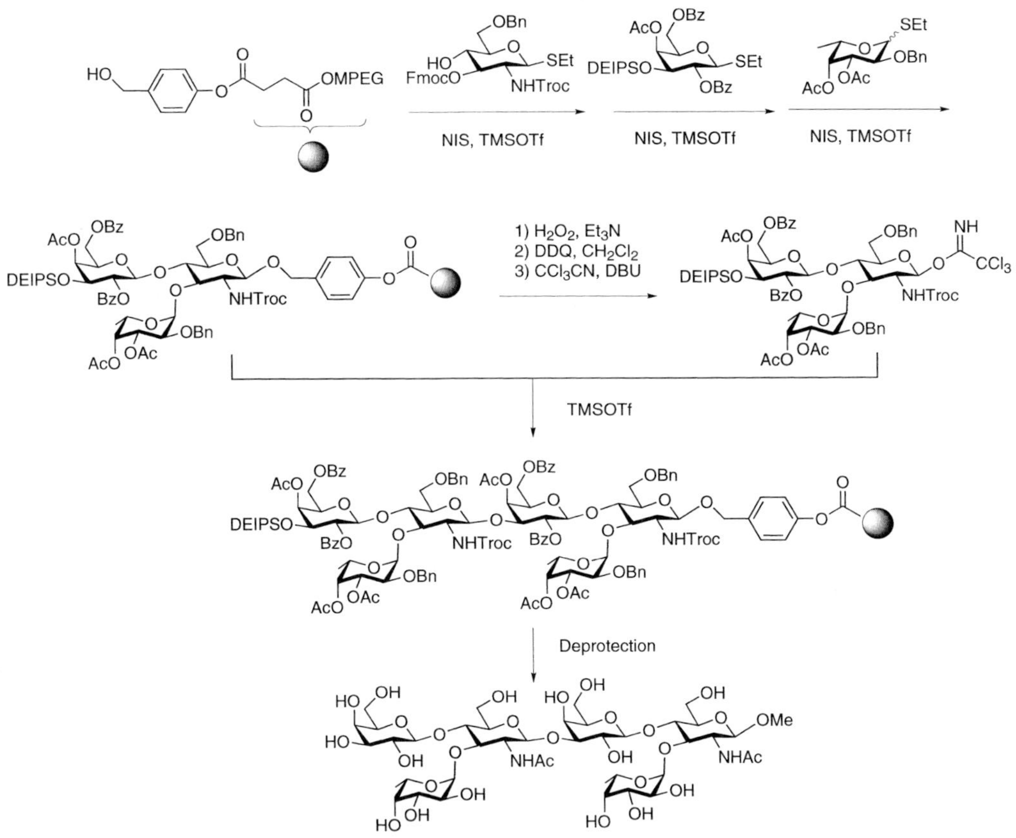

Scheme 4.4 Synthesis of a dimeric Le^x hexasaccharide on soluble support. ●: Methoxypolyethylene glycol (MPEG) resin.

4.5.2
Acid- and Base-Labile Linkers

A host of acid-labile linkers known from solid-phase peptide synthesis have been applied to solid-phase oligosaccharide assembly. Amino-functionalized Rink resin [27] and benzylidene acetal-type linkages to Wang aldehyde resin [28] or regular Wang resin [29] were readily cleaved with trifluoroacetic acid (TFA). The tris(alkoxy)benzyl amine (BAL) safety-catch linker, also developed for peptide synthesis [30], was used to anchor an amino sugar to a support- and cleaved by acylation and treatment with TFA.

The base-labile succinoyl linker, commonly used in automated DNA synthesis, has also been employed in oligosaccharide syntheses on soluble supports [23a,31] and on TentaGel and polystyrene supports (Scheme 4.6). Treatment with aqueous ammonia released the target from the support. A linker based on 9-hydroxy-

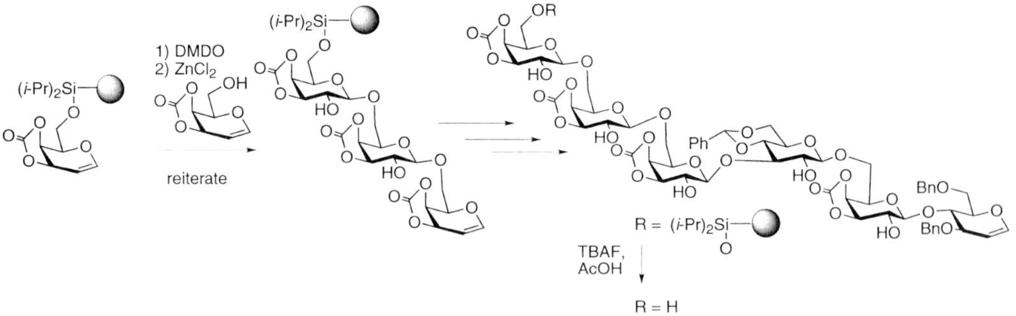

Scheme 4.5 Synthesis of a hexasaccharide by the glycal assembly approach. ●: Merrifield's resin.

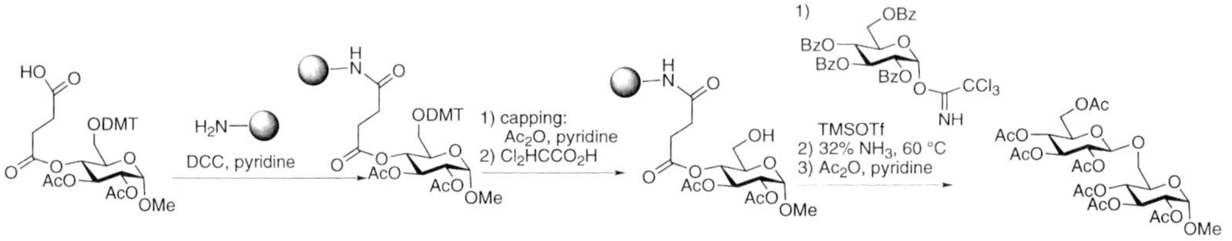

Scheme 4.6 Use of a base-labile succinoyl linker for oligosaccharide assembly. ●: Merrifield's or TentaGel resin.

methylfluorene-2-succinic acid could be cleaved under even milder conditions (20% triethylamine), but has not found widespread application [32].

4.5.3
Thioglycoside Linkers

The stability of thioglycosides to a wide range of activation conditions and their ability to be activated by thiophiles makes them an attractive mode of anomeric protection. This type of attachment to the carrier resin was utilized with trichloroacetimidate donors in the assembly of an oligomannoside [33], the product of the synthesis being cleaved by methanolysis induced by NBS and di-*tert*-butyl pyridine (DTBP). A *p*-hydroxythiophenyl glycoside was applied to the synthesis of oligosaccharides by the use of glycosyl sulfoxides and hydrolyzed at the end of the synthesis by treatment with trifluoroacetate [34].

4.5.4
Linkers Cleaved by Oxidation

Several linkers, related to the *p*-methoxybenzyl (PMB) group, that may be cleaved by oxidation have been introduced [35]. To overcome the inherent acid lability associated with the PMB group, an acyl moiety was introduced to provide an acid-stable protecting group [36]. Bound to ArgoPore™ resin, this protecting group can be cleaved under oxidative conditions by treatment with dichlorodicyanodiquinone (DDQ).

4.5.5
Photocleavable Linkers

Light is a mild reagent that can be used to effect cleavage of an oligosaccharide from the solid support and does not interfere with any other modes of protection. Several groups have revitalized the idea of employing a photocleavable linker [9] for chemical oligosaccharide synthesis. Photolabile o-nitrobenzylic linkers were used by Nicolaou and co-workers for the construction of very large, branched carbohydrate structures [37].–Since photolytic cleavage of primary o-nitrobenzyl linkers is often slow and incomplete, Fraser-Reid designed a new system based on a secondary o-nitrobenzyl ether linkage (Scheme 4.7) [38].

4.5.6
Linkers Cleaved by Olefin Metathesis

In addition to serving purely as a means of attachment to a polymer, a linker may also allow for further functionalization, glycosylation, or conjugation after completion of the synthesis. We introduced a new linker concept that fulfilled these requirements (Scheme 4.8a) [39]. The first carbohydrate moiety is connected through a glycosidic bond to octenediol-functionalized Merrifield's resin, and the loading is readily determined by colorimetry after cleavage of the DMT protecting group. The octenediol linker is stable to a wide range of reaction conditions and was quantitatively cleaved by olefin cross-metathesis in the presence of Grubbs' catalyst under an atmosphere of ethylene to afford fully protected oligosaccharides

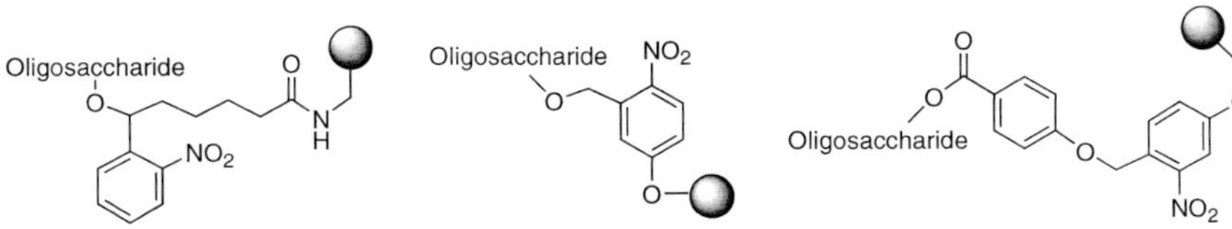

Scheme 4.7 Photocleavable linkers used for oligosaccharide synthesis. ●: Polymer resin.

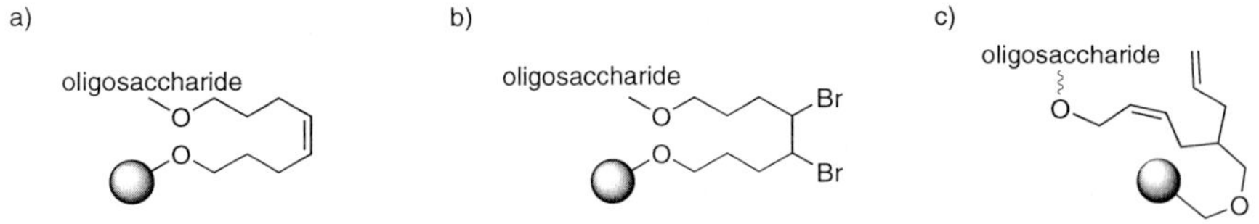

Scheme 4.8 Linkers removed by olefin cross metathesis (a, b) and ring-closing (c) metathesis. ●: Merrifield's resin.

in the form of *n*-pentenyl glycosides. These fragments can serve as glycosyl donors or as precursors for other anomeric functionalities and linking moieties [40]. To render the octenediol linker compatible with glycosylating agents that require electrophiles as activators, the double bond was converted into the corresponding dibromide (Scheme 4.8b) [41].

Ring-closing metathesis (RCM) with Grubbs' catalyst was applied to cleave two linkers introduced by Schmidt et al. (Scheme 4.8c) [42].

4.6
Synthesis of Oligosaccharides on Solid Support by Use of Different Glycosylating Agents

Glycosylating agents that react efficiently and selectively are the key to solid-phase oligosaccharide synthesis, since purification is not possible during elongation. The development of novel glycosyl donors carrying a variety of anomeric leaving groups for solution-phase syntheses also had a great impact on solid-phase oligosaccharide synthesis. This section summarizes the most successful approaches used for the synthesis of complex oligosaccharides on solid support. The advances in this area are grouped according to the type of glycosyl donors used in each synthesis.

4.6.1
1,2-Anhydrosugars – The Glycal Assembly Approach

1,2-Anhydrosugars are readily derived from glycal precursors and have been activated to fashion a variety of glycosidic linkages. A host of complex oligosaccharides, glycoconjugates, and glycosylated natural products have been prepared by the glycal assembly method [19]. Following the success in solution-phase synthesis, glycals were adapted to the solid-phase strategy by the donor-bound approach [20]. Glycals minimize protecting group manipulations, serve as glycosyl acceptors, and may readily be converted into different glycosylating agents. The first glycal was linked to Merrifield's resin through a diisopropylsilane. Support-bound glycals were readily converted into the corresponding anhydrosugars by epoxidation with dimethyldioxirane. After the feasibility of the approach had initially been demonstrated on the example of a linear tetrasaccharide (Scheme 4.2), a linear hexasaccharide containing β-(1 → 3)-glucosidic and β-(1 → 6)-galactosidic linkages was prepared [43].

Carbohydrate blood group determinants are important for binding events during inflammation and bacterial infection and are tumor-associated antigens. The Le^b blood group antigen has been identified as a mediator for the binding of the pathogen *Helicobacter pylori* to human gastric epithelium. Initially, parts of the assembly of a Lewis^b-(Le^b) hexasaccharide were carried out on solid support [44]. Synthetic access to the complete Le^b antigen on a solid support was achieved after the iodosulfonamidation reaction [45] was successfully adapted to solid-support

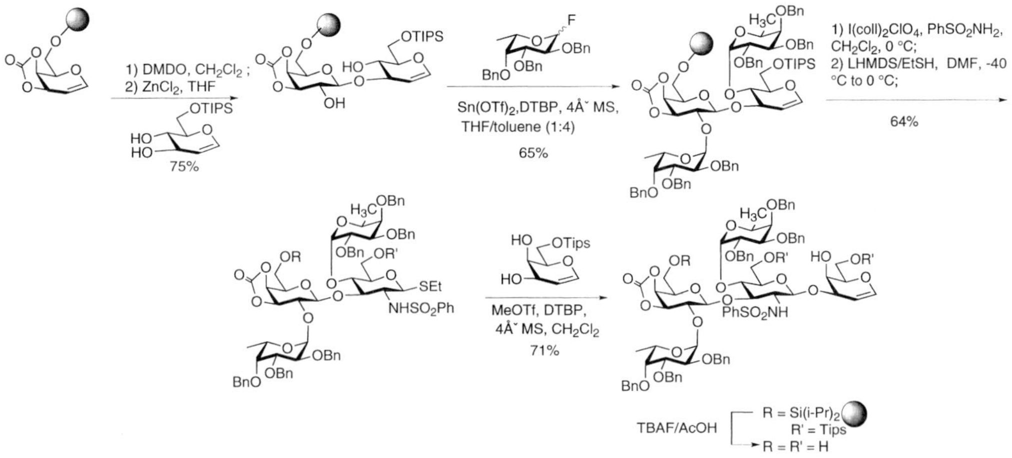

Scheme 4.9 Solid-phase synthesis of the Lewis b blood group determinant pentasaccharide glycal. ●: Merrifield's resin.

conditions to install a thioethyl glycosyl donor from a glycal precursor. The desired pentasaccharide was obtained in 20% overall yield after release from the support (Scheme 4.9) [46].

The reliable and selective installation of β-glucosidic linkages on the solid support by the glycal assembly method proved to be a challenge. The initially explored conformationally constrained galactal was found to be relatively stable to mild Lewis acids such as zinc chloride and even allowed for galactosylation of hindered C4-hydroxy acceptors. The lack of a constrained glucosyl epoxide rendered these donors highly reactive and prone to rapid decomposition upon treatment with Lewis acids. From solution-phase precedence [47], the glucosyl epoxides could reliably be converted into thioethyl glycosyl donors bearing a C2 pivaloyl group. By this methodology a tetrasaccharide containing exclusively β-(1 → 4)-glucosidic linkages was synthesized in 20% overall yield (84% per step) [48]. The glycal method was further extended to access N-linked glycopeptides by solid-phase synthesis [49].

4.6.2
Glycosyl Sulfoxides

Anomeric glycosyl sulfoxides act as highly reactive glycosylating agents upon activation with Lewis acidic promoters. Triflic anhydride has most commonly been used to induce facile reactions at –78 °C and to fashion even difficult linkages with hindered acceptors [50]. The use of glycosyl sulfoxides in solid-phase oligosaccharide synthesis was initiated by Kahne and explored in the context of the preparation of single compounds and of a combinatorial library of di- and trisaccharides [51]. In the absence of C2-participating groups, anomeric sulfoxides gave high α-selectivity when coupled with secondary alcohols. The selective formation of β-glycosidic linkages was achieved with the aid of a C2-pivaloyl participating group.

4.6.3
Glycosyl Trichloroacetimidates

Glycosyl trichloroacetimidates have become the most commonly used glycosyl donors among the wide array of glycosylating agents. Their versatility, high yields, and excellent selectivity in glycosylation reactions provided the basis for their outstanding success in solution-phase oligosaccharide synthesis [52]. Krepinsky reported the first successful use of trichloroacetimidate glycosyl donors for the synthesis of a disaccharide on a soluble PEG support [53], and many laboratories have since demonstrated their utility on various insoluble supports.

Schmidt et al. initially relied on a thioether linkage and Merrifield's resin [33]. The synthesis of β-linked linear oligosaccharides was followed by the solid-phase synthesis of a branched pentasaccharide common to most complex N-glycan structures (Scheme 4.10). Cleavage of the thioether linker with N-bromosuccinimide (NBS) in the presence of methanol released the pentasaccharide methyl glycoside in 20% overall yield [54].

Use of trichloroacetimidate donors with an octenediol linker was particularly successful. Repetitive α-mannosylations furnished a linear heptasaccharide. The desired n-pentenyl heptamannoside was obtained in 9% overall yield (84% per step), and the corresponding penta- and trimannosides were cleaved in 41% and 76% yields, respectively (91–95% per step) [39].

Trichloroacetimidate donors also performed extremely well with other solid support materials. Iadonisi et al. [23a] explored the performance of trichloroacetimidates in glycosylation reactions with acceptors bound to different polymeric supports. Coupling yields of up to 95% were reported with polystyrene or CPG, but PEG-containing polymers were found to perform significantly more poorly in

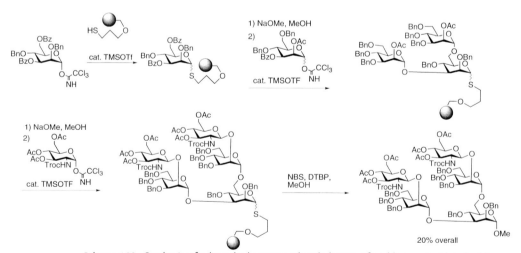

Scheme 4.10 Synthesis of a branched pentasaccharide by use of trichloroacetimidate building blocks. ●: Merrifield's resin.

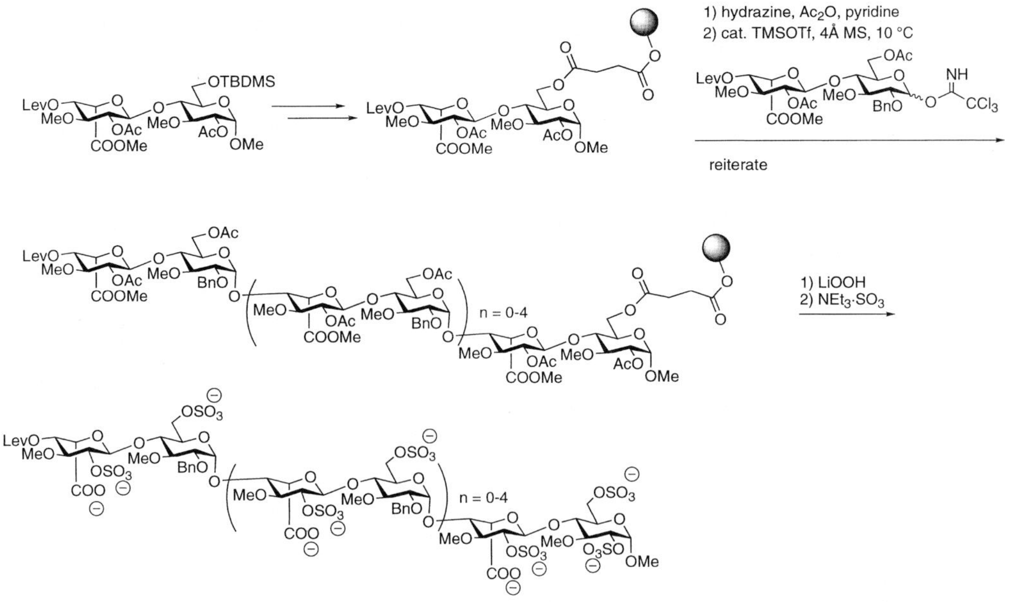

Scheme 4.11 Synthesis of a heparin mimetic. ●: MPEG resin.

these reactions. This group also explored glycosylations of CPG-bound oligonucleotides with trichloroacetimidates [55].

Heparan sulfate mimetics of varying length were prepared on a soluble MPEG resin with the aid of trichloroacetimidate donors [56]. Reaction temperature, excess of donor, and sometimes double glycosylation were found to be crucial if coupling efficiencies were to exceed 95%. A capping step was introduced after each glycosylation to acetylate any unreacted acceptor sites. Reiteration of the deprotection, glycosylation, and capping cycle provided oligosaccharides up to a dodecamer ($n=4$) (Scheme 4.11).

4.6.4
Thioglycosides

Thioglycosides can be prepared on large scale, stored over prolonged periods of time even at room temperature, and selectively activated with a range of thiophilic promoters such as dimethylthiosulfonium triflate (DMTST), methyl triflate, or NIS/triflic acid [57]. Drawbacks of thioglycoside donors on solid support are the high toxicity of the activators, but these have not prevented their widespread use in solid-phase syntheses of oligosaccharides.

The application of ethyl thioglycosides to the synthesis of highly branched oligosaccharides on a polymer support was investigated by van Boom et al. in the synthesis of a heptaglucoside exhibiting phytoalexin elicitor activity [58]. This heptasaccharide, containing β-(1 → 6)- and β-(1 → 3)-glucosidic linkages, had previously

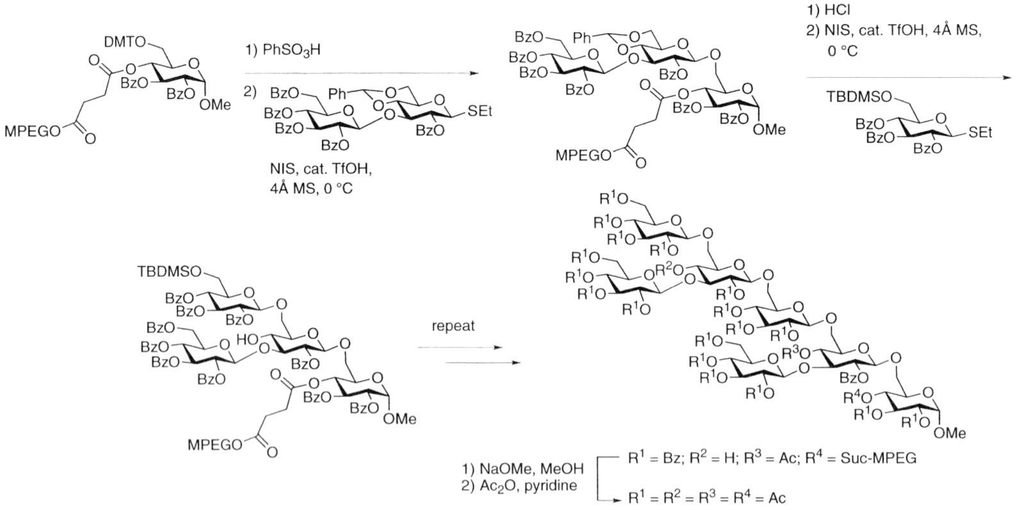

116 | 4 Solid-Phase Oligosaccharide Synthesis

Scheme 4.12 Synthesis of a phytoalexin elicitor heptasaccharide on soluble support.

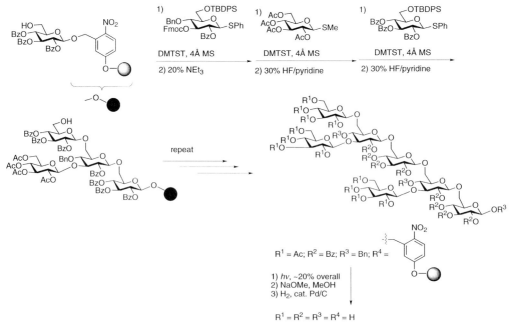

Scheme 4.13 Synthesis of a phytoalexin elicitor hexasaccharide on insoluble support. ●: Merrifield's resin.

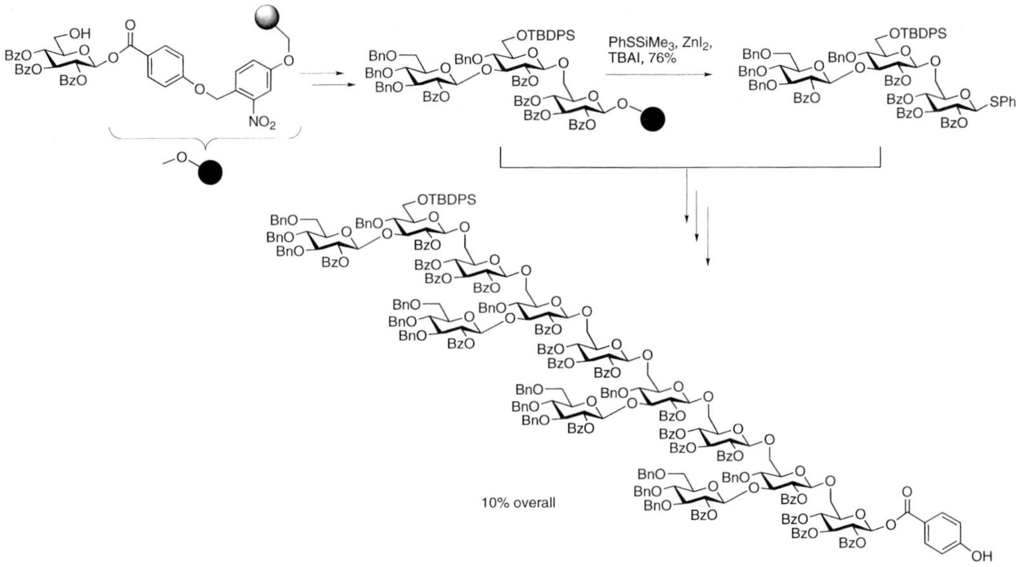

4.6 Synthesis of Oligosaccharides on Solid Support by Use of Different Glycosylating Agents

Scheme 4.14 Synthesis of a phytoalexin elicitor dodecasaccharide on solid support. ●: Merrifield's resin.

been synthesized by conventional solution-phase methodology [59]. Elaboration of the starting monomer by subsequent deprotection and peracetylation steps furnished homogeneous heptasaccharide in 18% overall yield after purification (Scheme 4.12). Full β-selectivity of all glycosidation reactions was achieved thanks to the C2-benzoyl groups.

The same phytoalexin elicitor heptasaccharide was later synthesized by Nicolaou and co-workers on a polystyrene support with a photolabile o-nitrobenzyl linker through use of thiomethyl and thiophenyl glycosides [37a]. This synthesis was achieved by subsequent coupling of monomers, with the use of a key 3,6-differentially protected glucose. The first monosaccharide was attached to the linker, which in turn was coupled to phenolic polystyrene. Acceptor sites were temporarily TBDPS- or Fmoc-protected, and glycosylations employed phenylthio donors. Photolytic cleavage followed by deacylation and hydrogenation in solution procured the fully deprotected heptasaccharide (Scheme 4.13).

Incorporation of a 4-hydroxybenzoic acid spacer between the photolabile linker and the anomeric position of the first monosaccharide allowed for the generation of fully protected oligosaccharide fragments that served in turn as glycosylating agents [37b]. Trisaccharide thioglycoside was obtained in 76% yield after cleavage with $PhSSiMe_3/ZnI_2/nBu_4NI$ and was used in fragment couplings to furnish a dodecamer in 10% yield (Scheme 4.14).

4.6.5
Glycosyl Fluorides

Anomeric halides, for many years the workhorse of carbohydrate assembly, have not seen much use on solid support. The requirement for heavy metal salt activation renders this method difficult on polymeric supports. Only fucosyl fluorides are commonly employed, for the installation of α-fucosidic linkages, as they provide the desired linkages in excellent yield and selectivity [60].

4.6.6
n-Pentenyl Glycosides

Fraser-Reid and co-workers introduced the use of NPGs as glycosylating agents for solution-phase synthesis [40a]. NPGs are activated by electrophilic reagents such as NIS/TESOTf and have also found applications in solid-phase synthesis. Polystyrene grafted "crowns" were applied to the construction of a trisaccharide by use of a photocleavable o-nitrobenzyl linker and n-pentenyl glycoside donors. The first aminoglucosyl moiety was attached to the linker through the anomeric position. Further elaboration furnished a trisaccharide. Global deprotection, peracetylation, and photolytic cleavage from the support provided unprotected trisaccharide, although no yield was reported [61].

4.6.7
Glycosyl Phosphates

A straightforward new route for the preparation of glycosyl phosphates from glycal precursors was the key to providing sufficient quantities of differentially protected building blocks in order to study this class of molecules as glycosylating agents [62]. Glycosyl phosphates are extremely reactive glycosyl donors that can be activated at low temperatures to form a variety of linkages in very high yields within minutes [63]. We have studied the use of glycosyl phosphates in depth and have demonstrated their versatility in solution-phase [63] and solid-phase approaches [64].

Manual solid-phase oligosaccharide synthesis, with use of the octenediol linker, allowed for a facile synthesis of β-$(1 \rightarrow 4)$-linked trisaccharide (53% overall yield, 7 steps). These findings provided the foundation for our program directed at automating solid-phase oligosaccharide synthesis (Scheme 4.15) [62].

4.7
Automated Solid-Phase Oligosaccharide Synthesis

In recent years, two different approaches to the automation of oligosaccharide assembly have been developed: the automated solid-phase synthesis method described in this chapter and the one-pot approach described elsewhere in this volume, focussing on automation of synthesis planning [4].

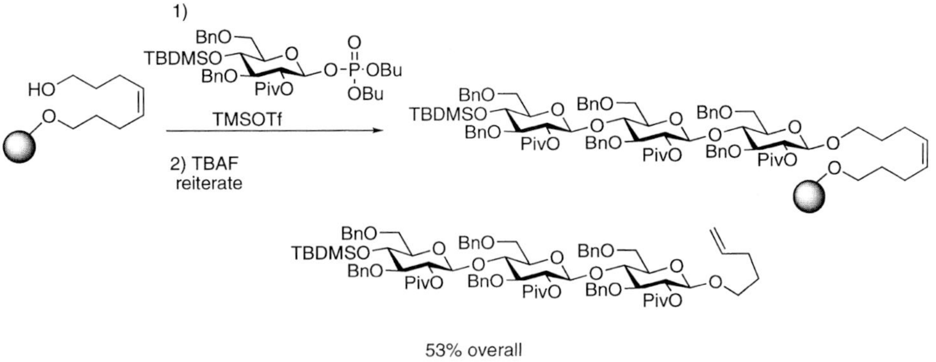

Scheme 4.15 Synthesis of oligosaccharides by the use of glycosyl phosphate building blocks. ●: Merrifield's resin.

4.7.1
Fundamental Considerations

The general strategy for automated carbohydrate synthesis was based on solid-phase techniques we had developed for manual solid-support oligosaccharide assembly (see Section 4.6.7). The acceptor-bound method was chosen, as side products, reagents, and unreacted starting material are washed away and the carbohydrate chain remains covalently attached to the polymer support. A unique protecting group is removed to provide another nucleophilic position for subsequent elongation.

In order to evaluate automated oligosaccharide assembly, we adapted an ABI-433 peptide synthesizer to carbohydrate synthesis (Fig. 4.1). A specially designed low-temperature reaction vessel was cooled by a commercially available cooling. The necessary reagents were loaded onto the instrument ports and reaction conditions were programmed on the computer, in a fashion similar to that used for the automated synthesis of peptides [65].

An automated solid-phase method necessitates a polymer support and linker that are compatible with the reagents used in carbohydrate synthesis. We utilized both swellable resins such as Merrifield's (1% crosslinked) resin and rigid macroreticular resins such as Argopore, functionalized with our octenediol linkers that are readily cleaved under neutral conditions at the end of a synthesis.

We demonstrated the utility of the automated method with glycosyl trichloroacetimidate and glycosyl phosphate building blocks. Temporary protecting groups such as levulinate esters, silyl ethers, and acetate esters were shown to be compatible with automation.

4 Solid-Phase Oligosaccharide Synthesis

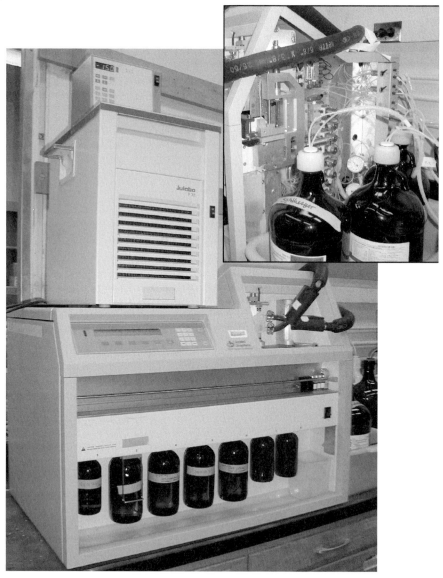

Fig. 4.1 The first automated solid-phase oligosaccharide synthesizer, based on an ABI 433 peptide synthesizer.

4.7.2
Automated Synthesis with Glycosyl Trichloroacetimidates

Initially, automation of the synthetic process was explored with the use of glycosyl trichloroacetimidate building blocks. A polystyrene support, functionalized with an olefinic linker, is loaded into a reaction vessel in the instrument. By use of a coupling cycle, the activating reagent (trimethylsilyl triflate (TMSOTf)/CH_2Cl_2) and deblocking reagent (e.g. NaOMe) and building blocks were applied (Scheme 4.16). The synthesis was performed in an iterative manner according to the programmed coupling cycle. Double glycosylations (95–98%) gave better coupling yields than single glycosylations (90–95%).

4.7.3
Automated Synthesis with Glycosyl Phosphates

Following the success with glycosyl trichloroacetimidate monomers, we developed a similar coupling cycle for glycosyl phosphate building blocks (Tab. 4.1). A reaction vessel designed for low temperatures and a cooling apparatus was used to provide a –15 °C reaction temperature ideal for phosphate couplings. The success of this approach was demonstrated by the assembly of a dodecasaccharide in automated fashion (Scheme 4.17).

4.7.4
Automated Oligosaccharide Synthesis by Use of Different Glycosylating Agents

The coupling protocols described in Sections 4.7.3 and 4.7.4 were combined in the synthesis of a branched tetrasaccharide antigen found on Leishmania [66]. Both glycosyl phosphate and glycosyl trichloroacetimidate building blocks were used, along with acetate and levulinate esters as temporary protecting groups (Scheme 4.18).

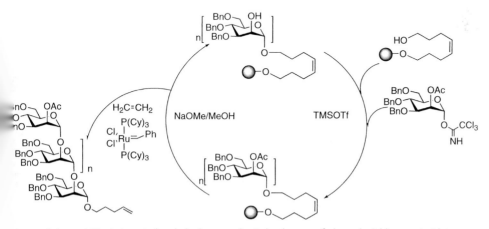

Scheme 4.16 Automated carbohydrate synthesis by the use of glycosyl trichloroacetimidates. ●: Polystyrene resin.

4 Solid-Phase Oligosaccharide Synthesis

Tab. 4.1 Automated coupling cycle used with glycosyl phosphates (25 mmol scale)

Step	Function	Reagent	Time (min)
1	couple	5 equiv. donor and 5 equiv. TMSOTf	15
2	wash	dichloromethane	6
3	couple	5 equiv. donor and 5 equiv. TMSOTf	15
4	wash	methanol:dichloromethane	4
5	wash	tetrahydrofuran	4
6	wash	pyridine:acetic acid	3
7	deprotection	2×20 equiv. hydrazine (pyridine:acetic acid)	30
8	wash	pyridine:acetic acid	3
9	wash	methanol:dichloromethane	4
10	wash	0.2 m acetic acid in tetrahydrofuran	4
11	wash	tetrahydrofuran	4
12	wash	dichloromethane	6

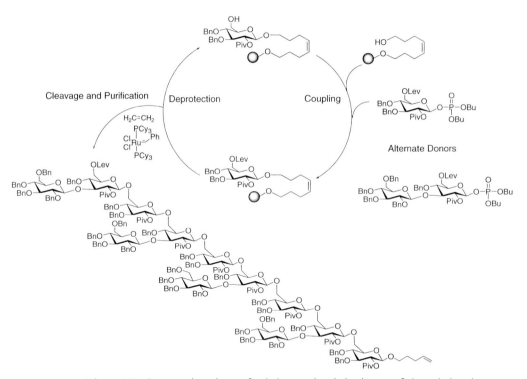

Scheme 4.17 Automated synthesis of a dodecasaccharide by the use of glycosyl phosphate building blocks. ●: Polystyrene resin.

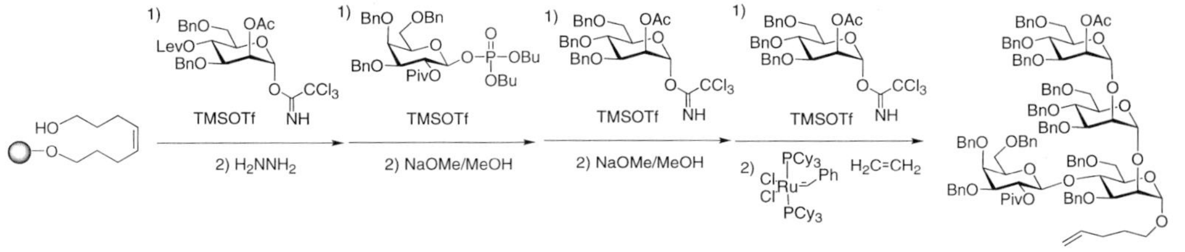

Scheme 4.18 Automated synthesis of a Leishmania tetrasaccharide. ●: Polystyrene resin.

4.7.5
"Cap-Tags" to Suppress Deletion Sequences

Deletion sequences ($n-1$, $n-2$, etc.) are the most common side-products in automated solid-phase synthesis and may complicate purification of the final product. To aid in the purification process, we developed a capping procedure that allows for the facile removal of deletion sequences [67]. After each coupling event, unreacted hydroxy groups that may result in deletion sequences were subjected to a capping reagent that rendered these sites silent in subsequent couplings (Scheme 4.19). The caps also functioned as handles for easy separation of all unwanted capped sequences from the desired (uncapped) products. Attachment of a polyfluorinated silyl ether F-tag onto unreacted hydroxy groups precluded further elongation of the deletion sequence. After cleavage from the resin, all fluorinated intermediates were easily removed by filtration through a pad of fluorinated silica gel. This modification of the automated coupling cycles greatly facilitates the purification of synthetic carbohydrates.

4.7.6
Current State of the Art of Automated Synthesis

Automated solid-phase carbohydrate synthesis utilizes instrumentation and reaction design in order to remove the most time-consuming aspect of carbohydrate assembly [65]. This process reduces the challenge of carbohydrate synthesis to the preparation of simple building blocks, by use of established methods. A complete set of building blocks has not been established, but the variety of possible building blocks should enable the production of diverse libraries of complex carbohy-

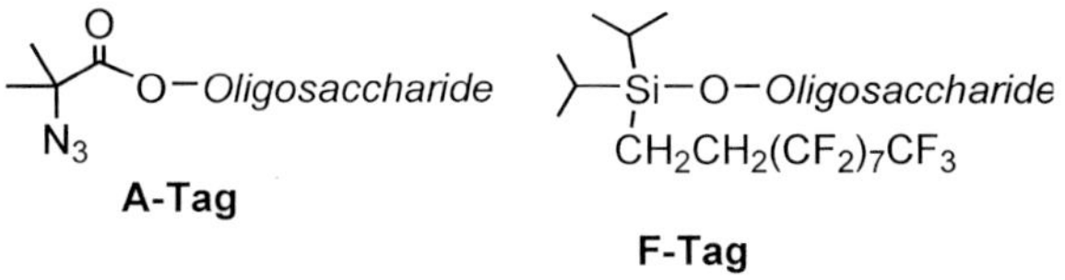

Scheme 4.19 Capping reagents used to minimize contamination of the oligosaccharide products with deletion sequences.

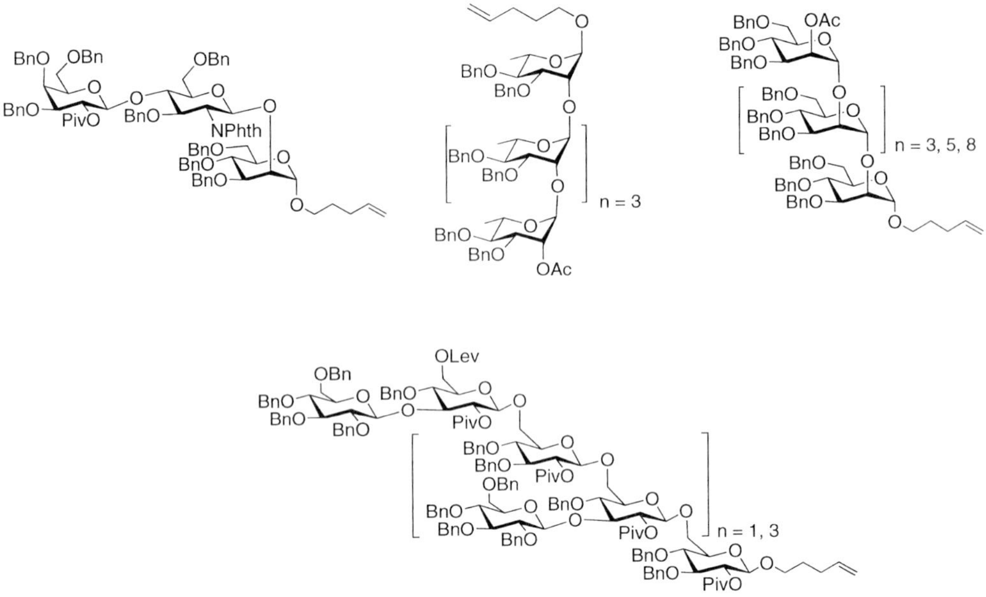

Fig. 4.2 Representative sequences prepared by automated solid-phase oligosaccharide synthesis.

drates. Some of the carbohydrate sequences prepared by automated solid-phase synthesis are shown in Fig. 4.2.

Automated carbohydrate synthesis allows for the production of complex carbohydrate orders of magnitude more rapidly than achievable by other approaches. This advance has the potential to parallel the breakthroughs achieved by researchers in the peptide and DNA fields that opened up the proteomics and genomic eras in biotechnology. When the selection of carbohydrate building blocks is further widened and the reaction conditions further streamlined, automated solid-phase carbohydrate synthesis should become the method of choice for carbohydrate production.

4.8
Conclusion and Outlook

Advances in glycobiology depend heavily upon the ability of synthetic organic chemists to provide defined compounds for biochemical and biophysical studies. Driven by the need to create more efficient methods for the synthesis of complex oligosaccharides, the past ten years have seen a major push toward a reliable procedure for the solid-phase synthesis of oligosaccharides, culminating in the past two years in the development of automated solid-phase oligosaccharide synthesis. As the coupling yields have improved to 95% and above, larger structures have come within reach. These efforts are crucial for providing molecular tools to elucidate biologically important interactions. Further development of this method should eventually allow even non-specialist synthetic chemists to create important molecular tools for biochemical, biophysical, and medical applications.

4.9 References

1 VARKI, A., *Glycobiology* **1993**, *3*, 97–130.
2 CARUTHERS, M.H., *Science* **1985**, *230*, 281–285.
3 a) ATHERTON, E.; SHEPPARD, R.C., *Solid phase peptide synthesis: A practical approach*, Oxford University Press, Oxford, **1989**. b) Nobel lecture: MERRIFIELD, R.B., *Angew. Chem. Int. Ed. Engl.* **1985**, *24*, 799–810.
4 a) ZHANG, Z.; OLLMANN, I.R.; YE, X.-S.; WISCHNAT, R.; BAASOV, T.; WONG, C.-H., *J. Am. Chem. Soc.* **1999**, *121*, 734–753. b) BURKHART, F.; ZHANG, Z.Y.; WACOWICH-SGARBI, S.; WONG, C.-H., *Angew. Chem. Int. Ed. Engl.* **2001**, *40*, 1274–1276.
5 a) OSBORN, H.M.I.; KHAN, T.H., *Tetrahedron* **1999**, *55*, 1807–1850. b) ITO, Y.; MANABE, S., *Curr. Opin. Chem. Biol.* **1998**, *2*, 701–708. c) SOFIA, M.J. in *Combinatorial Chemistry and Molecular Diversity in Drug Discovery*; GORDON, E.M.; KERWIN, J.F., Jr., Eds.; Wiley: New York, **1998**, pp. 243–269. d) SEEBERGER, P.H.; HAASE, W.-C., *Chem. Rev.* **2000**, *100*, 4349–4394.
6 MERRIFIELD, R.B., *J. Am. Chem. Soc.* **1963**, *85*, 2149–2152.
7 a) MALIK, A.; BAUER, H.; TSCHAKERT, J.; VOELTER, W., *Chemiker-Z.* **1990**, *114*, 371–375. b) For a more comprehensive review covering the initial studies on oligosaccharide synthesis on polymeric supports, see: FRÉCHET, J.M. in *Polymer-supported Reactions in Organic Synthesis*; HODGE, P.; SHERRINGTON, D.C., Eds.; Wiley, Chichester, **1980**; pp. 407–434.
8 FRÉCHET, J.M.J.; SCHUERCH, C., *J. Am. Chem. Soc.* **1971**, *93*, 492–496.
9 a) ZEHAVI, U.; AMIT, B.; PATCHORNIK, A., *J. Org. Chem.* **1972**, *37*, 2281–2285. b) ZEHAVI, U.; PATCHORNIK, A., *J. Am. Chem. Soc.* **1973**, *95*, 5673–5677.
10 a) PFÄFFLI, P.J.; HIXSON, S.H.; ANDERSON, L., *Carbohydr. Res.* **1972**, *23*, 195–206. b) CHIU, S.H.L.; ANDERSON, L., *Carbohydr. Res.* **1976**, *50*, 277–283.
11 EXCOFFIER, G.; GAGNAIRE, D.; UTILLE, J.-P.; VIGNON, M., *Tetrahedron* **1975**, *31*, 549–553.
12 EBY, R.; SCHUERCH, C., *Carbohydr. Res.* **1975**, *39*, 151–155.
13 FRÉCHET, J.M.J.; NUYENS, L.J.; SEYMOUR, E., *J. Am. Chem. Soc.* **1979**, *101*, 432–436.
14 GUTHRIE, R.D.; JENKINS, A.D.; STEHLÍCEK, J., *J. Chem. Soc. (C)* **1971**, 2690–2696.
15 a) BOCHKOV, A.F.; SNYATKOVA, V.I.; KOCHETKOV, N.K., *Izvest. Akad. Nauk S.S.S.R., Ser khim.* **1967**, 2684–2691. b) GUTHRIE, R.D.; JENKINS, A.D.; ROBERTS, G.A.F., *J. Chem. Soc., Perkin Trans. 1* **1973**, 2414–2417.
16 VEENEMAN, G.H.; NOTERMANS, S.; LISKAMP, R.M.J.; VAN DER MAREL, G.A.; VAN BOOM, J.H., *Tetrahedron Lett.* **1987**, *28*, 6695–6698.
17 TOSHIMA, K.; TATSUTA, K., *Chem. Rev.* **1993**, *93*, 1503–1531.
18 KHAN, S.H.; O'NEILL, R.A., Eds., *Modern Methods in Carbohydrate Synthesis*, Harwood Academic: Amsterdam, **1996**.
19 a) DANISHEFSKY, S.J.; BILODEAU, M.T., *Angew. Chem. Int. Ed. Engl.* **1996**, *35*, 1380–1419. b) SEEBERGER, P.H.; BILODEAU, M.T.; DANISHEFSKY, S.J., *Aldrichimica Acta* **1997**, *30*, 75–92.
20 SEEBERGER, P.H.; DANISHEFSKY, S.J., *Acc. Chem. Res.* **1998**, *31*, 685–695.
21 ZHU, T.; BOONS, G.-J., *Angew. Chem. Int. Ed. Engl.* **1998**, *37*, 1898–1900.
22 BAYER, E., *Angew. Chem. Int. Ed. Engl.* **1991**, *30*, 113–147.
23 a) ADINOLFI, M.; BARONE, G.; DE NAPOLI, L.; IADONISI, A.; PICCIALLI, G., *Tetrahedron Lett.* **1998**, *39*, 1953–1956. b) HECKEL, A.; MROSS, E.; JUNG, K.-H.; RADEMANN, J.; SCHMIDT, R.R., *Synlett* **1998**, 171–173.

24 Zhu, T.; Boons, G.-J., *J. Am. Chem. Soc.* **2000**, *122*, 10222–10223.
25 For comprehensive reviews on linker groups for solid-phase organic synthesis see: a) Gordon, K.; Balasubramanian, J., *Chem. Technol. Biotechnol.* **1999**, *74*, 835–851. b) James, I. W., *Tetrahedron* **1999**, *55*, 4855–4946.
26 Doi, T.; Sugiki, M.; Yamada, H.; Takahashi, T., *Tetrahedron Lett.* **1999**, *40*, 2141–2144.
27 Silva, D. J.; Wang, H.; Allanson, N. M.; Jain, R. K.; Sofia, M. J., *J. Org. Chem.* **1999**, *64*, 5926–5929.
28 Hanessian, S.; Huynh, H. K., *Synlett* **1999**, 102–104.
29 Shimizu, H.; Ito, Y.; Kanie, O.; Ogawa, T., *Bioorg. Med. Chem. Lett.* **1996**, *6*, 2841–2846.
30 a) Jensen, K. J.; Alsina, J.; Songster, M. F.; Vagner, J.; Albericio, F.; Barany, G., *J. Am. Chem. Soc.* **1998**, *120*, 5441–5452. b) Tolborg, J. F.; Jensen, K. J., *Chem. Commun.* **2000**, 147–148.
31 Leung, O. T.; Douglas, S. P.; Whitfield, D. M.; Dong, H. Y. S.; Krepinsky, J. J., *New J. Chem.* **1994**, *18*, 349–363.
32 Wang, Y.; Zhang, H.; Voelter, W., *Chem. Lett.* **1995**, 273–274.
33 Rademann, J.; Schmidt, R. R., *J. Org. Chem.* **1996**, *62*, 3650–3653.
34 a) Yan, L.; Taylor, C. M.; Goodnow, R.; Kahne, D., *J. Am. Chem. Soc.* **1994**, *116*, 6953–6954. b) Liang, R.; Yan, L.; Loebach, J.; Ge, M.; Uozumi, Y.; Sekanina, K.; Horan, N.; Gildersleeve, J.; Thompson, C.; Smith, A.; Biswas, K.; Still, W. C.; Kahne, D., *Science* **1996**, *274*, 1520–1522.
35 Fukase, K.; Nakai, Y.; Egusa, K.; Porco, Jr., J. A.; Kusumoto, S., *Synlett* **1999**, 1074–1078.
36 Fukase, K.; Egusa, K.; Nakai, Y.; Kusumoto, S., *Molecular Diversity* **1996**, *2*, 182–188.
37 a) Nicolaou, K. C.; Winssinger, N.; Pastor, J.; DeRoose, F., *J. Am. Chem. Soc.* **1997**, *119*, 449–450. b) Nicolaou, K. C.; Watanabe, N.; Li, J.; Pastor, J.; Winssinger, N., *Angew. Chem. Int. Ed. Engl.* **1998**, *37*, 1559–1561.
38 Rodebaugh, R.; Fraser-Reid, B.; Geysen, H. M., *Tetrahedron Lett.* **1997**, *38*, 7653–7656.
39 Andrade, R. B.; Plante, O. J.; Melean, L. G.; Seeberger, P. H., *Org. Lett.* **1999**, *1*, 1811–1814.
40 a) Fraser-Reid, B.; Udodong, U. E.; Wu, Z.; Ottosson, H.; Merritt, J. R.; Rao, S.; Roberts, C.; Madsen, R., *Synlett* **1992**, 927–942. b) Buskas, T.; Söderberg, E.; Konradsson, P.; Fraser-Reid, B., *J. Org. Chem.* **2000**, *65*, 958–963.
41 Melean, L. G.; Haase, W.-C.; Seeberger, P. H., *Tetrahedron Lett.* **2000**, *41*, 4329–4333.
42 a) Knerr, L.; Schmidt, R. R., *Synlett*. **1999**, 1802–1804. b) Mross, E.; Schmidt, R. R., *Poster Eurocarb X* **1999**, PA001.
43 Randolph, J. T.; McClure, K. F.; Danishefsky, S. J., *J. Am. Chem. Soc.* **1995**, *117*, 5712–5719.
44 Randolph, J. T.; Danishefsky, S. J. *Angew. Chem. Int. Ed. Engl.* **1994**, *33*, 1470–1473.
45 a) Griffith, D. A.; Danishefsky, S. J., *J. Am. Chem. Soc.* **1990**, *112*, 5811–5819. b) Hamada, T.; Nishida, A.; Yonemitsu, O., *J. Am. Chem. Soc.* **1986**, *108*, 140–145.
46 Zheng, C.; Seeberger, P. H.; Danishefsky, S. J., *Angew. Chem. Int. Ed. Engl.* **1998**, *37*, 789–792.
47 Seeberger, P. H.; Eckhardt, M.; Gutteridge, C. E.; Danishefsky, S. J., *J. Am. Chem. Soc.* **1997**, *119*, 10064–10072.
48 Zheng, C.; Seeberger, P. H.; Danishefsky, S. J., *J. Org. Chem.* **1998**, *63*, 1126–1130.
49 Roberge, J. Y.; Beebe, X.; Danishefsky, S. J., *Science* **1995**, *269*, 202–204.
50 Kahne, D.; Walker, S.; Cheng, Y.; Van Engen, D., *J. Am. Chem. Soc*, **1989**, *111*, 6881–6882.
51 Yan, L.; Taylor, C. M.; Goodnow, Jr., R.; Kahne, D., *J. Am. Chem. Soc.* **1994**, *116*, 6953–6954.
52 Schmidt, R. R.; Kinzy, W., *Adv. Carbohydr. Chem. Biochem.* **1994**, *50*, 21–123.
53 Douglas, S. P.; Whitfield, D. M.; Krepinsky, J. J., *J. Am. Chem. Soc.* **1991**, *113*, 5095–5097.
54 Rademann, J.; Geyer, A.; Schmidt, R.R., *Angew. Chem. Int. Ed. Engl.* **1998**, *37*, 1241–1245.
55 Adinolfi, M.; Barone, G.; De Napoli, L.; Guariniello, L.; Iadonisi, A.; Pic-

cialli, G., *Tetrahedron Lett.* **1999**, *40*, 2607–2610.
56 Dreef-Tromp, C. M.; Willems, H. A. M.; Westerduin, P.; van Veelen, P.; van Boeckel, C. A. A., *Bioorg. Med. Chem. Lett.* **1997**, *7*, 1175–1180.
57 For a review, see: Garegg, P. J., *Adv. Carbohydr. Chem. Biochem.* **1997**, *52*, 179–205.
58 Verduyn, R.; van der Klein, P. A. M.; Dowes, M.; van der Marel, G. A.; van Boom, J. H., *Recl. Trav. Chim. Pays-Bas* **1993**, *112*, 464–466.
59 Ossowski, B. P.; Pilotti, A.; Garegg, P. J.; Lindberg, B., *Angew. Chem. Int. Ed. Engl.* **1983**, *22*, 793–795.
60 Manabe, S.; Ito, Y.; Ogawa, T., *Synlett.* **1998**, 628–630.
61 Rodebaugh, R.; Joshi, S.; Fraser-Reid, B.; Geysen, H. M., *J. Org. Chem.* **1997**, *62*, 5660–5661.
62 Plante, O. J.; Andrade, R. B.; Seeberger, P. H., *Org. Lett.* **1999**, *1*, 211–214.
63 Plante, O. J.; Palmacci, E. R.; Andrade, R. B.; Seeberger, P. H., *J. Am. Chem. Soc.* **2001**, *123*, 9545–9554.
64 Palmacci, E. R.; Plante, O. J.; Seeberger, P. H., *Eur. J. Org. Chem.* **2002**, 595–606.
65 Plante, O. J.; Palmacci, E. R.; Seeberger, P. H., *Science* **2001**, *291*, 1523–1527.
66 Hewitt, M. C.; Seeberger, P. H., *Organic Letters* **2001**, *3*, 3699–3702.
67 Palmacci, E. R.; Hewitt, M. C.; Seeberger, P. H., *Angew. Chem. Int. Ed.* **2001**, *40*, 4433–4437.

5
Solution and Polymer-Supported Synthesis of Carbohydrates
SHIN-ICHIRO NISHIMURA

5.1
Introduction

Proteins are key and indispensable molecules for all of the biological recognition systems within cells. However, these biomacromolecules are genetically modified by carbohydrates in order to modulate their structures and specific functions. During the biosynthesis of glycoproteins in mammalian cells, immature proteins are subjected to glycan introduction, trimming, and further modifications. In the early biosynthetic pathways at the endoplasmic reticulum and Golgi apparatus, the glycans seem to have a common role in promoting protein folding, quality control, and certain sorting events. Moreover, carbohydrate chains attached to the mature glycoproteins displayed on the cell surfaces exhibit a variety of functions as specific signals in cell-cell interactions. In the cellular immune system, for example, specific glycoforms are involved in the folding, quality control, and assembly of peptide-loaded major histocompatibility complex antigens and the T cell receptor complex. Some typical oligosaccharide chains of glycoproteins and/or glycolipids found in cancer cells significantly influence tumor metastasis. In influenza virus infection, sialic acid-containing oligosaccharides are well known ligands for the envelope proteins (hemaglutinin and sialidase) located on virus membranes. Vero toxin of *E. coli* O-157 invades by recognizing the globotriaose trisaccharide chain of the sphingoglycolipid in the initial step of human infection. These events proceed through molecular recognition processes between cell surface glycoconjugates and microbe proteins.

The emerging understanding of the critical structural and functional roles of glycoconjugates in cellular biology and the promise of therapeutics based on carbohydrate-related compounds create an urgent need for an efficient synthetic methodology for these biomolecules. Although the progress made in carbohydrate chemistry in recent years has been remarkable, chemical synthesis of oligosaccharides still involves difficult and time-consuming processes. These problems are mainly due to the requirement for multistep transformations of sugar intermediates, involving iterative "protection → glycosylation → deprotection" reactions and tedious purification procedures of the products. Enzyme-assisted synthesis of carbohydrates [1] is a promising potential alternative method to conventional chemical synthesis, because these biocatalysts have effected highly stereo- and regioselective glycosylation reactions without any protective groups under aqueous and mild conditions. Recently, a large number of glycosyltransferases have been cloned [2] and some of their recombinant forms have been produced.

In this chapter, I would like to describe an efficient and practical method for enzymatic synthesis through the use of "high-performance polymers" as glycosyl acceptor substrates. It has been demonstrated that the use of glycosyltransferases with these polymer substrates in the synthetic schemes has accelerated practical syntheses of oligosaccharides and related compounds. Next, our interest should be focused on the systematic approach for achieving "automated glycosynthesis" based on the combined chemical and enzymatic strategy.

5.2
Mimicking Glycoprotein Biosynthetic Systems

As illustrated in Fig. 5.1, carbohydrate chains of proteins seem to be introduced in stepwise fashion at the Golgi apparatus (partially at the ER) by glycosyltransferases displayed on the surface of this organelle in the presence of suitable sugar nucleotides (sugar donor substrates) and proteins (acceptor substrates). We have recently found that a variety of synthetic water-soluble glycopolymers bearing multivalent sugar branches function as excellent glycosyl acceptor substrates for glycosyltransferases [2]. These polymers could be used as appropriate models of naturally occurring and immature glycoproteins to be modified by a series of glycosyltransferases.

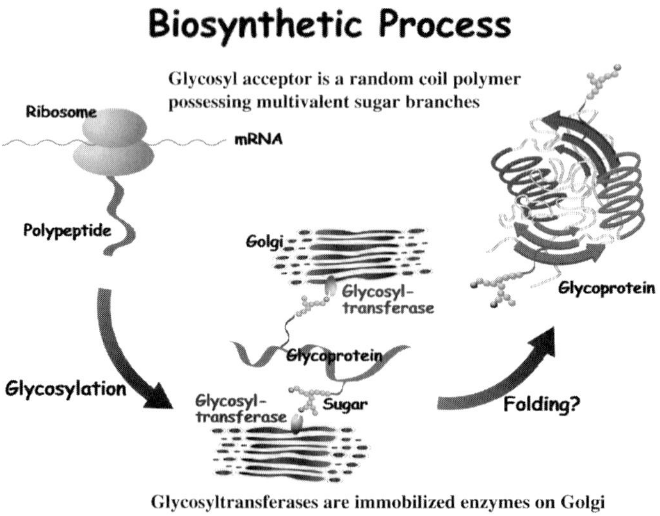

Fig. 5.1 Stepwise introduction of carbohydrate chains of proteins at the Golgi apparatus.

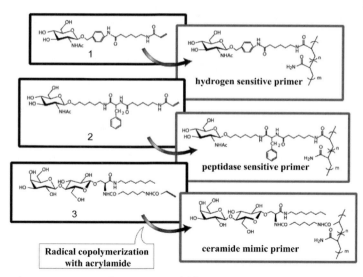

Scheme 5.1 Design of some water-soluble primers as acceptors.

Scheme 5.1 shows the chemical structures of typical glycomonomers employed for the preparation of water-soluble glycopolymers as acceptor substrates for glycosyltransferases. These three materials are composed of: (i) acrylamide for the polymerization, (ii) a simple sugar residue as primer (the starting point of the synthesis), and (iii) a flexible and functional linker. The reasons for our interest in polyacrylamide-type acceptor substrates are that all polymers should exhibit satisfactory water solubility and structural flexibility, providing good accessibility to the binding pocket of the glycosyltransferases. The merits of the use of water-soluble polymers are evident in that they can easily be isolated from the reaction mixtures and that monitoring of glycosylation may also be performed by conventional spectroscopic techniques such as NMR and fluorescence spectroscopy and MALDI-TOF MS. The linkers have specific functions for releasing or transferring oligosaccharide products from polymers. When specially designed linkers recognizable by specific enzymes are involved in the glycopolymers, the products can easily be released from the polymer supports by treatment with the particular enzymes after the sugar-elongation reactions. For example, compound **1** [3] contains a benzyl ether moiety that can be cleaved by conventional mild conditions of hydrogenolysis, compound **2** [4] has an L-phenylalanine residue in the middle of the linker, suitable for cleaving by treatment with α-chymotrypsin to supply oligosaccharide products as ω-aminoalkykyl glycosides, and the ceramide mimetic linker in compound **3** [5, 6] can be recognized by ceramide glycanases, oligosaccharides synthesized on the water-soluble polymers being transferred to natural ceramide or sphingosine analogues by the enzymatic transglycosylation reaction.

It has also been demonstrated that polymers bearing multivalent sugars showed drastically enhanced affinity for enzymes (glycosyltransferases) due to "cluster effects" [7]. As shown in Fig. 5.2, the affinity constant of a polymer bearing N-acetyl-

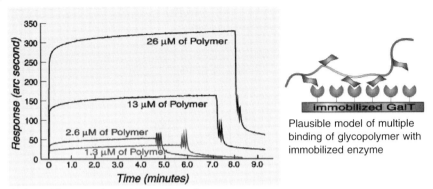

Plausible model of multiple binding of glycopolymer with immobilized enzyme

Ka (5.0 μM) was estimated by surface plasmon resonance analysis

Fig. 5.2 Enhanced affinity for enzymes due to cluster effects.

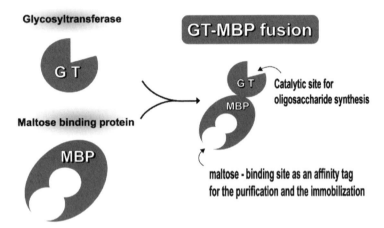

Fig. 5.3 Use of an "engineered catalyst".

D-glucosamine, prepared from **2** with recombinant galactosyltransferase, was estimated by surface plasmon resonance as 5 μM. Since the affinity constants of galactosyltransferase with popular N-acetyl-D-glucosamine derivatives of low molecular weight are known to be in the 1–100 mM range, this result suggests that stepwise glycosylation reactions on water-soluble polymer supports may have the potential to become a general concept for the efficient carbohydrate synthesis by enzymes.

On the other hand, we have succeeded in establishing an efficient recycling system for engineered glycosynthesis based on glycosyltransferases. Thus, some glycosyltransferases obtained as the fusion proteins with maltose-binding protein [8] (Figs. 5.3 and 5.4) were specifically immobilized on tailored polymer gels prepared from a polymerizable maltooligosaccharide derivative. As anticipated, the maltose-binding protein domain of the fusion enzymes proved to act as an affini-

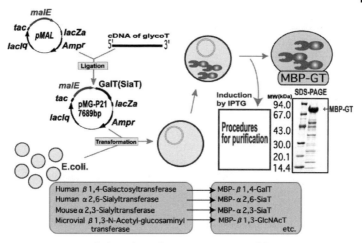

Fig. 5.4 Engineered glycosyltransferases: preparation of fusion protein.

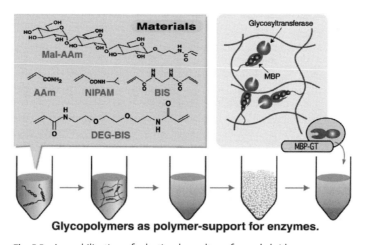

Fig. 5.5 Immobilization of a lectin glycosyltransferase hybrid.

ty tag for the immobilization of these engineered biocatalysts without any chemical treatment for crosslinking. Fig. 5.5 shows monomers and crosslinkers employed for the preparation of polymer gels to immobilize fusion proteins. Immobilization of these enzymes on sepharose-type gel was also carried out by conventional chemical coupling [7]. The immobilized glycosyltransferases showed both satisfactory activity and recyclability for further synthetic study.

These results prompted us to examine high-speed syntheses of glycoconjugates by combined use of the immobilized fusion enzymes with water-soluble polymers as acceptor substrates. For example, a sphingoglycolipid was synthesized as shown in Scheme 5.2. Here, the oligosaccharide was synthesized on the water-sol-

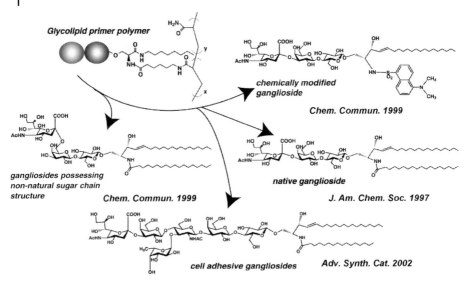

Fig. 5.6 Various sphingolipid syntheses of water-soluble polymer supports.

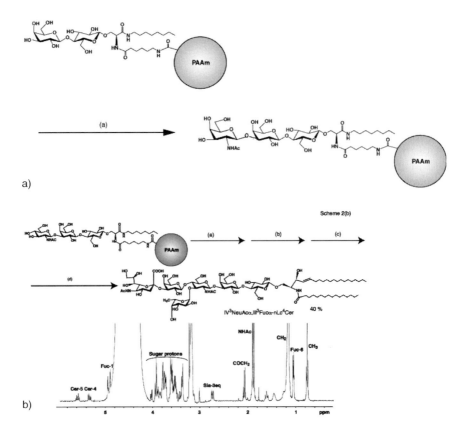

5.2 Mimicking Glycoprotein Biosynthetic Systems

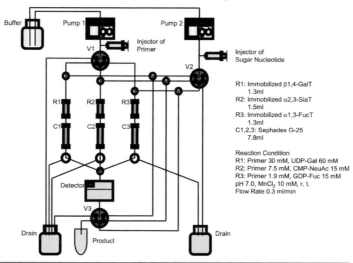

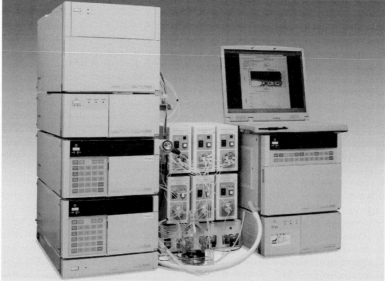

Fig. 5.7 Automated glycosynthesizer.

Scheme 5.2 a) Synthetic scheme for a polymer bearing a GlcNAcβ(1 → 3)Galβ(1 → 4) structure by use of MBP-β1,3-GlcNAcT in the presence of UDP-GlcNAc. b) Synthesis of a sphingoglycolipid, IV^{3}NeuAcα, III^3 Fucα-nLc^{4}Cer. i) Compound **9** (7 mg), β(1 → 4)GalT (1.0 unit), UDPGal disodium salt (1.2 equiv., 2.60 mg), α-lactalbumin (0.26 mg), 50 mM HEPES buffer (pH 6.0), 37 °C, 24 h; ii) α-SiaT (0.03 unit) CMP-NANA disodium salt (1.4 equiv., 3.27 mg), 50 mM sodium cacodylate buffer (pH 7.4), 10 mM MnCl$_2$, 1 mM NaN$_3$, CIAP (20 unit), Triton CF-54 (0.5% v/v), 37 °C, 72 h; iii) α-1,3-FucT (0.08 unit), GDP-Fuc disodium salt (1.4 equiv., 3.14 mg), 50 mM sodium cacodylate buffer (pH 6.5), 10 mM MnCl$_2$, 1 mM NaN$_3$, 37 °C, 72 h; iv) ceramide glycanase (0.03 unit), ceramide (5.0 equiv., 9.52 mg), 50 mM sodium citrate buffer (pH 6.0), Triton CF-54 (0.5% v/v), 37 °C, 17 h.

uble polymer in high yield by use of four recombinant glycosyltransferases and the product was transferred from polymer support to ceramide by treatment with ceramide glycanase to afford the target sphingoglycolipid in 40% yield. In addition, it was also demonstrated that GM3 and some non-natural sphingoglycolipids can be prepared by the same procedure, as shown in Fig. 5.6. Similar strategies might be followed for the synthesis of glycopeptides, and results will be reported in the near future.

Our research has shown that the combined use of water-soluble polymers and immobilized glycosyltransferases, an idea arising from glycoconjugate biosynthetic pathways in nature, can be used as a practical and general strategy for glycosynthetic system. In July 2001, at Hokkaido University, our research group demonstrated that this biomimetic concept could be applied for making a prototype computer-controlled "automated synthesizer", as shown in Fig. 5.7. Since glycobiologists are seeking to establish a genetic library of the human glycosyltransferases, it is hoped that our automated synthesizer will become a practically standard machine for producing "sweet compounds" in post-genome bioscience and biotechnologies.

5.3
References

1 KOELLER, K.M., WONG, C.-H., "Emerging themes in medicinal glycoscience", *Nat. Biotechnol.* **2001**, *18*, 835–841.

2 NISHIMURA, S.-I., "Combinatorial syntheses of sugar derivatives", *Curr. Opin. Chem. Biol.* **2001**, *5*, 325–335.

3 NISHIMURA, S.-I., MATSUOKA, K., LEE, Y.C., "Chemoenzymatic Oligosaccharide Synthesis on a Soluble Polymeric Carrier", *Tetrahedron Lett.* **1994**, *35*, 5657–5660.

4 YAMADA, K., NISHIMURA, S.-I., "An Efficient Synthesis of Sialoglycoconjugates on a Peptidase-Sensitive Polymer Support", *Tetrahedron Lett.* **1995**, *36*, 9493–9496.

5 NISHIMURA, S.-I., YAMADA, K., "Transfer of Ganglioside GM3 Oligosaccharide from a Water-Soluble Polymer to Ceramide by Ceramide Glycanase. A Novel Approach for the Chemical-Enzymatic Synthesis of Glycosphingolipids", *J. Am. Chem. Soc.* **1997**, *119*, 10555–10556.

6 YAMADA, K., FUJITA, E., NISHIMURA, S.-I., "High Performance Polymer Supports for Enzyme-assisted Synthesis of Glycoconjugates", *Carbohydr. Res.* **1998**, *305*, 443–461.

7 NISHIGUCHI, S., YAMADA, K., FUJI, Y., SHIBATANI, S., TODA, A., NISHIMURA, S.-I., "Highly efficient oligosaccharide synthesis on water-soluble polymeric primers by recombinant glycosyltransferases immobilized on solid supports", *Chem. Commun.* **2001**, 1944–1945.

8 TODA, A., YAMADA, K., NISHIMURA, S.-I., "An Engineered Biocatalyst for the Synthesis of Glycoconjugates: Utilization of β_1 3-N-Acetyl-D-glucosaminyltransferase from *Streptococcus agalactiae* Type Ia Expressed in *Escherichia coli* as a Fusion with Maltose-Binding Protein", *Adv. Synth. Catal.* **2002**, *344*, 61–69.

6
Enzymatic Synthesis of Oligosaccharides
Jianbo Zhang, Jun Shao, Prezemk Kowal, and Peng George Wang

6.1
Introduction

Glycobiology has become one of the fastest growing branches of biology [1, 2]. The tremendous interest in this area stems from the fact that oligosaccharides and oligosaccharide-decorated molecules pervade all biological systems. Oligosaccharide components of cells play essential roles in physiological and pathological processes such as molecular recognition, signal transduction, differentiation, and developmental events [3–16]. Further growth in research on the biological functions of the varied glycan structures will undoubtedly be closely tied to the availability of bioactive carbohydrates [17–19]. Most glycoconjugates (oligosaccharides, glycoproteins, and glycolipids) and their derivatives are difficult to obtain in large quantities, due partially to considerable limitations in oligosaccharide synthesis. Like proteins and nucleic acids, glycans consist of limited numbers of building blocks, the monosaccharides. However, the saccharide oligomers come in far greater structural diversity, because of varying glycosidic linkage patterns, stereochemistry, and branching. Unlike DNA or peptide syntheses, which are commonly performed on the solid phase with commercial instruments, comparable methodologies for oligosaccharides are still in their infancy [20, 21]. Chemical oligosaccharide synthetic procedures are plagued with often tedious protection and deprotection steps, making one-pot enzymatic systems with high regio- and stereoselectivities attractive alternatives [22–28].

Among the numerous enzymes associated with carbohydrate processing in cells, those used in glycoconjugate syntheses belong to three categories: glycosidases, glycosynthases, and glycosyltransferases (Tab. 6.1).

Glycosidases are enzymes that cleave oligosaccharides and polysaccharides *in vivo*. Under *in vitro* conditions they can form glycosidic linkages in which a carbohydrate hydroxyl moiety acts as a more efficient nucleophile than water. They have been of tremendous utility in the enzymatic synthesis of oligosaccharides, thanks to their availability, stability, organic solvent compatibility, and low cost [29–31]. Nevertheless, traditional glycosidase-catalyzed transglycosylations still suffer from low yields and poor regioselectivities. On the basis of structure/function relationship information and the mechanisms of glycosidase-catalyzed reactions,

Tab. 6.1 Enzymatic formation of glycosidic bonds.

Donor + Acceptor $\xrightarrow{\text{Enzymes}}$ Product

Enzyme	Glycosyl donor	Advantage	Disadvantage
Glycosidase	Sugar-NP[a]	– Easy to perform – Low cost	– Low yield – Low regioselectivity
Glycosynthase	Sugar-F[b]	– Higher yield	– Hard to obtain enzyme – Difficult to predict results
Leloir glycosyl- transferase	Sugar nucleotide	– High yield – High regio- and stereose- lectivity – Essential for important sequences	– High cost
Non-Leloir glyco- syltransferase	Sugar phosphate or glycoside	– High yield – High regio- and stereose- lectivity	– Not useful for important sugar sequences

a) Sugar-NP: Nitrophenyl glycoside
b) Sugar-F: Glycosyl fluoride

glycosynthases, a class of glycosidase mutants, have been developed to enhance enzymatic activities towards the synthesis of oligosaccharides, through the mutation of a single catalytic carboxylate nucleophile to a neutral amino acid residue (Ala or Ser). The resulting enzymes have no hydrolytic activity, but show increased activity towards the synthesis of oligosaccharides with use of glycosyl fluorides as activated donors [32–35].

Glycosyltransferases are enzymes that transfer a sugar moiety to a defined acceptor to construct a specific glycosidic linkage. The "one enzyme-one linkage" concept makes glycosyltransferases useful and important in the construction of glycosidic linkages in carbohydrates [36–38]. They can be further divided into two groups: transferases of the Leloir pathway and those of non-Leloir pathways. The Leloir pathway enzymes need sugar nucleotides as glycosylation donors, while non-Leloir glycosyltransferases typically utilize glycosyl phosphate or glycosides. The Leloir transferases are responsible for the synthesis of most glycoconjugates in cells, especially in mammalian systems, and are the focus of this chapter [39–41].

Leloir glycosyltransferase-catalyzed syntheses start with the conversion of monosaccharides into activated sugar nucleotides (donors), which then donate the sugars to various acceptors through the action of specific transferases (Scheme 6.1).

Glycosyltransferases form highly regiospecific and stereoselective glycosidic linkages. Though they are generally substrate-specific, minor modifications on donor and acceptor structures can be tolerated [42, 43]. Despite these merits, the preparative application of glycosyltransferases is limited by their inadequate availability. The amount of glycosyltransferases that can be isolated from natural resources is often limited by the low concentrations of these enzymes in most tissues. Furthermore, their purification procedures are quite complicated by their relative

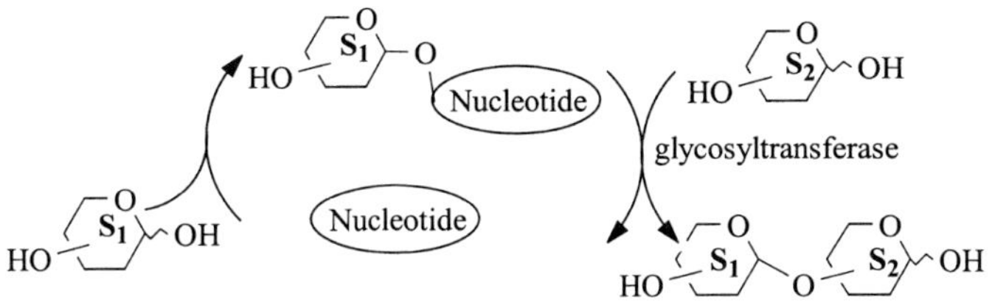

Scheme 6.1 Enzymatic glycosylation and regeneration of sugar nucleotide.

instability, for glycosyltransferases are often membrane-bound or membrane-associated. For these reasons, many efforts have been geared toward genetic engineering and recombinant sources of glycosyltransferases [44–48]. Historically, reflecting the major focus of glycobiology researchers, most glycosyltransferases investigated have been from mammalian sources. They can be expressed at high levels in mammalian systems, such as Chinese hamster ovary (CHO) cells. However, this expression procedure is too tedious and expensive to be applied in practical transferase production. Efforts toward the expression of mammalian enzymes in insect, plant, yeast, and bacterial cells have been made, but high-level expression remains difficult. Fortunately, glycosyltransferases from bacterial sources can be easily cloned and expressed in large quantities in *E. coli* [49]. They also have a broader range of substrates than mammalian glycosyltransferases, and furthermore, some bacterial transferases have been found to produce mammalian-like oligosaccharide structures, which makes these enzymes quite promising for syntheses of biologically important oligosaccharides [50–52]. The recent phenomenal expansion of genomic sequencing has allowed for many glycosyltransferases to be characterized and expressed in recombinant form.

The second obstacle to enzymatic carbohydrate synthesis is the sugar nucleotides. Although all the common sugar nucleotides are now commercially available, these materials are prohibitively expensive. Since a sugar nucleotide only serves as an intermediate in the enzymatic glycosylation, the most efficient synthetic approach is to regenerate it in situ. In addition, the low concentration of sugar nucleotide regenerated can avoid its inhibitory effect on the glycosyltransferase and increase the synthetic efficiency. The idea of in situ regeneration of sugar nucleotides was first demonstrated in 1982 by Wong and Whitesides, with their work on UDP-Gal regeneration (Scheme 6.2) [53]. Since then, this revolutionary concept has been adopted in other regeneration systems and further developed in glycoconjugate syntheses [54–65]. Our discussion here is centered on the production and regeneration systems for common sugar nucleotides and their applications in glycosyltransferase-catalyzed reactions.

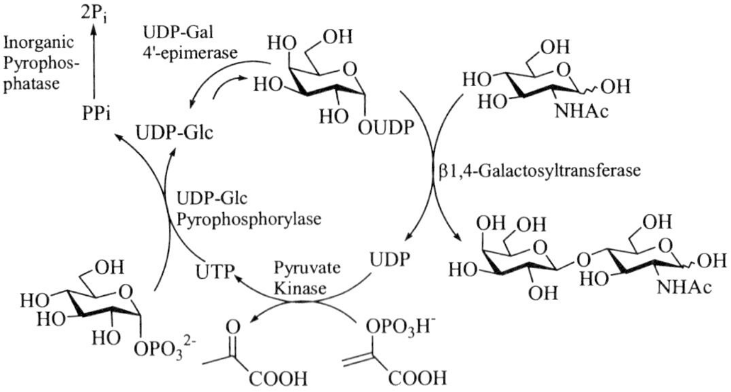

Scheme 6.2 Whiteside and Wong's first sugar nucleotide regeneration cycle.

6.2
Sugar Nucleotide Biosynthetic Pathways

6.2.1
Basic Principle

The structures of oligosaccharides can be quite complex, but the number of important oligosaccharide sequences is limited. There are only nine common sugar monomers: glucose (Glc), galactose (Gal), glucuronic acid (GlcA), xylose (Xyl), N-acetylglucosamine (GlcNAc), N-acetylgalactosamine (GalNAc), mannose (Man), fucose (Fuc), and sialic acid (Neu5Ac). In order to construct the sugar chains, the monosaccharides have to be activated by attachment to nucleoside phosphates. L. F. Leloir, 1950 Nobel laureate, showed that a nucleotide triphosphate such as UTP reacts with a glycosyl-1-phosphate to form a high-energy donor capable of participating in glycoconjugate synthesis. Half a century later, the biosynthetic pathways of nine common sugar nucleotides (UDP-Glc, UDP-Gal, UDP-GlcA, UDP-Xyl, UDP-GlcNAc, UDP-GalNAc, GDP-Man, GDP-Fuc, and CMP-NeuAc) are now well established [66]. They are also known as primary sugar nucleotides, for they are generated in vivo from sugar-1-phosphates. Other sugar nucleotides are known as secondary sugar nucleotides because they are synthesized by modification of a primary sugar nucleotide [67]. The general biosynthetic pathway for common sugar nucleotides provides the basic guidelines on how to construct them in vitro (Fig. 6.1).

Regardless of the sugar and its origin, all monosaccharides must be activated by a kinase (reaction 1 below) or generated from a previously synthesized sugar nucleotide (reactions 2 and 3 below):

1. Sugar + ATP → Sugar-P + NTP → Sugar-NDP + PPi
2. Sugar(A)-NDP ↔ Sugar(B)-NDP
3. Sugar(A)-NDP + Sugar(B)-1-P ↔ Sugar(B)NDP + Sugar(A)-1-P

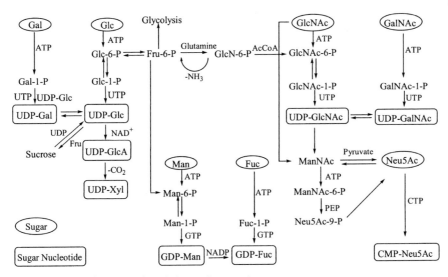

Fig. 6.1 Integrated sugar nucleotide biosynthetic pathways.

For Glc, GlcNAc, and Man, routes to the corresponding sugar nucleotides include (1) phosphorylation to sugar-6-P by a kinase, (2) conversion of sugar-6-P into sugar-1-P by a mutase, and (3) condensation of sugar-1-P with an NTP by a pyrophosphorylase.

No C6 sugar kinases have been reported for Gal and GalNAc and Fuc. In fucose there is no 6-OH group, while in Gal and GalNAc the 6-OH is quite sterically hindered for enzymatic phosphorylation, due to the axial 4-OH group. These three monosaccharides are therefore phosphorylated on the anomeric position, converting them directly into sugar-1 phosphates.

There are several enzymes that interconvert different sugar nucleotides, including UDP-Gal 4′-epimerase (GalE), and UDP-GalNAc 4′-epimerase (GalNAcE). These enzymes are very useful in sugar nucleotide regeneration systems since they can be employed to convert one sugar nucleotide into another or provide both sugar nucleotides at the same time. It should be noted that UDP-GlcA universally comes from the oxidation of UDP-Glc by UDP-Glc 6′-dehydrogenase (UDPGDH, UGD) and UDP-Xyl comes from the decarboxylation of UDP-GlcA by UDP-GlcA decarboxylase.

The biosynthesis of CMP-NeuAc is different from those of the rest of the common sugar nucleotides. It can be synthesized directly from Neu5Ac and CTP without a sugar-1-phosphate intermediate.

6.2.2
Regeneration Systems for nine Common Sugar Nucleotides

To prepare carbohydrate structures efficiently, in situ regeneration systems for common sugar nucleotides have been developed by mimicking the natural biosynthetic pathways. In the experimental design, the convenience and efficiency of biosynthetic pathways are carefully evaluated and exemplified in practical syntheses [68, 69].

6.2.2.1 Regeneration Systems for UDP-Gal, UDP-Glc, UDP-GlcA and UDP-Xyl

UDP-Glc, as the central sugar nucleotide in cells, can be prepared from UTP and glucose-1-phosphate (Glc-1-P) in the presence of UDP-glucose pyrophosphorylase (Route a in Fig. 6.2) or from sucrose and UDP by sucrose synthase (Route b in Fig. 6.2). UDP-Gal can be prepared from UDP-Glc by C4 epimerization, from galactose-1-phosphate (Gal-1-P) and UTP (Route c in Fig. 6.2) or from UDP-Glc and Gal-1-P by use of Gal-1-P uridyltransferase (GalPUT) (Route d in Fig. 6.2). UDP-GlcA can readily be prepared from UDP-Glc and NAD by UGD (Route e in Fig. 6.2). UDP-Xyl can be generated from UDP-GlcA and NAD by use of UDP-GlcA carboxylase (Route f in Fig. 6.2).

UDP-Glc UDP-Glc can be regenerated from Glc-1-P through the action of UDP-Glc pyrophosphorylase (EC 2.7.7.9) [70, 71]. As shown in Fig. 6.2, there is another efficient way to generate UDP-Glc, with the help of sucrose synthase (SusA) (Route b) [59, 72–74]. Use of SusA allows UDP-Glc to be regenerated and applied to glucosylation as shown in Scheme 6.3. This system involves only two enzymes: sucrose synthase and a glucosyltransferase.

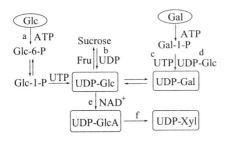

Fig. 6.2 Biosynthetic pathways of UDP-Glc, UDP-Gal, and UDP-GlcA.

Scheme 6.3 Regeneration cycle of UDP-Glc by SusA.

UDP-Gal The first UDP-Gal regeneration system, in which UDP-Gal is epimerized from UDP-Glc by GalE (EC 5.1.3.2), was constructed by Whitesides and Wong (Scheme 6.2) [53]. UDP-Gal can also be simply regenerated by addition of GalE into the sucrose-catalyzed UDP-Glc regeneration cycle (Scheme 6.4A) [65, 75]. A key simplification of this approach by Wang's group followed the discovery of sucrose synthase from cyanobacterium *Anabaena* sp. PCC 7119 [76–78]. This enzyme is more readily expressed in *E. coli* than the plant homologue [78, 79]

Another approach starts from direct phosphorylation at the 1-position of galactose, to give Gal-1-P, which can be converted to UDP-Gal by means of a uridyltransferase-catalyzed exchange reaction with UDP-Glc or a pyrophosphorylase-catalyzed reaction with UTP [60, 80, 81]. Besides Whitesides, Wong, and their collaborators' pioneering work, Wang's group have applied recombinant enzymes to regenerate UDP-Gal. There is no known bacterial enzyme able to catalyze the condensation between UTP and Gal-1-P, and so Route d in Fig. 6.2 was chosen as the viable option to obtain UDP-Gal. As shown in Scheme 6.4B, the *E. coli* biosynthetic pathway for UDP-Gal regeneration involves five enzymes [82, 83]. Galactose kinase (galactokinase, GalK, EC 2.7.7.6) phosphorylates galactose to Gal-1-P, consuming the first molecule of phosphoenol pyruvate (PEP). A combination of GalPUT (EC 2.7.1.10) and glucose-1-phosphate uridylyltransferase (GalU, EC 2.7.1.9) then converts the Gal-1-P and UTP to UDP-Gal. In these two steps, UDP-Glc and Glc-1-P function as transient intermediates. To complete the regeneration cycle,

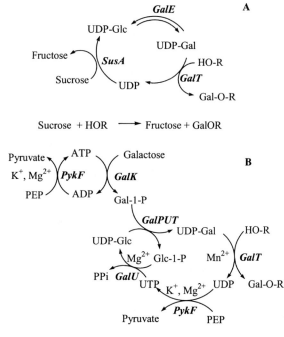

Scheme 6.4 Regeneration cycles of UDP-Gal.

PykF (pyruvate kinase, EC 2.7.1.40) recycles UDP, the byproduct of the galactosyltransferase-catalyzed reaction, back to UTP with the consumption of the second molecule of PEP.

UDP-GlcA UDP-GlcA can also be produced easily, by adding UGD (EC 1.1.1.22) in the UDP-Glc regeneration cycle [84, 85]. Gygax et al. have reported a UDP-GlcA regeneration for β-glucuronide synthesis from Glc-1-P [86]. Wong et al. also applied a recombinant UGD to regenerate UDP-GlcA for the synthesis of hyaluronic acid (Scheme 6.5) [87, 88].

UDP-Xyl The UDP-Xyl regeneration cycle can be established by extending the UDP-GlcA regeneration. The UDP-GlcA carboxylase (EC 4.1.1.35) will convert UDP-GlcA into UDP-Xyl with NAD^+ (Scheme 6.6) [89–91]. To our knowledge, no regeneration system for UDP-Xyl has been reported [92].

6.2.2.2 Regeneration Systems for UDP-GlcNAc and UDP-GalNAc

Biosynthesis of UDP-GlcNAc begins with the formation of glucosamine-6-phosphate (GlcN-6-P) from Fru-6-P by transamination with glutamine as the $-NH_2$ do-

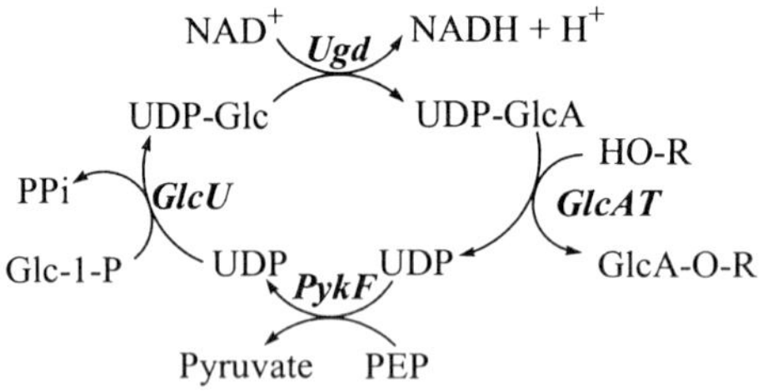

Glc-1-P + PEP + HOR + NAD^+ → GlcAOR + Pyruvate + PPi + NADH + H^+

Scheme 6.5 UDP-GlcA regeneration cycle.

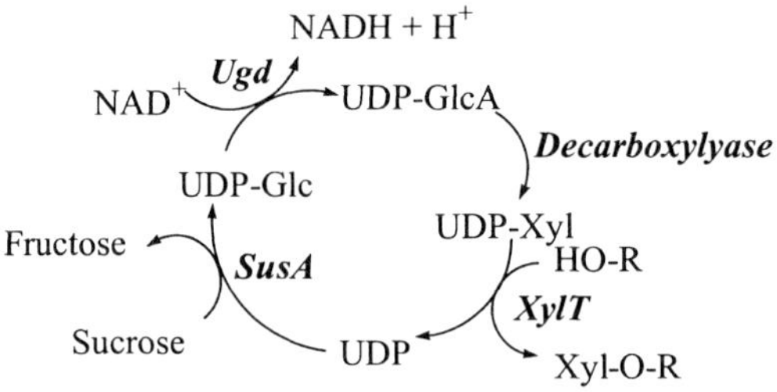

Sucrose + HOR + NAD^+ → Fructose + XylOR + NADH + H^+ + CO_2

Scheme 6.6 UDP-Xyl regeneration cycle.

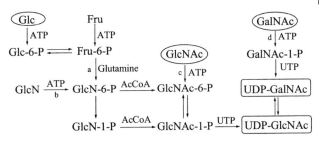

Fig. 6.3 Biosynthetic pathways of UDP-GlcNAc and UDP-GalNAc.

nor (Route a in Fig. 6.3). GlcN-6-P is then N-acetylated, with acetyl-CoA serving as the acetate donor. N-Acetylglucosamine-6-phosphate (GlcNAc-6-P) is then isomerized to N-acetylglucosamine-1-phosphate (GlcNAc-1-P) via a 1,6-bisphosphate intermediate. As in the other activation reactions, GlcNAc-1-P reacts with UTP to form UDP-GlcNAc and pyrophosphate. Alternatively, GlcN-6-P can be generated from glucosamine (GlcN) directly (Route b in Fig. 6.3). GlcNAc-6-P can be made by direct phosphorylation of GlcNAc (Route c in Fig. 6.3). UDP-GalNAc can be obtained by two routes. One is the direct reaction between GalNAc-1-P and UTP (Route d in Fig. 6.3). GalNAc-1-P is formed by a specific kinase that is distinct from GalK. This route is present only in eukaryotic organisms as a means of recycling GalNAc generated by the degradation of glycosaminoglycans and glycoproteins. UDP-GalNAc can also be formed by epimerization of UDP-GlcNAc using a GlcNAcE or GalE in eukaryotes. However, bacterial GalE cannot interconvert UDP-GlcNAc and UDP-GalNAc [93–95].

UDP-GlcNAc Regeneration Wong et al. have applied UDP-GlcNAc pyrophosphorylase (GlmU, EC 2.7.7.32) in enzymatic regeneration of UDP-GlcNAc from GlcNAc-1-P 87. Wang's group has extended this research and applied recombinant enzymes in UDP-GlcNAc regeneration.

Regeneration of UDP-GlcNAc by Following the Biosynthetic Pathway in Prokaryotes
Bacteria utilize external GlcN to produce UDP-GlcNAc (Route b in Fig. 6.3) [96]. The glucokinase-catalyzed (GlcK, EC 2.7.1.2) phosphorylation of GlcN is essential for this pathway. The GlcN-6-P product is converted into GlcN-1-P by phosphoglucosamine mutase (GlmM, EC 5.4.2.10). UDP-GlcNAc pyrophosphorylase (GlmU, EC 2.7.7.32), a bifunctional enzyme, then catalyzes two consequent steps: acetylation and coupling (Scheme 6.7). Since GlcN-6-P is relatively cheap, the whole cycle of UDP-GlcNAc regeneration starts from GlcN-6-P and consists of two separate regeneration cycles, in order to solve the cost problems associated with AcCoA and UTP. In this route, GlcN was acetylated with AcCoA and phosphorylated with PEP by phosphotransacetylase (PTA, EC 2.3.1.8) and PykF [97]. The by-product pyrophosphate (PPi) is hydrolyzed by inorganic pyrophosphatase (PPA) [98].

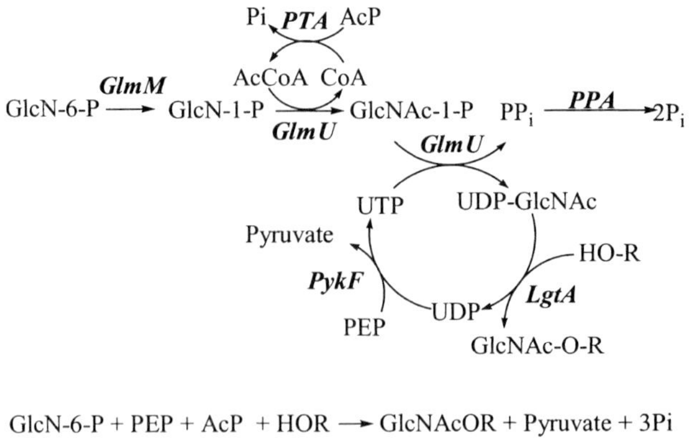

Pi PTA AcP
GlmM
AcCoA CoA
GlcN-6-P
GlcN-1-P
GlcNAc-1-P
PPi
PPA
2Pi
GlmU
GlmU
UTP
UDP-GlcNAc
Pyruvate
HO-R
PykF
LgtA
UDP
PEP
GlcNAc-O-R
GlcN-6-P + PEP + AcP + HOR → GlcNAcOR + Pyruvate + 3Pi

Scheme 6.7 Regeneration cycle of UDP-GlcNAc from GlcN-6-P.

Regeneration of UDP-GlcNAc by Following the Biosynthetic Pathway in Eukaryotes Alternatively, GlcNAc can be directly phosphorylated to form GlcNAc-6-P by use of an N-acetylglucosamine kinase (GlcNAcK, EC 2.7.1.59) [99, 100]. Phosphoacetylglucosamine mutase (Agm1, EC 5.4.2.3) then converts this into GlcNAc-1-P, which is further converted into the corresponding sugar nucleotide with GlmU 101. Therefore, we can obtain a simplified regeneration pathway that begins with GlcNAc as the starting material (Scheme 6.8) [102].

UDP-GalNAc Regeneration UDP-GalNAc can be generated by addition of GalNAcE (EC 5.1.3.2) to the regeneration cycle for UDP-GlcNAc. GalNAcE catalyzes the conversion of UDP-GlcNAc into UDP-GalNAc, with the equilibrium shifted towards the glucose derivative [94, 95, 103]. Our attention has been focused on finding the gene encoding the GlcNAcE in microbial organisms, although the functional assignment of putative GalNAcE genes is often complicated by their close relationship to other enzymes of this group. BLAST protein sequence similarity searches and amino acid sequence alignment were conducted using the known

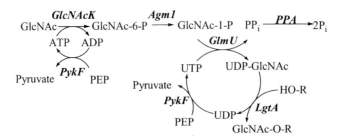

GlcNAcK
Agm1
GlcNAc
GlcNAc-6-P
GlcNAc-1-P
PPi
PPA
2Pi
ATP ADP
GlmU
UTP
UDP-GlcNAc
PykF
Pyruvate
PEP
Pyruvate
HO-R
PykF
LgtA
UDP
PEP
GlcNAc-O-R
GlcNAc + 2PEP + HOR → GlcNAcOR + 2Pyruvate + 2Pi

Scheme 6.8 Regeneration cycle of UDP-GlcNAc from GlcNAc.

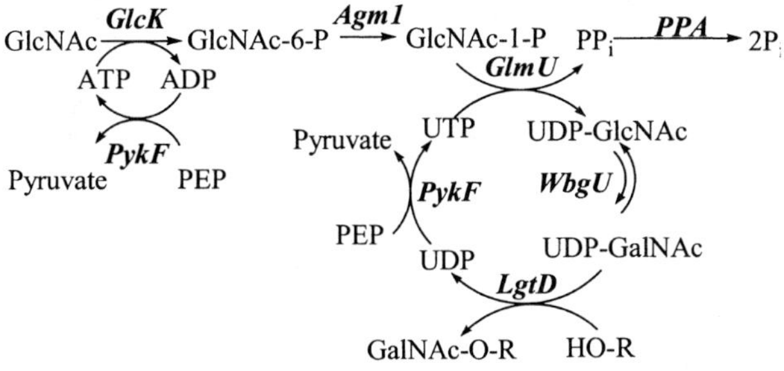

6.2 Sugar Nucleotide Biosynthetic Pathways

$$\text{GlcNAc} \xrightarrow[\text{ATP} \quad \text{ADP}]{GlcK} \text{GlcNAc-6-P} \xrightarrow{Agm1} \text{GlcNAc-1-P} \xrightarrow[\text{PP}_i]{GlmU, \text{UTP}} \text{UDP-GlcNAc} \xrightarrow{PPA} 2\text{P}_i$$

$$\text{Pyruvate} \xleftarrow[PykF]{} \text{PEP} \quad\quad \text{Pyruvate} \xleftarrow[PykF]{} \text{PEP} \quad \text{UDP} \xleftarrow[LgtD]{WbgU} \text{UDP-GalNAc}$$

$$\text{GalNAc-O-R} \quad \text{HO-R}$$

$$\text{GalNAc} + 2\text{PEP} + \text{HOR} \longrightarrow \text{GalNAcOR} + 2\text{Pyruvate} + 2\text{Pi}$$

Scheme 6.9 Regeneration cycle of UDP-GalNAc from GlcNAc.

amino acid sequence of the UDP-GalNAc 4-epimerase, WbpP, from *Pseudomonas aeruginosa* serogroup O6 [104]. A search of protein databanks identified two proteins from food-borne human pathogens *Shigella sonnei* serotype O17 (ORF2S) and *Plesiomonas shigelloides* (ORF2P, *wbgU*) [105]. Because these two proteins are virtually identical, the *wbgU* gene from *Plesiomonas shigelloides* was cloned and expressed in *E. coli*. Biochemical characterization confirmed that it is a GalNAcE 106.

The establishment of the UDP-GalNAc regeneration cycle allows it to be combined with any N-acetylgalactosaminyltransferase (GalNAcT) to synthesize oligosaccharides. In the first model study, Wang's group used efficient UDP-GalNAc regeneration systems in the synthesis of globotetraose (Gb₄) and a series of derivatives (Scheme 6.9) [107–109].

6.2.2.3 Regeneration Systems for GDP-Man and GDP-Fuc

GDP-Man can be produced in one of two ways. The first is by direct phosphorylation of mannose by a specific mannokinase in some invertebrates (Route a in Fig. 6.4). The second, and so far the best known, way is by conversion of Fru-6-P into Man-6-P through the action of phosphomannose isomerase (PMI) (Route b in Fig. 6.4). Two pathways to generate GDP-fucose from GDP-mannose are known. The first involves the epimerization and oxido-reduction of GDP-αD-man-

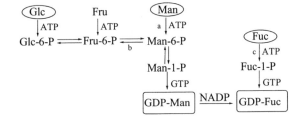

Fig. 6.4 Biosynthetic pathways of GDP-Man and GDP-Fuc.

nose to GDP-β-L-fucose. The second pathway is an alternative and salvage route in certain mammalian organs such as liver and kidney (Route c in Fig. 6.4). It involves a specific fucose kinase that catalyzes the phosphorylation of L-fucose, and a GDP-L-fucose pyrophosphorylase that catalyzes the reaction between β-L-fucose-1-P and GTP to generate GDP-L-fucose [110].

GDP-Man Regeneration Wong et al. have reported GDP-Man regeneration from mannose-1-phosphate (Man-1-P), and this was successfully applied in an enzymatic synthesis of mannoside [61]. Since Fru-6-P is inexpensive, Wang's group developed a new regeneration pathway directly from Fru-6-P with all recombinant enzymes. This cycle is constructed with three key enzymes: PMI (EC 5.3.1.8) for the conversion of Fru-6-P to Man-6-P [111–114], phosphomannomutase (PMM, EC 5.4.2.8) for the conversion of Man-6-P to Man-1-P [115], and GDP-mannose pyrophosphorylase (GMP, EC 2.7.7.13) for condensation of Man-1-P with GTP to form GDP-Man [116–121]. The system was then coupled with a truncated mannosyltransferase (ManT-catalytic domain) to synthesize the mannosyloligosaccharides (Scheme 6.10).

GDP-Fucose Regeneration Wong et al. have reported the regeneration of GDP-Fuc from Fuc-1-P or through the GDP-Man pathway from Man-1-P [60]. Because the DNA sequence of fucose kinase in Route c in Fig. 6.4 has not yet been reported, Wehmeier et al. have accomplished a practical fucosyllactose synthesis through regeneration of GDP-Fuc from GDP-Man with recombinant enzymes [122, 123]. Recently, Wang's group have extended the GDP-fucose regeneration system with recombinant enzymes, based on the same biosynthetic pathway of GDP-Man from Fru-6-P [124]. As shown in Scheme 6.11, starting from Fru-6-P, fucosylation with GDP-Fuc regeneration requires seven enzymes: namely, PMI, PMM, GMP, GDP-mannose 4,6-dehydratase (GMD, EC 4.2.1.37), GDP-fucose synthase (GFS), PykF, and the corresponding fucosyltransferase (FucT). In this cycle, PMI, PMM, and GMP catalyze the conversion of fructose-6-phosphate into GDP-Man as discussed in the GDP-Man regeneration section. Then, two enzymes – GMD and GFS – cat-

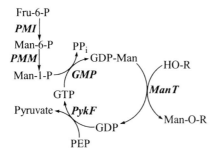

Fru-6-P + PEP + HOR → ManOR + Pyruvate + PPi

Scheme 6.10 GDP-Man regeneration cycle.

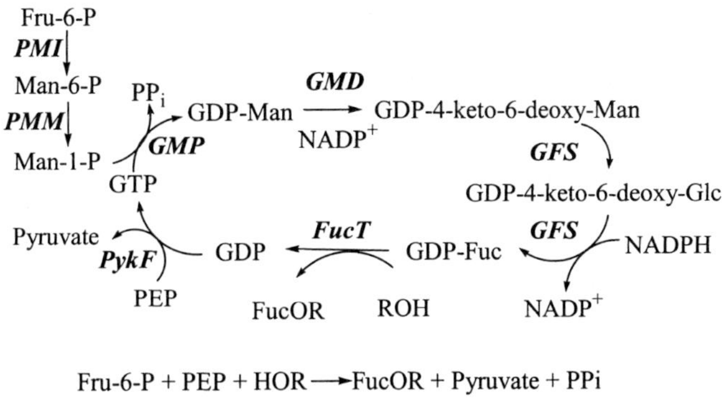

Fru-6-P
PMI ↓
Man-6-P   PPi   GMD
PMM ↓ ↗ GDP-Man ⟶ GDP-4-keto-6-deoxy-Man
Man-1-P GMP   NADP⁺   GFS ↘
         GTP   GDP-4-keto-6-deoxy-Glc
Pyruvate ↗   FucT   GFS ↙
      PykF ⤻ GDP ⇌ GDP-Fuc ⤻ NADPH
         PEP   FucOR ROH   NADP⁺

Fru-6-P + PEP + HOR ⟶ FucOR + Pyruvate + PPi

Scheme 6.11 GDP-Fuc regeneration cycle.

alyze three transformations in the conversion of GDP-Man into GDP-Fuc, involving 4,6-dehydrogenation, 3,5-epimerization, and 4-reduction [125–129].

6.2.2.4 CMP-Neu5Ac Regeneration

CMP-Neu5Ac, synthesized from CTP and Neu5Ac, is a common substrate for the synthesis of sialylated oligosaccharides [130–132]. Wong et al. reported the CMP-Neu5Ac regeneration system from Neu5Ac [133, 134], and Wang's group extended this regeneration to allow the use of cheaper starting material. As can be seen in Fig. 6.5, Neu5Ac may originate from two source pathways. The first pathway involves three steps from ManNAc to Neu5Ac and consumes one molecule of ATP and one molecule of PEP. The second pathway produces Neu5Ac directly from ManNAc through an aldolase-catalyzed condensation with pyruvate. From the viewpoint of pathway engineering, the pathway in mammalian cells involves three steps and consumes two molecules of high-energy phosphates, so it is technically difficult and energetically inefficient (Route c in Fig. 6.5). In bacteria, on the other hand, Neu5Ac is synthesized by the reversible condensation of ManNAc with pyruvate, catalyzed by Neu5Ac aldolase (NanA, EC 4.1.3.3) [135, 136]. The second pathway has only one step from ManNAc to NeuAc and no ATP or PEP requirement, which makes it a far better choice [132].

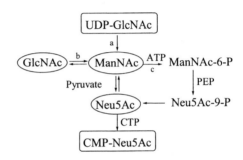

Fig. 6.5 Biosynthetic pathway of CMP-Neu5Ac.

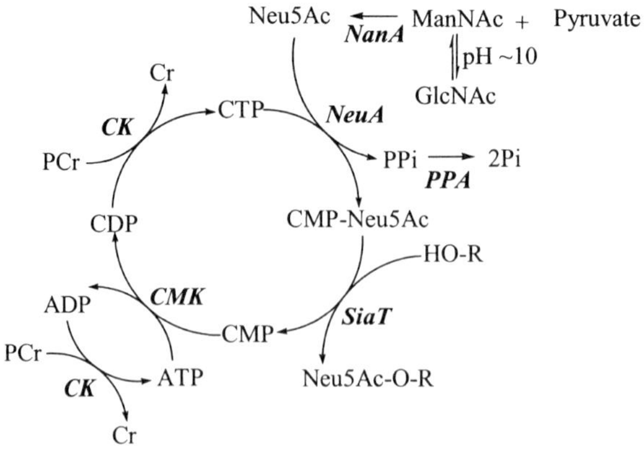

GlcNAc + Pyruvate + 2PCr + HOR ⟶ Neu5AcOR + 2Cr + 2Pi

Scheme 6.12 CMP-Neu5Ac regeneration cycle.

CMP-NeuAc regeneration can therefore be achieved with five enzymes (Scheme 6.12): NanA synthesizes Neu5Ac from ManNAc and pyruvate. Neu5Ac is activated with CTP by CMP-Neu5Ac synthetase (NeuA, EC 2.7.7.43). CMP-Neu5Ac serves as the substrate for sialyltransferase (SiaT) to make sialyloligosaccharide. The by-product, CMP, was regenerated with cytidine 5′-monophosphate kinase (CMK, EC 2.7.4.14) and creatine kinase (CK, EC 2.7.3.2). Here, the NanA-catalyzed reaction is reversible, but the presence of pyruvate completely inhibits the lyase activity and the coupled transfer reaction will also drive the equilibrium toward Neu5Ac synthesis. Overall, this system requires only ManNAc, pyruvate, and creatine phosphate to generate CMP-Neu5Ac for the synthesis of sialylated oligosaccharides. There are two biosynthetic pathways to obtain ManNAc. The first is from UDP-GlcNAc by use of UDP-GlcNAc 2′-epimerase (Route a in Fig. 6.5). The second is from GlcNAc by use of GlcNAc 2-epimerase (Route b in Fig. 6.5). UDP-GlcNAc is quite expensive and there are no known bacterial enzymes that catalyze these reactions. ManNAc can also be epimerized chemically from GlcNAc under basic conditions (pH 10). We can therefore use GlcNAc as starting material to synthesize sialyloligosaccharides.

6.2.3
Novel Energy Source in Sugar Nucleotide Regeneration

As we have seen, most biosynthetic pathways for sugar nucleotide regeneration use PEP as the energy source. Although the associated kinase (PykF) can be expressed in *E. coli* in large quantities, the cost of PEP is still one of the most inhibitory factors in sugar nucleotide regeneration. Therefore, to solve the energy supply problem, the creatine phosphate-creatine kinase (CP-CK) system has been integrated into the regeneration cycles (Scheme 6.13) [137]. It is a more efficient,

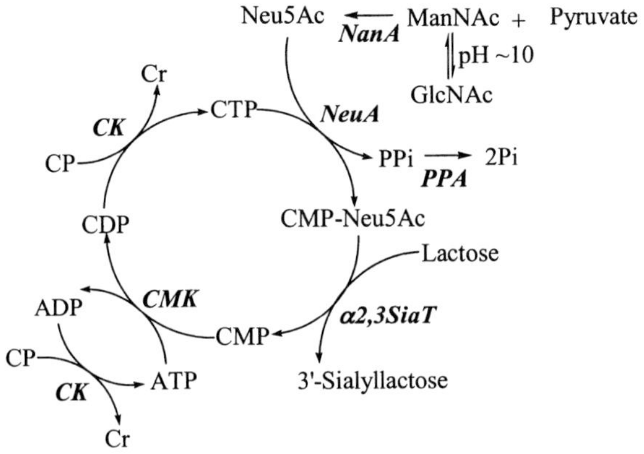

Scheme 6.13 Syntheses of 3'-sialyllactose with use of the CP-CK energy system.

$$\text{GlcNAc} + \text{Pyruvate} + 2\text{CP} + \text{Lac} \longrightarrow \text{Neu5Ac } \alpha 2,3\text{Lac} + 2\text{Cr} + 2\text{Pi}$$

cheap, and convenient energy source than PEP-PK and acetyl phosphate-acetyl kinase systems, especially in large-scale synthesis of glycoconjugates [137].

6.3
Enzymatic Oligosaccharide Synthesis Processes

The greater availability of recombinant carbohydrate biosynthetic enzymes is making enzymatic oligosaccharide synthesis increasingly competitive with organic synthesis. In addition, attachment of sugar acceptors on solid phases has significantly simplified product recovery [138–151]. However, the recombinant proteins remain too expensive and unstable to be used as versatile catalysts in synthesis. Moreover, they cannot be recovered from the reaction mixture by conventional workup processes, and so these biocatalysts are at present not suited for large-scale or combinatorial syntheses of glycoconjugates. Although glycosyltransferases are extremely powerful in small-scale synthesis, the high cost of sugar nucleotides and product inhibition caused by the released nucleoside mono- or diphosphates present major obstacles to scale-up efforts. Here we discuss recent developments in enzymatic oligosaccharide synthesis that aim to overcome these limitations.

6.3.1
Cell-Free Oligosaccharide Synthesis

Immobilized enzyme systems have advantages such as ease of product separation, increased stability, and reusability of the catalysts [152–164]. Immobilization of glycosyltransferases and other enzymes has therefore been used to overcome the difficulties in common solution-phase enzymatic glycosylation.

Tab. 6.2 Immobilization systems for enzymatic oligosaccharide synthesis.

Acceptor	Enzyme	Comments	Reference
Free	Solution phase	Easy to manipulate, but high cost	26–28
	Glycosyltransferase immobilized	Increased stability and reusability	152–161
	Enzymes for sugar nucleotide regeneration immobilized	Increased stability, reusability, and low cost	82, 180
	All enzymes immobilized	Increased stability, reusability, and low cost	82
Solid phase	Solution phase	Easy product separation, long cleavable tether needed	138–147
Water-soluble polymer-bound	Solution phase	Easy product separation	170, 172
	Glycosyltransferase immobilized	Easy product separation, increased stability and reusability	173–175

6.3.1.1 Immobilized Glycosyltransferases and Water-Soluble Glycopolymer

Immobilization of glycosyltransferases seems to have become a common solution to defeat problems associated with large-scale enzymatic syntheses. However, enzymes immobilized on solid materials cannot be used for glycosylation of solid-supported sugars. Therefore, the feasibility of reactions between glycosyltransferases immobilized on solid phases and sugars bound on flexible polymers showing satisfactory water solubility has been explored. This new technology combines the advantages of enzymatic accessibility of solution-phase glycosylation and the convenience of easy product recovery [150, 165–171]. Recently, Wong et al. described a thermoresponsive polymer support that can be used to immobilize glycosyltransferases and carbohydrate acceptors. After the reactions, the enzymes and products can easily be recovered by thermal precipitation of the polymer [172]. Nishimura et al. have also described a practical synthesis of some oligosaccharides that makes use of immobilized recombinant glycosyltransferases and a sugar-carrying primer [173]. The success of this new strategy is dependent on drastically enhanced affinity of glycosyltransferases for multivalent glycopolymers. The glycosyltransferases expressed as fusion proteins with maltose-binding protein (MBP) in *E. coli* were directly subjected to coupling reactions with CNBr-activated Sepharose 4B (MBP-GalT) or NHS-activated Sepharose 4FF (MBP-SiaT). The MBP portion in the fusion enzyme is used for enzyme purification or immobilization. As anticipated, immobilized GalT exhibited satisfactory elongation of the sugar primer and the polymer-GalT beads were found to be recyclable.

In these experiments, the versatility of oligosaccharide synthesis based on the immobilized glycosyltransferases was demonstrated by the construction of a simple trisaccharide derivative that can be applied to further conjugation studies to

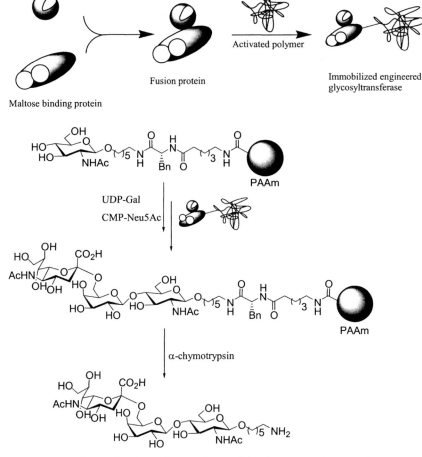

Scheme 6.14 Oligosaccharide synthesis with immobilized enzyme and water-soluble glycopolymer.

produce glycodrugs or glycomaterials with interesting bioactivities [173]. For example, Sialylα(2–6)-N-acetyllactosamine (72%) was finally obtained by treating the peptide primer with α-chymotrypsin (Scheme 6.14).

It should also be noted that glycosylation of polymer primers by glycosyltransferases conjugated with a macromolecular support affords product even in the presence of high concentrations of inhibitors (nucleotides). This result indicates that the inhibitory effect of a nucleotide (UDP) was significantly reduced by immobilization of the enzyme [173]. The availability of the immobilized glycosyltransferases reported here should greatly accelerate the development of enzyme-based automated glycosynthesizers [20, 174, 175].

6.3.1.2 "Superbeads"

In large-scale oligosaccharide synthesis, sugar nucleotide regeneration systems have to be introduced to solve the cost and product inhibition problems. As we have seen, the regeneration systems are complicated, since many enzymes are involved in each cycle. We have prepared oligosaccharide synthesis "superbeads", which are essentially agarose resin-immobilized enzymes of the biosynthetic pathway for glycoconjugate synthesis. The immobilization principle is based on immobilized metal ion affinity chromatography (IMAC), the His_6-tagged recombinant enzymes binding specifically to the Ni-nitrilotriacetic acid (NTA) agarose beads [176–179]. This technology involves the following steps: (1) cloning and overexpression of individual N-terminal His_6-tagged enzymes, (2) stepwise testing of the combined activity of the recombinant enzymes, (3) co-immobilization on the NTA beads, and (4) combination of the superbeads with glycosyltransferases (either on the beads or in solution) for glycoside synthesis. As an example, we introduce our UDP-Gal superbeads here (Scheme 6.15).

All the related enzymes for UDP-Gal regeneration – GalK, GalPUT, GalU, and PykF – were expressed in *E. coli* BL21 (DE3) with N-terminal His_6-tag. Their ex-

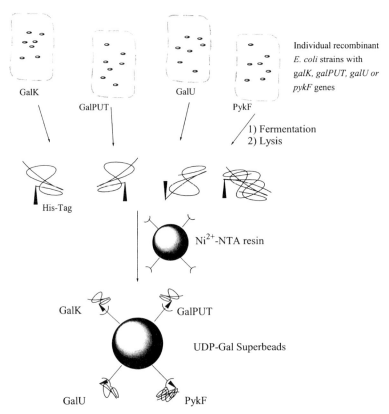

Scheme 6.15 UDP-Gal superbeads.

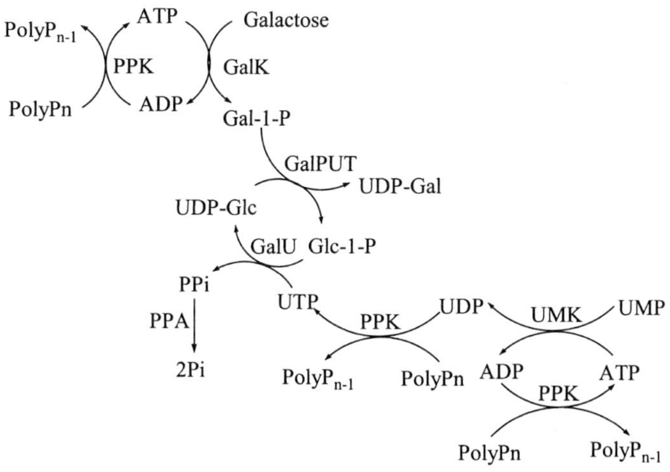

```
PolyP_{n-1}    ATP      Galactose
         \   /    \    /
          PPK      GalK
         /   \    /    \
PolyPn    ADP      Gal-1-P
                    \
                     GalPUT
                         → UDP-Gal
         UDP-Glc    /
              \  GalU  Glc-1-P
               \ /
           PPi  
      PPA|      UTP      UDP    UMK    UMP
         |       \  PPK  /  \   /
       2Pi        \ /   \    \ /
              PolyP_{n-1}  PolyPn  ADP    ATP
                                \  PPK  /
                              PolyPn   PolyP_{n-1}
```

$$3\ \text{PolyPn} + \text{Gal} + \text{UMP} \rightleftharpoons 3\ \text{PolyP}_{n-1} + 2\text{Pi} + \text{UDP-Gal}$$

Scheme 6.16 UDP-Gal production with superbeads.

pression levels were high and their activities were confirmed in a solution-phase syntheses of α-Gal oligosaccharides. UDP-Gal regeneration beads were then obtained by incubating cell lysates consisting of same number of activity units of His$_6$-tagged GalK, GalT, GalU, and PykF with the Ni^{2+}-NTA resin [82]. With this UDP-Gal superbeads, a variety of oligosaccharides, including deoxy and labeled derivatives, were synthesized on preparative scales with galactosyltransferases either on the beads or in solution.

Sugar nucleotide regeneration systems contain only low concentrations of sugar nucleotides. To obtain high concentrations of sugar nucleotides, especially when trying to accumulate them, our laboratory has developed UDP-Gal production superbeads (Scheme 6.16) [180]. We found that the UDP-Gal regeneration cycle could readily be converted into the biosynthetic pathway of UDP-Gal by removing the glycosyltransferase. Instead, the necessary driving force to complete the cycle was provided by inorganic pyrophosphorylase-catalyzed hydrolysis of the pyrophosphate byproduct.

Through this approach, we were able to synthesize UDP-Gal on gram scales by recirculating the reaction solution through the UDP-Gal regeneration superbead column [180]. The cell-free UDP-Gal production system proved efficient for the production of UDP-Gal, with a yield of more than 90% based on PEP. To reduce PEP-associated costs, we replaced PEP with polyphosphate. The replacement reduced the synthetic efficiency, but a 50% UDP-Gal yield (based on UMP) could still be obtained after 48 h incubation at 37 °C. We have also prepared superbeads to produce UDP-GlcNAc, as shown in Scheme 6.17 [102].

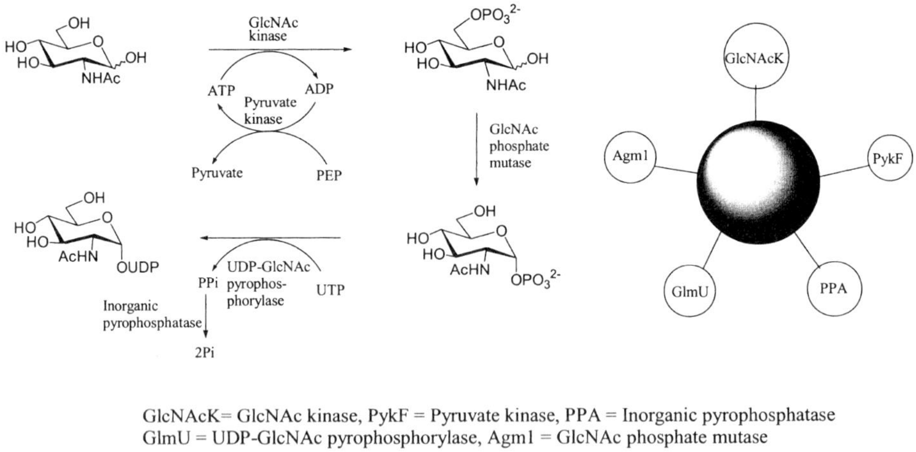
GlcNAc kinase
ATP
ADP
Pyruvate kinase
Pyruvate
PEP
GlcNAc phosphate mutase
UDP-GlcNAc pyrophos-phorylase
PPi
UTP
Inorganic pyrophosphatase
2Pi
GlcNAcK
Agm1
PykF
GlmU
PPA
GlcNAcK= GlcNAc kinase, PykF = Pyruvate kinase, PPA = Inorganic pyrophosphatase
GlmU = UDP-GlcNAc pyrophosphorylase, Agm1 = GlcNAc phosphate mutase

Scheme 6.17 UDP-GlcNAc superbeads.

6.3.2
Large-Scale Syntheses of Oligosaccharides with Whole Cells

Without wishing to negate the impressive results obtained with purified enzymes, the real promise of oligosaccharide synthesis may reside in the recently developed whole-cell biocatalysis [56, 58, 62, 181–183]. In enzymatic synthesis, the isolation and purification of multiple enzymes is laborious and the purification process may result in decreased enzymatic activity. To avoid these obstacles, scientists have applied metabolic pathway engineering techniques to construct carbohydrate-producing organisms [184–187]. In 1998, for instance, Kyowa Hakko Inc. in Japan introduced the large-scale synthesis of carbohydrates by metabolic engineering of microorganisms [183]. Wang et al. have developed the 'superbug' technology independently to address this problem [83]. Though many different cell types have been used to make oligosaccharides, here we discuss only metabolically engineered bacteria [188–193].

Tab. 6.3 Comparison of whole-cell systems for oligosaccharide synthesis

Technology	*Kyowa Hakko's*	*Wang's*	*Samain's*
Plasmid	Multiple	Single, pLDR20	Single
Strains	Multiple	Single *E. coli* strain	Single bacteria strain
Fermentation	Multiple batches	One batch	One batch
Reaction	During fermentation	Two-step	During fermentation
Sugar nucleotides	High concentration	Low concentration, regenerated	Low concentration, using internal pool
Reference	56, 58, 181–183	78, 83	194–197

6.3.2.1 Kyowa Hakko's Technology

As shown in Scheme 6.18, the key component in Kyowa Hakko's technology for the large-scale production of UDP-Gal and Gb$_3$ was a *C. ammoniagenes* strain engineered to convert cheap orotic acid efficiently into uridine 5′-triphosphate (UTP) [183]. Through combination with an engineered *E. coli* strain overexpressing the UDP-Gal biosynthetic genes *galK* (galactokinase), *galT* (galactose-1-phosphate uridyltransferase), *galU* (glucose-1-phosphate uridyltransferase), and *ppa* (pyrophosphatase), UDP-Gal was accumulated in the reaction solution (72 mM). Addition of another recombinant *E. coli* strain, overexpressing the α1,4-galactosyltransferase gene (*lgtC*) of *Neisseria gonorrhoeae*, allowed accumulation of the trisaccharide product, globotriose, at a high concentration (372 mM). The same UDP-Gal production system was also successfully applied in the large-scale synthesis of the disaccharide LacNAc [279 mM (107 g L^{-1})] by replacement of the *E. coli* strain harboring the *lgtC* gene with a strain expressing *lgtB* (β1, 4-GalT) [56].

By the same bacterial coupling concept, Kyowa Hakko achieved large-scale production of other sugar nucleotides and related oligosaccharides, such as UDP-GlcNAc, CMP-NeuAc and sialylated oligosaccharides, GDP-Fuc, and Sialyl Lewis X.

Their success initiates a new era in large-scale enzymatic synthesis of carbohydrates. However, this methodology still suffers from several problems that need to be addressed: (1) multiple fermentations of several bacterial strains are required, including one engineered *C. ammoniagenes* and two or more engineered *E. coli* strains, and (2) the transport of substrates between different bacterial strains.

6.3.2.2 Wang's "Superbug"

After our synthetic success with purified recombinant enzymes and superbeads, we attempted the use of whole *E. coli* cells as biocatalysts. Synthesis of the galacto-

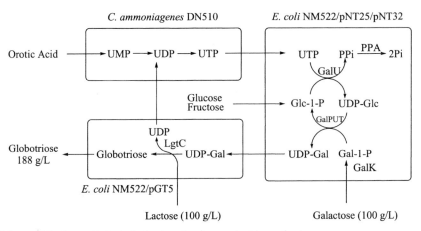

Scheme 6.18 Kyowa Hakko's technology in oligosaccharide synthesis.

sylated oligosaccharides, shown in Scheme 6.19, requires five enzymes. The genes encoding these proteins were cloned in tandem into a single vector and transformed into an expression host. A bacterial strain capable of simultaneous α-Gal production and UDP-Gal regeneration was obtained. Such a multi-enzyme producing strain was named a "superbug" in our laboratory. In this approach, it is unnecessary to purify and immobilize individual enzymes, and the proteins may be expressed without tags [122]. For example, *Helicobacter pylori* β1,4GalT is inactive when expressed with the His_6-tag in *E. coli* cells and hence cannot be immobilized in our superbead system. However, it is active when expressed in its native form [181].

α-Gal Superbug (Scheme 6.19) We have used the pET15b expression vector to express individual enzymes of the α-Gal biosynthetic pathway. However, in superbug construction, this expression system will generate β-galactosidase, which can hydrolyze the β-galactosidic bond in the acceptor. It is necessary to utilize another expression vector, pLDR20, with a temperature-controlled promoter and a β-galactosidase-deficient *E. coli* host strain.

The pLDR20 vector contains an ampicillin resistance gene, a P_R promoter, and a C_I repressor gene. The DNA fragment containing the ribosomal binding site, the *N*-terminal His_6-tag codons, and the open reading frame for each of the five enzymes were cloned one by one into the pLDR20 vector to form the final plasmid pLDR20-αKTUF. Since GalK and GalT exist in the same *gal* operon and close to each other, they were cloned together into the pET15b vector and then subcloned into the pLDR20 vector. The expression of the target genes could be induced above 37 °C.

SDS-PAGE of the superbug indicated that all five enzymes were expressed. The activity of these enzymes was further confirmed by the synthesis of α-Gal by use of the enzymes purified from the superbug. The superbug was then used as a whole-cell catalyst in the production of α-Gal oligosaccharides. After growth, the

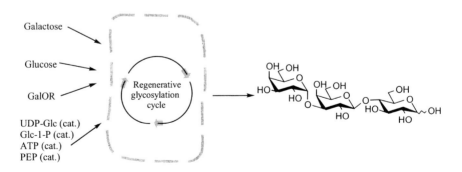

Scheme 6.19 α-Gal superbug.

cells were collected from the medium by centrifugation, and added to a reaction mixture containing glucose, galactose, and lactose acceptor.

One of the most important findings in the use of the superbug was that only catalytic amounts of high-energy phosphate were necessary, there being no need to use stoichiometric amounts of PEP or ATP as required for the cell-free in vitro synthesis. This indicated that both PEP and ATP were generated and recycled, although the bacteria did not grow in the reaction system. Obviously, this makes the superbug-based production of oligosaccharide the most cost-effective approach. Another feature of the superbug reaction was that both the starting mono- and disaccharides and the product trisaccharides were membrane-permeable.

Other Superbugs The superbug technology can be expanded if one or more glycosyltransferases are incorporated in the recombinant *E. coli*. Replacement of α-1,3-GalT with other transferases, such as LgtC, resulted in another superbug, CKTUF. This new strain was successfully utilized in synthesis of globotriose and its derivatives (Scheme 6.20).

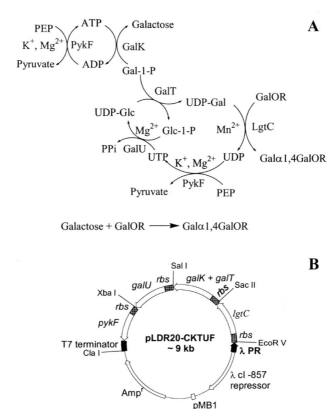

Scheme 6.20 Globotriose superbug.

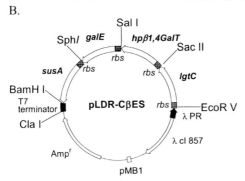

Scheme 6.21 P₁ antigen superbug.

The oligosaccharide sequence Galα1, 4Galβ1, 4GlcNAc (P$_1$ antigen) can be synthesized by insertion of β1, 4GalT in the globotriose regeneration cycle. Such a superbug, named CβES, was constructed and shown to be an efficient catalyst of P$_1$ antigen (Scheme 6.21).

Comparison with Kyowa Hakko's Technology The superbug technology and Kyowa Hakko's technology have their respective advantages and disadvantages (see Tab. 6.3). The most striking feature of Kyowa's technology is that it makes use of a special bacterial strain that converts orotic acid into UTP. This makes the production of sugar nucleotides cost-effective. In Wang's technology, however, all of the enzymes essential for oligosaccharide syntheses, including glycosyltransferases and sugar nucleotide regeneration, are contained in one bacterial strain, so the superbug can be easily adapted and used in a variety of synthetic and biochemical procedures.

6.3.2.3 Other Whole Cell-Based Technologies

A complementary strategy for the synthesis of oligosaccharides is in vivo production in recombinant microorganisms expressing glycosyltransferases. The idea of a "living factory" is to use growing bacterial cells as natural mini-reactors for the regeneration of the sugar nucleotides and to utilize the intracellular pool of sugar nucleotides as substrates for in vivo synthesis of "recombinant" carbohydrate [194–199]. Although comparable practices in the production of exopolysaccharides by lactic acid bacteria had been tried for some time [200], they had not gained much attention until the large-scale syntheses by Samain and co-workers. In their work, chitooligosaccharides were produced by cultivation of E. coli cells harboring heterologous genes only involved in oligosaccharide synthesis. For example, penta-N-acetylchitopentaose (2.5 g L^{-1}) was produced by cultivation of E. coli expressing the nodC gene from Azorhizobium caulinodans, encoding chitooligosaccharide synthase. O-Acetylated and sulfated chitooligosaccharides were also produced on a gram scale by coexpression of nodC or nodBC with nodH and/or nodL, enzymes that encode chitooligosaccharide sulfotransferase and chitooligosaccharide O-acetyltransferase, respectively. In addition, when E. coli cells coexpressing nodC from A. caulinodans and lgtB from N. meningitidis were cultivated, more than 1.0 g L^{-1} of a hexasaccharide, identified as βGal (1,4)[βGlcNAc (1,4)]$_4$ GlcNAc, was accumulated (Scheme 6.22). Obviously, this method possesses some advantages over the common chemoenzymatic methods, as there is no need for the isolation and purification of recombinant glycosyltransferases, and the cells already possess the machinery required for sugar nucleotide synthesis. On the other hand, the synthetic efficiency of this system is critically dependent on the intracellular pool of sugar nucleotides.

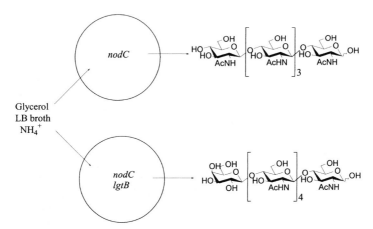

Scheme 6.22 Production of carbohydrates in vivo by use of engineered bacteria.

6.4
Future Directions

Recent progress in glycobiology and glycochemistry has allowed enormous development in the syntheses of oligosaccharides. In the near future, large number of enzymes, strains, and vectors for oligosaccharide syntheses should become available either through academic collaborations such as the glycomics consortium (http://glycomics.scripps.edu) or through commercial suppliers. Although traditional cell-free enzymatic syntheses of oligosaccharides will continue to be used in laboratory scale synthesis, the future direction will be large-scale preparation of oligosaccharides with whole-cell systems. Thus, further investment in industrial-scale carbohydrate preparation should become feasible once the market for these products has expanded. Moreover, the increasing availability of genomic information should make the production of many previously inaccessible complex carbohydrates possible. These include complex structures of *N*-glycans, *O*-glycans, lipopolysaccharides, and others, to support the increasing demands of the carbohydrate-based drug industry.

6.5
References

1 KOBATA, A. Acc. Chem. Res. 1993, 26, 319–24.
2 DWEK, R. A. Chem. Rev. 1996, 96, 683–720.
3 DENNIS, J. W.; GRANOVSKY, M.; WARREN, C. E. Bioessays 1999, 21, 412–21.
4 APWEILER, R.; HERMJAKOB, H.; SHARON, N. Biochim. Biophys. Acta 1999, 1473, 4–8.
5 RUDD, P. M.; WORMALD, M. R.; STANFIELD, R. L.; HUANG, M.; MATTSSON, N.; SPEIR, J. A.; DIGENNARO, J. A.; FETROW, J. S.; DWEK, R. A.; WILSON, I. A. J. Mol. Biol. 1999, 293, 351–66.
6 HAKOMORI SI, S. I. Proc. Natl. Acad. Sci. USA 2002, 99, 225–32.
7 SEITZ, O. Chem. Bio. Chem. 2000, 1, 214–46.
8 COOPER, D. K. C.; ORIOL, R. Glycosciences 1997, 531–545.
9 CHEN, X.; ANDREANA, P. R.; WANG, P. G. Curr. Opin. Chem. Biol. 1999, 3, 650–658.
10 GALILI, U. Biochimie 2001, 83, 557–63.
11 LAI, L.; KOLBER-SIMONDS, D.; PARK, K. W.; CHEONG, H. T.; GREENSTEIN, J. L.; IM, G. S.; SAMUEL, M.; BONK, A.; RIEKE, A.; DAY, B. N.; MURPHY, C. N.; CARTER, D. B.; HAWLEY, R. J.; PRATHER, R. S. Science 2002, 295, 1089–92.
12 KARLSSON, K. A. Curr. Opin. Struct. Biol. 1995, 5, 622–35.
13 LINGWOOD, C. A. Biochim. Biophys. Acta 1999, 1455, 375–86.
14 PETRI, W. A., JR.; HAQUE, R.; MANN, B. J. Annu. Rev. Microbiol. 2002, 56, 39–64.
15 HAKOMORI, S.; ZHANG, Y. Chem. Biol. 1997, 4, 97–104.
16 DANISHEFSKY, S. J.; ALLEN, J. R. Angew. Chem. Int. Ed. Engl. 2000, 39, 836–863.
17 WONG, C.-H.; HALCOMB, R. L.; ICHIKAWA, Y.; KAJIMOTO, T. Angew. Chem. Int. Ed. Engl. 1995, 34, 412–32.
18 WONG, C.-H.; HALCOMB, R. L.; ICHIKAWA, Y.; KAJIMOTO, T. Angew. Chem. Int. Ed. Engl. 1995, 34, 521–46.
19 KOELLER, K. M.; WONG, C.-H. Glycobiology 2000, 10, 1157–1169.
20 SEARS, P.; WONG, C.-H. Science 2001, 291, 2344–2350.
21 PLANTE, O. J.; PALMACCI, E. R.; SEEBERGER, P. H. Science 2001, 291, 1523–7.
22 KOELLER, K. M.; WONG, C.-H. Nature (London) 2001, 409, 232–240.

23 Wymer, N.; Toone, E. J. Curr. Opin. Chem. Biol. 2000, 4, 110–9.
24 Seto, N. O. L.; Compston, C. A.; Evans, S. V.; Bundle, D. R.; Narang, S. A.; Palcic, M. M. Eur. J. Biochem. 1999, 259, 770–775.
25 Crout, D. H.; Vic, G. Curr. Opin. Chem. Biol. 1998, 2, 98–111.
26 Whitesides, G. M.; Wong, C. H. Aldrichimica Acta 1983, 16, 27–34.
27 Drueckhammer, D. G.; Hennen, W. J.; Pederson, R. L.; Barbas, C. F., III; Gautheron, C. M.; Krach, T.; Wong, C. H. Synthesis 1991, 499–525.
28 Wong, C. H.; Schuster, M.; Wang, P.; Sears, P. J. Amer. Chem. Soc. 1993, 115, 5893–901.
29 Crout, D. H. G.; Critchley, P.; Muller, D.; Scigelova, M.; Singh, S.; Vic, G. Special Publication – Royal Society of Chemistry 1999, 246, 15–23.
30 van Rantwijk, F.; Woudenberg-van Oosterom, M.; Sheldon, R. A. J. Mol. Catal. B: Enzymatic 1999, 6, 511–532.
31 Vocadlo, D. J.; Withers, S. G. Glycosidase-Catalysed Oligosaccharide Synthesis, in Carbohydrates in Chemistry and Biology, Vol. 2, 723–844, Ernst, B.; Hart, G. W.; Sinaÿ, P. (Eds.), Wiley-VCH, Weinheim, 2000.
32 Withers, S. G. Can. J. Chem. 1999, 77, 1–11.
33 Ly, H. D.; Withers, S. G. Ann Rev. Biochem.1999, 68, 487–522.
34 Jakeman, D. L.; Withers, S. G. Trends Glycosci. and Glycotechnol. 2002, 14, 13–25.
35 Tolborg, J. F.; Petersen, L.; Jensen, K. J.; Mayer, C.; Jakeman, D. L.; Warren, R. A.; Withers, S. G. J. Org. Chem. 2002, 67, 4143–9.
36 Roseman, S. Chem. Phys. Lipids 1970, 5, 270–97.
37 Watkins, W. M. Carbohydr. Res. 1986, 149, 1–12.
38 Hehre, E. J. Carbohydr. Res. 2001, 331, 347–368.
39 Unligil, U. M.; Rini, J. M. Curr. Opin. Struct. Biol. 2000, 10, 510–517.
40 Davies, G. J.; Henrissat, B. Biochem. Soc. Trans. 2002, 30, 291–297.
41 Kaneko, M.; Nishihara, S.; Narimatsu, H.; Saitou, N. Trends Glycosci. and Glycotechnol. 2001, 13, 147–155.
42 Palcic, M. M.; Sujino, K.; Qian, X. Enzymatic Glycosylations with Non-Natural Donors and Acceptors, in Carbohydrates in Chemistry and Biology, Vol. 2, 685–703, Ernst, B.; Hart, G. W.; Sinaÿ, P. (Eds.), Wiley-VCH, Weinheim, 2000.
43 Sujino, K.; Uchiyama, T.; Hindsgaul, O.; Seto, N. O. L.; Wakarchuk, W. W.; Palcic, M. M. J. Amer. Chem. Soc. 2000, 122, 1261–1269.
44 Fukuda, M.; Bierhuizen, M. F.; Nakayama, J. Glycobiology 1996, 6, 683–9.
45 Fang, J.; Li, J.; Chen, X.; Zhang, Y.; Wang, J.; Guo, Z.; Zhang, W.; Yu, L.; Brew, K.; Wang, P. G. J. Amer. Chem. Soc. 1998, 120, 6635–6638.
46 Palacpac, N. Q.; Yoshida, S.; Sakai, H.; Kimura, Y.; Fujiyama, K.; Yoshida, T.; Seki, T. Proc. Natl. Acad. Sci. USA 1999, 96, 4692–7.
47 Shibatani, S.; Fujiyama, K.; Nishiguchi, S.; Seki, T.; Maekawa, Y. J. Biosci. Bioeng. 2001, 91, 85–87.
48 Malissard, M.; Zeng, S.; Berger, E. G. Glycoconjugate J. 1999, 16, 125–39.
49 Johnson, K. F. Glycoconj. J. 1999, 16, 141–6.
50 Blixt, O.; van Die, I.; Norberg, T.; van den Eijnden, D. H. Glycobiology 1999, 9, 1061–71.
51 Izumi, M.; Shen, G.-J.; Wacowich-Sgarbi, S.; Nakatani, T.; Plettenburg, O.; Wong, C.-H. J. Amer. Chem. Soc. 2001, 123, 10909–10918.
52 DeAngelis, P. L. Glycobiology 2002, 12, 9R–16R.
53 Wong, C. H.; Haynie, S. L.; Whitesides, G. M. J. Org. Chem. 1982, 47, 5416–18.
54 Ichikawa, Y.; Wang, R.; Wong, C.-H. Methods Enzymol. 1994, 247, 107–27.
55 Wong, C. H.; Wang, R.; Ichikawa, s. J. Org. Chem. 1992, 57, 4343–4.
56 Endo, T.; Koizumi, S.; Tabata, K.; Kakita, S.; Ozaki, A. Carbohydr. Res. 1999, 316, 179–183.
57 Bulter, T.; Elling, L. Glycoconj. J. 1999, 16, 147–159.
58 Endo, T.; Koizumi, S. Curr. Opin. Struct. Biol. 2000, 10, 536–541.

59 ELLING, L.; GROTHUS, M.; KULA, M.R. Glycobiology 1993, 3, 349–55.
60 ICHIKAWA, Y.; LIN, Y.C.; DUMAS, D.P.; SHEN, G.J.; GARCIA-JUNCEDA, E.; WILLIAMS, M.A.; BAYER, R.; KETCHAM, C.; WALKER, L.E.; et al. J. Am. Chem. Soc. 1992, 114, 9283–98.
61 WANG, P.; SHEN, G.J.; WANG, Y.F.; ICHIKAWA, Y.; WONG, C.H. J. Org. Chem. 1993, 58, 3985–90.
62 HERRMANN, G.F.; WANG, P.; SHEN, G.-J.; WONG, C.-H. Angew. Chem., Int. Ed. Engl., 1994, 33(12), 1241–2.
63 LOOK, G.C.; ICHIKAWA, Y.; SHEN, G.J.; CHENG, P.W.; WONG, C.H. J. Org. Chem. 1993, 58, 4326–30.
64 GYGAX, D.; SPIES, P.; WINKLER, T.; PFAAR, U. Tetrahedron 1991, 47, 5119–22.
65 HOKKE, C.H.; ZERVOSEN, A.; ELLING, L.; JOZIASSE, D.H.; VAN DEN EIJNDEN, D.H. Glycoconj. J. 1996, 13, 687–92.
66 FREEZE, H.H. Monosaccharide Metabolism, in *Essentials of Glycobiology*, 69–84, VARKI, A.; CUMMINGS, R.; ESKO, J.; FREEZE, H.H., HART, G.; MARTH J. (Eds.), CSHL Press, Cold Spring Harbor, NY, 1999.
67 HE, X.M.; LIU, H.W. Annu. Rev. Biochem. 2002, 71, 701–54.
68 HEIDLAS, J.E.; WILLIAMS, K.W.; WHITESIDES, G.M. Acc. Chem. Res. 1992, 25, 307–14.
69 BULTER, T.; ELLING, L. Glycoconj. J. 1999, 16, 147–59.
70 MA, X.; STOCKIGT, J. Carbohydr. Res. 2001, 333, 159–63.
71 HAYNIE, S.L.; WHITESIDES, G.M. Appl. Biochem. Biotechnol. 1990, 23, 155–70.
72 WINTER, H.; HUBER, S.C. Crit. Rev. Biochem. Mol. Biol. 2000, 35, 253–89.
73 NGUYEN-QUOC, B.; FOYER, C.H. J. Exp. Bot. 2001, 52, 881–9.
74 FERNIE, A.R.; WILLMITZER, L.; TRETHEWEY, R.N. Trends. Plant. Sci. 2002, 7, 35–41.
75 CHEN, X.; KOWAL, P.; HAMAD, S.; FAN, H.; WANG, P.G. Biotechnol. Lett. 1999, 21, 1131–1135.
76 HASELKORN, R.; BUIKEMA, W.J.; BAUER, C.C. Anabaena sucrose synthase, its gene sequence, and applications to alter starch and sucrose content and nitrogen fixation in transgenic plants. In: *Anabaena sucrose synthase, its gene sequence, and applications to alter starch and sucrose content and nitrogen fixation in transgenic plants*, p. 109. HASELKORN, R.; BUIKEMA, W.J.; BAUER, C.C. (Eds.). Arch Development Corp., USA 1998.
77 PORCHIA, A.C.; CURATTI, L.; SALERNO, G.L. Planta 1999, 210, 34–40.
78 CHEN, X.; ZHANG, J.; KOWAL, P.; LIU, Z.; ANDREANA, P.R.; LU, Y.; WANG, P.G. J. Amer. Chem. Soc. 2001, 123, 8866–8867.
79 CURATTI, L.; PORCHIA, A.C.; HERRERA-ESTRELLA, L.; SALERNO, G.L. Planta 2000, 211, 729–735.
80 HEIDLAS, J.E.; LEES, W.J.; WHITESIDES, G.M. J. Org. Chem. 1992, 57, 152–7.
81 CHACKO, C.M.; MCCRONE, L.; NADLER, H.L. Biochim. Biophys. Acta 1972, 268, 113–20.
82 CHEN, X.; FANG, J.; ZHANG, J.; LIU, Z.; SHAO, J.; KOWAL, P.; ANDREANA, P.; WANG, P.G. J. Amer. Chem. Soc. 2001, 123, 2081–2082.
83 CHEN, X.; LIU, Z.; ZHANG, J.; ZHANG, W.; KOWAL, P.; WANG, P.G. ChemBioChem 2002, 3, 47–53.
84 STEWART, D.C.; COPELAND, L. Plant Physiology 1998, 116, 349–355.
85 JOHANSSON, H.; STERKY, F.; AMINI, B.; LUNDEBERG, J.; KLECZKOWSKI, L.A. Biochim. Biophys. Acta 2002, 1576, 53–8.
86 GYGAX, D.; SPIES, P.; WINKLER, T.; PFAAR, U. Tetrahedron 1991, 47, 5119–22.
87 DE LUCA, C.; LANSING, M.; MARTINI, I.; CRESCENZI, F.; SHEN, G.-J.; O'REGAN, M.; WONG, C.-H. J. Amer. Chem. Soc. 1995, 117, 5869–70.
88 DE LUCA, C.; LANSING, M.; CRESCENZI, F.; MARTINI, I.; SHEN, G.-J.; O'REGAN, M.; WONG, C.-H. Bioorg. Med. Chem. 1996, 4, 131–42.
89 BAR-PELED, M.; GRIFFITH, C.L.; DOERING, T.L. Proc. Natl. Acad. Sci. USA 2001, 98, 12,003–8.
90 BREAZEALE, S.D.; RIBEIRO, A.A.; RAETZ, C.R. J. Biol. Chem. 2002, 277, 2886–96.
91 MORIARITY, J.L.; HURT, K.J.; RESNICK, A.C.; STORM, P.B.; LAROY, W.; SCHNAAR, R.L.; SNYDER, S.H. J. Biol. Chem. 2002, 277, 16968–75.
92 BDOLAH, A.; FEINGOLD, D.S. Biochem. Biophys. Res. Commun. 1965, 21, 543–6.

93 Wang, L.; Huskic, S.; Cisterne, A.; Rothemund, D.; Reeves, P. R. J. Bacteriol. 2002, 184, 2620–5.
94 Thoden, J. B.; Wohlers, T. M.; Fridovich-Keil, J. L.; Holden, H. M. J. Biol. Chem. 2001, 276, 15131–6.
95 Thoden, J. B.; Henderson, J. M.; Fridovich-Keil, J. L.; Holden, H. M. J. Biol. Chem. 2002, 277, 27528–34.
96 Milewski, S. Biochim. Biophys. Acta 2002, 1597, 173–92.
97 Knorr, R.; Ehrmann, M. A.; Vogel, R. F. J. Basic Microbiol. 2001, 41, 339–49.
98 Lahti, R.; Pitkaranta, T.; Valve, E.; Ilta, I.; Kukko-Kalske, E.; Heinonen, J. J. Bacteriol. 1988, 170, 5901–7.
99 Yamada-Okabe, T.; Sakamori, Y.; Mio, T.; Yamada-Okabe, H. Eur. J. BioChem. 2001, 268, 2498–505.
100 Berger, M.; Chen, H.; Reutter, W.; Hinderlich, S. Eur. J. BioChem. 2002, 269, 4212–8.
101 Hofmann, M.; Boles, E.; Zimmermann, F. K. Eur. J. BioChem. 1994, 221, 741–7.
102 Shao, J.; Zhang, J.; Nahalka, J.; Wang, P. G. Chem. Commun. 2002, 2586–2587.
103 Bengoechea, J. A.; Pinta, E.; Salminen, T.; Oertelt, C.; Holst, O.; Radziejewska-Lebrecht, J.; Piotrowska-Seget, Z.; Venho, R.; Skurnik, M. J. Bacteriol 2002, 184, 4277–87.
104 Creuzenet, C.; Belanger, M.; Wakarchuk, W. W.; Lam, J. S. J. Biol. Chem. 2000, 275, 19,060–7.
105 Shepherd, J. G.; Wang, L.; Reeves, P. R. Infect. Immun 2000, 68, 6056–61.
106 Kowal, P.; Wang, P. G. Biochemistry 2002, 41, 15410–15414.
107 Shao, J.; Zhang, J.; Kowal, P.; Lu, Y.; Wang, P. G. Biochem. Biophys. Res. Commun. 2002, 295, 1–8.
108 Shao, J.; Zhang, J.; Kowal, P.; Wang, P. G. Appl. Environ. Microbiol. 2002, 68, 5634–40.
109 Shao, J.; Zhang, J.; Kowal, P.; Lu, Y.-Q.; Wang, P. G. Chem. Commun. 2003, 1422–1423.
110 Park, S. H.; Pastuszak, I.; Drake, R.; Elbein, A. D. J. Biol. Chem. 1998, 273, 5685–91.
111 Davis, J. A.; Freeze, H. H. Biochim. Biophys. Acta 2001, 1528, 116–26.
112 Wills, E. A.; Roberts, I. S.; Del Poeta, M.; Rivera, J.; Casadevall, A.; Cox, G. M.; Perfect, J. R. Mol. Microbiol. 2001, 40, 610–20.
113 Davis, J. A.; Wu, X. H.; Wang, L.; DeRossi, C.; Westphal, V.; Wu, R.; Alton, G.; Srikrishna, G.; Freeze, H. H. Glycobiology 2002, 12, 435–42.
114 Privalle, L. S. Ann. N Y. Acad. Sci. 2002, 964, 129–38.
115 Regni, C.; Tipton, P. A.; Beamer, L. J. Structure (Camb.) 2002, 10, 269–79.
116 Ning, B.; Elbein, A. D. Eur. J. BioChem. 2000, 267, 6866–74.
117 Ning, B.; Elbein, A. D. Arch. Biochem. Biophys. 1999, 362, 339–45.
118 Keller, R.; Renz, F. S.; Kossmann, J. Plant J. 1999, 19, 131–41.
119 Yoda, K.; Kawada, T.; Kaibara, C.; Fujie, A.; Abe, M.; Hitoshi; Hashimoto; Shimizu, J.; Tomishige, N.; Noda, Y.; Yamasaki, M. Biosci. Biotechnol. BioChem. 2000, 64, 1937–41.
120 Agaphonov, M. O.; Packeiser, A. N.; Chechenova, M. B.; Choi, E. S.; Ter-Avanesyan, M. D. Yeast 2001, 18, 391–402.
121 Garami, A.; Mehlert, A.; Ilg, T. Mol. Cell. Biol. 2001, 21, 8168–83.
122 Albermann, C.; Distler, J.; Piepersberg, W. Glycobiology 2000, 10, 875–81.
123 Albermann, C.; Piepersberg, W.; Wehmeier, U. F. Carbohydr. Res. 2001, 334, 97–103.
124 Wu, B.; Zhang, Y.; Wang, P. G. Biochem. Biophys. Res. Commun. 2001, 285, 364–71.
125 Somoza, J. R.; Menon, S.; Schmidt, H.; Joseph-McCarthy, D.; Dessen, A.; Stahl, M. L.; Somers, W. S.; Sullivan, F. X. Structure Fold Des. 2000, 8, 123–35.
126 Mattila, P.; Rabina, J.; Hortling, S.; Helin, J.; Renkonen, R. Glycobiology 2000, 10, 1041–7.
127 Bonin, C. P.; Reiter, W. D. Plant J. 2000, 21, 445–54.
128 Bonin, C. P.; Potter, I.; Vanzin, G. F.; Reiter, W. D. Proc. Natl. Acad. Sci. USA 1997, 94, 2085–90.
129 Ohyama, C.; Smith, P. L.; Angata, K.; Fukuda, M. N.; Lowe, J. B.; Fukuda, M. J. Biol. Chem. 1998, 273, 14,582–7.

130 Liu, J.L.C.; Shen, G.J.; Ichikawa, Y.; Rutan, J.F.; Zapata, G.; Vann, W.F.; Wong, C.H. J. Amer. Chem. Soc. 1992, 114, 3901–10.

131 Karwaski, M.F.; Wakarchuk, W.W.; Gilbert, M. Protein Expr. Purif. 2002, 25, 237–40.

132 Lee, S.G.; Lee, J.O.; Yi, J.K.; Kim, B.G. Biotechnol. Bioeng. 2002, 80, 516–24.

133 Ichikawa, Y.; Shen, G.J.; Wong, C.H.J. Amer. Chem. Soc. 1991, 113, 4698–700.

134 Ichikawa, Y.; Liu, J.L.C.; Shen, G.J.; Wong, C.H. J. Am. Chem. Soc. 1991, 113, 6300–2.

135 Rodriguez-Aparicio, L.B.; Ferrero, M.A.; Reglero, A. Biochem. J. 1995, 308. (Pt 2), 501–5.

136 Baumann, W.; Freidenreich, J.; Weisshaar, G.; Brossmer, R.; Friebolin, H. Biol. Chem. Hoppe Seyler 1989, 370, 141–9.

137 Zhang, J.; Wu, B.; Zhang, Y.; Kowal, P.; Wang, P.G. Org. Lett. 2003, in press.

138 Halcomb, R.L.; Huang, H.; Wong, C.-H. J. Amer. Chem. Soc. 1994, 116, 11315–22.

139 Nunez, H.A.; Barker, R. Biochemistry 1980, 19, 489–95.

140 Zehavi, U.; Herchman, M. Carbohydr. Res. 1986, 151, 371–8.

141 Zehavi, U.; Herchman, M.; Kopper, S. Carbohydr. Res. 1992, 228, 255–63.

142 Blixt, O.; Norberg, T. J. Org. Chem. 1998, 63, 2705–2710.

143 Blixt, O.; Norberg, T. Carbohydr. Res. 1999, 319, 80–91.

144 Yan, F.; Wakarchuk, W.W.; Gilbert, M.; Richards, J.C.; Whitfield, D.M. Carbohydr. Res. 2000, 328, 3–16.

145 Yan, F.; Gilbert, M.; Wakarchuk, W.W.; Brisson, J.R.; Whitfield, D.M. Org. Lett 2001, 3, 3265–8.

146 Schuster, M.; Wang, P.; Paulson, J.C.; Wong, C.-H. J. Amer. Chem. Soc. 1994, 116, 1135–6.

147 Kopper, S. Carbohydr. Res. 1994, 265, 161–6.

148 Seeberger, P.H.; Haase, W.C. Chem. Rev. 2000, 100, 4349–94.

149 Witte, K.; Seitz, O.; Wong, C.-H. J. Amer. Chem. Soc. 1998, 120, 1979–1989.

150 Brinkmann, N.; Malissard, M.; Ramuz, M.; Romer, U.; Schumacher, T.; Berger, E.G.; Elling, L.; Wandrey, C.; Liese, A. Bioorg. Med. Chem. Lett 2001, 11, 2503–6.

151 Auge, C.; Le Narvor, C.; Lubineau, A. Solid-Phase Synthesis with Glycosyltransferases, in Carbohydrates in Chemistry and Biology, Vol. 2, 705–722, Ernst, B.; Hart, G.W.; Sinaÿ, P. (Eds.), Wiley-VCH, Weinheim, 2000.

152 Smiley, K.L.; Strandberg, G.W. Adv. Appl. Microbiol. 1972, 15, 13–38.

153 Klibanov, A.M. Anal. BioChem. 1979, 93, 1–25.

154 Liang, J.F.; Li, Y.T.; Yang, V.C.J. Pharm. Sci 2000, 89, 979–90.

155 Turkova, J. J. Chromatogr. B Biomed. Sci. Appl 1999, 722.

156 Tischer, W.; Kasche, V. Trends Biotechnol. 1999, 17, 326–35.

157 Saleemuddin, M. Adv. Biochem. Eng. Biotechnol. 1999, 64, 203–26.

158 Balcao, V.M.; Paiva, A.L.; Malcata, F.X. Enzyme Microb. Technol. 1996, 18, 392–416.

159 Clark, D.S. Trends Biotechnol. 1994, 12, 439–43.

160 Katchalski-Katzir, E. Trends Biotechnol. 1993, 11, 471–8.

161 Beeckmans, S.; Van Driessche, E.; Kanarek, L. J. Mol. Recognit. 1993, 6.

162 David, S.; Auge, C.; Gautheron, C. Adv. Carbohydr. Chem. BioChem. 1991, 49, 175–237.

163 Dols-Lafargue, M.; Willemot, R.M.; Monsan, P.F.; Remaud-Simeon, M. Biotechnol. Bioeng. 2001, 75.

164 Spagna, G.; Barbagallo, R.N.; Pifferi, P.G.; Blanco, R.M.; Guisan, J.M. J. Mol. Catal. B: Enzymatic 2000, 11, 63–69.

165 Yamada, K.; Nishimura, S.I. Tetrahedron Lett. 1995, 36, 9493–9496.

166 Gravert, D.J.; Janda, K.D. Curr. Opin. Chem. Biol. 1997, 1, 107–13.

167 Krepinsky, J.J.; Douglas, S.P.; Whitfield, D.M. Methods Enzymol. 1994, 242, 280–93.

168 Douglas, S.P.; Whitfield, D.M.; Krepinsky, J.J. J. Amer. Chem. Soc. 1991, 113, 5095–5097.

169 Lu, J.A.; Hartley, R.C.; Chan, T.H. Chem. Commun. 1996, 2193–2194.

170 Nishimura, S.-I.; Matsuoka, K.; Lee, Y.C. Tetrahedron Lett. 1994, 35, 5657–60.
171 Toda, A.; Yamada, K.; Nishimura, S.-I. Adv. Synth. Catal. 2002, 344, 61–69.
172 Huang, X.; Witte, K.L.; Bergbreiter, D.E.; Wong, C.-H. Adv. Synth. Catal. 2001, 343, 675–681.
173 Nishiguchi, S.; Yamada, K.; Fuji, Y.; Shibatani, S.; Toda, A.; Nishimura, S.-I. Chem. Commun. 2001, 1944–1945.
174 Nishimura, S.-I. Curr. Opin. Chem. Biol. 2001, 5, 325–335.
175 Nishimura, S.-I. Abstracts of Papers, 224th ACS National Meeting, Boston, MA, United States, August 18–22, 2002. 2002, CARB-032.
176 Byra, A.; Dworniczak, K.; Szumilo, T. Acta Biochim. Pol. 1991, 38, 101–5.
177 Kastner, M.; Neubert, D. J. Chromatogr. 1991, 587.
178 Lindeberg, G.; Bennich, H.; Engstrom, A. Int.J. Pept. Protein. Res. 1991, 38.
179 Roos, P.H. J. Chromatogr. 1991, 587.
180 Liu, Z.; Zhang, J.; Chen, X.; Wang, P.G. ChemBioChem 2002, 3, 348–355.
181 Endo, T.; Koizumi, S.; Tabata, K.; Ozaki, A. Glycobiology 2000, 10, 809–13.
182 Endo, T.; Koizumi, S.; Tabata, K.; Kakita, S.; Ozaki, A. Carbohydr. Res. 2001, 330.
183 Koizumi, S.; Endo, T.; Tabata, K.; Ozaki, A. Nat. Biotechnol. 1998, 16.
184 Hang, H.C.; Bertozzi, C.R. J. Amer. Chem. Soc. 2001, 123, 1242–1243.
185 Yarema, K.J.; Mahal, L.K.; Bruehl, R.E.; Rodriguez, E.C.; Bertozzi, C.R. J. Biol. Chem.1998, 273, 31168–31179.
186 Saxon, E.; Bertozzi, C.R. Annu. Rev. Cell. Dev. Biol. 2001, 17, 1–23.
187 Saxon, E.; Bertozzi, C.R. Science 2000, 287, 2007–10.
188 Seo, N.S.; Hollister, J.R.; Jarvis, D.L. Protein. Expr. Purif. 2001, 22, 234–41.
189 Jarvis, D.L.; Howe, D.; Aumiller, J.J. J. Virol. 2001, 75, 6223–7.
190 Breitbach, K.; Jarvis, D.L. Biotechnol. Bioeng. 2001, 74, 230–9.
191 Hollister, J.R.; Jarvis, D.L. Glycobiology 2001, 11, 1–9.
192 Jenkins, N.; Parekh, R.B.; James, D.C. Nat. Biotechnol. 1996, 14, 975–81.
193 Weikert, S.; Papac, D.; Briggs, J.; Cowfer, D.; Tom, S.; Gawlitzek, M.; Lofgren, J.; Mehta, S.; Chisholm, V.; Modi, N.; Eppler, S.; Carroll, K.; Chamow, S.; Peers, D.; Berman, P.; Krummen, L. Nat. Biotechnol. 1999, 17, 1116–21.
194 Samain, E.; Drouillard, S.; Heyraud, A.; Driguez, H.; Geremia, R.A. Carbohydr. Res. 1997, 302, 35–42.
195 Bettler, E.; Samain, E.; Chazalet, V.; Bosso, C.; Heyraud, A.; Joziasse, D.H.; Wakarchuk, W.W.; Imberty, A.; Geremia, A.R. Glycoconj.J. 1999, 16, 205–12.
196 Dumon, C.; Priem, B.; Martin, S.L.; Heyraud, A.; Bosso, C.; Samain, E. Glycoconj. J. 2001, 18, 465–74.
197 Priem, B.; Gilbert, M.; Wakarchuk, W.W.; Heyraud, A.; Samain, E. Glycobiology 2002, 12, 235–40.
198 Samain, E.; Debeire, P.; Touzel, J.P. J. Biotechnol 1997, 58, 71–8.
199 Levander, F.; Svensson, M.; Radstrom, P. Appl. Environ. Microbiol. 2002, 68, 784–90.
200 Kawaguchi, S. Trends Glycosci. and Glycotechnol. 1997, 9, 473–474.

7
Glycopeptides and Glycoproteins: Synthetic Chemistry and Biology
OLIVER SEITZ

7.1
Introduction

Protein glycosylation is an omnipresent process, introducing enormous structural diversity. Why is nature bothering with this most complicated type of protein modification? Comparatively little is known about how the attachment of carbohydrates affects the activity of peptides and proteins, one reason being the lack of suitable methods with which to access biologically relevant glycopeptides and glycoproteins. A given protein can exist as variant glycoforms, which complicates the isolation of well defined glycoconjugates. Another difficulty arises when glycoproteins are produced by heterologous expression techniques. The protein glycosylation pattern provided by a particular host cell might not match that of "natural" biosynthesis. In addition, host cells might respond to the artificial environment of cell culture media by displaying aberrant glycosylation.

Chemists are meeting the challenges involved in solving this availability problem. Rapid progress in "glycoscience" can, however, only be achieved through cooperative efforts. This requires information transfer within and between the biological and chemical disciplines. The current state of the art of glycopeptide synthesis is presented here in these terms, together with selected examples pointing out the biological role of protein glycosylation in T cell stimulation and cancer immunotherapy. The examples presented in the succeeding sections were selected with the aim of outlining some principles both of current synthetic methodology and of biological investigations, and their purpose is instructive rather than comprehensive. For more detailed information the reader is guided to some excellent review articles [1–9].

7.2
The Glycosidic Linkage

The majority of naturally occurring carbohydrates are linked through N-glycosidic bonds to the side chain of asparagine or through O-glycosidic bonds to the side chains of serine and threonine [10, 11]. Less common connections are O-glycosidic linkages to tyrosine, hydroxyproline, and hydroxylysine and C-glycosidic attach-

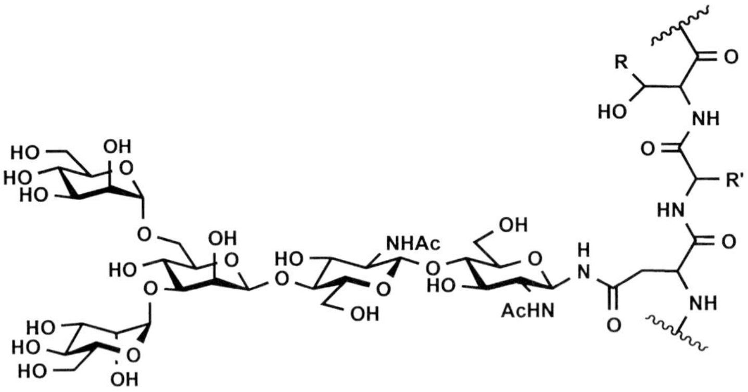

(Manα1→6[Manα1→3]Manβ1→4GlcNacβ1→4GlcNacβ)Asn-Xxx-Thr/Ser-

Fig. 7.1 Pentasaccharide core fragment and consensus sequence Asn-Xxx-Ser/Thr of N-glycosidically linked oligosaccharides.

ment to tryptophan. In the biosynthesis of N-glycosides, one common saccharide unit is co-translationally transferred from a dolichol phosphate to the consensus sequence Asn-Xxx-Ser/Thr, Xxx being any amino acid other than proline [12]. Various glycosidases shear the oligosaccharide to a common pentasaccharide core (Fig. 7.1) on which all further glycosylations occur.

The biosynthesis of O-glycosides does not follow a common pathway and so is more diverse. A D-2-acetamido-2-deoxygalactose (GalNAc) residue α-O-linked to serine and threonine – also termed T_N-antigen – is the most commonly found bridgehead (Fig. 7.2). A host of glycosyltransferases processes this core structure and creates a diverse set of so-called mucin-type O-glycosides. Mucins are heavily O-glycosylated proteins expressed on the surfaces of epithelial cells and the subjects of considerable research efforts, due to their roles as potential targets in tumor diagnosis and tumor therapy [13]. Nuclear pore proteins, transcription factors, and cytoskeletal proteins have been found to contain 2-acetamido-2-deoxyglucose (GlcNAc) residues linked to serine and threonine side chains [14, 15]. The introduction of the β-O-GlcNAc moiety seems to be a regulatory modification involved in signal transduction cascades [16, 17]. Structural proteins such as collagens contain hydroxylysine and hydroxyproline. Galactose residues or the Glcα1→2Gal disaccharide are β-O-glycosidically attached to these amino acids and appear to play a role in a mouse model of rheumatoid arthritis [18, 19]. Many more glycosylation types are known; for more detailed coverage the reader is guided to the review literature [20–29].

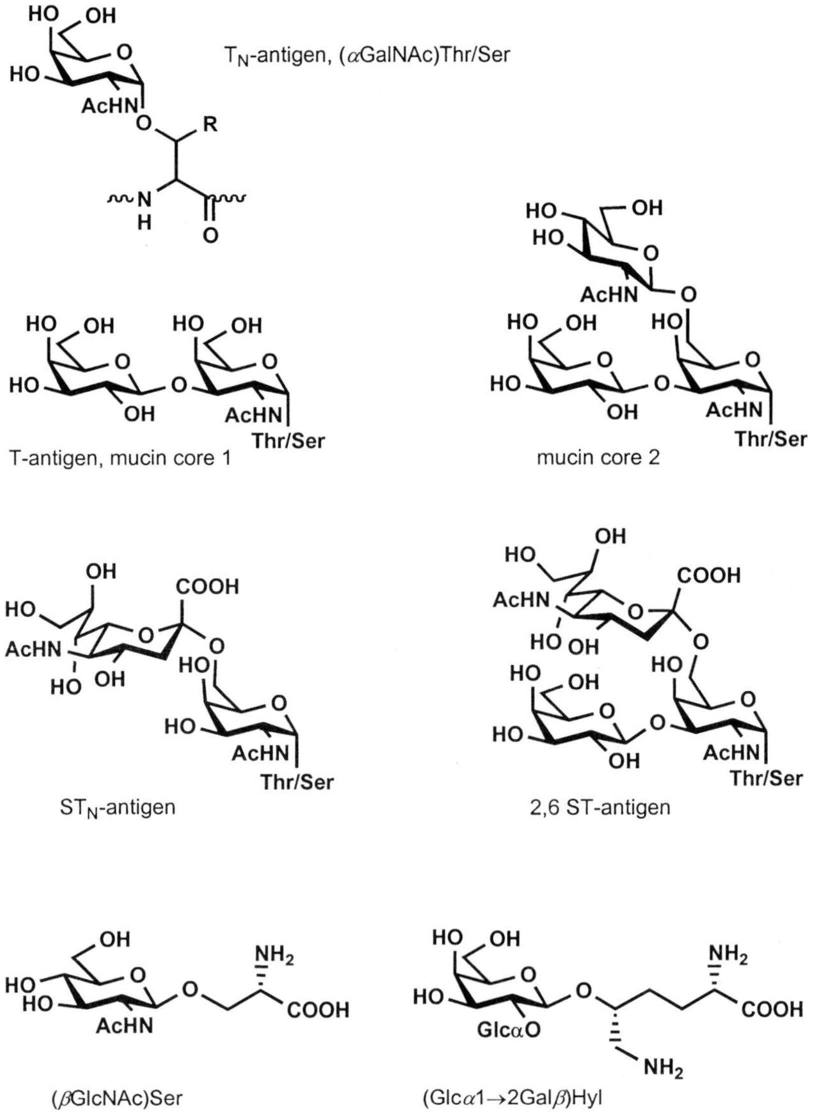

T_N-antigen, (αGalNAc)Thr/Ser

T-antigen, mucin core 1

mucin core 2

ST_N-antigen

2,6 ST-antigen

(βGlcNAc)Ser

(Glcα1→2Galβ)Hyl

Fig. 7.2 Selected examples of O-glycosidically linked carbohydrates.

7.3
The Challenges of Glycopeptide Synthesis

In the planning of a glycopeptide synthesis the additional complexity and lability conferred by the attachment of the carbohydrate group must be taken into account. The crucial step is the introduction of the glycan moiety. The establishment of N-glycosidic bonds usually proceeds through activation of the side chain

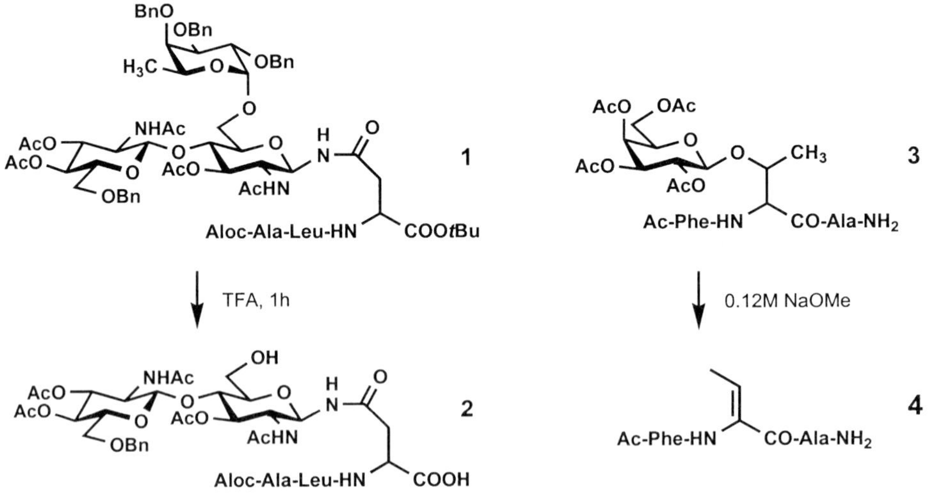

Scheme 7.1 Evidence of the acid lability (**1 → 2**) and the base lability (**3 → 4**) of some O-glycosidic linkages.

carboxyl group of aspartic acid. When performed with C-terminally elongated aspartic acid derivatives, this reaction can be complicated by the formation of aspartimide- and isoaspartic acid-containing by-products. Even more complex is the formation of O-glycosidic bonds. Stereoselective glycosylation has to be achieved, which is difficult when complex targets are involved. Furthermore, the polyfunctionality of carbohydrates necessitates an additional set of protecting groups, which have to be removable without harming the acid- and base-sensitive glycoside structures. Scheme 7.1 provides evidence of the acid-lability of certain O-glycosidic bonds. During the acidolytic cleavage of the *tert*-butyl ester in **1** a concomitant cleavage of the fucosidic bond occurred (→ **2**) [30]. It was found that acetylation, instead of benzylation, rendered the fucosidic linkage less labile, and global acetyl protection of carbohydrate hydroxy groups is now a standard technique. Acyl-type protection, however, requires a final base treatment. Because of the base-sensitivity of the O-glycosidic linkage, the conditions for acyl group removal have to be carefully adjusted. One possible side-reaction is β-elimination. This reaction was observed during the treatment of **3** with a 0.12 M solution of sodium methoxide in methanol, conditions typically applied for the removal of O-benzoyl groups [31]. The removal of O-acetyl groups usually proceeds smoothly when performed with highly diluted solutions of sodium methoxide. Recently, it has been shown that fluorobenzoyl groups have deprotection characteristics similar to those of acetyl groups [32].

7.4
Synthesis of Preformed Glycosyl Amino Acids

7.4.1
N-Glycosides

The usual tactics for the construction of N-glycosidic bonds involve peptide coupling between an aspartic acid derivative and a glycosylamine, so the preparation of glycosylamines is a key requirement. The most common procedure involves the synthesis of glycosylazides such as **6**, which can be obtained by treatment of glycosyl donors such as the halide **5** with azide salts (Scheme 7.2) [33–38].

An alternative route to glycosyl azides was reported by McDonald and Danishefsky [39]. The glycal **7** was allowed to react with iodocollidinium perchlorate in the presence of phenylsulfonamide. The resulting iodosulfonamide **8** was treated with sodium azide, which afforded the desired glycosyl azide **9**. Prior to the construction of the N-glycosidic bond the glycosyl azides are reduced to the glycosyl amines **10** by hydrogenation, hydride reduction, or Staudinger reaction (Scheme 7.3) [30, 39–44]. Glycosylamines can be prepared directly from unprotected sugars (reaction b) by Kochetkov's procedure [45, 46]. Both protected and unprotected glycosyl amines have been used in peptide coupling with a partially protected aspartic acid. Virtually all protecting group combinations appear to be possible [37, 41, 46–52].

A critical step in the synthesis of complex N-glycosyl asparagines is the stereoselective construction of the β-mannose linkage. For example, in Ogawa's elegant synthesis of an Asn-linked core pentasaccharide, the preparation of the key intermediate trisaccharide **16** began with the use of the mannosyl bromide **12** as a glycosyl donor and the glycosyl fluoride **13** as an acceptor (Scheme 7.4) [53]. The desired β isomer was obtained in 53% yield, the α isomer in 38% yield. At this early stage of the synthesis the rather unselective formation of the β-mannoside **14** was

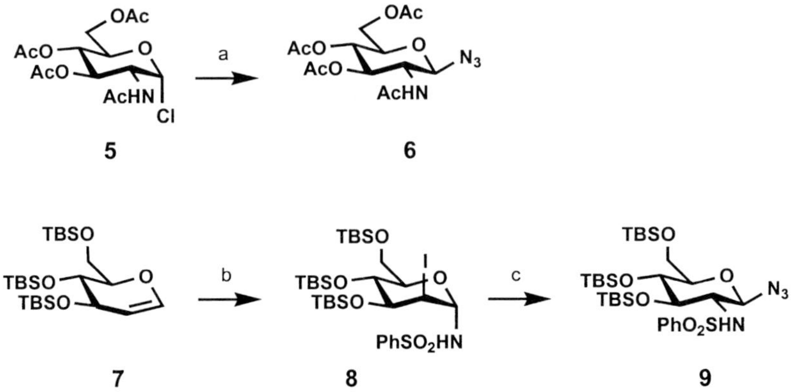

Scheme 7.2 Typical glycosyl azide syntheses. a) NaN_3, CH_2Cl_2, $NaHCO_3$, H_2O, Bu_4NHSO_4, 98%; b) $I(coll)_2ClO_4$, H_2NSO_2Ph, CH_2Cl_2, $-10\,°C$, 60%; c) NaN_3, DMF, 92%.

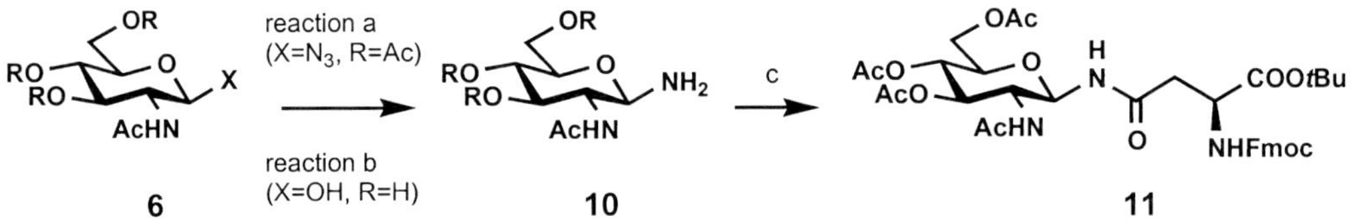

Scheme 7.3 Typical examples of: a) reduction of a glycosyl azide: H_2, Raney-Ni; b) conversion of an unprotected carbohydrate into the glycosylamine: sat. NH_4HCO_3, and c) coupling between a glycosylamine and Fmoc/tBu-protected aspartic acid: DCC, HOBt, THF, 59% (from **6**).

tolerated, especially since the disaccharide product **14** could then be employed directly as glycosyl donor for the glycosylation of glycosyl azide **15**. Danishefsky and Seeberger coped with the β-mannose problem by using a β-glucose C2-epimerization approach [54]. Firstly, the use of an acetate group at the C2-position of the trisaccharide donor **17** secured the stereoselective β-glycosylation of acceptor **18**. Deacetylation of **19** yielded the free C2-alcohol, which was subjected to an oxidation/reduction sequence to deliver the β-mannoside product **21**. In Unverzagt's impressive synthesis of the sialylated undecasaccharide-asparagine conjugate **24** [40], the β-manno configuration was installed by use of a procedure reported by Günther and Kunz [55, 56]. Heating of the trisaccharide building block **22** in DMF and pyridine resulted in inversion at C2 through cyclization. The resulting β-mannoside **23** was chemically extended to a heptasaccharide-asparagine conjugate, which was subjected to a series of enzymatic glycosyl-transfer reactions.

The effort involved in glycan assembly can be avoided by releasing naturally occurring N-linked oligosaccharides from natural glycoproteins. For example, mild hydrazinolysis of bovine fetuin **25**, the major glycoprotein of fetal calf serum, provided the triantennary complex-type oligosaccharidyl hydrazine **26** (Scheme 7.5) [57]. After N-acetylation, the acetohydrazide was treated with copper acetate to yield the oligosaccharide **27**. The conversion to the glycosyl amine by Kochetkov's procedure was followed by coupling with Fmoc/tBu-protected aspartic acid. Subsequent O-acetylation and acidolysis furnished building block **28** ready for usage in solid-phase synthesis (see Scheme 7.19).

Scheme 7.4 a) Ag,Al silicate, MS (4 Å), CH_2Cl_2, 0 °C to rt, 91% (α:β=1:1.4); b) MS (4 Å), Cp_2HfCl_2, $AgClO_4$, CH_2Cl_2, –20 °C to 0 °C, 76%; c) MS (4 Å), MeOTf, DTBP, CH_2Cl_2, 64%; d) LAH, Et_2O, 0 °C, 86%; e) Dess-Martin, CH_2Cl_2; f) L-selectride, THF, –40 °C, 83%; g) (i) DMF, Pyr, 60 °C; (ii) AcOH, dioxane, H_2O, 0 °C; $BF_3 \cdot Et_2O$, CH_2Cl_2, 85%; h) NaOMe, MeOH, CH_2Cl_2, 70%.

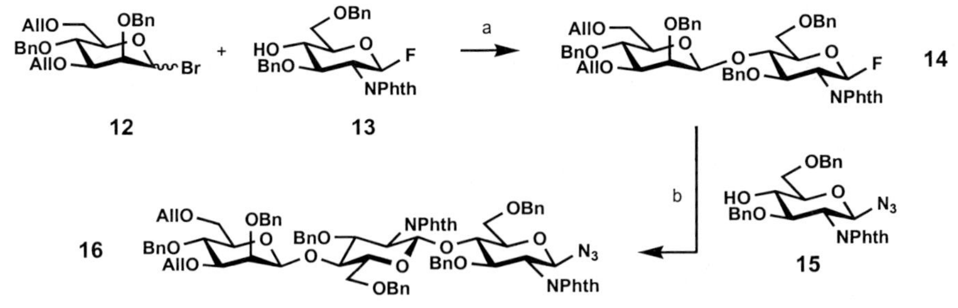

7.4 Synthesis of Preformed Glycosyl Amino Acids

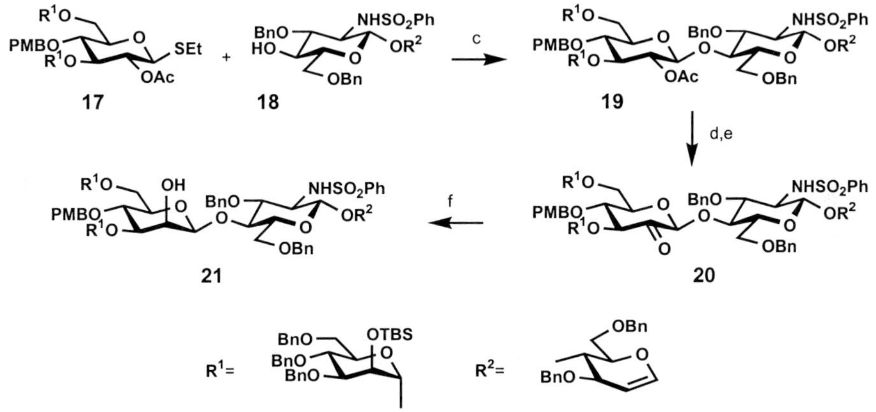

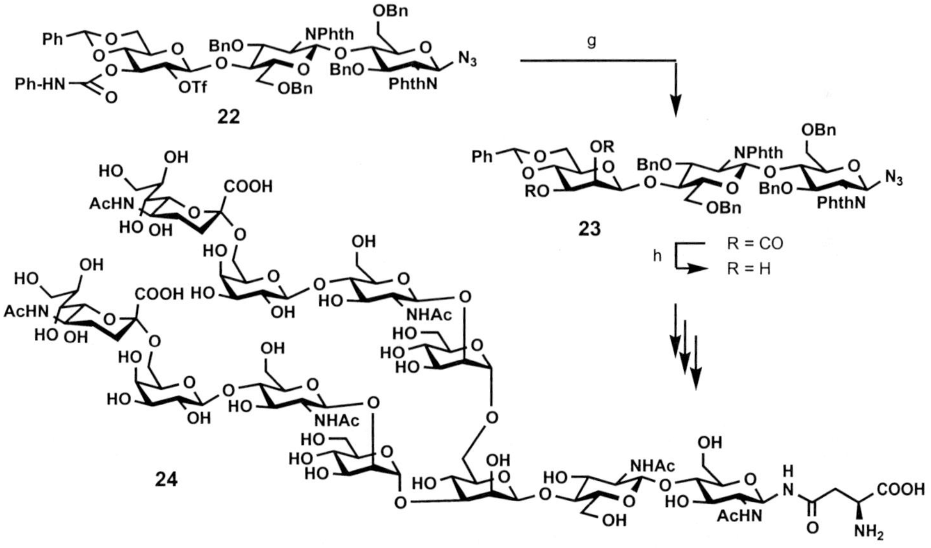

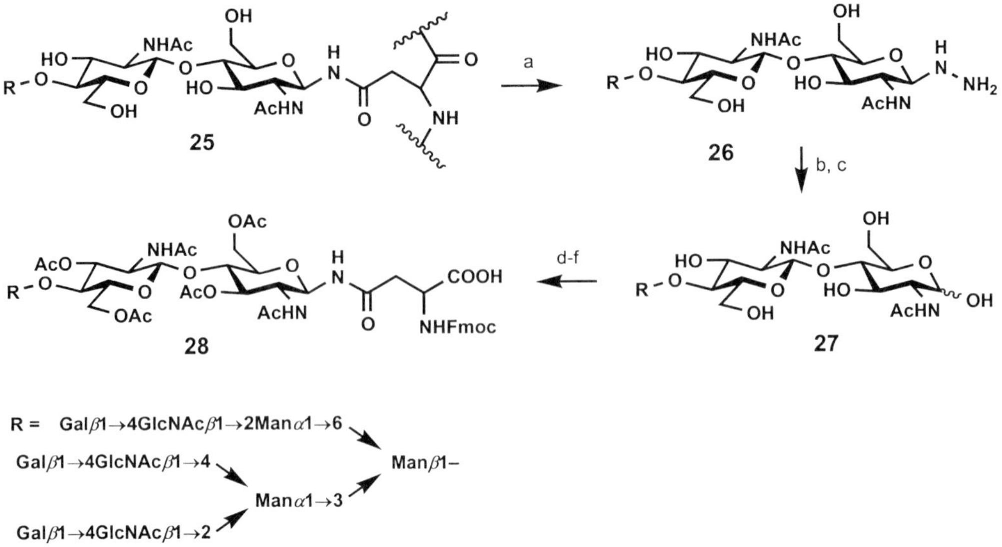

Scheme 7.5 a) H$_2$NNH$_2$, 85 °C; b) Ac$_2$O, sat. NaHCO$_3$, 0 °C, Dowex AG-50X12; c) Cu(OAc)$_2$, 0.1 M AcOH; d) sat. NaHCO$_3$, 45 °C, 95%; e) Fmoc-Asp(ODhbt)-O*t*Bu, Et-N*i*Pr$_2$, DMSO; f) (*i*) Ac$_2$O, Pyr; (*ii*) TFA.

7.4.2
O-Glycosides

7.4.2.1 O-Glycosyl Amino Acids bearing Mono- or Disaccharides

A great deal of work has been directed towards the synthesis of suitably protected serine/threonine building blocks bearing the most abundant α-*O*-GalNAc modification. Nearly all syntheses have relied on the 2-azido group as a non-participating precursor of the 2-acetamido moiety [58]. Most commonly, 1-bromo and 1-chloro sugars such as **29** have been employed in the glycosylation of Fmoc-protected serine/threonine esters and active esters (Scheme 7.6) [59–62]. The use of

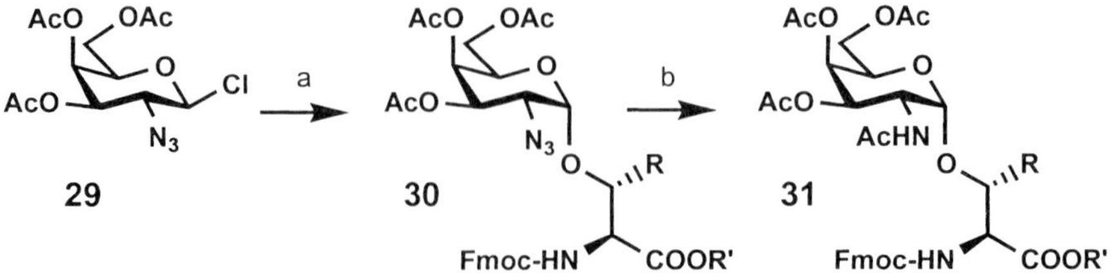

Scheme 7.6 a) Fmoc-NH-CH(CHROH)-COOR′, Ag$_2$CO$_3$/AgClO$_4$ (R′=Bn: R=CH$_3$, 60%, pure α anomer; R=H, 65% pure α anomer; R′=*t*Bu: R=CH$_3$, 81%, α:β=9:1; R=H, 70%, α:β=8:1; R′=All: R=CH$_3$, 50%, pure α anomer; R=H, 55%, pure α anomer; b) R′=Bn:(*i*) NaBH$_4$, NiCl$_2$; (*ii*) Ac$_2$O, Pyr, 85% (R=CH$_3$), 86% (R=H); R′=*t*Bu: (*i*) H$_2$S, Pyr; (*ii*) Ac$_2$O, Pyr, 81% (R=CH$_3$), 75% (R=H); R′=All: CH$_3$COSH, 82% (R=CH$_3$), 81% (R=H).

7.4 Synthesis of Preformed Glycosyl Amino Acids

insoluble promoters such as silver perchlorate usually gives high α selectivity. To establish the 2-acetamido group, the azido group in **30** is subjected to a reduction/acetylation sequence or reductively acetylated with thioacetic acid. Recently, Schmidt and co-workers reported a conceptually different route [61–64]. A 2-nitrogalactal was used as a Michael-type acceptor in the reaction with Boc/tBu-protected serine or threonine [65].

Interest in the synthesis of β-O-GlcNAc-Ser/Thr building blocks has been increased by their putative roles as regulatory protein modifications [16]. In principle, treatment of the peracetylated glucosamine **32** with N-protected serine or threonine provides the quickest route to GlcNAc-substituted amino acids such as **34** (Scheme 7.7) [66]. This reaction is induced by the presence of a Lewis acid such as $BF_3 \cdot Et_2O$ and proceeds through the oxazoline **33**. Interestingly, protection of the amino acid carboxyl group is not required, although the need for laborious purification procedures can be a serious limitation [67]. Very recently, it has been shown that $CuCl_2$ can serve as mild activator of the oxazoline **33**, enabling efficient coupling to Fmoc/All-protected serine [68]. Suppression of oxazoline formation can be achieved by replacement of the N-acetyl group with electron-withdrawing groups; Aloc, Troc, and Dts protecting groups were introduced and shown to support high-yielding glycosylations [69–73]. The DMTST-promoted reaction between the thioglycoside **35** and Fmoc/Bn-protected threonine, for example, furnished conjugate **36** in high yield [74]. Reductive cleavage of the Troc-group, N-acetylation, and hydrogenolysis completed the building block synthesis.

For studies of the T-cell response involved in the induction of rheumatoid arthritis, β-galactose and β-(glucosyl-(α1 → 2)-galactose) were attached to hydroxylysine. A convenient synthesis of the (Galβ)-Hyl building block **39** proceeded

Scheme 7.7 a) MS (4 Å), CH_2Cl_2, $BF_3 \cdot Et_2O$, PG^N=Fmoc: R=H, 55%, R=CH_3, 53%; PG^N=Z: R=H, 49%, R=CH_3, 41%; PG^N=PhacOZ: 54%; b) DMTST, CH_2Cl_2, 78%; c) (i) Zn, AcOH, (ii) Ac_2O, Pyr, 77%; d) Pd/C (5%), H_2, EtOAc, EtOH, 88%.

through the copper complex **37**, which was used as acceptor in the glycosylation with peracetylated 1-bromogalactose (Scheme 7.8) [75]. Decomplexation of **38** and subsequent introduction of the Fmoc group furnished the desired conjugate **39** in five steps in total and in 29% overall yield. Kihlberg and co-workers reported the synthesis of the (Glcα1 → 2Galβ)-hydroxylysine building block **45** [76]. The glycosylation of the hydroxylysine **41** was achieved in the presence of $ZnCl_2$ by use of α-1,2-anhydrogalactose **40** as glycosyl donor. This reaction resulted in opening of the

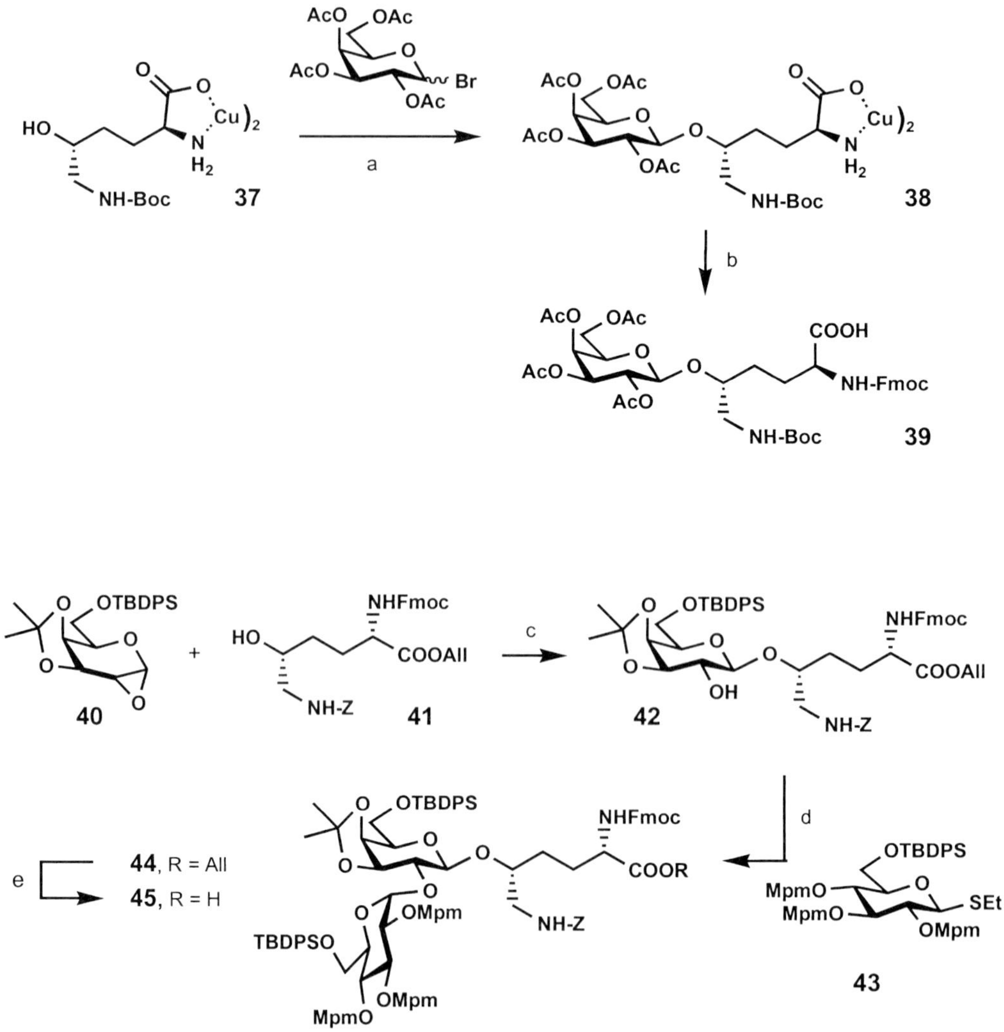

Scheme 7.8 a) (*i*) NaH, MeCN, 77%; b) MeOH, H_2O, chelex 100 (H^+ form); (*ii*) Fmoc-OSu, $NaHCO_3$, H_2O, Me_2CO, 50%; c) $ZnCl_2$, THF, MS (AW-300), –50 °C to rt., 30%; d) MS (4 Å), NIS, AgOTf, CH_2Cl_2, –45 °C, 47%; e) [Pd(PPh_3)$_4$], PhNHMe, THF, 92%.

epoxide structure and afforded the unprotected C2-hydroxy group. Treatment of **42** with the benzyl-type protected thioglucoside **43** and N-iodosuccinimide and silver triflate promoted the formation of the disaccharide conjugate **44**. The subsequent Pd(0)-catalyzed allyl transfer to N-methylaniline yielded the diglycosylated hydroxylysine **45** as a building block for solid-phase peptide synthesis.

7.4.2.2 O-Glycosyl Amino Acids bearing Complex Carbohydrates

Tactics for the synthesis of complex O-glycosyl amino acids follow two categories: the block glycosylation approach and the cassette approach. In the former, a fully elaborated oligosaccharide is attached to a hydroxy group-containing amino acid. The glycosylation chemistry and conditions have to be carefully optimized, as shown in the synthesis of the sialyl-(2→6)-T-antigen building blocks **47** (Scheme 7.9) [77]. Activation as a trichloroacetimidate gave the highest α selectivity for the glycosylation of serine, whereas the Königs-Knorr reaction was found to be optimal for threonine. The trichloroacetimidate method appears to have a broad applicability for the activation of complex oligosaccharides. A bissialyl-T-antigen building block, for example, has been obtained by trichloroacetimidate activation of the corresponding tetrasaccharide [78].

Despite some success, the block glycosylation is often plagued by modest yields and poor selectivity, which is tedious at this advanced stage of synthesis. The cassette approach, in which a preformed α-GalNAc-Thr/Ser glycoside is used as acceptor for further glycosylations, offers an appealing alternative. In the laboratories of Bock, Meldal, and Paulsen and Kunz, the 4,6-benzylidene-protected **48** was used as universal precursor of mucin core 1, core 2, core 4, and core 6 building blocks and a sialyl-T_N threonine (Scheme 7.10) [60, 79].

The broad applicability of the cassette approach was demonstrated in Danishefsky's syntheses of building blocks containing T, sialyl-T_N, sialyl-T, and bissialyl-T antigen structures, as well as in the synthesis of a Lewis Y-containing pentasaccharide-serine conjugate [80]. In a sialyl-T antigen synthesis, the common precur-

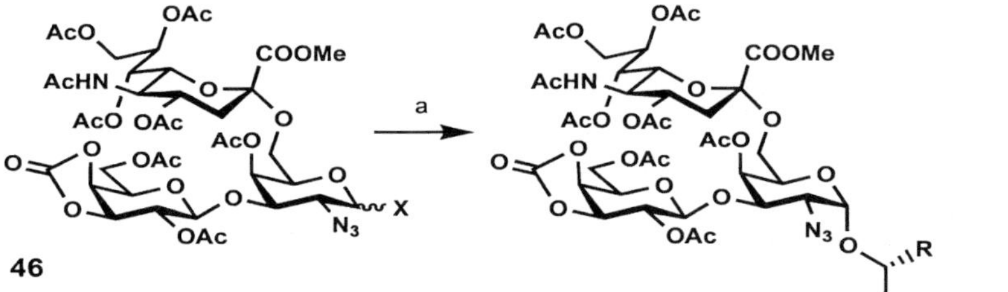

X=	47, R = H α:β (%)	R = CH$_3$ α:β (%)
Br	2.6:1 (70%)	pure α (74%)
O(CNH)CCl$_3$	12:1 (65%)	pure α (63%)

Scheme 7.9 a) X=Br: AgClO$_4$, CH$_2$Cl$_2$; X=O(CNH)CCl$_3$: BF$_3$·Et$_2$O, THF, −30 °C.

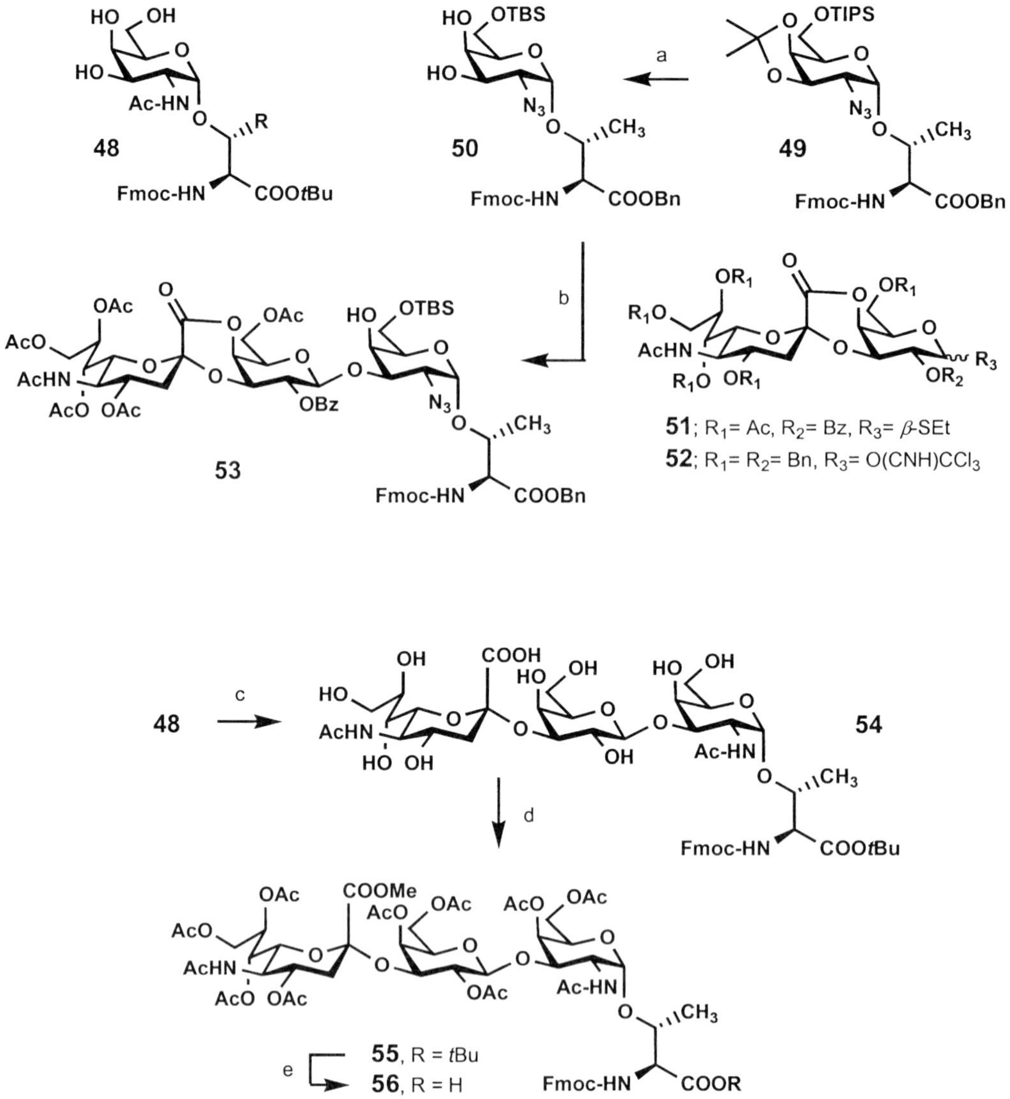

Scheme 7.10 a) (i) I₂, MeOH; (ii) TBS-Cl, imidazole, DMF, 85%; b) **51**, NIS, TfOH, CH₂Cl₂, MS (4 Å), 10 min, 62%; c) (i) lactose, β-galactosidase, BSA, 2,6-dimethyl-β-cyclodextrin, pH 6.5; (ii) CMP-NeuNAc, sialyltransferase, BSA, 2,6-dimethyl-β-cyclodextrin, pH 6.5, 50%; d) (i) Ac₂O, Pyr; (ii) MeOH, EtNiPr₂; (iii) Ac₂O, Pyr, 61%; e) TFA, 84%.

sor 2-azidogalactosyl threonine **49** was converted into the 3-acceptor **50** [59]. Activation of the thioethyl glycoside **51** with NIS in the presence of trifluoromethanesulfonic acid afforded the trisaccharide **53** in high yield and with excellent β selectivity. A similar lactone-protected disaccharide donor **52** was used by Nakahara and co-workers, who favored trichloroacetimidate activation and O-benzyl protection [81]. Kunz and co-workers reported a short, chemoenzymatic synthesis of the sialyl-T threonine building block **56** [82]. The sialyl-T threonine derivative **54** was obtained in a one-pot reaction sequence in which Fmoc/*t*Bu-protected (GalNAc*α*)-threonine **48** was subjected to a *β*-galactosidase-catalyzed transgalactosylation employing lactose as donor, and a subsequent sialyl transfer [83]. The resulting *O*-unprotected conjugate **54** was converted into the fully protected sialyl T threonine **55** before acidolysis of the *t*Bu-ester completed the building block synthesis.

7.5
Synthesis of Glycopeptides

7.5.1
N-Glycopeptide Synthesis in Solution

A typical example of the solution-phase assembly of *N*-glycopeptides is shown in Scheme 7.11. In their synthesis of the *N*-glycopeptide cluster **62**, containing two Lewis X residues, von dem Bruch and Kunz demonstrated that the Boc group was removable without detriment to the labile *α*-fucoside linkage (**57** → **58**), thereby highlighting the beneficial effect of *O*-acetyl protection [84]. After elaboration to the dipeptide **60**, coupling with the *C*-terminally unprotected glycosyl asparagine **59** was performed. Deallylation of the tripeptide cluster **61** was followed by a fragment condensation to give the fully protected form of conjugate **62**. Successive treatment with formic acid and with highly diluted sodium methylate in methanol cleaved the *t*Bu ester and the *O*-acetyl groups, respectively. A convergent method was reported by Cohen-Anisfield and Lansbury [45], who optimized the acylation of *O*-unprotected glycosylamines, synthesized by Kochetkov's procedure, with aspartic acid-containing peptides. Danishefsky and co-workers recently applied this methodology in their impressive syntheses of the *N*-glycopeptide **66**, bearing a 15mer oligosaccharide with H-type 2 blood group determinants (Scheme 7.12) [85]. The use of the Kochetkov-Cohen-Lansbury procedure was deemed necessary because it was observed that the pentasaccharide glycosyl amine **63**, prepared by reduction of the corresponding glycosyl azide, suffered from anomerization during the azide reduction/acylation sequence, resulting in a mixture of *α-N-* and *β-N*-glycosyl linkages. It was found that cleaner acylations could be achieved by use of OH-unprotected glycosyl amines such as **64** or **65** and potent coupling reagents such as HBTU or HATU.

Despite the great synthetic utility of the approach, particularly when only small amounts of glycosyl amines are available, it has to be noted that the formation of aspartimides can be a side-reaction (see Scheme 7.20). The aspartimide-forming reactions are dependent on steric factors, convergent coupling of protected glycosylamines to conformationally constrained cyclopeptides usually proceeding smoothly, as shown in the synthesis of the trivalent sialyl-Lewis X conjugate **68** (Scheme 7.13) [86].

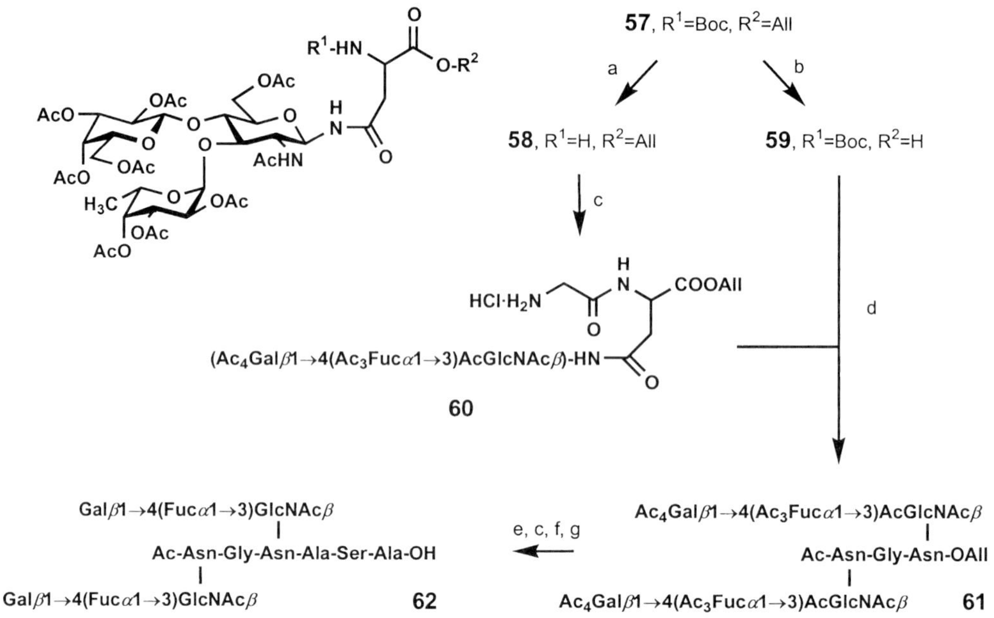

Scheme 7.11 a) HCl, Et$_2$O, 92%; b) [(PPh$_3$)$_3$RhCl], EtOH/H$_2$O (9:1), 70 °C, 93%; c) (i) Boc-Gly-OH, IIDQ, CH$_2$Cl$_2$; (ii) HCl, Et$_2$O, 86%; d) EDC, DIPEA, HOBt, CH$_2$Cl$_2$, 64%; e) (i) HCl, Et$_2$O; (ii) Ac$_2$O, Pyr, 82%; f) TBTU, HOBt, H-Ala-Ser-Ala-O*t*Bu, MeCN, 73%; g) (i) HCOOH; (ii) NaOMe, MeOH, pH 8.5, 87%.

Enzymatic synthesis has been demonstrated to be a very powerful approach for the construction of glycopeptides and glycoproteins. One promising technique is the use of endoglycosidases in transglycosylation reactions [87]. Inazu and co-workers applied endoglycosidase M from *Mucor hiemalis*, which shows a broad substrate tolerance [88]. The preparation of eel calcitonin analogues containing natural *N*-linked glycans began with the synthesis of the *N*-GlcNAc peptide **69**, which was obtained by a combination of solid-phase synthesis and fragment condensation in solution (Scheme 7.14). In the central step, the glycosylasparagine **70** was used as glycosyl donor. Endo-β-*N*-acetylglucosaminidase from *Mucor hiemalis* then catalyzed the transfer of the disialo complex-type oligosaccharide from human transferrin to the *N*-glycopeptide **69**. After 6 h the desired product **71** was obtained in 9% yield. It should be noted that the pursuit of a transglycosylation reaction requires preformed glycosyl asparagines. However, it adds to the attractiveness of this approach that bioactive peptides or proteins that might be difficult to synthesize by chemical means can be used as glycosyl acceptors.

Single-glycoform proteins can also be obtained by the use of glycosyltransferases. Wong and co-workers removed the natural *N*-glycans from ribonuclease B with the aid of glycosidases, leaving only the innermost *N*-linked GlcNAc-residue [89]. Successive treatment with galactosyl, sialyl, and fucosyl transferases together with the corresponding sugar donors yielded a well defined sialyl-Lewis X-contain-

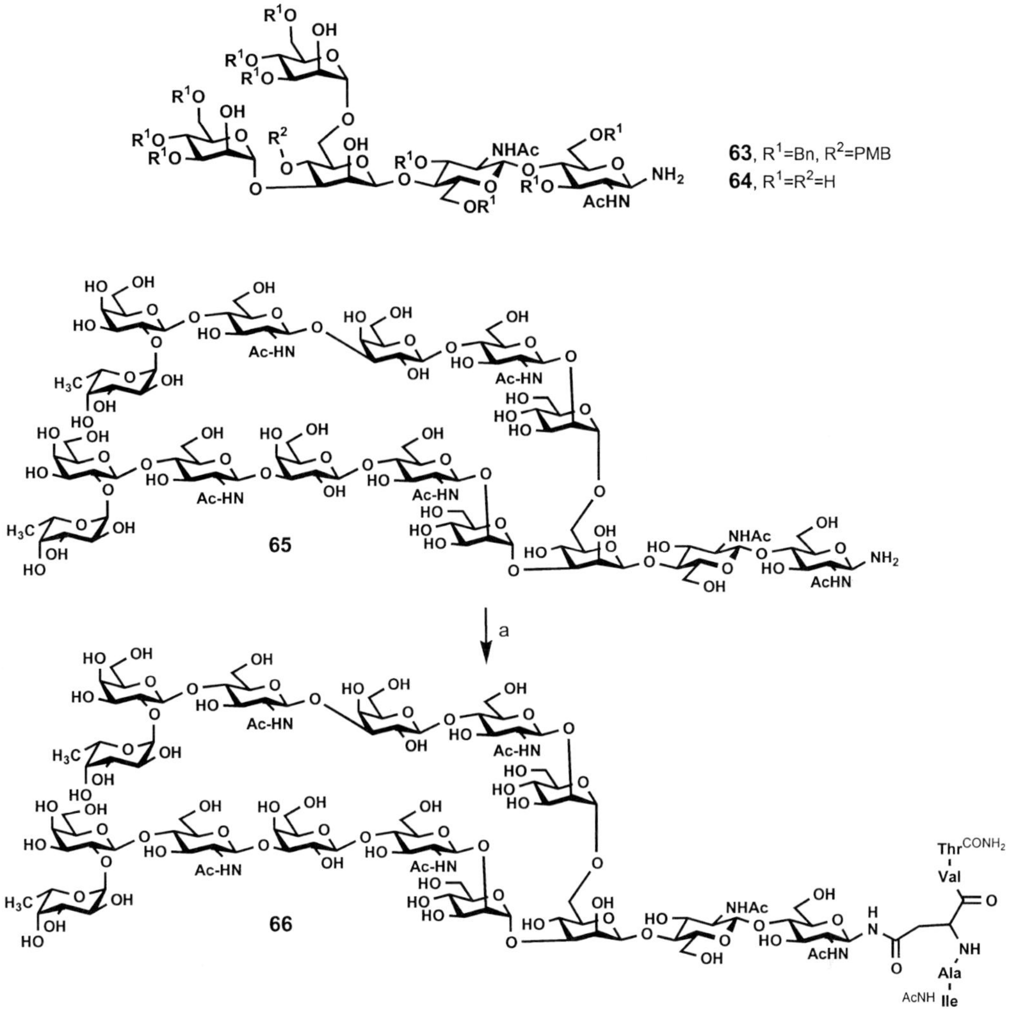

Scheme 7.12 a) HBTU, HOBt, DIPEA, DMSO, 20%.

ing ribonuclease, thus converting a potentially heterogeneous population of glycoproteins into a single glycoform. An alternative approach is the use of intein-mediated ligation [90, 91] of synthetic glycopeptides to larger biosynthesized proteins (Scheme 7.15). This technique utilizes a natural splicing event, so that a C-terminal protein thioester such as **73** is created. Addition of mercaptoethylsulfonic acids results in the formation of a second thioester. Thioesters react with peptides or with a glycopeptide such as **75**, containing an N-terminal cysteine, to form ligation products (→**76**) by a mechanism known as Native Chemical Ligation (see also Scheme 7.29) [92].

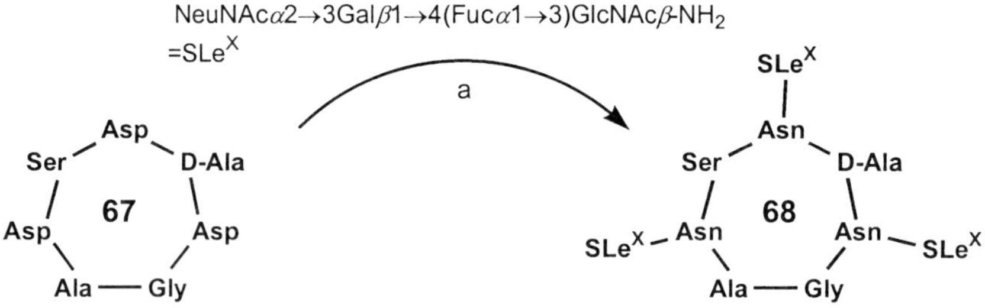

Scheme 7.13 a) (*i*) HATU, HOAt, DIPEA, DMF, 48%; (*ii*) H₂, Pd/C, MeOH/dioxane/AcOH (5:1:1), 88%; (*iii*) NaOH in H₂O/MeOH, pH 10.4, 96%.

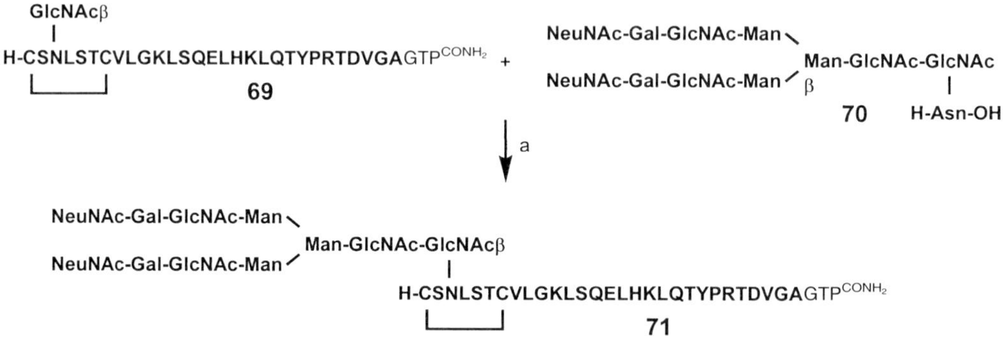

Scheme 7.14 a) Phosphate buffer (pH 6.25), *endo*-β-GlcNAc-ase, 9%.

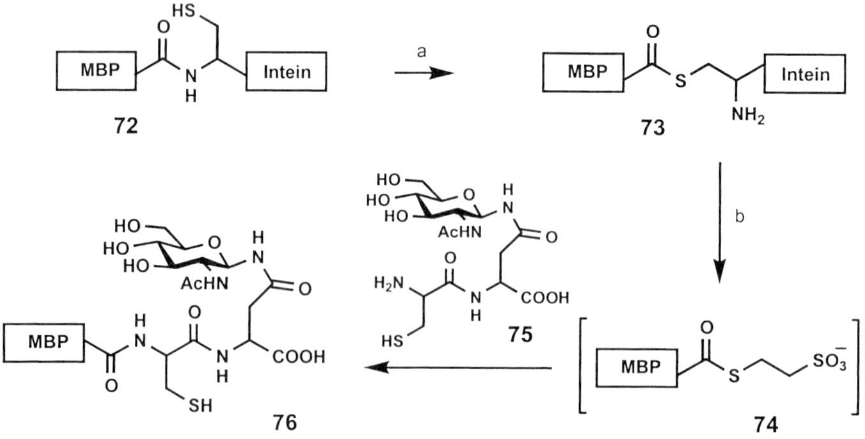

Scheme 7.15 a) Intein-catalyzed thioester formation; b) 1 mM HS(CH₂)₂SO₃H, 30 mM **75**, pH 7.5.

7.5.2
O-Glycopeptide Synthesis in Solution

Danishefsky's group has employed solution-phase techniques for the synthesis of various O-glycopeptides with mucin-type carbohydrate structures. The sialyl T building blocks **77**, for example, were C-terminally elongated and used in iterative peptide coupling to yield pentapeptide **78** (Scheme 7.16) [80]. A similar strategy al-

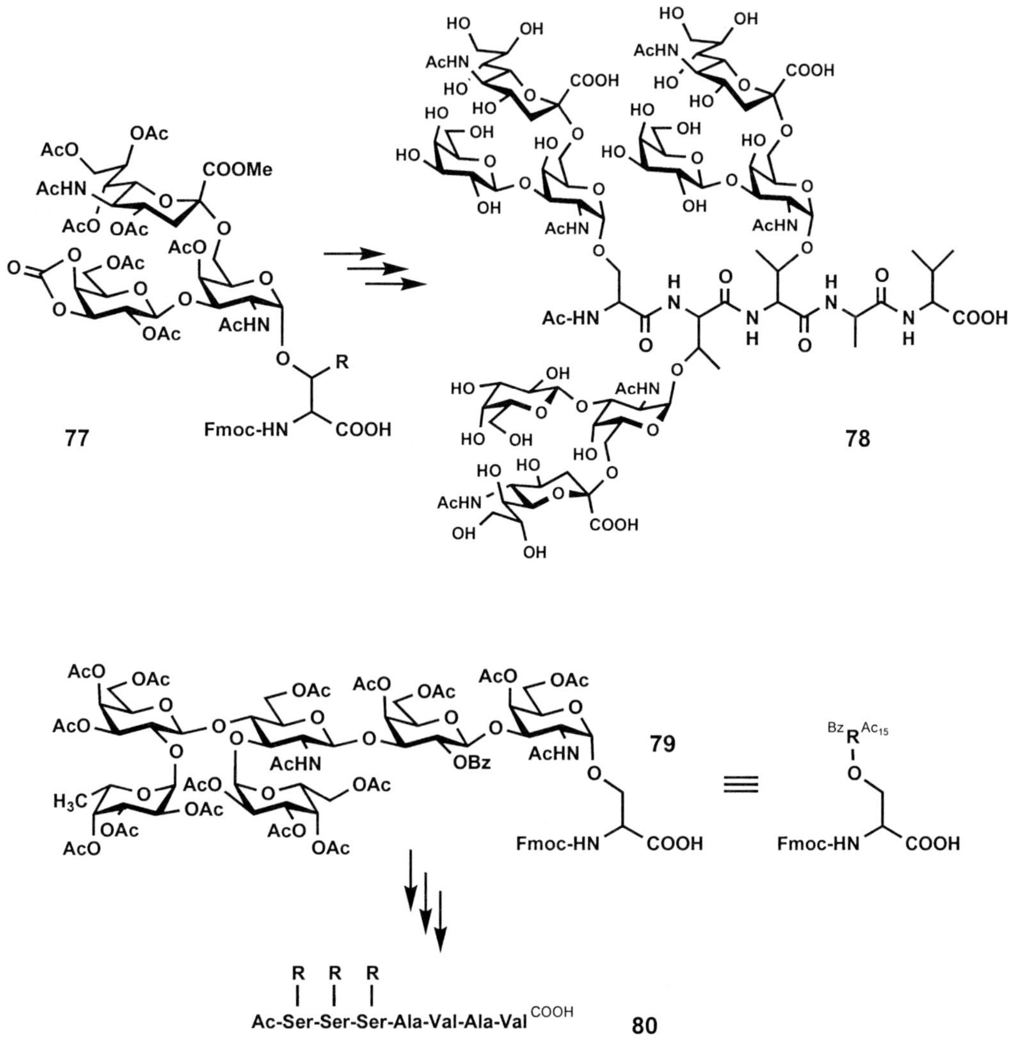

Scheme 7.16 Use of the sialyl T building blocks **77** in the synthesis of sialyl T cluster **78**. The Lewis Y-bearing mucin-type serine building blocks **79** were applied in the solution assembly of the hexasaccharide cluster **80**.

lowed the synthesis of a nonapeptide clustering three sialyl-T structures and a tripeptide with three T antigen units [59, 77]. The synthesis of the hexasaccharide cluster **80** was based on the Lewis Y-bearing mucin-type serine building block **79** and glycopeptide assembly in solution [93, 94]. During the deprotection of the clustered hexasaccharide, removal of the benzoyl esters proved difficult. Mild hydrazinolysis, however, cleaved acetates and benzoates in a clean reaction, affording the fully deprotected glycopeptide for subsequent evaluation of its immunological properties.

Significant progress has been made in the enzymatic oligosaccharide assembly of O-glycopeptides. An early example was reported by Schultz and Kunz, who employed $\beta 1 \rightarrow 4$-galactosyltransferase to achieve a regioselective galactosylation of a short O-glycopeptide [73]. More recently, an O-linked sialyl-Lewis X-containing glycopeptide was synthesized by drawing on a set of three glycosyltransferases [74]. Kihlberg and co-workers used recombinant mouse GalNAc $a2 \rightarrow 6$ sialyltransferase and human $a2 \rightarrow 3$ sialyltransferase to sialylate the T_N and T antigen-containing glycopeptides **81** and **83**, respectively (Scheme 7.17) [95].

Wong and co-workers applied a microbial $a2 \rightarrow 3$ sialyltransferase from *Neisseria gonorrheae* and reported a remarkably broad acceptor substrate tolerance [96]. As an example, attempts to sialylate the trisaccharide-sulfopeptide **86a** with a mamma-

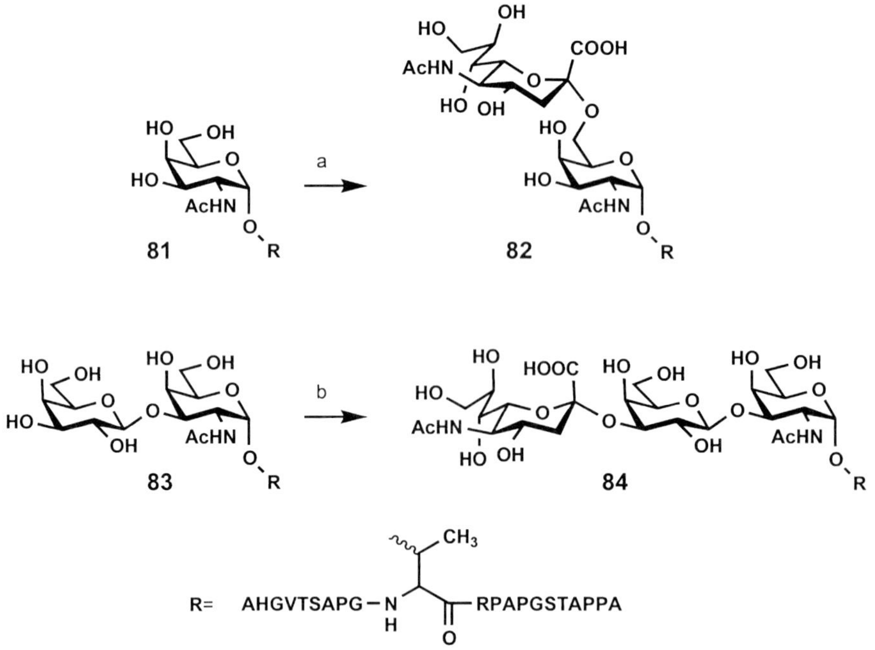

Scheme 7.17 a) CMP-NeuNAc, $a2,6$-sialyltransferase, Bis-Tris buffer (20 mM, pH 6.0), 37°C, 74%; : b) CMP-NeuNAc, $a2,3$-sialyltransferase, calf intestinal phosphatase, Tris-HCl buffer (25 mM, pH 6.5), 37°C, 94%.

lian sialyltransferase had failed (Scheme 7.18) [97]. The microbial sialyltransferase, however, proved reactive and catalyzed the reaction with the sulfated glycopeptide **86a**, giving sialylated product **87a**. During the chemoenzymatic synthesis of PSGL-1 glycopeptides it was noted that the activity of the $\beta 1 \rightarrow 4$ GalNAc transferase was also influenced by sulfation of the distant tyrosine residue in **85a**. The reaction between $\beta 1 \rightarrow 4$ GalNAc transferase and sulfated glycopeptide **85a** proceeded slowly

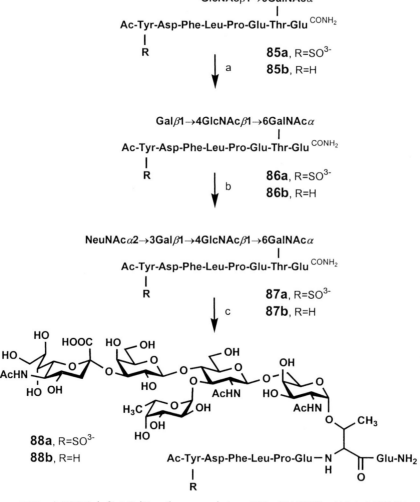

Scheme 7.18 a) UDP-Gal, β1,4-GalTase (bovine), alkaline phosphatase, 130 mM HEPES, pH 7.4, 0.25% Triton X-100, MnCl$_2$, protease inhibitor cocktail ("PIC" Sigma) (**85a**: 75%, **85b**: 72%); b) CMP-NeuNAc, α2,3-SiaTase (**86a**: microbial, **86b**: rat liver), alkaline phosphatase, 130 mM HEPES, pH 7.4, 0.25% Triton X-100, MnCl$_2$ (**86a**: +PIC, 69%, **86b**: –PIC, 48%); c) GDP-Fuc, α1,3-FucTase V (human), alkaline phosphatase, 100 mM MES, pH 6.0, 0.25% Triton X-100, MnCl$_2$ (**87a**: +PIC, 87%, **87b**: –PIC, 65%).

relative to the reaction of unsulfated **85b**. The discussed sialylation and subsequent $\alpha 1 \rightarrow 3$ fucosyltransferase-catalyzed fucosylation completed the synthesis of the PSGL-1 glycopeptides **88** with α-O-linked sialyl-Lewis X moiety [98].

Cummings and co-workers applied a similar strategy for the synthesis of various PSGL-1 glycosulfopeptides [99]. The approaches described above require the synthesis of a preformed glycosyl amino acid. A de novo glycosylation is feasible by employment of a polypeptide:GalNAc transferase preparation from microsomal membranes of colorectal cancer cells [100]. These enzymes catalyzed the O-glycosylation of synthetic peptides corresponding to the human MUC2 tandem-repeat domain. A recent study presented evidence that the glycosylation specificity is controlled both by the peptide sequence and by the structure of previously introduced O-glycans [101].

7.5.3
Solid-Phase Synthesis of N-Glycopeptides

Solid-phase synthesis offers the opportunity to automate the repetitive process of building block coupling. The most important advantages of solid-phase-based methods are the opportunity to use large excesses of building blocks that can drive reactions to completion, the high speed of automated synthesis, and the possibility to implement parallel or combinatorial synthesis formats easily.

The most commonly applied strategy employs the iterative coupling of preformed building blocks. With a few alterations, which usually concern the removal of carbohydrate protecting groups, the techniques of modern peptide chemistry can be applied [102–104]. An example illustrating the current state of the art is shown in Scheme 7.19. The resin-bound tripeptide **89** was assembled by the Fmoc strategy. In the subsequent coupling of the impressive glycosylasparagine **28**, only 1.1 equivalents were used. Chain extension furnished **91**, and acidolytic cleavage from the hydroxymethylphenyl acetic acid (HMPA) linker [105] gave the O-acetylated peptide **92**. Methoxide-promoted O-deacetylation then allowed the synthesis of the triantennary N-glycopeptide **93**. Kihlberg and co-workers, and later Imperiali and co-workers, experienced problems during the final O-deacetylation of N-linked chitobiose peptides [106, 107]. A chitobiosylasparagine in which the carbohydrate hydroxy groups were protected as acid-labile TBDMS ethers was therefore prepared. The main advantage of this method is that concomitant cleavage of both peptide side chain-protecting groups and the resin linkage can be achieved. Nakahara and Ogawa preferred O-benzyl protection in the synthesis of an N-linked core pentasaccharide CD52-glycopeptide [53, 108].

The use of preformed glycosyl asparagines is usually a reliable method for N-glycopeptide synthesis. The synthesis of the corresponding building blocks, however, is time-consuming. Rapid and convergent access is provided by methods that utilize on-resin coupling of glycosylamines to aspartic acid-containing peptides. Albericio's group incorporated Fmoc-protected aspartic acid β allyl ester and performed the deallylation and a subsequent coupling to a glycosylamine on the resin-bound full-length peptide **95a** (Scheme 7.20) [109]. As a severe side-reaction,

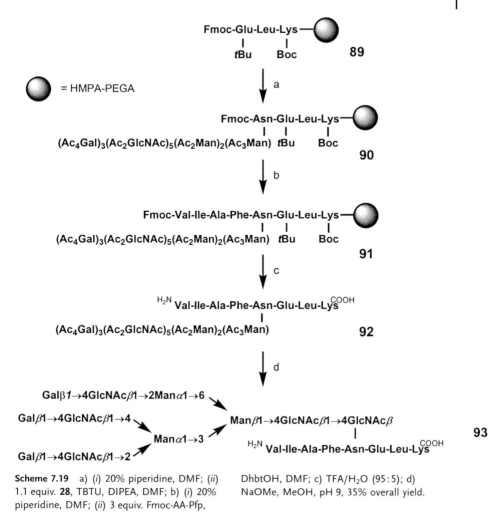

Scheme 7.19 a) (i) 20% piperidine, DMF; (ii) 1.1 equiv. **28**, TBTU, DIPEA, DMF; b) (i) 20% piperidine, DMF; (ii) 3 equiv. Fmoc-AA-Pfp, DhbtOH, DMF; c) TFA/H$_2$O (95:5); d) NaOMe, MeOH, pH 9, 35% overall yield.

the formation of the aspartimide **97** occurred during both peptide assembly and aspartic acid activation. Offer and co-workers presented a solution to this problem by using the N-(2-hydroxy-4-methoxybenzyl) (Hmb) group as a protecting group for the amide backbone [110]. In preparation for the on-resin coupling, O-acetylation of the Hmb group in **95b** was followed by deallylation. Activation of the aspartic acid-containing peptide **98** and subsequent treatment with O-unprotected glucosamine afforded the resin-bound N-glycopeptide **99**. O-Deacetylation reestablished the acid-lability of the Hmb group. The final TFA-promoted cleavage removed all protecting groups, including the Hmb group, and liberated the GlcNAc-substituted peptide **101**. A potential disadvantage of this strategy arises from the order of the deprotection steps. Labile linkages such as the fucosidic bond survive TFA treatment only in the acyl-protected form. However, prior to acidic cleavage

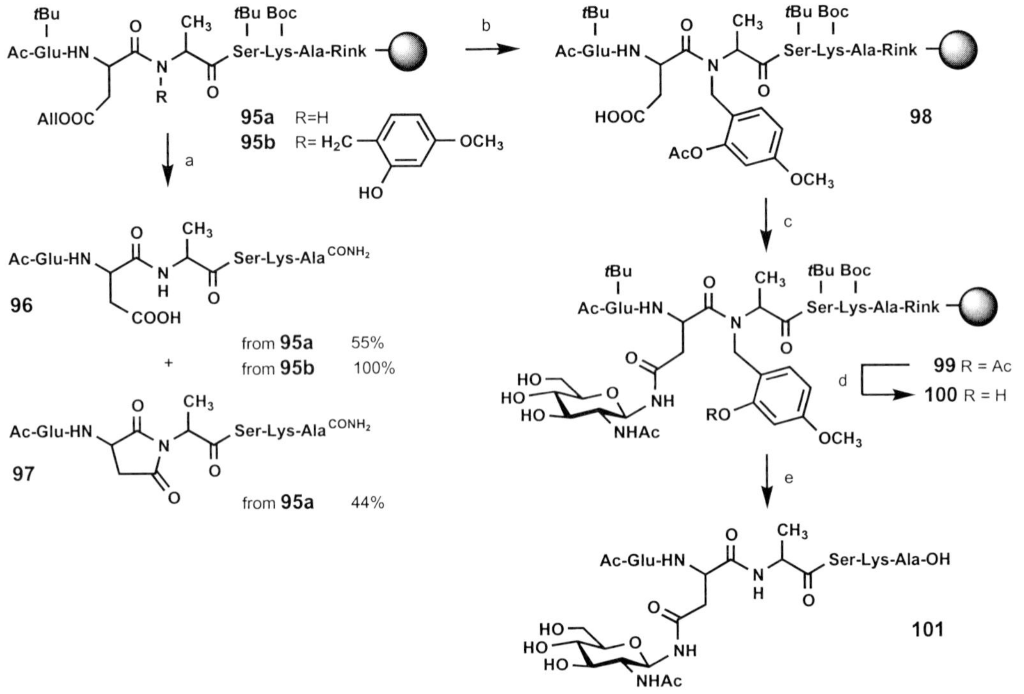

Scheme 7.20 a) (i) [Pd(PPh₃)₄], NMM, AcOH, DMF, CHCl₃; (ii) TFA/H₂O/Et₃SiH (90:5:5); b) (i) Ac₂O, DIPEA; (ii) [Pd(PPh₃)₄], NMM, AcOH, DMF, CHCl₃; c) glucosylamine; BOP, HOBt, DIPEA, DMSO, DMF; d) N₂H₄, DMF; e) TFA/H₂O/Et₃SiH (90:5:5), 30%.

the AcHmb group has to be deacetylated, which inevitably results in deacetylation and concomitant enhancement of the acid lability of labile glycosyl linkages.

Danishefsky's group reported a conceptually different method for convergent solid-phase glycopeptide synthesis (Scheme 7.21) [111]. The glycopeptide was anchored to the solid support through the carbohydrate part rather than the peptide C-terminus. Firstly, oligosaccharide glycal **102** was assembled on a solid phase, and was converted to the resin-bound glycosyl amine **103** by the known sequence of iodosulfonamidation, sulfonamide rearrangement, and subsequent azide reduction (see Scheme 7.2). Coupling with a tripeptide yielded conjugate **104**. Liberation of **106** was achieved by treatment with HF · pyridine. For global deprotection, the allyl, Troc, and carbohydrate protecting groups were removed by Pd(0)-catalyzed allyl transfer, Zn-mediated reductive cleavage, and KCN-induced methanolysis. Aspartimide formation is likely to restrict the general applicability of this approach, however. This side reaction can be completely prevented by synthesis of a supported glycopeptide with the *N*-linked GlcNAc already in place. Wong and co-workers employed glycosyltransferases to extend the glycan of a resin-bound glycopeptide [112]. The pursuit of enzymatic on-resin glycosylations requires a careful choice of the solid support, which has to enable steric accessibility and ef-

Scheme 7.21 a) (*i*) I(coll)₂ClO₄, THF, −10 to 0 °C, Anth-SO₂NH₂; (*ii*) Bu₄NN₃, THF; (*iii*) Ac₂O, DMAP, THF; (*iv*) propanedithiol, DIPEA, DMF; b) Troc-Asn-Leu-Ser(OBn)-OAll, IIDQ, CH₂Cl₂; c) HF·Pyr, PhOMe, CH₂Cl₂, −10 °C, 35%; d) (*i*) [Pd(PPh₃)₄], dimethylbarbituric acid, THF; (*ii*) Zn, AcOH, MeOH; (*iii*) Pd(OAc)₂, H₂, AcOH, MeOH; (*iv*) KCN, MeOH, 65%.

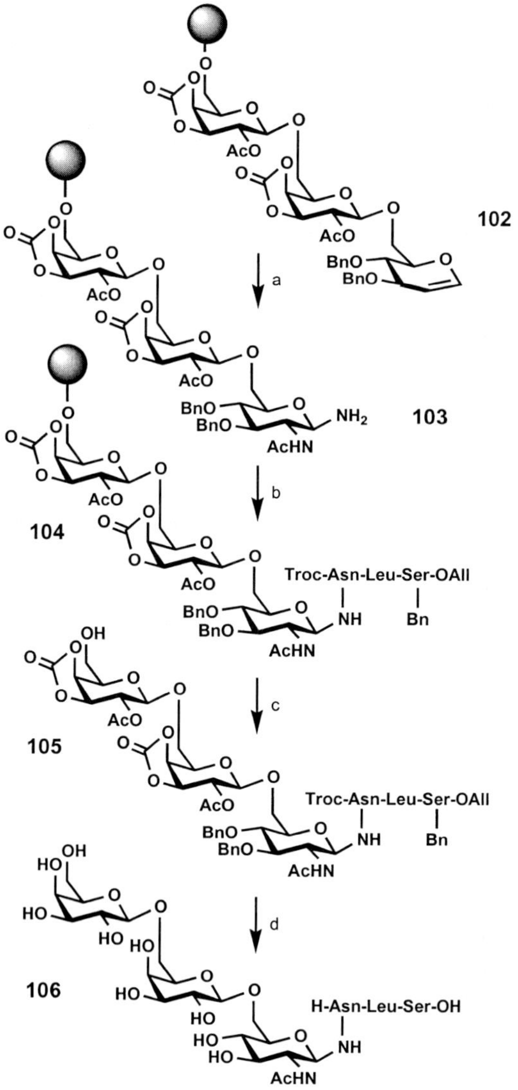

ficient mass transfer of the bulky biocatalyst. Non-swellable supports with high surface areas, such as silica gel or controlled pore glass (CPG), fulfil these criteria [74]. Alternatively, resins that swell in aqueous solutions, such as acrylamide resins or the PEGA support, have also been used [113, 114].

7.5.4
Solid-Phase Synthesis of O-Glycopeptides

Most synthetic efforts have been directed towards the preparation of the most frequently occurring mucin-type glycopeptides. The groups of Bock, Meldal, and Paulsen were amongst the first to advocate parallel glycopeptide synthesis and to demonstrate its utility in the assembly of MUC 2 and MUC 3 peptides with T_N antigen core 1 (T antigen), core 2, core 3, core 4, and core 6 structures [79]. Acid-labile resins such as the Wang resin were chosen. The O-acyl-protected carbohydrates remained intact during the TFA cleavage, and final treatment with sodium methylate removed the O-acetyl and O-benzoyl groups. Danishefsky described a case in which extensive experimentation was required in order to achieve the removal of an O-benzoyl group [94]. The use of peracetylated building blocks allows the application of milder deprotection conditions that leave the base-labile O-glycosidic linkage unaffected. Kihlberg and Kunz, for example, independently reported the incorporation of O-acetyl-protected sialyl-T_N building blocks [60, 115]. Liebe and Kunz used the allylic HYCRON-linker and liberated the protected glycopeptide **108** by applying Pd(0)-catalyzed allyl transfer to weak nucleophiles such as N-methylaniline (Scheme 7.22) [60]. Global deprotection was achieved by treatment with TFA and aqueous sodium hydroxide to give the sialyl-T_N glycopeptide **109**.

It seems that O-acetylation is the most versatile means of protecting carbohydrates during glycopeptide synthesis. In rare cases, however, side-reactions occur during the removal of O-acetyl groups from O-linked glycopeptides [116, 117]. Protection by O-benzyl groups, as demonstrated by Nakahara and co-workers, is an alternative [118]. It has to be taken into account, however, that O-benzylation enhances the acid lability of glycosidic linkages. Furthermore, complete benzyl ether cleavage sometimes requires long reaction times. Kihlberg's group reported the use of silyl-type protecting groups such as TBDMS and TBDPS ethers, as well as protection with 4-methoxybenzyl ethers [119]. The disaccharide conjugate **45**, for example, was used for coupling during the synthesis of the type II collagen glycopeptide **111** (Scheme 7.23) [76]. TFA treatment of the resin-bound peptide **110** cleaved the Wang linker and removed both the peptide side chain and the carbohydrate protecting groups.

A lot of research has been devoted to minimization of the number of steps necessary for building block preparation. Glycosyltransferases have been shown to accept resin-bound glycopeptide substrates, which removes the need to preform the complex glycosyl amino acid in solution [112–114]. Typically, glycosyltransferases work upon unprotected substrates and in aqueous environments. Accordingly, a linker that will allow all protecting groups to be removed without detaching the supported substrates is required. Seitz and Wong applied the acid- and moderately base-stable HYCRON linker [120, 121] (see also Scheme 7.22) [74]. The synthesis of the MAdCAM 1 peptide **115**, containing an O-linked sialyl-Lewis X saccharide, began with the assembly of the fully protected GlcNAc-peptide **112** on the CPG-support by conventional Fmoc techniques (Scheme 7.24). Treatment of **112** with TFA removed all side chain-protecting groups, while the unprotected

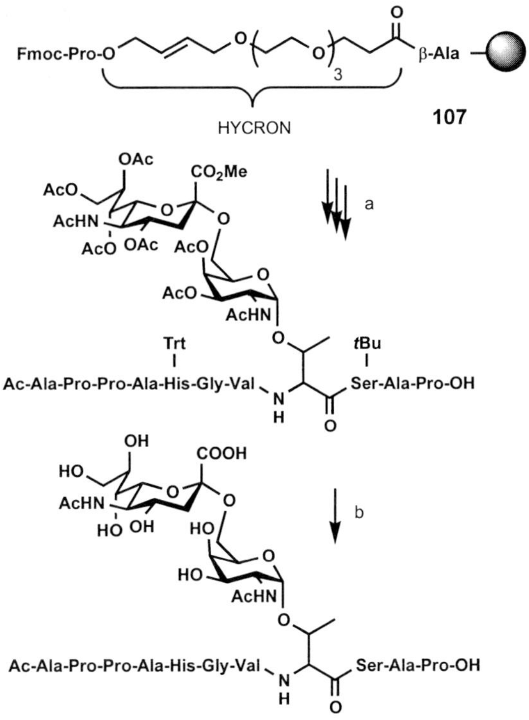

Scheme 7.22 a) (*i*) 50% morpholine, DMF; (*ii*) Fmoc-AA-OH, TBTU, HOBt, NMM, DMF; (*iii*) repetitive cycles of *i–ii*; (*iv*) [Pd(PPh₃)₄], morpholine, DMF/DMSO (1:1), 42%; b) (*i*) TFA, PhOMe, EtSMe; (*ii*) NaOH, MeOH, 76%.

conjugate **113** remained on the solid support. The series of enzymatic galactosyl, sialyl, and fucosyl transfer reactions was followed by the Pd(0)-catalyzed cleavage of the HYCRON linker, which released sialyl-Lewis X peptide **115** in a smooth reaction. A chemical alternative to this approach was developed in Paulsen's group [122]. Selectively masked glycopeptides were subjected to on-resin glycosylations, and remarkable resin effects were noted.

The direct glycosylation of resin-bound peptides would eliminate the bottleneck of glycosyl amino acid synthesis. Several rather unsuccessful attempts to achieve on-resin glycosylation of peptide acceptors have been reported [121, 123, 124]. It was suggested that the low yields could be due to an effect of the resin [125], and indeed, the development of the polyether-based inert POEPOP resin allowed the glycosylation of serine residues of supported tetra- and pentapeptides through the use of acyl protected galactose, fucose, and glucosamine trichloroacetimidate donors [125]. The glycosylation yields amounted to 48–71% when TMSOTf was applied as promoter (see, for example, **116 → 117**, Scheme 7.25). Recently, Meldal and co-workers have shown that the TMSOTf treatment can result in the undesired cleavage of *t*Bu protecting groups [126]. High yields and purity, however,

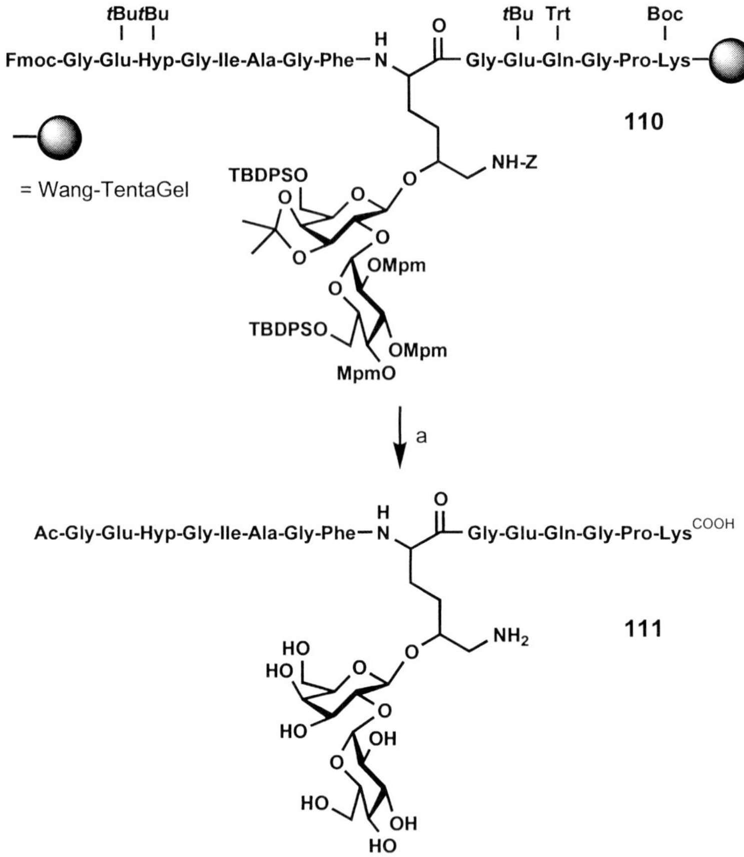

Scheme 7.23 a) TFA/H$_2$O/PhSMe/ethanedithiol (87.5:5:5:2.5).

were obtained with a BF$_3 \cdot$ Et$_2$O-activation of the glycosyl trichloroacetimidates. This investigation introduced a novel photolabile linker, which facilitated reaction monitoring by MALDI-TOF analysis. Relatively simple peptides were used in both reports, and it remains to be seen whether the outlined methodology will also support on-resin glycosylation of peptides containing nucleophilic tryptophan, cysteine, or methionine residues.

A high degree of convergence with respect to the peptide part is provided by on-resin fragment couplings (convergent solid-phase synthesis). Chemical condensations are usually performed with protected peptide segments. Habermann and Kunz, for example, employed the allylic HYCRON linkage to synthesize the protected lipopeptide **118** (Scheme 7.26, see also Scheme 7.22) [127]. This segment was coupled to the HYCRON-linked glycopeptide resin **119**. Subsequent TFA treatment removed all side chain protecting groups from the resin-bound lipo-glycopeptide **120**. Pd(0)-catalyzed cleavage of the allylic linkage was followed by a rather unselective saponification of the acetate esters to furnish conjugate **121**.

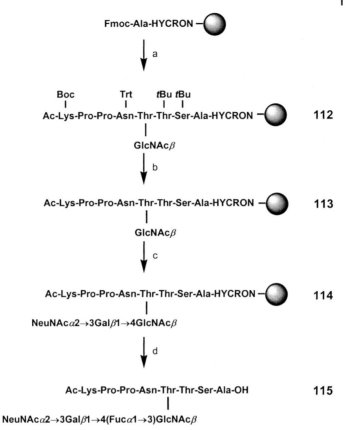

Scheme 7.24 a) (i) 50% morpholine, DMF; (ii) Fmoc-AA-OH, HBTU, HOBt, NMM, DMF; (iii) N-acetylation: AcOH, HBTU, NMM, HOBt, DMF; b) TFA/H$_2$O/ethanedithiol (40:1:1); c) (i) UDG-Gal, GalTase, 0.1 M HEPES (pH 7), 5 mM MnCl$_2$, 37 °C; (ii) CMP-NeuNAc, α-2,3-sialyltransferase, 50 M HEPES (pH 7), 5 mM MnCl$_2$, alkaline phosphatase, 37 °C; d) (i) [Pd(PPh$_3$)$_4$], morpholine, DMF/DMSO (1:1), 9%; (ii) FucTase, GDP-Fuc, 0.1 M HEPES (pH 7), 37 °C, 59%.

Similar tactics were applied for the synthesis of the sialyl-T$_N$ glycopeptide conjugate **122** (Fig. 7.3) [128]. In the key step, the T cell epitope peptide was coupled to the tumor-associated MUC1 glycopeptide by convergent solid-phase synthesis. These syntheses demonstrate the high versatility of the allylic HYCRON linker, which provides protected peptide segments as well as the opportunity to perform on-resin deprotections.

Wagner and Kunz have recently introduced the PTMSEL linker **123** and demonstrated its usefulness for the synthesis of the protected glycopeptide **124** (Scheme 7.27) [129]. The PTMSEL linkage was shown to suppress the sometimes problematic diketopiperazine formation. Cleavage occurred upon treatment with a mildly basic fluoride source such as TBAF in dichloromethane. Nakahara and co-workers

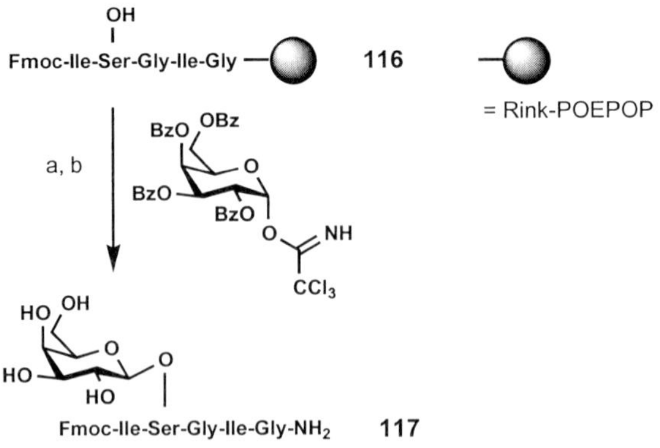

7 Glycopeptides and Glycoproteins: Synthetic Chemistry and Biology

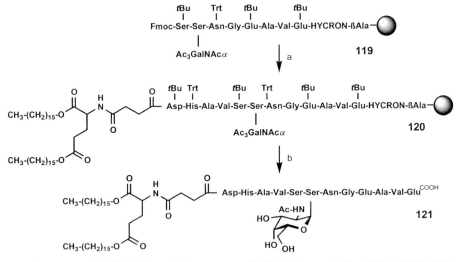

Scheme 7.25 a) TMSOTf, CH₂Cl₂; b) TFA/H₂O.

Scheme 7.26 a) (*i*) Morpholine, DMF; (*ii*) **118**, PfPyU, DIPEA, *s*-collidine, DMF, DMSO; (*iii*) Ac₂O, Pyr; b) (*i*) TFA, TIS, H₂O; (*ii*) [(PPh₃)₄], morpholine, DMF, DMSO, 13%; (*iii*) NaOMe, MeOH (pH 9.0), 32%.

7.5 Synthesis of Glycopeptides | 197

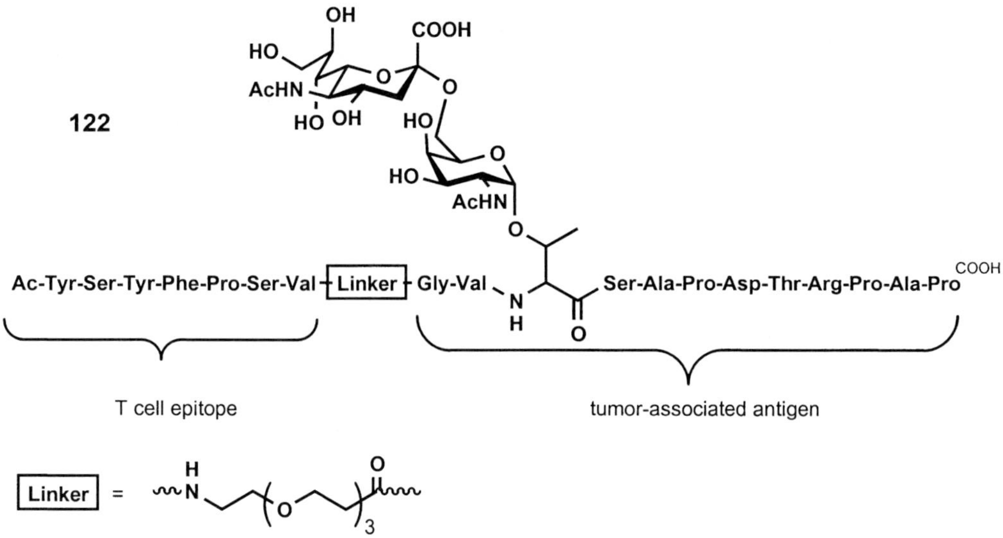

Fig. 7.3 Convergent solid-phase peptide synthesis provided the sialyl-T_N glycopeptide conjugate **122**, made up of a T cell epitope peptide and a tumor-associated MUC1 B cell epitope.

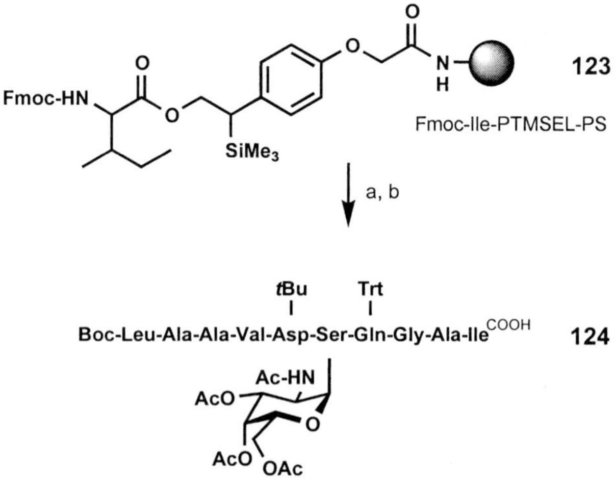

Scheme 7.27 a) Fmoc solid-phase synthesis; b) cleavage: 2 equiv. TBAF·3H$_2$O, CH$_2$Cl$_2$, 25 min, rt., 66%.

have shown that fragment condensations can be performed on both the N- and the C-termini when the glycopeptide is silyl-anchored through a carbohydrate or a peptide side chain such as a threonine hydroxy group [130, 131]. In Wong's labs, an acetal linkage was employed for the attachment of a fucosyl threonine [132]. Subsequent N- and C-terminal elongations furnished fucopeptides as sialyl-Lewis X mimetics.

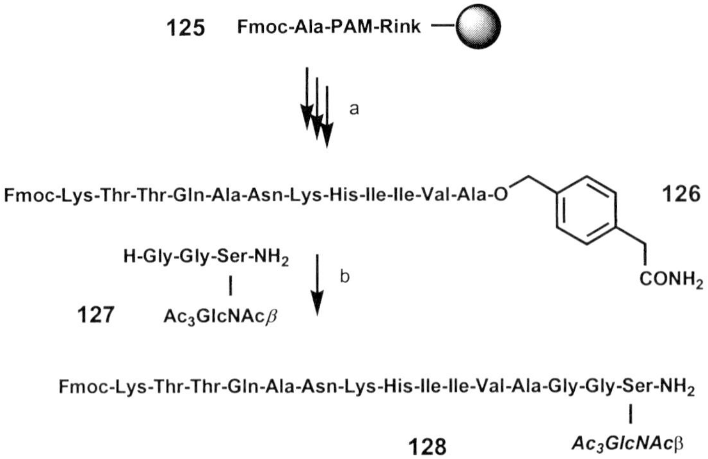

125 Fmoc-Ala-PAM-Rink

↓↓↓ a

Fmoc-Lys-Thr-Thr-Gln-Ala-Asn-Lys-His-Ile-Ile-Val-Ala-O-CH₂-C₆H₄-CONH₂ **126**

H-Gly-Gly-Ser-NH₂
|
127 Ac₃GlcNAcβ

↓ b

Fmoc-Lys-Thr-Thr-Gln-Ala-Asn-Lys-His-Ile-Ile-Val-Ala-Gly-Gly-Ser-NH₂
|
128   Ac₃GlcNAcβ

Scheme 7.28 a) Standard Fmoc solid-phase peptide synthesis, cleavage: TFA/Et$_3$SiH/H$_2$O (95:2.5:2.5), 89%; b) subtilisin 8397 K256Y, 50 M triethanolamine/DMF (1:9), 84%.

Protected peptides often exhibit poor solubility in commonly used solvents, complicating both purification and usage in peptide couplings. It adds to the difficulties that couplings of the relatively unreactive segments suffer from racemization of the activated amino acid. As a solution to these problems, Witte, Seitz, and Wong utilized enzyme-catalyzed fragment condensations relying on water-soluble unprotected segments [133]. Solid-phase synthesis provided unprotected glycopeptide and peptide esters such as **126**, which were shown to serve as acyl donors in subtilisin-mediated peptide couplings (Scheme 7.28). In one reported case, glycopeptide **127** was ligated with **126** in 84% yield.

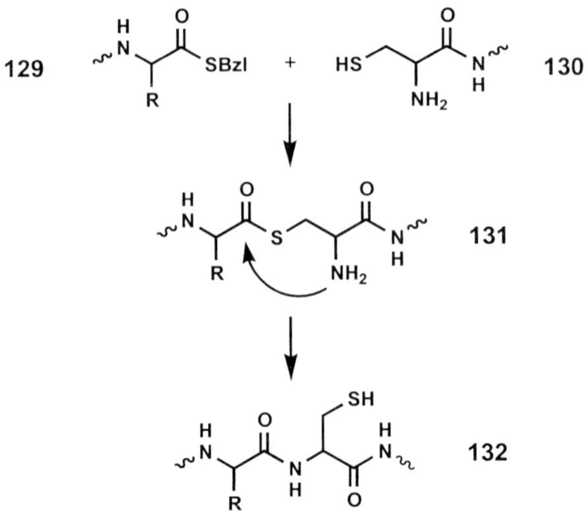

Scheme 7.29 The mechanism of "native chemical ligation".

Chemical ligation of two unprotected peptide segments can be achieved by use of the "native chemical ligation" approach [134]. This reaction makes use of the distinct reactivity of an N-terminal cysteine (**130** in Scheme 7.29). A peptide thioester such as **129** undergoes a thioester exchange to form the secondary thioester intermediate **131**, which is subject to a spontaneous $S \rightarrow N$ acyl shift to deliver the final ligation product **132** [135].

Crucial for native chemical ligation is access to peptide thioesters. Bertozzi and Ellman demonstrated the solid-phase synthesis of glycopeptide thioesters such as **135** (Scheme 7.30) [136]. In this study, the N-terminal fragment **133**, containing one O-linked GalNAc moiety, was synthesized with Fmoc-protected building blocks on Ellman's modification of Kenner's sulfonamide resin [137]. Alkylation with iodoacetonitrile furnished **134** and prepared the sulfonamide linkage for sub-

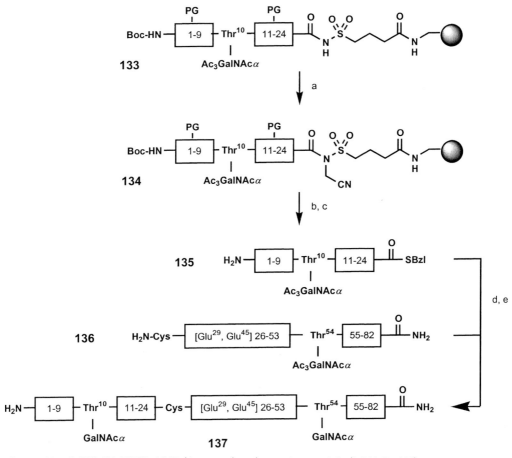

Scheme 7.30 a) ICH$_2$CN, DIPEA, NMP; b) BnSH, THF; c) TFA/PhOH/H$_2$O/PhSMe/EDT (82.5:5:5:5:2.5), 4 h, 21% overall yield (based on resin capacity); d) 6 M Gn·HCl, 100 mM NaH$_2$PO$_4$, pH 7.5, 4% PhSH, 55%; e) 5% equiv. N$_2$H$_4$, DTT, 53%.

sequent thiolytic cleavage. After removal of the acid-labile protecting groups, the unprotected acyl donor **135** was employed in native chemical ligation with segment **136**. The final O-deacetylation afforded the 82mer glycoprotein **137**. A similar technique was applied in the synthesis of lymphotactin, a 93mer glycoprotein with eight O-linked GalNAc residues [138].

7.6
Biological and Biophysical Studies

7.6.1
Conformations of Glycopeptides

In principle, there are two types of glycosylation pattern. One is characterized by a single attachment of a carbohydrate and the other by the so-called clustered mode of glycosylation [139]. Several studies have shown that a single glycan can influence the secondary structure of the peptide backbone when appropriately attached. In the early days, conformational analyses were mainly carried out by means of CD spectroscopy. With the recent improvements in NMR spectroscopy and computational methods it is now feasible to gain more detailed insights in local structures [140]. O'Connor and Imperiali, for example, have studied the synthetic hemaglutinin nonapeptide **138** (Fig. 7.4) [141]. The N-glycosylation site was found to adopt a turn structure, the Asx-turn, which appears to be a prerequisite for natural N-glycosylation [142]. The appendage of the N-linked chitobiosyl residue in **139** induced a major alteration of the peptide backbone conformation. The

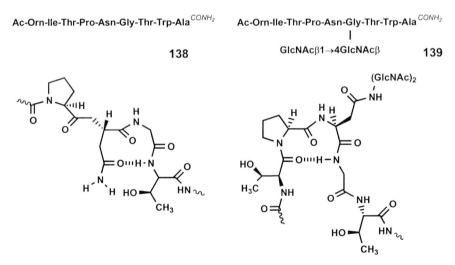

Fig. 7.4 Peptide **138** adopts a turn structure, the Asx-turn, around the N-glycosylation site. The chitobiosyl residue in **139** induces a β-turn structure.

intense NOE between the amide protons of asparagine and glycine, the very low temperature dependence of the glycine amide signal, and a simulated annealing procedure with 109 NOE-derived distance constraints suggested the presence of a β-turn structure. Interestingly, conformational analyses of GlcGlc-, GlcGlcNAc-, GlcGlcNAc-, and GlcNAc-containing peptides revealed that the 2-acetamido group at the proximal sugar unit is required for maintenance of the β-turn [143].

A similar analysis was carried out with the glycopeptide Ala-Leu-(Glcβ1 → 6Glcβ1 → 6GlcNAcβ)Asn-Leu-Thr [144] and a glycosylated segment of the repeating unit of RNA polymerase II [145]. In each case a conformational equilibrium between an ordered structure and the random coil conformation was detected, while the unglycosylated peptide failed to show any secondary structure.

A recent study explored the conformational interplay between the carbohydrate and the peptide backbone of the O-SLe^X-MAdCAM-1 peptide 115 and its synthetic intermediates, the O-GlcNAc, O-LacNAc and O-sialyl-LacNAc peptides (Fig. 7.5) [146]. The conformation of the peptide was significantly altered upon introduction of the SLe^X group. All carbohydrates stabilized a turn-like structure. However, ROESY peaks between the 2-acetamido group and the methyl group of Thr6 disappeared after introduction of the fucosyl residue. Concomitantly, ROESY peaks between GlcNAc-H1 and Thr5-Hγ decreased. It was concluded that glycosylation of the O-GlcNAc-peptide 142 produced a reorientation of the carbohydrate and, through the 2-acetamido group, an alteration of the peptide backbone conformation.

The examples presented indicate that glycosylation can induce turn-like structures, and this was reported to be feasible even for the O-glycosidic attachment of a single GlcNAc residue [145]. Nevertheless, strong influences of singular glycans are rare. In contrast, multiple attachment of carbohydrates seems to confer a remarkable stabilization of secondary structures. As an example, Bächinger and coworkers compared the triple helix-forming propensities of the model peptides Ac-(Gly-Pro-Thr)$_{10}$-NH$_2$ and Ac-(Gly-Pro-(βGal)Thr)$_{10}$-NH$_2$ [147], and it was found

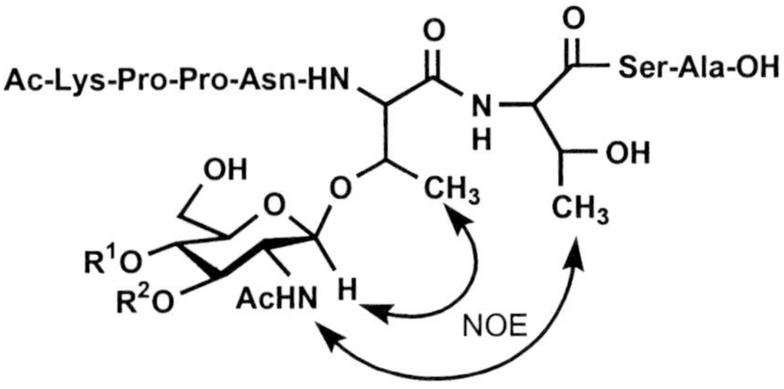

Fig. 7.5 Conformational interplay between the carbohydrate and the peptide backbone. Distal saccharides affected the NOEs between the peptide backbone and the proximal sugar.

115, R^1 = NeuNAcα2→3Galβ; R^2 = Fucα
140, R^1 = NeuNAcα2→3Galβ; R^2 = H
141, R^1 = Galβ; R^2 = H
142, R^1 = R^2 = H

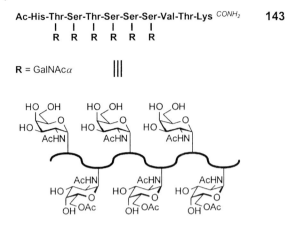

Fig. 7.6 Multiple attachment of monosaccharides to the mucin **143** forced the peptide backbone to adopt a "wave-type" conformation. NMR and computer modeling of glycopeptides **144** suggested that α-O-glycosidic but not β-O-glycosidic linkages are able to stabilize the extended backbone conformation.

that glycosylation was required to achieve a stable collagen triple helix. Mucins, proteins expressed on the surfaces of various epithelial cells, form rod-like structures. The extended conformation of the mucin peptide backbone seems to be stabilized by the multiple attachment of O-linked carbohydrates. This stiffening effect has been demonstrated in studies with the hexaglycosylated glycophorin A decapeptide **143** (Fig. 7.6) [148]. The lack of NH_i–NH_{i-1} mid-range and long-range contacts and a constrained molecular dynamics simulation suggested the presence of a "wave-type" conformation. NMR and computer modeling revealed that the sequential attachment of α-O-linked T_N-, T-, and 2,6-bissialyl-T-antigen moieties in **144** induced a conformationally highly stable structure [149]. The NOE patterns of the three α-linked carbohydrates were virtually identical and, analogously with **143**, the appendage of the GalNAc monosaccharide was sufficient to stabilize the extended conformation of the peptide backbone. One striking observation was that the β-O-linked sugar failed to confer such a stabilizing effect, which emphasizes the specific role of the α-O-GalNAc bridgehead.

7.6.2
Glycopeptides as Substrates of Enzymes and Receptors

Glycosylation serves as a means of maintaining the structural integrity of a given protein [150]. The glycans can protect the peptide backbone from proteolytic attack, as a result of which the introduction of carbohydrates can increase the biological half-life of a putative peptide drug. Glycosylation can also enhance delivery and target the aglycon to specific cells or tissues. The intestinal absorption of peptide drugs, for example, has been improved by glycosylation [151]. Extensive investigations have been carried out on enkephalin glycosylation and its effects on antinociceptive activity [152, 153]. It was shown that an appended carbohydrate can destroy binding of the enkephalin peptide to the opioid receptor [154, 155]. However, suitably positioned glycans do not interfere with receptor binding and, as an added advantage, increase penetration through endothelial barriers such as the blood-brain barrier [156, 157]. In a recent study, peptide **145** and glycopeptide **146** (Fig. 7.7) were radioiodinated at the tyrosine residues in order to assess their blood-brain barrier permeability [158]. It became apparent that glycopeptide **146** showed a significantly higher accumulation in the brain than peptide **145**. Further analysis suggested that the glycopeptide **146** produced analgesic effects similar to morphine, with a reduced dependence liability. Carbohydrate-increased cellular uptake has been demonstrated for various medicinally interesting compounds such as toxophores [159–161], oligonucleotides [162], proteins [163], and even polymers [164].

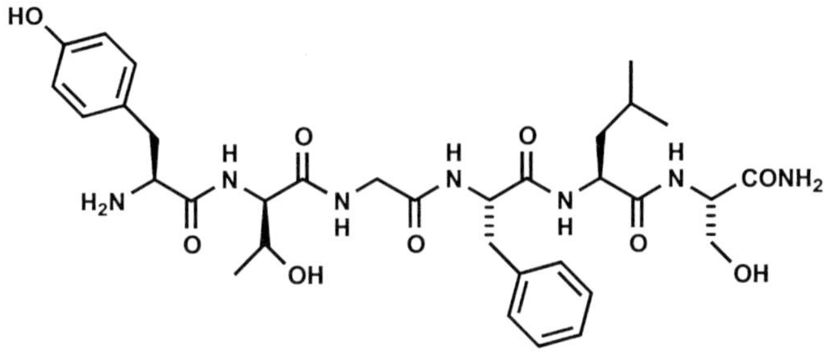

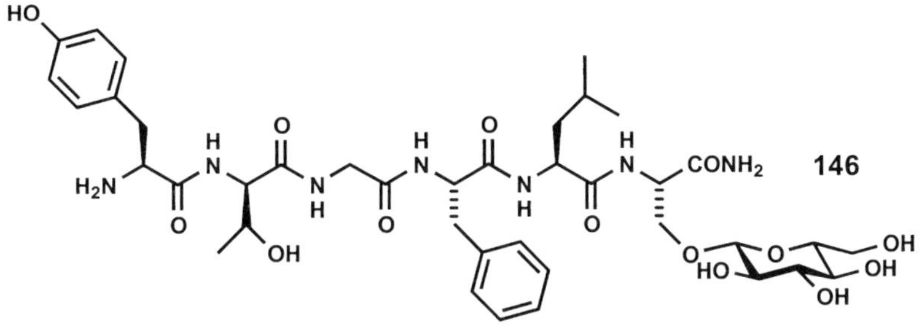

Fig. 7.7 Glycopeptide **146** was shown to pass through the blood-brain barrier to produce significantly higher accumulation in the brain than seen in the case of peptide **145** and produced analgesic effects similar to morphine.

7.6.3
Glycopeptides and Cancer Immunotherapy

Classical cancer therapy is based on surgery, irradiation, and chemotherapy. The first is a mechanical means of tumor removal, while the last two destroy – more or less selectively – proliferating cells by damaging DNA or by inhibiting proteins essential for cell growth. Immunotherapy is a new approach in which the goal is to direct the immune defense to the tumor tissue. The malignant phenotype of cells is characterized by a dramatic transformation of the glycosylation machinery. As a result, aberrantly glycosylated proteins are amongst the most frequently occurring tumor antigens [165]. The carbohydrates identified comprise the T_N-, sialyl-T_N-, and T-antigens and Lewis-X and Lewis-A structures. In the active immunization approach, synthetic tumor-specific structures are used as vaccines and it is hoped that the elicited antibodies and T cells will cross-react with the cancer cells [166]. Synthetic antigens are usually of low molecular weight and so are poorly immunogenic. As a solution to this problem the antigen is commonly conjugated to immunogenic carrier proteins such as BSA or KLH.

Danishefsky's group has synthesized the trimeric T_N-peptide **147** (Fig. 7.8) as a partial structure of mucin-related tumor antigens [59]. Immunization studies revealed that conjugation to KLH induced high IgM and moderate IgG antibody titers in mice. It was shown that the formed antibodies cross-reacted with T_N-positive LS-C colon cancer cells and induced their complement-mediated lysis. The immune response was, however, dominated by high IgM titers. The absence of the so-called class switch from IgM to IgG antibodies, along with the lack of a secondary response, indicated a T cell-independent antibody response. Unfortunately, this usually less effective immune response is characteristic of many carbohydrate antigens. Similar results were obtained for a sialyl-T_N-KLH conjugate and for a hexaglycosidic globo H-KLH conjugate, the most complex vaccine to have been synthesized and evaluated in clinical studies [167]. Lloyd and co-workers have studied the influence of epitope clustering, carrier structure, and adjuvant on the antibody response to Lewis-Y conjugates in mice [168]. It was found that the clustered Lewis-Y structure **148** was more efficient than the non-clustered antigen **149** in eliciting antibodies against Lewis-Y mucins and the OVCAR-3 ovarian cancer cell line. The response was, however, still dominated by IgM antibodies. Addition of the adjuvant QS-21 produced a mixed IgM and IgG response, but the reactivity against cancer cells was weaker.

Evidence for a T cell response has been obtained in immunization studies with the THERATOPE vaccine **150**, a sialyl-T_N-KLH conjugate currently being studied in phase III clinical trials [169, 170]. The high IgG titers suggest T cell participation. Furthermore, sialyl-T_N-specific T cell proliferation and killer cell activity was demonstrated in vitro [171]. A recent study demonstrated that a T cell-dependent immune response can be elicited by use of a synthetic vaccine with a dendrimer structure, thereby removing the need for a carrier protein. The multiple antigenic peptide **151** was designed to contain a well known T cell epitope from the type I poliovirus and four T_N residues as B cell epitopes [172, 173]. The sera obtained

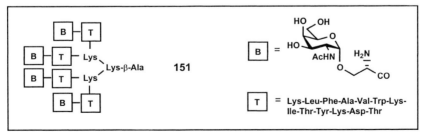

Fig. 7.8 Selected examples of synthetic glycopeptide conjugates used as putative tumor vaccines.

from immunized mice were able to recognize the native T_N-antigen on human Jurakt T-lymphoma and LS180 adenocarcinoma cell lines. The humoral response was dominated by IgG antibodies, supporting the notion that a T cell-dependent response was induced. Interestingly, stimulation of a T cell hybridoma cell line specific for the unglycosylated poliovirus peptide (**T** in Fig. 7.8) was achieved with a dose 10 000 times lower than required for a construct in which the T_N-antigen was omitted. Clearly, presentation of **151** by MHC was enhanced either through favorable intracellular processing or through increased endocytosis of **151** by the antigen-presenting cells.

7.6.4
Glycopeptides and T Cell Recognition

Highly immunogenic antigens have to be endowed with a B cell epitope and a T cell epitope, which provide the recognition structures for B cell receptors (membrane-bound antibodies) and T cell receptors. Multivalent presentation as provided by clustering, conjugation to carriers, or use of adjuvants is beneficial, since uptake by professional antigen-presenting cells is facilitated. These specialized antigen-presenting cells (APCs), such as macrophages or dendritic cells, internalize proteins by endocytosis. After passage through an acidic compartment, these proteins are degraded and the resulting peptides are bound to MHC class II molecules located on the cell surface and presented to helper T cells, provided that certain residues anchor the peptide to the MHC-binding cleft. The binding of the T cell receptor (TCR) to non-self peptides on the MHC II-peptide binding groove is a prerequisite for T cell activation. Ultimately, helper T cell activation results in the production of cytokines, which are necessary for B cell activation. A second and possibly more effective line of immune defense employs cytolytic T cells for recognition and lysis of antigen-charged cells. Every cell constantly converts endogenous proteins to small-size peptides by means of its proteasome. The fragments are transported into the endoplasmic reticulum, in which the 10–20mer peptides bind to MHC class I molecules. After transfer through the Golgi network, the peptide-MHC I complex is located on the cell surface and presented to $CD8^+$-cytolytic T cells. Only if the T cell receptor (TCR) recognizes non-self peptides on the MHC I peptide-binding groove can a cytolytic response towards the antigen-presenting cell be triggered.

In the examples discussed in the previous section, T cell epitopes were derived from peptide structures. It should be possible to elicit a specific and possibly highly effective immune response against glycoprotein tumor antigens by selecting protein-linked carbohydrates as T cell epitopes. Binding studies have revealed that synthetic glycopeptides can bind well to MHC I and MHC II molecules [102]. T cell proliferation assays confirmed that glycopeptides are able to induce a carbohydrate-specific T cell response [174, 175]. Werdelin and co-workers, for example, raised T cell hybridomas that proliferated and secreted interleukin 2 upon activation with the synthetic glycopeptide **152b** (Fig. 7.9) [176]. Glycopeptides **152a–h** were used to investigate the fine-specificity of the hybridomas against the glycan structure. Remarkably, 19

152, H-Val-Ile-Thr-Ala-Phe-Xxx-Glu-Gly-Leu-Lys

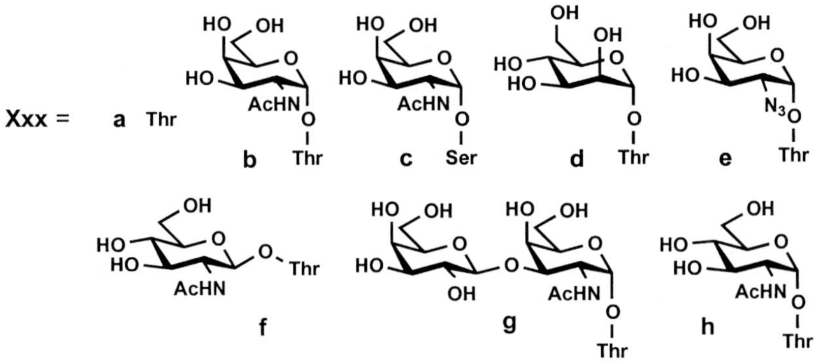

Fig. 7.9 Peptide **152a** and glycopeptides **152b–h** were used for evaluation of the carbohydrate specificity of MHC class II restricted T cell hybridomas raised against the O-glycosylated self-peptide **152b**.

of the 22 tested hybridomas responded only to the glycopeptide **152b**, displaying a total lack of cross-reactivity with the unglycosylated peptide **152a**. Of the 19 glycopeptide-responsive hybridomas, 17 were able to distinguish between the αGalNAc-peptide **152b** and the αGlcNAc peptide **152c**, which indicates that the glycan is the entity recognized by the T cell receptor. In a recent study it was shown that the αGlcNAc peptide **152c** failed to trigger phosphorylation of the T cell receptor ζ chain, which comprises the initial signaling event [177].

Specific T cells have been demonstrated to become stimulated upon presentation of glycopeptide fragments of type II collagen, causing rheumatoid arthritis in a mouse model. The observation that peptides **153a** and **153b**, spanning the immunodominant sequence of native collagen, activated only a few T cell hybridomas produced the conclusion that the T cell recognition structure was a glycan attached to the hydroxylysine residue (Fig. 7.10) [178, 179]. Indeed, the majority of the hybridomas were activated upon incubation with glycopeptide **153c**, in which the central hydroxylysine carried the β-galactosyl moiety. These investigations presented the first demonstration that immunization with a natural glycopeptide can elicit carbohydrate-specific T cells. A subsequent study provided evidence that glycopeptides such as **153c** were bound by MHCII molecules and that the glycan residue is accessible to the T cell receptor [180]. Similar conclusions were drawn from analysis of crystal structures of glycopeptides with H-2D^b MHC [181].

Cytosolic proteins often carry the GlcNAc monosaccharide β-linked to serine. It has been shown that these GlcNAc-containing glycopeptides are transported into the endoplasmic reticulum, bind to MHCI molecules, and elicit glycopeptide-specific T cell responses [182, 183]. Isolated MHC-derived peptides were incubated with ^{3}H-labeled

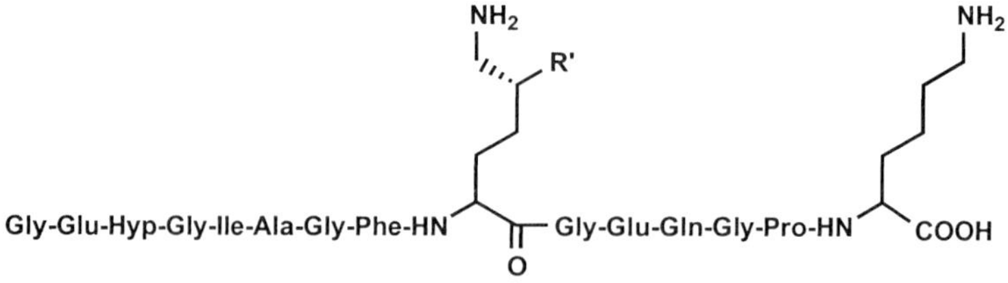

Gly-Glu-Hyp-Gly-Ile-Ala-Gly-Phe-HN—C(=O)—Gly-Glu-Gln-Gly-Pro-HN-CH(CH2CH2CH2R')(NH2)...CH(CH2CH2CH2CH2NH2)COOH

153a; R'=H
153b; R'=OH
153b; R'=βGal-O

Fig. 7.10 The majority of T cells obtained after immunization with native type II collagen specifically recognized glycopeptide **153c** with no cross-reactivity to the unglycosylated peptides **153a** and **153b**.

UDP-galactose and a GlcNAc-specific galactosyltransferase. Incorporation of radioactivity proved the presence of the (βGlcNAc)Ser modification, indicating that presentation of glycosylated peptides by MHC class I molecules occurs in vivo.

7.7
Summary and Outlook

The methodology of glycopeptide synthesis has been improved considerably. The increased understanding of glycosylation reactions has facilitated the preparation of glycosyl amino acid building blocks for subsequent use in peptide assembly. At present it appears that the building block approach provides the most reliable means of glycopeptide synthesis. The usual tactics for the construction of complex N-glycosyl-asparagine building blocks is peptide coupling between an aspartic acid derivative and a full-length glycosylamine. For the preparation of complex O-glycosyl amino acids, an approach in which a preformed amino acid glycoside is used as acceptor for further glycosylations offers an appealing alternative. With appropriately protected glycosyl amino acid building blocks to hand, solid-phase synthesis allows rapid access to various glycopeptides. On the other hand, solution-based methods suit the needs for the synthesis of very complex structures, particularly when performed by enzymatic techniques. The examples described in this review demonstrate that enzymatic glycosylations provide facile routes to complex glycoconjugates. A key requirement, however, is the availability of the corresponding glycosyltransferases or, when glycosidases are used, of the oligosaccharide substrates.

It is very encouraging that the synthetic methodology available today allows the synthesis of glycopeptides with degrees of complexity that meet many of the needs of biological and medicinal research. The synthesis of very complex targets, however, is still a challenge and often a research project in its own right. It appears that rapid progress can only be achieved through the combined use of chemical and enzymatic methods, along with selective ligation techniques such as native chemical ligation, to allow the total synthesis of glycoproteins.

Biological studies have revealed that glycosylation is a biological means to regulate the structure and activity of a given protein. For example, a single glycan can induce a turn-like structure, whereas multiple attachment of carbohydrates can force the peptide backbone to accommodate an extended conformation. The introduction of carbohydrates can protect the peptide backbone from proteolytic attack. In contrast, glycosylation was found to enhance delivery to specific cells or tissues, thereby increasing in vivo clearance. Some glycans are tumor antigens and it has been shown that a carbohydrate-specific immune response can be elicited. Both B cells and T cells can be activated, which supports the notion that vaccination with tumor-specific carbohydrate-based antigens can allow cancer immunotherapy by eradicating circulating metastases.

7.8 References

1 A. Varki, *Glycobiol.* **1993**, *3*, 97–130.
2 H. Kunz, *Angew. Chem. Int. Ed.* **1987**, *26*, 294–308.
3 M. Meldal, P. M. Sthilaire, *Curr. Op. Chem. Biol.* **1997**, *1*, 552–563.
4 O. Seitz, *ChemBiochem.* **2000**, *1*, 215–246.
5 O. Seitz, I. Heinemann, A. Mattes, H. Waldmann, *Tetrahedron* **2001**, *57*, 2247–2277.
6 P. M. St Hilaire, M. Meldal, *Angew. Chem. Int. Ed.* **2000**, *39*, 1163–1179.
7 H. Herzner, T. Reipen, M. Schultz, H. Kunz, *Chem. Rev.* **2000**, *100*, 4495–4537.
8 P. Sears, C. H. Wong, *Science* **2001**, *291*, 2344–2350.
9 B. G. Davis, *Chem. Rev.* **2002**, *102*, 579–601.
10 For naturally occurring C-glycosylation: J. F. G. Vliegenthart, F. Casset, *Curr. Op. Struct. Biol.* **1998**, *8*, 565–571.
11 For naturally occurring S-glycosylation; C. J. Lote, J. B. Weiss, *FEBS Lett.* **1971**, 16.
12 G. W. Hart, K. Brew, G. A. Grant, R. A. Bradshaw, W. J. Lennarz, *J. Biol. Chem.* **1979**, *254*, 9747–9753.
13 J. Hilkens, M. J. L. Ligtenberg, H. L. Vos, S. V. Litvinov, *Trends Biochem. Sci.* **1992**, *17*, 359–363.
14 G. W. Hart, *Annu. Rev. Biochem.* **1997**, *66*, 315–335.
15 G. W. Hart, R. S. Haltiwanger, G. D. Holt, W. G. Kelly, *Annu. Rev. Biochem.* **1989**, *58*, 841–874.
16 N. E. Zachara, G. W. Hart, *Chem. Rev.* **2002**, *102*, 431–438.
17 L. Wells, K. Vosseller, G. W. Hart, *Science* **2001**, *291*, 2376–2378.
18 R. G. Spiro, *J. Biol. Chem.* **1967**, *242*, 4813–&.
19 W. T. Butler, L. W. Cunningh, *J. Biol. Chem.* **1966**, *241*, 3882–&.
20 A. Gunnarsson, B. Svensson, B. Nilsson, S. Svensson, *Eur. J. Biochem.* **1984**, *145*, 463–467.
21 D. H. Williams, B. Bardsley, *Angew. Chem. Int. Ed.* **1999**, *38*, 1173–1193.
22 M. Gohlke, G. Baude, R. Nuck, D. Grunow, C. Kannicht, P. Bringmann, P. Donner, W. Reutter, *J. Biol. Chem.* **1996**, *271*, 7381–7386.
23 H. Nishimura, T. Takao, S. Hase, Y. Shimonishi, S. Iwanaga, *J. Biol. Chem.* **1992**, *267*, 17520–17525.
24 C. Smythe, P. Cohen, *Eur. J. Biochem.* **1991**, *200*, 625–631.
25 R. J. Harris, C. K. Leonard, A. W. Guzzetta, M. W. Spellman, *Biochemistry* **1991**, *30*, 2311–2314.
26 S. Hase, H. Nishimura, S. I. Kawabata, S. Iwanaga, T. Ikenaka, *J. Biol. Chem.* **1990**, *265*, 1858–1861.
27 C. Smythe, F. B. Caudwell, M. Ferguson, P. Cohen, *Embol J.* **1988**, *7*, 2681–2686.

28 I. R. Rodriguez, W. J. Whelan, *Biochem. Biophys. Res. Commun.* **1985**, *132*, 829–836.
29 D. T. A. Lamport, *Biochemistry* **1969**, *8*, 1155–&.
30 H. Kunz, C. Unverzagt, *Angew. Chem. Int. Ed.* **1988**, *27*, 1697–1699.
31 P. Sjölin, M. Elofsson, J. Kihlberg, *J. Org. Chem.* **1996**, *61*, 560–565.
32 P. Sjölin, J. Kihlberg, *J. Org. Chem.* **2001**, *66*, 2957–2965.
33 C. Li, T. L. Shih, J. U. Jeong, A. Arasappan, P. L. Fuchs, *Tetrahedron Lett.* **1994**, *35*, 2645–2646.
34 C. Li, A. Arasappan, P. L. Fuchs, *Tetrahedron Lett.* **1993**, *34*, 3535–3538.
35 C. Unverzagt, H. Kunz, *J. Prakt. Chem.* **1992**, *334*, 570–578.
36 S. Sabesan, S. Neira, *Carbohydrate Res.* **1992**, *223*, 169–185.
37 C. Auge, C. Gautheron, H. Pora, *Carbohydrate Res.* **1989**, *193*, 288–293.
38 C. H. Bolton, R. W. Jeanloz, *J. Org. Chem.* **1963**, *28*, 3228–&.
39 F. E. McDonald, S. J. Danishefsky, *J. Org. Chem.* **1992**, *57*, 7001–7002.
40 C. Unverzagt, *Angew. Chem. Int. Ed.* **1996**, *35*, 2350–2353.
41 R. S. Clark, S. Banerjee, J. K. Coward, *J. Org. Chem.* **1990**, *55*, 6275–6285.
42 M. A. E. Shaban, R. W. Jeanloz, *Bull. Chem. Soc. Jpn.* **1981**, *54*, 3570–3576.
43 H. G. Garg, R. W. Jeanloz, *Carbohydrate Res.* **1980**, *86*, 59–68.
44 G. S. Marks, A. Neuberger, *J. Chem. Soc. Perkin Trans. 1* **1961**, 4872–&.
45 S. T. Cohen-Anisfeld, P. T. Lansbury, *J. Am. Chem. Soc.* **1993**, *115*, 10531–10537.
46 L. M. Likhosherstov, O. S. Novikova, V. A. Derevitskaja, N. K. Kochetkov, *Carbohydrate Res.* **1986**, *146*, C1–C5.
47 G. Arsequell, J. S. Haurum, T. Elliott, R. A. Dwek, A. C. Lellouch, *J. Chem. Soc. Perkin Trans. 1* **1995**, 1739–1745.
48 C. H. Wong, M. Schuster, P. Wang, P. Sears, *J. Am. Chem. Soc.* **1993**, *115*, 5893–5901.
49 I. Christiansen-Brams, M. Meldal, K. Bock, *J. Chem. Soc. Perkin Trans. 1* **1993**, 1461–1471.
50 H. Kunz, J. März, *Angew. Chem. Int. Ed.* **1988**, *27*, 1375–1377.

51 H. Waldmann, H. Kunz, *Liebigs Ann.* **1983**, 1712–1725.
52 S. Lavielle, N. C. Ling, R. C. Guillemin, *Carbohydrate Res.* **1981**, *89*, 221–228.
53 Z. W. Guo, Y. Nakahara, T. Ogawa, *Bioorg. Med. Chem.* **1997**, *5*, 1917–1924.
54 S. J. Danishefsky, S. Hu, P. F. Cirillo, M. Eckhardt, P. H. Seeberger, *Chem. Eur. J.* **1997**, *3*, 1617–1628.
55 H. Kunz, W. Günther, *Angew. Chem. Int. Ed.* **1988**, *27*, 1086–1087.
56 W. Günther, H. Kunz, *Angew. Chem. Int. Ed.* **1990**, *29*, 1050–1051.
57 E. Meinjohanns, M. Meldal, H. Paulsen, R. A. Dwek, K. Bock, *J. Chem. Soc. Perkin Trans. 1* **1998**, 549–560.
58 R. U. Lemieux, R. R. M., *Can. J. Chem.* **1979**, *57*, 1244–1251.
59 S. D. Kuduk, J. B. Schwarz, X. T. Chen, P. W. Glunz, D. Sames, G. Ragupathi, P. O. Livingston, S. J. Danishefsky, *J. Am. Chem. Soc.* **1998**, *120*, 12474–12485.
60 B. Liebe, H. Kunz, *Angew. Chem. Int. Ed.* **1997**, *36*, 618–621.
61 H. Paulsen, K. Adermann, *Liebigs Ann.* **1989**, 751–769.
62 H. Kunz, S. Birnbach, *Angew. Chem. Int. Ed.* **1986**, *25*, 360–362.
63 H. Paulsen, T. Bielfeldt, S. Peters, M. Meldal, K. Bock, *Liebigs Ann.* **1994**, 369–379.
64 T. Bielfeldt, S. Peters, M. Meldal, K. Bock, H. Paulsen, *Angew. Chem. Int. Ed.* **1992**, *31*, 857–859.
65 G. A. Winterfeld, R. R. Schmidt, *Angew. Chem. Int. Ed.* **2001**, *40*, 2654–2657.
66 G. Arsequell, L. Krippner, R. A. Dwek, S. Y. C. Wong, *J. Chem. Soc. Chem. Commun.* **1994**, 2383–2384.
67 L. A. Salvador, M. Elofsson, J. Kihlberg, *Tetrahedron* **1995**, *51*, 5643–5656.
68 V. Wittmann, D. Lennartz, *Eur. J. Org. Chem.* **2002**, 1363–1367.
69 U. K. Saha, R. R. Schmidt, *J. Chem. Soc. Perkin Trans. 1* **1997**, 1855–1860.
70 K. J. Jensen, P. R. Hansen, D. Venugopal, G. Barany, *J. Am. Chem. Soc.* **1996**, *118*, 3148–3155.
71 E. Meinjohanns, M. Meldal, K. Bock, *Tetrahedron Lett.* **1995**, *36*, 9205–9208.
72 A. Vargas-Berenguel, M. Meldal, H. Paulsen, K. Bock, *J. Chem. Soc. Perkin Trans. 1* **1994**, 2615–2619.

73 M. Schultz, H. Kunz, *Tetrahedron Lett.* **1992**, *33*, 5319–5322.
74 O. Seitz, C. H. Wong, *J. Am. Chem. Soc.* **1997**, *119*, 8766–8776.
75 N. B. Malkar, J. L. Lauer-Fields, G. B. Fields, *Tetrahedron Lett.* **2000**, *41*, 1137–1140.
76 J. Broddefalk, M. Forsgren, I. Sethson, J. Kihlberg, *J. Org. Chem.* **1999**, *64*, 8948–8953.
77 D. Sames, X. T. Chen, S. J. Danishefsky, *Nature* **1997**, *389*, 587–591.
78 H. Iijima, T. Ogawa, *Carbohydrate Res.* **1989**, *186*, 107–118.
79 N. Mathieux, H. Paulsen, M. Meldal, K. Bock, *J. Chem. Soc. Perkin Trans. 1* **1997**, 2359–2368.
80 J. B. Schwarz, S. D. Kuduk, X. T. Chen, D. Sames, P. W. Glunz, S. J. Danishefsky, *J. Am. Chem. Soc.* **1999**, *121*, 2662–2673.
81 Y. Nakahara, Y. Ito, T. Ogawa, *Tetrahedron Lett.* **1997**, *38*, 7211–7214.
82 N. Bezay, G. Dudziak, A. Liese, H. Kunz, *Angew. Chem. Int. Ed.* **2001**, *40*, 2292–2295.
83 G. Dudziak, N. Bezay, T. Schwientek, H. Clausen, H. Kunz, A. Liese, *Tetrahedron* **2000**, *56*, 5865–5869.
84 K. von dem Bruch, H. Kunz, *Angew. Chem. Int. Ed.* **1994**, *33*, 101–103.
85 Z. G. Wang, X. F. Zhang, M. Visser, D. Live, A. Zatorski, U. Iserloh, K. O. Lloyd, S. J. Danishefsky, *Angew. Chem. Int. Ed.* **2001**, *40*, 1728–1732.
86 U. Sprengard, M. Schudok, W. Schmidt, G. Kretzschmar, H. Kunz, *Angew. Chem. Int. Ed.* **1996**, *35*, 321–324.
87 K. Yamamoto, *J. Biosci. Bioeng.* **2001**, *92*, 493–501.
88 M. Mizuno, K. Haneda, R. Iguchi, I. Muramoto, T. Kawakami, S. Aimoto, K. Yamamoto, T. Inazu, *J. Am. Chem. Soc.* **1999**, *121*, 284–290.
89 K. Witte, P. Sears, R. Martin, C. H. Wong, *J. Am. Chem. Soc.* **1997**, *119*, 2114–2118.
90 S. R. Chong, Y. Shao, H. Paulus, J. Benner, F. B. Perler, M. Q. Xu, *J. Biol. Chem.* **1996**, *271*, 22159–22168.
91 M. Q. Xu, D. G. Comb, M. W. Southworth, F. B. Mersha, M. E. Scott, F. B. Perler, *Prot. Engin.* **1995**, *8*, 96.
92 T. J. Tolbert, C. H. Wong, *J. Am. Chem. Soc.* **2000**, *122*, 5421–5428.
93 P. W. Glunz, S. Hintermann, J. B. Schwarz, S. D. Kuduk, X. T. Chen, L. J. Williams, D. Sames, S. J. Danishefsky, V. Kudryashov, K. O. Lloyd, *J. Am. Chem. Soc.* **1999**, *121*, 10636–10637.
94 P. W. Glunz, S. Hintermann, L. J. Williams, J. B. Schwarz, S. D. Kuduk, V. Kudryashov, K. O. Lloyd, S. J. Danishefsky, *J. Am. Chem. Soc.* **2000**, *122*, 7273–7279.
95 S. K. George, T. Schwientek, B. Holm, C. A. Reis, H. Clausen, J. Kihlberg, *J. Am. Chem. Soc.* **2001**, *123*, 11117–11125.
96 M. Izumi, G. J. Shen, S. Wacowich-Sgarbi, T. Nakatani, O. Plettenburg, C. H. Wong, *J. Am. Chem. Soc.* **2001**, *123*, 10909–10918.
97 K. M. Koeller, M. E. B. Smith, C. H. Wong, *J. Am. Chem. Soc.* **2000**, *122*, 742–743.
98 K. M. Koeller, M. E. B. Smith, R. F. Huang, C. H. Wong, *J. Am. Chem. Soc.* **2000**, *122*, 4241–4242.
99 A. Leppänen, S. P. White, J. Helin, R. P. McEver, R. D. Cummings, *J. Biol. Chem.* **2000**, *275*, 39569–39578.
100 M. Inoue, I. Yamashina, H. Nakada, *Biochem. Biophys. Res. Commun.* **1998**, *245*, 23–27.
101 F. G. Hanisch, C. A. Reis, H. Clausen, H. Paulsen, *Glycobiol.* **2001**, *11*, 731–740.
102 J. Kihlberg, M. Elofsson, *Curr. Med. Chem.* **1997**, *4*, 85–116.
103 T. Norberg, B. Luning, J. Tejbrant, *Neoglycoconjugates, Pt B* **1994**, *247*, 87–106.
104 M. Meldal, *Methods Enzymol.* **1997**, *289*, 83–104.
105 R. C. Sheppard, B. J. Williams, *Int. J. Pept. Protein Res.* **1982**, *20*, 451–454.
106 B. Holm, S. Linse, J. Kihlberg, *Tetrahedron* **1998**, *54*, 11995–12006.
107 C. J. Bosques, V. W. F. Tai, B. Imperiali, *Tetrahedron Lett.* **2001**, *42*, 7207–7210.
108 Z. W. Guo, Y. Nakahara, T. Ogawa, *Angew. Chem. Int. Ed.* **1997**, *36*, 1464–1466.
109 S. A. Kates, B. G. Delatorre, R. Eritja, F. Albericio, *Tetrahedron Lett.* **1994**, *35*, 1033–1034.

110 J. Offer, M. Quibell, T. Johnson, *J. Chem. Soc. Perkin Trans. 1* **1996**, Perkin Transactions 1, 175–182.
111 J.Y. Roberge, X. Beebe, S.J. Danishefsky, *Science* **1995**, *269*, 202–204.
112 M. Schuster, P. Wang, J.C. Paulson, C.H. Wong, *J. Am. Chem. Soc.* **1994**, *116*, 1135–1136.
113 M. Meldal, F.I. Auzanneau, O. Hindsgaul, M.M. Palcic, *J. Chem. Soc. Chem. Commun.* **1994**, 1849–1850.
114 T. Wiemann, N. Taubken, U. Zehavi, J. Thiem, *Carbohydrate Res.* **1994**, *257*, C1–C6.
115 M. Elofsson, L.A. Salvador, J. Kihlberg, *Tetrahedron* **1997**, *53*, 369–390.
116 S. Peters, T.L. Lowary, O. Hindsgaul, M. Meldal, K. Bock, *J. Chem. Soc. Perkin Trans. 1* **1995**, Perkin Transactions 1, 3017–3022.
117 K.J. Jensen, M. Meldal, K. Bock, *J. Chem. Soc. Perkin Trans. 1* **1993**, 2119–2129.
118 Y. Nakahara, Y. Ito, T. Ogawa, *Carbohydrate Res.* **1998**, *309*, 287–296.
119 J. Broddefalk, K.E. Bergquist, J. Kihlberg, *Tetrahedron* **1998**, *54*, 12047–12070.
120 O. Seitz, H. Kunz, *Angew. Chem. Int. Ed.* **1995**, *34*, 803–805.
121 O. Seitz, H. Kunz, *J. Org. Chem.* **1997**, *62*, 813–826.
122 H. Paulsen, A. Schleyer, N. Mathieux, M. Meldal, K. Bock, *J. Chem. Soc. Perkin Trans. 1* **1997**, 281–293.
123 D.M. Andrews, P.W. Seale, *Int. J. Pept. Protein Res.* **1993**, *42*, 165–170.
124 M. Hollosi, E. Kollat, I. Laczko, K.F. Medzihradszky, J. Thurin, L. Otvos, *Tetrahedron Lett.* **1991**, *32*, 1531–1534.
125 A. Schleyer, M. Meldal, R. Manat, H. Paulsen, K. Book, *Angew. Chem. Int. Ed.* **1997**, *36*, 1976–1978.
126 K.M. Halkes, C.H. Gotfredsen, M. Grotli, L.P. Miranda, J.O. Duus, M. Meldal, *Chem. Eur. J.* **2001**, *7*, 3584–3591.
127 J. Habermann, H. Kunz, *Tetrahedron Lett.* **1998**, *39*, 4797–4800.
128 S. Keil, C. Claus, W. Dippold, H. Kunz, *Angew. Chem. Int. Ed.* **2001**, *40*, 366–369.
129 M. Wagner, H. Kunz, *Angew. Chem. Int. Ed.* **2001**, *41*, 317–321.
130 A. Ishii, H. Hojo, A. Kobayashi, K. Nakamura, Y. Nakahara, Y. Ito, *Tetrahedron* **2000**, *56*, 6235–6243.
131 K. Nakamura, N. Hanai, M. Kanno, A. Kobayashi, Y. Ohnishi, Y. Ito, Y. Nakahara, *Tetrahedron Lett.* **1999**, *40*, 515–518.
132 T.F.J. Lampe, G. Weitzschmidt, C.H. Wong, *Angew. Chem. Int. Ed.* **1998**, *37*, 1707–1711.
133 K. Witte, O. Seitz, C.H. Wong, *J. Am. Chem. Soc.* **1998**, *120*, 1979–1989.
134 P.E. Dawson, T.W. Muir, I. Clark-Lewis, S.B.H. Kent, *Science* **1994**, *266*, 776–779.
135 T. Wieland, E. Bokelmann, L. Bauer, H.U. Lang, H. Lau, *Ann. Chem.* **1953**, *583*, 129–149.
136 Y. Shin, K.A. Winans, B.J. Backes, S.B.H. Kent, J.A. Ellman, C.R. Bertozzi, *J. Am. Chem. Soc.* **1999**, *121*, 11684–11689.
137 B.J. Backes, A.A. Virgilio, J.A. Ellman, *J. Am. Chem. Soc.* **1996**, *118*, 3055–3056.
138 L.A. Marcaurelle, L.S. Mizoue, J. Wilken, L. Oldham, S.B.H. Kent, T.M. Handel, C.R. Bertozzi, *Chem. Eur. J.* **2001**, *7*, 1129–1132.
139 M.R. Wormald, A.J. Petrescu, Y.L. Pao, A. Glithero, T. Elliott, R.A. Dwek, *Chem. Rev.* **2002**, *102*, 371–386.
140 B. Imperiali, S.E. O'Connor, *Curr. Op. Chem. Biol.* **1999**, *3*, 643–649.
141 S.E. O'Connor, B. Imperiali, *J. Am. Chem. Soc.* **1997**, *119*, 2295–2296.
142 B. Imperiali, K.L. Shannon, *Biochemistry* **1991**, *30*, 4374–4380.
143 S.E. O'Connor, B. Imperiali, *Chem. Biol.* **1998**, *5*, 427–437.
144 D.H. Live, R.A. Kumar, X. Beebe, S.J. Danishefsky, *Proc. Natl. Acad. Sci. USA* **1996**, *93*, 12759–12761.
145 E.E. Simanek, D.H. Huang, L. Pasternack, T.D. Machajewski, O. Seitz, D.S. Millar, H.J. Dyson, C.H. Wong, *J. Am. Chem. Soc.* **1998**, *120*, 11567–11575.
146 W.G. Wu, L. Pasternack, D.H. Huang, K.M. Koeller, C.C. Lin, O. Seitz, C.H. Wong, *J. Am. Chem. Soc.* **1999**, *121*, 2409–2417.
147 J.G. Bann, D.H. Peyton, H.P. Bächinger, *FEBS Lett.* **2000**, *473*, 237–240.

148 O. Schuster, G. Klich, V. Sinnwell, H. Kranz, H. Paulsen, B. Meyer, *J. Biomol. NMR* **1999**, *14*, 33–45.

149 D. H. Live, L. J. Williams, S. D. Kuduk, J. B. Schwarz, P. W. Glunz, X. T. Chen, D. Sames, R. A. Kumar, S. J. Danishefsky, *Proc. Natl. Acad. Sci. USA* **1999**, *96*, 3489–3493.

150 H. Lis, N. Sharon, *Eur. J. Biochem.* **1993**, *218*, 1–27.

151 M. Nomoto, K. Yamada, M. Haga, M. Hayashi, *J. Pharm. Sci.* **1998**, *87*, 326–332.

152 R. Polt, M. M. Palian, *Drug Future* **2001**, *26*, 561–576.

153 N. Elmagbari, D. Notyce, W. Schmid, M. Palian, R. Polt, E. J. Bilsky, *FASEB J.* **2001**, *15*, A915–A915.

154 J. Horvat, S. Horvat, C. Lemieux, P. W. Schiller, *Int. J. Pept. Protein Res.* **1988**, *31*, 499–507.

155 S. Horvat, J. Horvat, L. Varga-Defterdarovic, K. Pavelic, N. N. Chung, P. W. Schiller, *Int. J. Pept. Protein Res.* **1993**, *41*, 399–404.

156 R. Tomatis, M. Marastoni, G. Balboni, R. Guerrini, A. Capasso, L. Sorrentino, V. Santagada, G. Caliendo, L. H. Lazarus, S. Salvadori, *J. Med. Chem.* **1997**, *40*, 2948–2952.

157 R. Polt, F. Porreca, L. Z. Szabo, E. J. Bilsky, P. Davis, T. J. Abbruscato, T. P. Davis, R. Horvath, H. I. Yamamura, V. J. Hruby, *Proc. Natl. Acad. Sci. USA* **1994**, *91*, 7114–7118.

158 E. J. Bilsky, R. D. Egleton, S. A. Mitchell, M. M. Palian, P. Davis, J. D. Huber, H. Jones, H. I. Yamamura, J. Janders, T. P. Davis, F. Porreca, V. J. Hruby, R. Polt, *J. Med. Chem.* **2000**, *43*, 2586–2590.

159 H. G. Lerchen, J. Baumgarten, N. Piel, V. Kolb-Bachofen, *Angew. Chem. Int. Ed.* **1999**, *38*, 3680–3683.

160 M. Veyhl, K. Wagner, C. Volk, V. Gorboulev, K. Baumgarten, W. M. Weber, M. Schaper, B. Bertram, M. Wiessler, H. Koepsell, *Proc. Natl. Acad. Sci. USA* **1998**, *95*, 2914–2919.

161 B. Rihova, *Cit. Rev. Biotechnol.* **1997**, *17*, 149–169.

162 J. J. Hangeland, J. T. Levis, Y. C. Lee, P. O. P. Tso, *Bioconj. Chem.* **1995**, *6*, 695–701.

163 K. Sarkar, H. S. Sarkar, L. Kole, P. K. Das, *Mol. Cell. Biochem.* **1996**, *156*, 109–116.

164 M. S. Wadhwa, K. G. Rice, *J. Drug Targeting* **1995**, *3*, 111–127.

165 Y. J. Kim, A. Varki, *Glycoconjugate J.* **1997**, *14*, 569–576.

166 R. R. Koganty, M. A. Reddish, B. M. Longenecker, *Drug Discovery Today* **1996**, *1*, 190–198.

167 S. F. Slovin, G. Ragupathi, S. Adluri, G. Ungers, K. Terry, S. Kim, M. Spassova, W. G. Bornmann, M. Fazzari, L. Dantis, K. Olkiewicz, K. O. Lloyd, P. O. Livingston, S. J. Danishefsky, H. I. Scher, *Proc. Natl. Acad. Sci. USA* **1999**, *96*, 5710–5715.

168 V. Kudryashov, P. W. Glunz, L. J. Williams, S. Hintermann, S. J. Danishefsky, K. O. Lloyd, *Proc. Natl. Acad. Sci. USA* **2001**, *98*, 3264–3269.

169 G. D. MacLean, M. A. Reddish, R. R. Koganty, B. M. Longenecker, *J. Immunother.* **1996**, *19*, 59–68.

170 L. A. Holmberg, D. V. Oparin, T. Gooley, K. Lilleby, W. Bensinger, M. A. Reddish, G. D. MacLean, B. M. Longenecker, B. M. Sandmaier, *Bone Marrow Transplant.* **2000**, *25*, 1233–1241.

171 B. M. Sandmaier, D. V. Oparin, L. A. Holmberg, M. A. Reddish, G. D. MacLean, B. M. Longenecker, *J. Immunother.* **1999**, *22*, 54–66.

172 R. Lo-Man, S. Bay, S. Vichier-Guerre, E. Deriaud, D. Cantacuzene, C. Leclerc, *Cancer Res.* **1999**, *59*, 1520–1524.

173 R. Lo-Man, S. Vichier-Guerre, S. Bay, E. Dériaud, D. Cantacuzène, C. Leclerc, *J. Immunol.* **2001**, *166*, 2849–2854.

174 E. Lisowska, *Cell. Mol. Life Sci.* **2002**, *59*, 445–455.

175 A. Corthay, J. Backlund, R. Holmdahl, *Ann. Med.* **2001**, *33*, 456–465.

176 T. Jensen, P. Hansen, L. Galli-Stampino, S. Mouritsen, K. Frische, E. Meinjohanns, M. Meldal, O. Werdelin, *J. Immunol.* **1997**, *158*, 3769–3778.

177 T. Jensen, M. Nielsen, M. Gad, P. Hansen, S. Komba, M. Meldal, N. Odum, O. Werdelin, *Eur. J. Immunol.* **2001**, *31*, 3197–3206.

178 E. Michalsson, V. Malmström, S. Reis, A. Engström, H. Burkhardt, R. Holmdahl, *J. Exp. Med.* **1994**, *180*, 745–749.

179 J. Broddefalk, J. Backlund, F. Almqvist, M. Johansson, R. Holmdahl, J. Kihlberg, *J. Am. Chem. Soc.* **1998**, *120*, 7676–7683.

180 P. Kjellen, U. Brunsberg, J. Broddefalk, B. Hansen, M. Vestberg, I. Ivarsson, A. Engstrom, A. Svejgaard, J. Kihlberg, L. Fugger, R. Holmdahl, *Eur. J. Immunol.* **1998**, *28*, 755–767.

181 A. Glithero, J. Tormo, J. S. Haurum, G. Arsequell, G. Valencia, J. Edwards, S. Springer, A. Townsend, Y. L. Pao, M. Wormald, R. A. Dwek, E. Y. Jones, T. Elliott, *Immunity* **1999**, *10*, 63–74.

182 J. S. Haurum, I. B. Høier, G. Arsequell, A. Neisig, G. Valencia, J. Zeuthen, J. Neefjes, T. Elliott, *J. Exp. Med.* **1999**, *190*, 145–150.

183 I. B. Kastrup, S. Stevanovic, G. Arsequell, G. Valencia, J. Zeuthen, H. G. Rammensee, T. Elliott, J. S. Haurum, *Tissue Antigens* **2000**, *56*, 129–135.

8
Synthesis of Complex Carbohydrates: Everninomicin 13,384-1
K.C. Nicolaou, Helen J. Mitchell, and Scott A. Snyder

8.1
Introduction

Carbohydrate chemistry is a discipline that has occupied the hearts and minds of innumerable scientists for well over a century, due principally to the relative abundance of carbohydrates in nature and their diverse and significant roles in biological systems. As we enter the twenty-first century, this field continues to be both vigorous and challenging. In particular, one of the most exciting aspects of organic synthesis over the course of the last few decades has been the symbiotic interplay between the specialized sub-discipline of carbohydrate chemistry and total synthesis, each enabling and advancing the other in new directions and toward greater heights. While the characteristic structural features of carbohydrates confer certain unique chemical, physical, and biological properties to their carrier molecules, making them interesting targets in their own right, the increasing complexity of carbohydrate-containing natural products isolated from nature's seemingly limitless library of chemical diversity serves as a vibrant engine driving forward the field of organic synthesis by requiring the development of improved and novel synthetic strategies and methodologies to effect their construction.

Over the course of the past twenty-five years, our group has engaged in numerous synthetic adventures in the realm of carbohydrate chemistry, enticed for the most part by objectives in chemical synthesis and chemical biology. Using natural products such as those shown in Fig. 8.1 as sources of inspiration, we have pursued and invented numerous synthetic technologies for carbohydrate synthesis, enabling the construction of complex oligosaccharides both in solution and on solid support for diverse applications; additionally, we have employed carbohydrate templates as scaffolds for peptide mimetics and for molecular diversity construction. Rather than review our extensive efforts in this area [1], we have instead decided in this chapter to focus in depth on a recently achieved total synthesis of the antibiotic everninomicin 13,384-1 (**1**, Fig. 8.1), a molecule that perhaps represents the most complex oligosaccharide-based structure synthesized to date. Our goal in this analysis is not only to indicate the current state of the art of carbohydrate synthesis, particularly in regards to total synthesis, but also to highlight

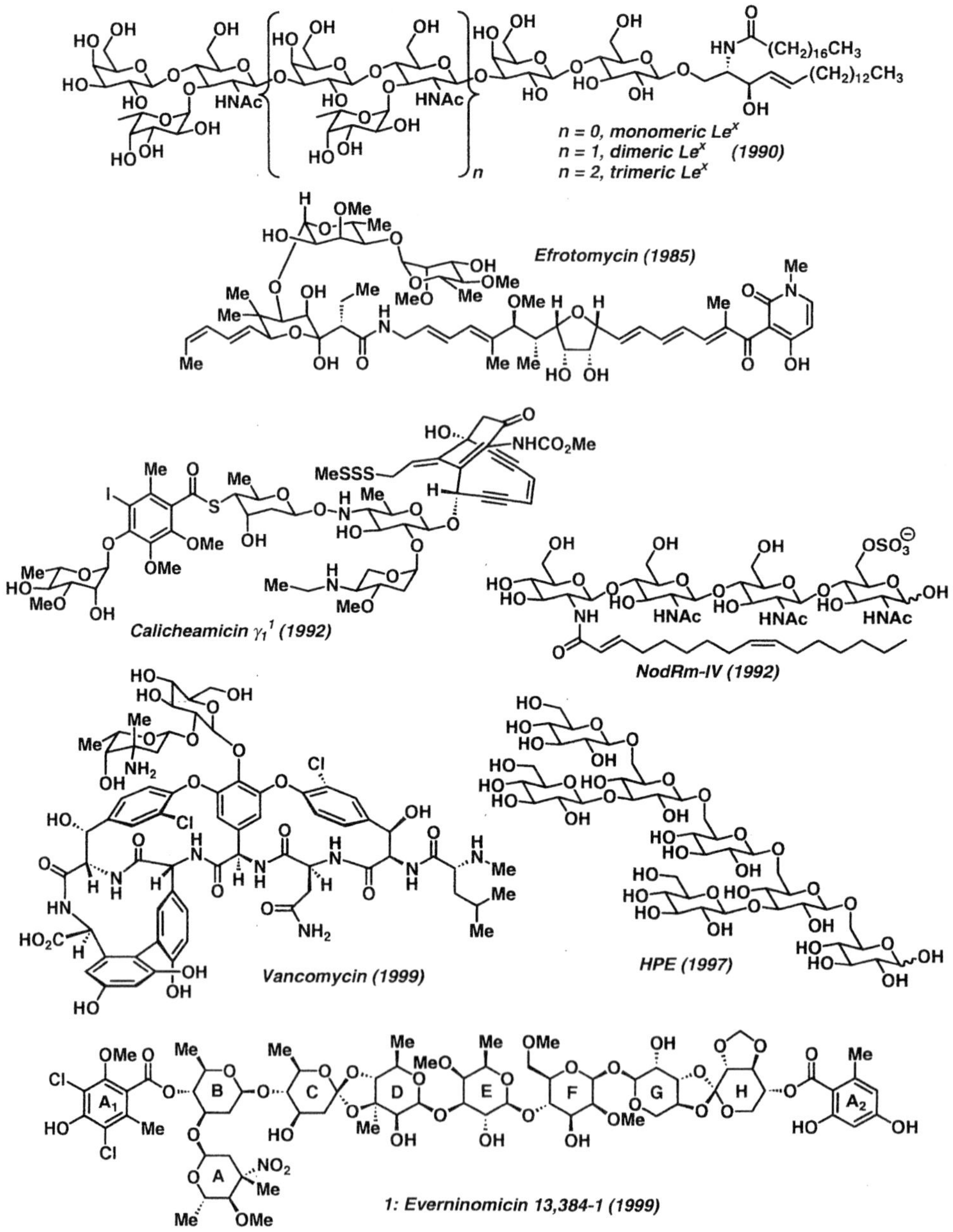

216 | *8 Synthesis of Complex Carbohydrates: Everninomicin 13,384-1*

Fig. 8.1 Selected glycoside-containing natural products synthesized by the Nicolaou group (year completed is in parentheses).

methodological advances, a benefit that can often result when one is confronted with unique and challenging molecular architectures.

Everninomicin 13,384-1 (**1**) constitutes a valuable new weapon in the battle against drug-resistant bacteria, specifically methicillin-resistant *Staphylococci* and vancomycin-resistant *Streptococci* and *Enterococci* [2]. First isolated in the 1960s, thanks to their strong activities against Gram-positive bacteria, the everninomicins are members of the orthosomycin class of natural products, a group of compounds produced as secondary metabolites from a wide variety of *Actinomycetes* of the genera *Streptomyces* and *Micromonospora* [3]. Architecturally, these molecules are unified by two unusual features: (1) a chain of three to eight carbohydrate residues, and (2) the replacement of at least one glycoside bond by a spiro-orthoester linkage. Although the structural and pharmacological properties of these early isolates, such as everninomicin D and flambamycin, were investigated in detail [4, 5], they were not considered for clinical use, due to severe adverse reactions [3]. Everninomicin component 13,384-1 (**1**) [2] was isolated more recently as one of several active compounds found in the fermentation broth of *Micromonospora carbonacea var. africana*, grown from a sample of soil collected from the banks of the Nyiro River in Kenya. In a number of studies, everninomicin 13,384-1 (**1**) demonstrated excellent in vitro activity, against resistant strains of Gram-positive bacteria, with MIC_{90} values similar to, or two to four times lower than, those possessed by vancomycin, the antibiotic currently regarded as the last line of defense against methicillin-resistant strains of *Staphylococcus aureus* [2]. With improved formulation techniques and a better activity profile than the older everninomicins [6], 13,384-1 (**1**) underwent initial stages of development and entered advanced clinical trials under the tradename of Ziracin.

To date, relatively few studies have been devoted to exploring and fully elucidating the chemical biology of the orthosomycins, including everninomicin 13,384-1 (**1**). One member of the family related to **1**, everninomicin B, was reported to cause alteration of the cytoplasmic membrane resulting in the inhibition of metabolic uptake, followed by subsequent disruption of DNA and protein synthesis [4]. A second orthosomycin, avilamycin C, appears to act as a specific inhibitor of protein synthesis by binding to the 30S subunit of ribosomes, where it interferes with attachment of aminoacyl-tRNA to the ribosome [7]. More recently, RNA probing has demonstrated that everninomicin 13,384-1 (**1**) interacts at a novel site on the large ribosomal subunit [8a]. In other studies, variations in ribosomal protein L16 corresponding to everninomicin susceptibility have been examined, also leading to the hypothesis that everninomicin 13,384-1 (**1**) acts as an inhibitor of protein biosynthesis [8b]. Early studies [9] on the structure-activity relationships of the everninomicins and flambamycin demonstrated that either hydrolysis of the central CD orthoester or derivatization of the phenolic hydroxy group of ring A_1 resulted in complete loss of biological activity. Additionally, chemical modification of the nitro sugar portion (ring A) of the everninomicins resulted in a number of highly active derivatives with improved pharmacokinetic properties. These and other modifications were also applied to 13,384-1 (**1**); however, none proved better than the original lead compound [9f].

8.2
Retrosynthetic Analysis and Strategy

8.2.1
Overview of Synthetic Strategies and Methodologies

One of the most critical issues which must be addressed in any attempted total synthesis of a complex oligosaccharide, such as everninomicin 13,384-1 (**1**), is control of anomeric stereochemistry during glycosidation reactions. Therefore, before embarking upon a description of the everninomicin synthesis [10], we shall first briefly describe the selected glycosylation methods [11] which proved invaluable in the schemes that follow. In most glycosidations, the nature of the C-2 substituent controls the resulting anomeric stereochemistry. Thus, when the C-2 position is occupied by an alkoxy, benzyloxy, or 2-deoxy substituent, the anomeric effect dominates and the a anomer is formed preferentially (Scheme 8.1a). Conversely, when the C-2 position is occupied by a participating group such as an ester (Scheme 8.1b) or a phenylthio group, the stereochemical outcome is a *trans* configuration between the anomeric and C-2 substituents [12]. Of the numerous types of donors currently known in the chemical literature, glycosyl fluorides, trichloroacetimidates, phenylsulfides, phenylselenides, and phenylsulfoxides have proven to be the most successful glycosidation partners for total synthesis applications, a statement whose veracity is demonstrated throughout the everninomicin synthesis. Glycosyl fluorides, first introduced as glycosyl donors by the Mukaiyama group [13] in 1981 (Scheme 8.1c), are usually more stable than other glycosyl halides, often exhibiting thermal and chemical stability sufficient to enable chromatographic purification. Glycosyl fluorides, activated by reagents such as Lewis acids [14, 15], silver salts, and various metallocenes [16], were used in the everninomicin 13,384-1 (**1**) synthesis to stereoselectively create linkages between the AB, BC, CD, and GH ring systems.

The trichloroacetimidate-mediated glycosidation procedure (Scheme 8.1d), first reported by Schmidt [17] in 1980, proved successful for the formation of the EF and FG systems. Trichloroacetimidate donors are easily prepared from lactols and trichloroacetonitrile in the presence of a base such as NaH or DBU (for abbreviations, see legends in schemes), and are typically activated by Lewis acids such as $BF_3 \cdot Et_2O$ [17] or TMSOTf [18].

Thioglycosides, particularly phenylthioglycosides (Scheme 8.1e), played a central role in the construction of the BC glycoside system of **1**, due to their stability to diverse reaction conditions and convenient activation with electrophilic [19] or oxidizing reagents. Glycosyl phenylsulfoxides act as mild glycosyl donors upon activation with Tf_2O, as first demonstrated by Kahne [20a], with later advances made by Crich providing a selective method for the construction of β-mannosides [20b, 20c]. This procedure proved to be an invaluable tool for successful construction of the target molecule's DE bond (Scheme 8.1f).

Several additional methodologies also merit some discussion at this juncture, as they played critical roles in the synthesis of everninomicin 13,384-1 (**1**). In 1986,

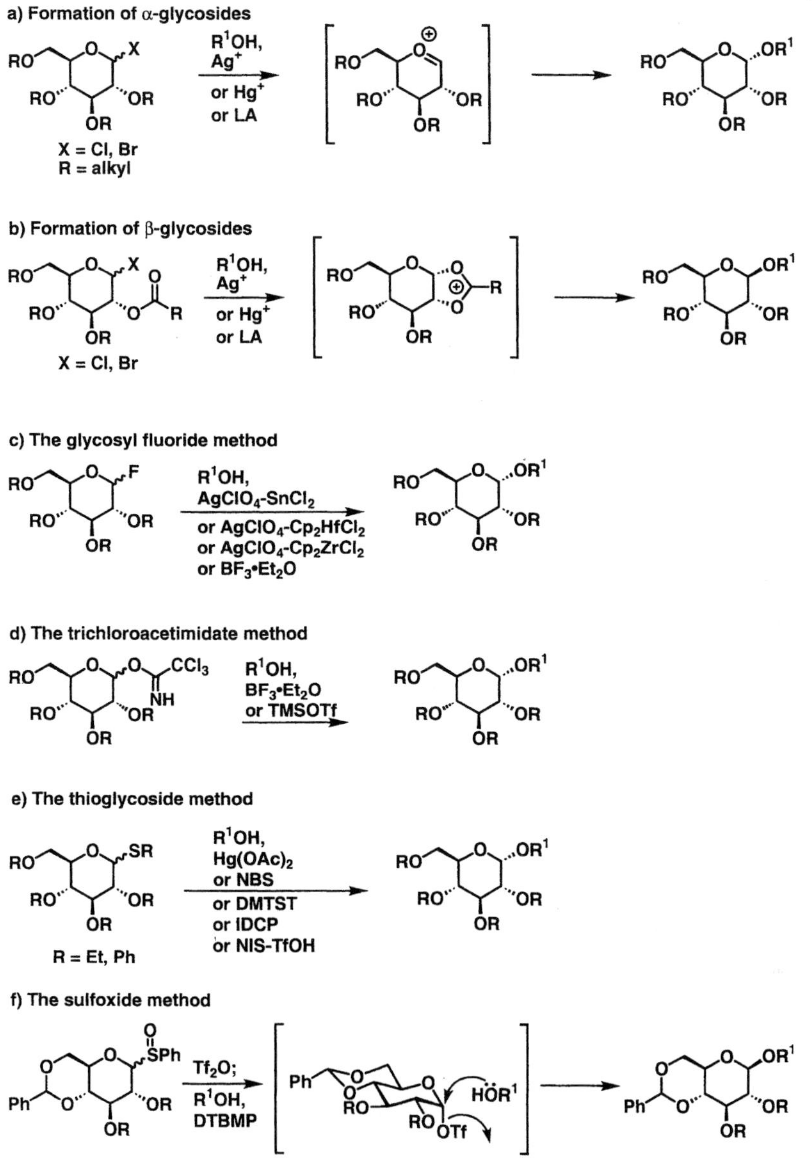

Scheme 8.1 Selected glycosylation methods employed in the everninomicin 13,384-1 total synthesis. LA = Lewis acid, TMS = trimethylsilyl, Cp = cyclopentadienyl, NBS = N-bromosuccinimide, DMTST = (dimethylthio)methyl- sulfonium trifluoromethanesulfonate, IDCP = iodobis(collidine) perchlorate, NIS = N-iodosuccinimide, Tf = trifluoromethanesulfonyl, DTBMP = 2,6-di-*tert*-butyl-4-methylpyridine.

for example, it was observed that treatment of certain carbohydrates with DAST promoted stereospecific 1,2-migration of the anomeric substituent, affording glycosyl fluorides in high yield with concomitant installation of a variety of useful functional groups at C-2 and inversion of configuration relative to the starting material at both centers, as illustrated in Scheme 8.2 [21]. One useful application of this procedure is subsequent reductive removal of C-2 thiophenyl groups, affording 2-deoxyglycosides, a process illustrated in the construction of the BC unit of **1**.

An additional topic that must be addressed to achieve a synthesis of everninomicin 13,384-1 (**1**) is the formation of spirocyclic esters (orthoesters). As such, extension of the 1,2-thiophenyl migration to include the selenophenyl moiety proved essential in the development of methodology to construct this unique molecular feature, as shown in Scheme 8.3. Thus, treatment of a 2-hydroxy-1-selenoglycoside (**I**) with DAST facilitates smooth 1,2-migration, affording the reactive 2-seleno-1-fluoro derivative (**III**) ready for coupling. Addition of a suitable alcohol and a Lewis acid such as $SnCl_2$ furnishes **IV** with a high degree of stereocontrol, due to participation of the phenylseleno group and the anomeric effect. Removal of the protecting group then sets the stage for the Sinaÿ orthoester formation, which proceeds by initial $NaIO_4$-mediated oxidation of the selenium residue to the selenoxide in a $MeOH/CH_2Cl_2/H_2O$ (3:2:1) solvent mixture, followed by heating at 140 °C in toluene/vinyl acetate/diisopropylamine (2:2:1) in a sealed tube [22] to effect *syn* elimination and cyclization to afford the desired spiro-orthoester **VIII** in high yield. This procedure proved to be the method of choice in the everninomicin synthesis for generating both the CD and the GH spirocyclic systems.

The construction of a 1,1′-bridge linking two carbohydrate units in a β-mannoside fashion requires control of the stereochemistry at both anomeric centers while the new bond is formed, a synthetic issue for the FG ring system in everninomicin 13,384-1 (**1**). After experimentation with various techniques, we developed a new technology for the construction of 1,1′-disaccharides and 1,1′:1″,2-trisaccharides to address this challenge [23]. As illustrated in Scheme 8.4, treatment of a lactol such as **2** with an activated donor (**3**) in the presence of TMSOTf furnished the undesired disaccharide **4**, containing the 1a,1′a stereochemistry, in 87%

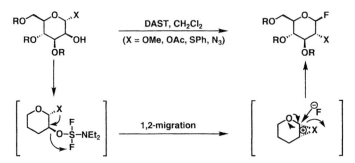

Scheme 8.2 DAST Promoted 1,2-migrations in carbohydrates (DAST = (diethylamino)sulfur trifluoride).

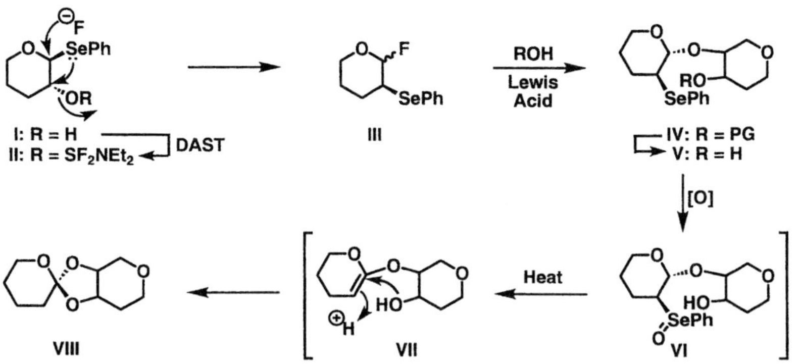

Scheme 8.3 Orthoester formation by phenylseleno 1,2-migration followed by glycosylation (I → II → III → IV) and ring-closure after *syn*-elimination (V → VI → VII → VIII). PG = protecting group.

yield. However, conversion of **2** into the five-membered ring stannane **5** fixed the anomeric configuration of **2**, and delivered the desired 1β,1′α-disaccharide **6** in 66% yield upon subsequent treatment with trichloroacetimidate **3** in the presence of TMSOTf.

Finally, we should highlight the most efficient methods utilized for the regioselective functionalization of polyol-containing mono- and oligosaccharides. As shown in Scheme 8.5, regioselective acylation and alkylation of tin acetals [24] prove extremely useful for the selective protection of a number of carbohydrate scaffolds, including equatorial over axial hydroxy groups (Scheme 8.5a), and for differentiating between two equatorial hydroxy groups in different steric environments (Scheme 8.5b). The first process is illustrated in the synthesis of ring D (see Scheme 8.27) of everninomicin 13,384-1 (**1**), while the second is featured in the construction of rings B and C (Schemes 8.11 and 8.12, respectively).

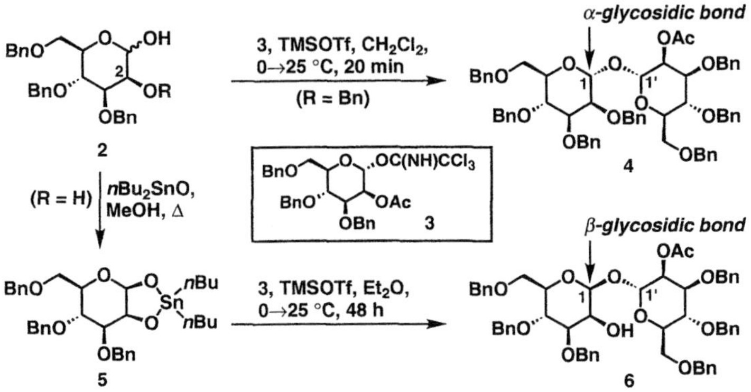

Scheme 8.4 Synthesis of model 1,1′-disaccharides **4** and **6**. Bn = benzyl, Ac = acetyl.

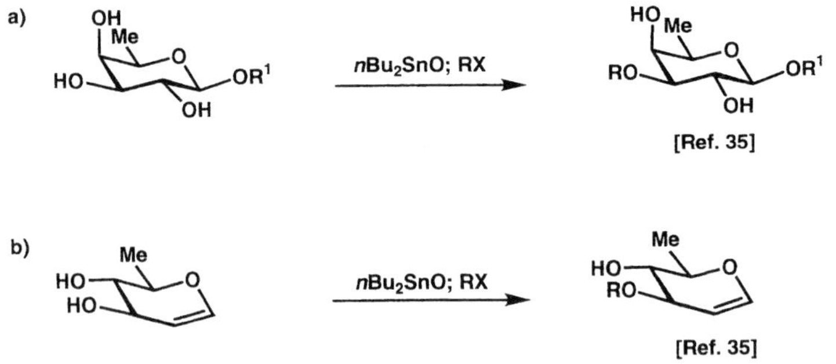

Scheme 8.5 Regioselective methods for protection of hydroxy groups: a) equatorial over axial protection of hydroxy groups with nBu₂SnO; b) selective protection of hydroxy groups with nBu₂SnO, based on steric grounds.

8.2.2
Retrosynthetic Analysis: Overall Approach

With these fundamental methods described, we are now ready to engage in a retrosynthetic analysis of everninomicin 13,384-1 (**1**). In addition to the novel oligosaccharide structure focused around the unusual connectivity of two sensitive orthoester moieties, everninomicin 13,384-1 (**1**) contains within its structure a 1,1′-disaccharide bridge, a nitrosugar (evernitrose), two highly substituted aromatic esters, a 2-deoxyglycoside, and two β-mannoside bonds, features highlighted in Fig. 8.2. In total, there are thirty-five stereogenic centers and thirteen rings within its impressive and imposing structure.

Scheme 8.6 outlines, in retrosynthetic format, the overall strategy which guided our total synthesis approach. While the modular nature of everninomicin's structure presents a seemingly limitless array of options for retrosynthetic simplification, the CD orthoester moiety was disassembled first, due to its well-precedented sensitivity to acid [3], giving the 2-phenylseleno fluoride **7** [A₁B(A)C fragment] and diol **8** [DEFGHA₂ fragment]. Close examination of the larger fragment **8** suggested that the EF glycosidic linkage would be the most straightforward to con-

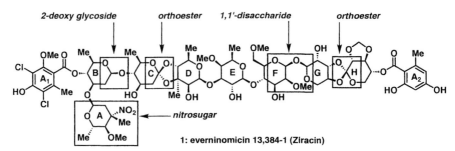

Fig. 8.2 Structure of everninomicin 13,384-1 (**1**) highlighting challenging and unique carbohydrate units and linkages.

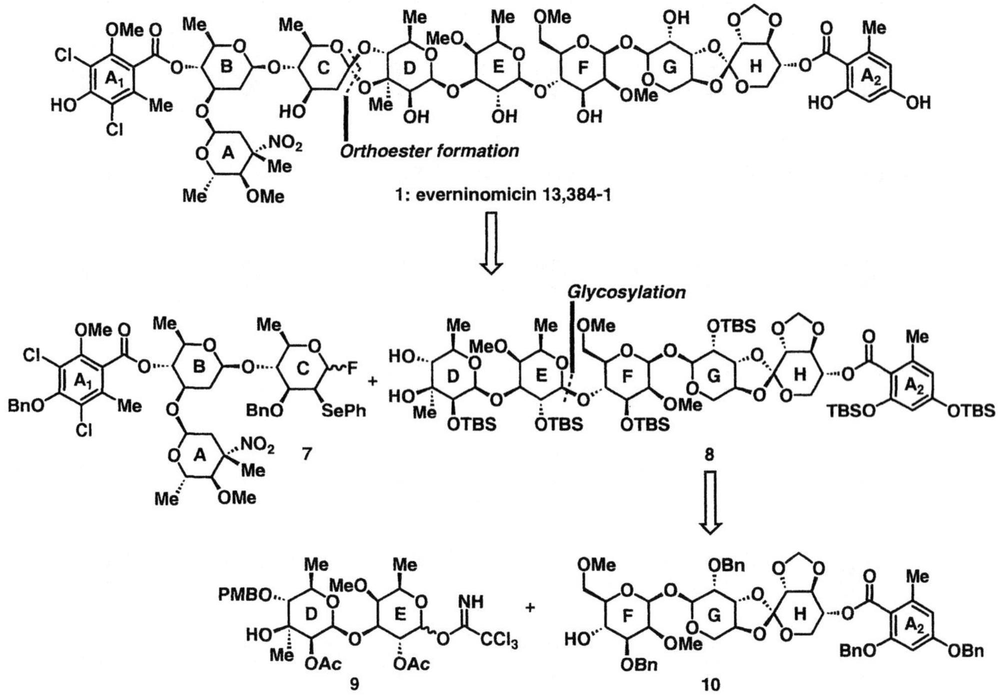

Scheme 8.6 Retrosynthetic analysis of everninomicin 13,384-1 (**1**). TBS = *tert*-butyldimethylsilyl, PMB = *p*-methoxybenzyl.

struct, and so this bond was retrosynthetically cleaved, revealing fragments **9** [DE] and **10** [FGHA$_2$] as target intermediates. Significantly, by adopting this particular approach, we ensured added flexibility in the late stages of the synthesis such that we would be able (after modification of the DE fragment) to assemble either the A$_1$B(A)CDE or the DEFGHA$_2$ fragment, which could then be joined with the remaining portion to complete the synthesis of everninomicin 13,384-1 (**1**).

8.3
Total Synthesis of Everninomicin 13,384-1 (1)

8.3.1
Approaches Towards the A$_1$B(A)C Fragment

8.3.1.1 Initial Model Studies
Containing the A-ring nitrosugar, a 2-deoxy-β-glycoside, and a fully substituted aromatic ring (A, ring), the A$_1$B(A)C portion of everninomicin 13,384-1 (**1**) has stimulated a number of synthetic studies. Notable among these explorations are the formation of the A$_1$B(A) and A$_1$BC systems by Scharf [25] and a BCDE model

system by Sinaÿ [22]. We first synthesized the A₁B(A)C model system (**11**) [26] in an effort to confirm the viability of our initial approach to this fragment, as depicted retrosynthetically in Scheme 8.7. Although we will not go into this successful synthesis in great detail here, suffice it to say that our preparation of this model system utilized a number of modern strategies, including: (a) a ring-closing olefin metathesis [27] approach to a common precursor for carbohydrate systems B and C (**20 → 19**, Scheme 8.7), (b) control of the 2-deoxy-β-anomeric stereochemistry based on the 1,2-phenylsulfeno migration/sulfur-directed glycosidation procedure [21], (c) use of an acyl fluoride to effect the formation of the sterically demanding ester bond between rings A₁ and B, and (d) an efficient synthesis of the unusual ring A nitrosugar (**14**).

Unfortunately, upon completing this model system (**11**), we had to reevaluate our strategy in order to incorporate functionality which we deemed necessary for

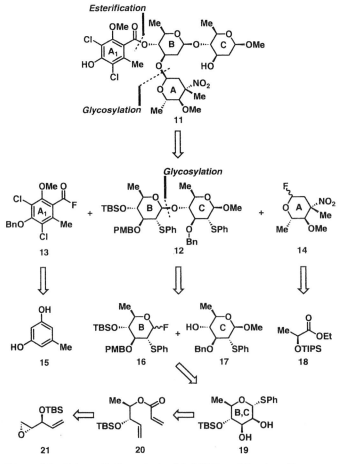

Scheme 8.7 Retrosynthetic analysis of A₁B(A)C model system **11**.

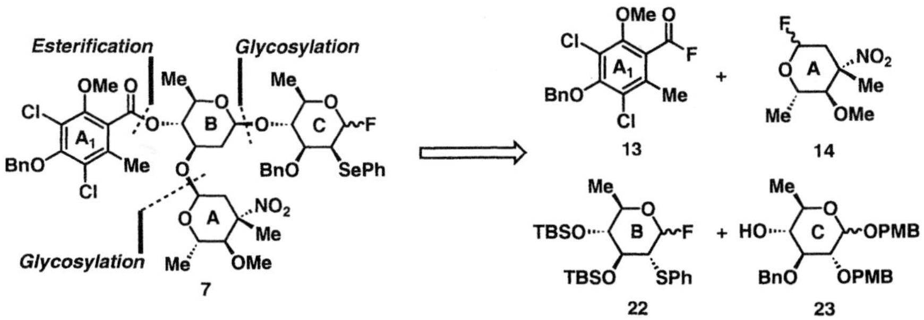

Scheme 8.8 Retrosynthetic analysis of A₁B(A)C fragment **7**.

the eventual construction of the CD orthoester. Crucial to this new plan were 1,2-phenylseleno- and 1,2-phenylthio-migrations on both rings B and C, for which we determined that a different set of protecting groups and glycosylation tactics would be required. Upon closer examination of the $A_1B(A)C$ fragment, while concurrently establishing a need to incorporate a C-2 phenylseleno moiety in ring C, we anticipated that the functionality shown in compound **7** would be suitable. Scheme 8.8 therefore indicates the revised retrosynthetic analysis of this fragment, with disconnection at the glycosidic and ester bonds defining the building blocks **13**, **14**, **22**, and **23** as the requisite starting materials. One should note that on the basis of our synthetic strategy to access the BC disaccharide, in which a sequential method for glycosidation, removal of the thiophenyl group from ring B, and introduction of the selenoglycoside into ring C was deemed necessary, we required a suitable combination of protecting groups. As such, the defined structures of **22** and **23** reflect the optimized set of TBS and PMB groups suitable for achieving these goals on the basis of several abortive attempts not described here.

8.3.1.2 Construction of the Building Blocks
A key objective of the initial strategy toward the nitrogen-containing, C-3-branched 2,6-dideoxy-L-sugars (nitrosugar **14**) was the synthesis of an advanced common intermediate from which the aminosugars of both vancomycin and everninomicin [28] could be constructed. Retrosynthetic analysis of this unit revealed that a nucleophilic chain-extension of an oxime [29] and a stereocontrolled *anti* addition [30] of an acyl anion equivalent to an aldehyde could be employed to install the C-3 and C-4 stereocenters, respectively. Efforts towards the key building block **14**, as depicted in Scheme 8.9, therefore began with the TIPS-derivative of ethyl L-lactate (**24**), which was sequentially reduced with DIBAL followed by addition of EVE-Li [31] at $-100\,^\circ$C to afford enol ether **25** as a mixture of diastereoisomers (85% *de*, 63% overall yield from **24**). Following separation, the *anti* isomer was methylated with MeI and NaH, hydrolyzed to the corresponding ketone with aqueous HCl, and converted into the oxime **26** by condensation with O-benzylhydroxylamine in pyridine (~4:1 ratio of $E:Z$ isomers, 87% overall yield for three

steps). Addition of allylmagnesium bromide to **26** in Et_2O at $-35\,°C$, followed by removal of the silyl group with TBAF, furnished alcohol **27** in 76% yield. It was anticipated that exposure of **27** to ozone would simultaneously generate the required aldehyde and nitro groups [32]; however, ozonolysis of **27** (CH_2Cl_2, $-78\,°C$), followed by Me_2S workup and silica gel chromatography, afforded a stable intermediate ozonide (**28** without the TMS group) as a 1:1 mixture of diastereoisomers. In contrast, Ph_3P workup led smoothly to the desired nitrosugar. The yield of the latter transformation was significantly improved by initial protection of the alcohol in **27** as its TMS ether [$(TMS)_2NH$, TMSCl] and carrying out the ozonolysis in a 2:1 mixture of isooctane and CCl_4, smoothly affording **28**. After in situ TMS cleavage, treatment with Ph_3P, and exposure to DAST [14a], a mixture of glycosyl fluorides **14** was obtained in 75% overall yield ($\alpha:\beta \sim 8:1$) from **27**.

Acyl fluoride **13** (Scheme 8.10) was evaluated as a potential coupling partner, based on previously reported [25] difficulties in forming the ester bond between rings A_1 and B. Thus, orcinol (**15**) was exposed to $Zn(CN)_2 \cdot AlCl_3$, facilitating a Gattermann formylation [33], and the resulting aldehyde was oxidized to carboxylic acid **29** ($NaClO_2$, 74% overall yield). Methylation (MeI, K_2CO_3) and subsequent chlorination with SO_2Cl_2 provided dichloride **30** in 86% yield. Sequential protection of the two phenolic groups proceeded smoothly and regioselectively through initial treatment with TIPSOTf and 2,6-lutidine (90% yield) followed by Ag_2O and MeI (91% yield), to afford compound **31**. The TIPS group was then ex-

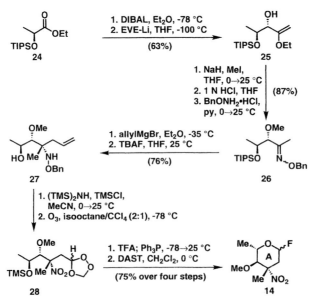

Scheme 8.9 Synthesis of nitrosugar A (**14**).
DIBAL=diisobutylaluminum hydride, EVE-Li=2-lithio-ethylvinyl-ether, py=pyridine, TBAF=tetra-*n*-butylammonium fluoride, THF=tetrahydrofuran, TFA=trifluoroacetic acid.

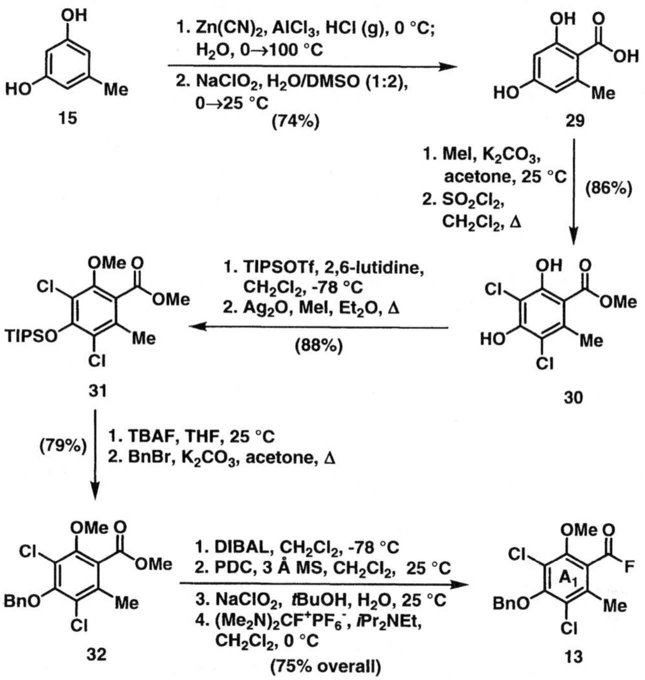

Scheme 8.10 Synthesis of dichloroisoeverninic acyl fluoride **13**. DMSO = dimethylsulfoxide, TIPS = triisopropylsilyl, PDC = pyridinium dichromate, MS = molecular sieves.

changed for a benzyl group (TBAF; K$_2$CO$_3$, BnBr, 79% overall yield), furnishing **32**. The inability to saponify this methyl ester directly resulted in a lengthy, but necessary, three-step procedure to transform it into the carboxylic acid through DIBAL reduction to give the primary alcohol, PDC oxidation to afford the aldehyde, and finally NaClO$_2$ oxidation. Treatment of this acid with (Me$_2$N)$_2$CF$^+$PF$_6^-$ [34] in the presence of iPr$_2$NEt gave the targeted acyl fluoride **13** in 75% overall yield from **32**.

With the A and A$_1$ building blocks in hand, we next turned our attention to the B and C ring portions. The synthesis of building block B (**22**) is summarized in Scheme 8.11. The primary hydroxy group of the known intermediate **33** [35] was tosylated with pTsCl and silylated with TIPSOTf and 2,6-lutidine to afford **34** in 88% yield over these two steps. Reduction with LiAlH$_4$, followed by methanolysis of the acetonide group in acidic media, provided diol **35** (72% yield). Selective protection of the resultant C-2 hydroxy group was then achieved by treatment with nBu$_2$SnO [24] followed by exposure of the resulting tin acetal with PMBCl/ nBu$_4$NI, furnishing the C-2 PMB ether. The likely source of the observed C-2 regioselectivity in this particular transformation was the bulkiness of the TIPS group. To decrease this steric bulk, for reasons that were to prove critical in later steps, the TIPS group had to be replaced with a slightly smaller protecting group.

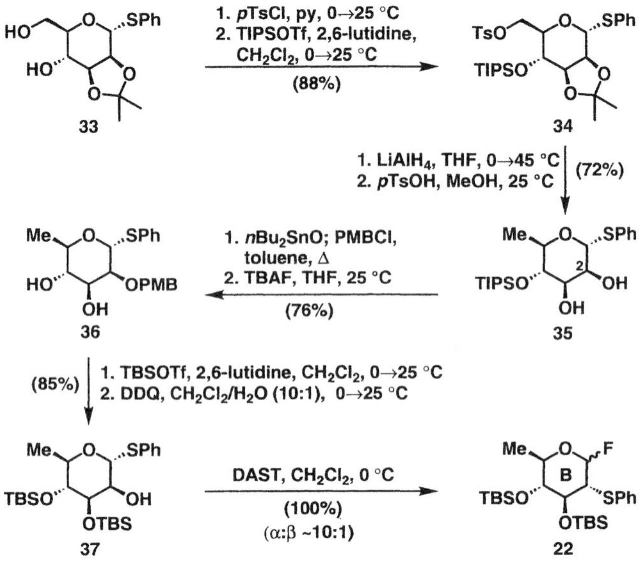

Scheme 8.11 Synthesis of carbohydrate building block B (**22**).
Ts = toluenesulfonyl, DDQ = 2,3-dichloro-5,6-dicyano-1,4-benzoquinone.

As such, TBAF-induced removal of the TIPS group provided diol **36** (76% overall yield), and treatment of this compound with TBSOTf and 2,6-lutidine furnished the bis-TBS derivative, which was then subjected to oxidative removal of the PMB group by use of DDQ to afford alcohol **37** in 85% yield. Finally, reaction of **37** with DAST facilitated the desired 1,2-migration of the thiophenyl group (accompanied by the inversion of the C-2 stereochemistry) and formation of glycosyl fluoride **22** (100% yield, $\alpha:\beta \sim 10:1$).

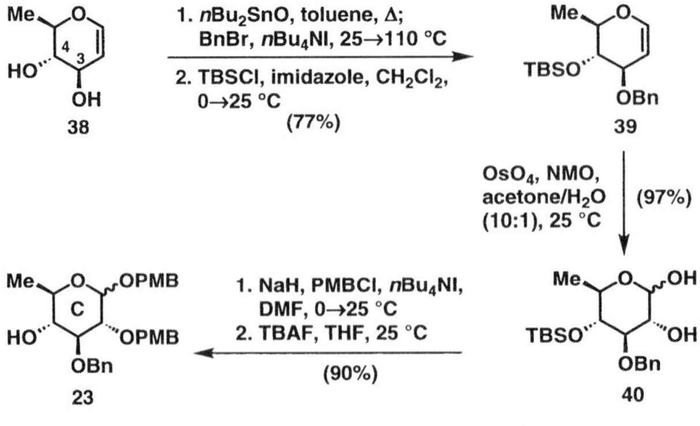

Scheme 8.12 Synthesis of carbohydrate building block C (**23**).
NMO = 4-methylmorpholine-N-oxide, DMF = dimethylformamide.

Ring C building block **23** was constructed in five steps from the readily available glucal **38** [36], as shown in Scheme 8.12. Regioselective tin acetal-mediated benzylation of **23** (nBu$_2$SnO; BnBr/nBu$_4$NI) afforded C-3 protection, which was followed by silylation of the C-4 alcohol with TBSCl to provide compound **39** in 77% yield overall for the two steps. Subsequent treatment of **39** with OsO$_4$/NMO furnished diol **40** in 97% yield ($\sim$1:1 mixture of anomers). Protection of both hydroxy groups of **40** as PMB ethers (NaH, PMBCl, nBu$_4$NI, $\sim$1:1 mixture of anomers) and treatment of the resulting compound with TBAF generated the desired building block **23** in 90% yield. The obtained mixture of C-1 stereochemistry was inconsequential as it would be destroyed later in the sequence.

8.3.1.3 Assembly and Completion of the A$_1$B(A)C Fragment

With the requisite four building blocks (**13, 14, 22,** and **23**) in hand, their stereoselective coupling was then undertaken in an attempt to form the key A$_1$B(A)C fragment **7**, as delineated in Scheme 8.13. Thus, SnCl$_2$-mediated coupling of glycosyl fluoride **22** with alcohol **23** in an equal volume mixture of CH$_2$Cl$_2$/Et$_2$O/Me$_2$S at $-10\,^\circ$C [21] gave the desired disaccharide as a single stereoisomer in 71% yield. This particular combination of solvents was selected in order to avoid cleavage of the PMB groups and to improve the solubility of the starting materials. The 2-thiophenyl group, necessary to achieve β-stereocontrol in the glycosidation but whose presence was no longer required, was then reductively extruded with Raney Ni, furnishing 2-deoxy-β-disaccharide **41**. In preparation for coupling with acyl fluoride **13**, bis-TBS ether **41** was exposed to excess TBAF, and regioselective tin acetal-mediated allylation of the resulting diol with nBu$_2$SnO and allyl bromide furnished **42** in 73% yield for these three steps. Addition of acyl fluoride **13** to the activated hydroxy group of **42** (nBuLi, THF, $-78 \to 0\,^\circ$C) provided the corresponding ester, from which the allyl group was then removed by a two-step procedure involving initial treatment with Wilkinson's catalyst [(Ph$_3$P)$_3$RhCl] and catalytic DABCO followed by exposure to OsO$_4$/NMO, affording alcohol **43** in 80% overall yield. This critical juncture in the synthesis having been reached, two paths were now available: (a) initial coupling with the nitrosugar (ring A), followed by introduction of the phenylseleno group onto ring C, or (b) the reverse sequence. The first scenario having been chosen, alcohol **43** was coupled with glycosyl fluoride **14** in the presence of SnCl$_2$ to furnish trisaccharide **44** in 77% yield and as the desired α anomer. The PMB ethers on ring C were then removed by treatment with PhSH under the activating influence of BF$_3\cdot$Et$_2$O at $-35\,^\circ$C, and the resultant diol was peracetylated (Ac$_2$O, Et$_3$N) to furnish **45** in 81% yield. Several attempts were now made to introduce the phenylseleno group onto diacetate **45**, or its respective lactol, by a variety of different activation methods (i.e., glycosyl fluoride, trichloroacetimidate, glycosyl bromide). In all cases, these endeavors unfortunately resulted only in decomposition of the trisaccharide. As a result, the alternative second approach was then attempted, as shown in Scheme 8.14.

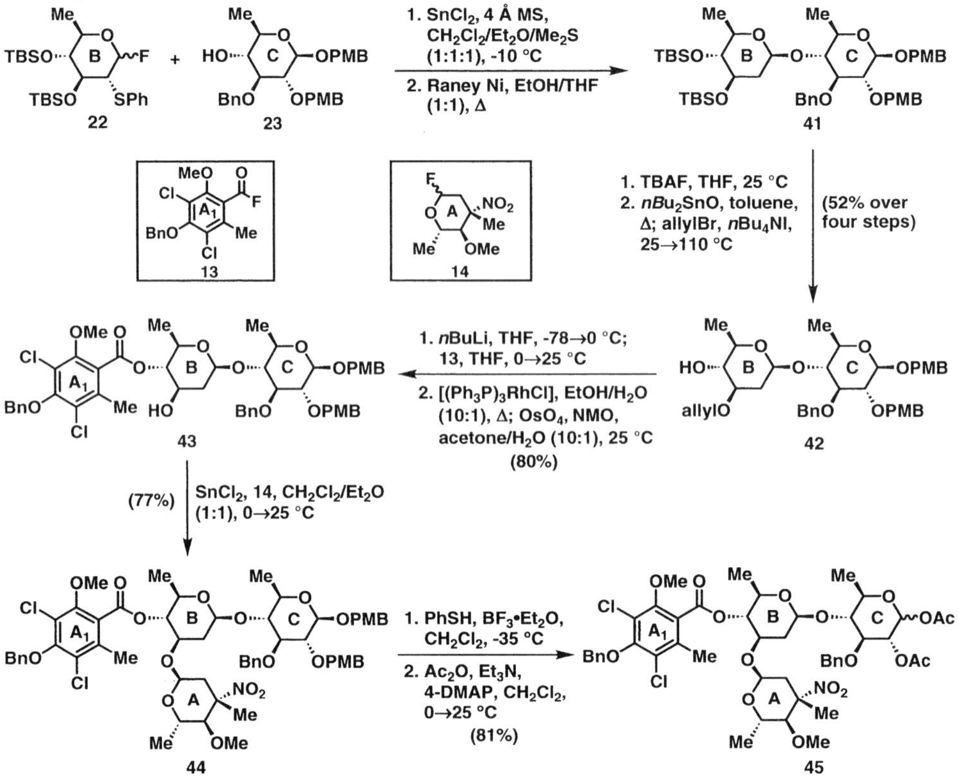

Scheme 8.13 Construction of $A_1B(A)C$ fragment **45**. 4-DMAP=4-dimethylaminopyridine.

Starting with protection of alcohol **43** as its corresponding TIPS ether (TIP-SOTf, 2,6-lutidine), the PMB groups were then removed as before by exposure to PhSH and $BF_3 \cdot Et_2O$, furnishing diol **46** in 77% yield. Acetate groups were installed (Ac_2O, Et_3N, catalytic 4-DMAP) and subsequent treatment of the resulting diacetate with $nBuNH_2$ gave selective cleavage of the C-1 acetate, liberating lactol **47** in 89% overall yield. This compound was then converted into its trichloroacetimidate derivative by treatment with Cl_3CCN and DBU, followed by addition of PhSeH [37] in the presence of $BF_3 \cdot Et_2O$ to afford the desired β-phenylseleno glycoside selectively, as would be expected from the participation of the C-2 acetate ($\alpha:\beta \sim 1:9$, 78% over two steps). The B ring TIPS group was then removed with TBAF, and attachment of evernitrose glycosyl fluoride **14** proceeded smoothly under the activating influence of $SnCl_2$, furnishing the desired $A_1B(A)C$ assembly (**49**) in 73% yield. The remaining cursory touches included: 1) basic hydrolysis of the acetate group from **49**, furnishing the C-2 hydroxy compound, and 2) treatment of this product with DAST, affording the targeted 2-phenylseleno glycosyl fluoride **7** in excellent overall yield (91%) with inversion of stereochemistry at C-2 of ring C, giving a 8:1 mixture of $\alpha:\beta$ anomers.

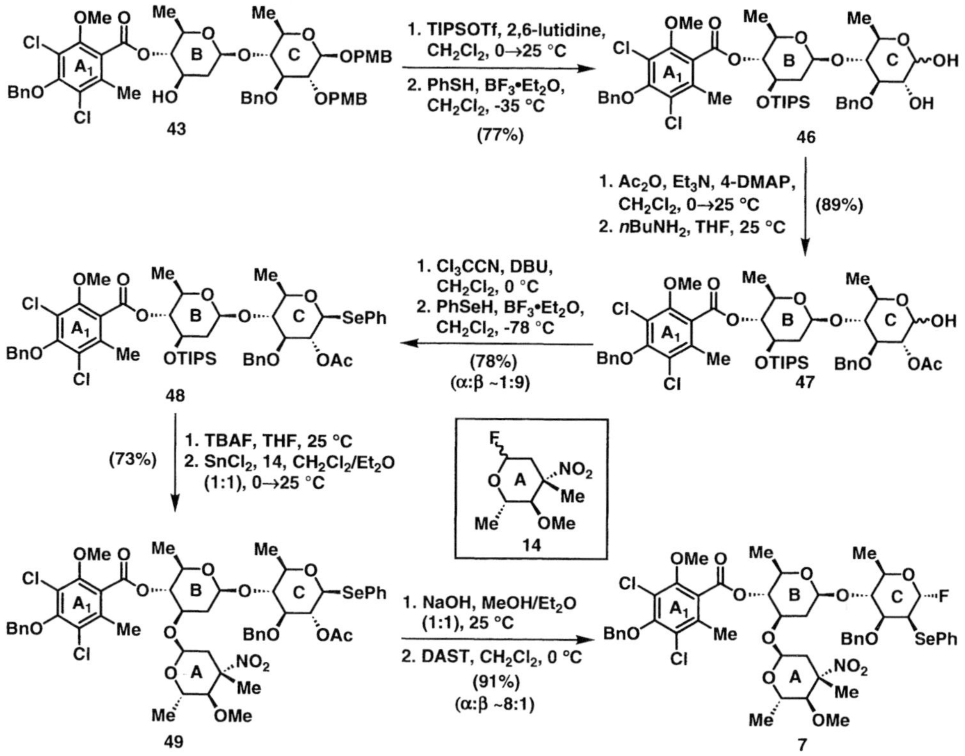

Scheme 8.14 Construction of everninomicin's $A_1B(A)C$ fragment **7**. DBU = 1,8-diazabicyclo[5.4.0]undec-7-ene.

8.3.2
Construction of the $FGHA_2$ Fragment

8.3.2.1 First Generation Approach to the $FGHA_2$ Fragment

Initial examination of $FGHA_2$ fragment **10** revealed several possible retrosynthetic options. As outlined in Scheme 8.15, our initial approach involved disconnection of the 1,1′-disaccharide bridge linking rings F and G and removal of the aromatic ester, to provide tin acetal **51**, acyl fluoride **52**, and GH orthoester **50**. Additional simplification of GH orthoester **50** gave the acyclic orthoester system **53**, which was further disconnected at the orthoester site to reveal xylose lactone **55** and threitol derivative **54**. Unfortunately, attempts to functionalize or trap the synthesized lactol **50** in situ failed, as did attempts to lactonize the corresponding carboxylic acid. Eventually, it was concluded that the strain caused by the formed orthoester would probably be insurmountable, and, as such, a new plan was devised on the principle that the FG 1,1′-disaccharide moiety would have to be constructed prior to formation of the GH orthoester. During this period, other methods to construct the GH orthoester were being explored, culminating in the decision to use and modify the Sinaÿ orthoester formation procedure.

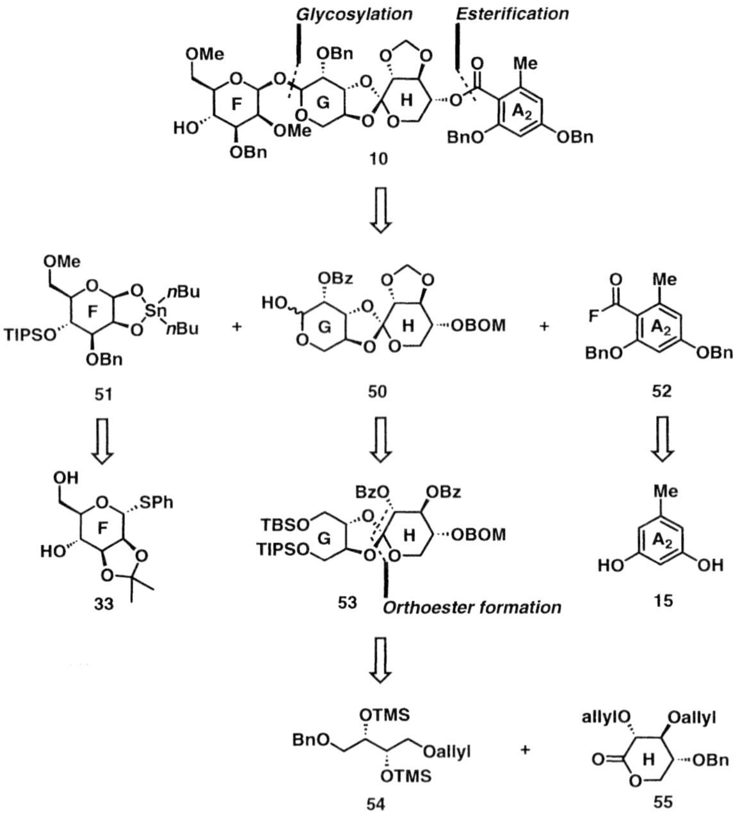

Scheme 8.15 First generation retrosynthetic analysis of FGHA$_2$ fragment **10**. BOM = benzyloxymethyl, Bz = benzoyl.

8.3.2.2 Second Generation Strategy Towards the FGHA$_2$ Fragment

Scheme 8.16 outlines our second generation retrosynthetic analysis of the FGHA$_2$ fragment **10**, based on the findings described above. Thus, disconnection of the indicated aromatic ester and orthoester bonds in **10** revealed components **56** (FGH fragment) and **52** (acyl fluoride). The FGH fragment **56** was further disconnected between rings G and H, furnishing FG diol **58** and 2-phenylselenoglycosyl fluoride **57** as potential precursors. A projected selenium-assisted coupling of **57** with **58** was expected to furnish trisaccharide **56** regio- and stereoselectively, its functionality poised for a Sinaÿ-type orthoester formation. Final disassembly of **58** at the 1,1′-disaccharide linkage gave the tin acetal **60** and the trichloroacetimidate **59** as desired building blocks. The stereoselective construction of the 1,1′-disaccharide bridge from **59** and **60** was assured from the previous studies on model systems to obtain **50**, as briefly described in the preceding section.

The construction of building blocks **57**, **58**, **59**, and **60** is presented in Schemes 8.17 through 8.20. The synthesis of ring F (**60**, Scheme 8.17) began with selective

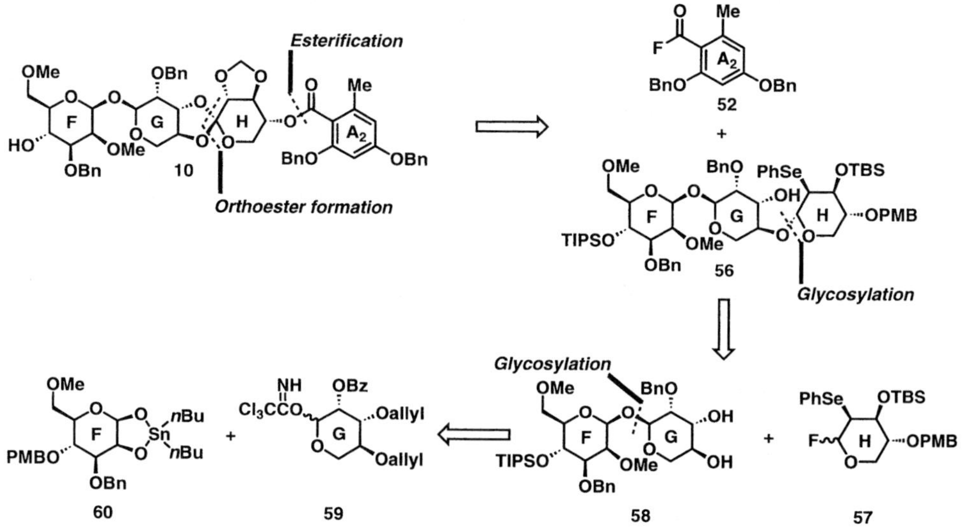

Scheme 8.16 Revised retrosynthetic analysis of FGHA$_2$ fragment **10**.

silylation of mannose diol **33** at the primary alcohol (TBSOTf, 2,6-lutidine, −78 °C), followed by PMB protection at C-4 (NaH, PMBCl) and removal of the TBS group with TBAF, affording primary alcohol **61** in 88% yield over these three steps. Methylation of **61** (NaH, MeI) and subsequent removal of the acetonide under acidic conditions (pTsOH, MeOH) afforded diol **62** in 81% overall yield. In this instance, application of tin acetal-mediated benzylation (nBu$_2$SnO, BnBr) provided the desired C-3 benzyl ether. The thioglycoside was then oxidatively released by treatment with NBS in aqueous acetone, and the resulting diol was heated at reflux with nBu$_2$SnO in MeOH to furnish tin acetal **60** in 86% overall yield, ready for coupling with ring G once the synthesis of that compound was achieved.

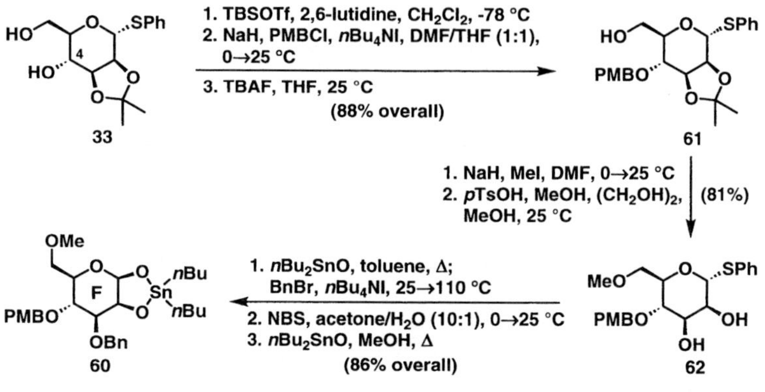

Scheme 8.17 Synthesis of tin acetal building block F (**60**).

Scheme 8.18 illustrates the final approach to G ring trichloroacetimidate **59** which featured Dondoni's TMS-thiazole chemistry [38]. Bis-allylation of diisopropyl L-tartrate (**63**) with NaH and allyl bromide was followed by reduction with LiAlH$_4$ and subsequent monosilylation (NaH, TBDPSCl), providing alcohol **64** in 81% overall yield, setting the stage for a one-carbon homologation. Swern oxidation [(COCl)$_2$, DMSO, Et$_3$N] followed by in situ treatment with TMS-thiazole afforded alcohols **65** and **66** in 91% combined yield, but as a mixture of diastereoisomers. While the lack of selectivity was disappointing, the undesired diastereoisomer could readily be recycled by Swern oxidation followed by reduction of the resultant ketone with LiAlH$_4$ in 67% overall yield, achieving a 2:1 ratio of **65:66**. Benzoylation of the correct diastereoisomer **65** (BzCl, Et$_3$N) afforded **67** in 98% yield. Completion of the sequence involved Dondoni's three-step, one-pot thiazole cleavage (MeOTf; NaBH$_4$; CuCl$_2$) [38] via the intermediates shown. Desilylation (TBAF) then afforded the lactol which was finally treated with Cl$_3$CCN and DBU to furnish the desired trichloroacetimidate **59** in 68% yield ($\alpha:\beta \sim 3:1$) from **67**.

Scheme 8.19 summarizes the synthesis of 2-phenylselenoglycosyl fluoride **57** (ring H) starting from peracetylated xylose (**68**). Treatment of **68** with PhSeH in the presence of BF$_3 \cdot$Et$_2$O gave the desired selenoglycoside in an approximately 5:1 anomeric ratio in favor of the β isomer. After chromatographic separation, the β-selenoglycoside was subjected to basic methanolysis (K$_2$CO$_3$, MeOH), and the resulting triol was converted into a 2,3-acetonide (**69**) in 69% overall yield for

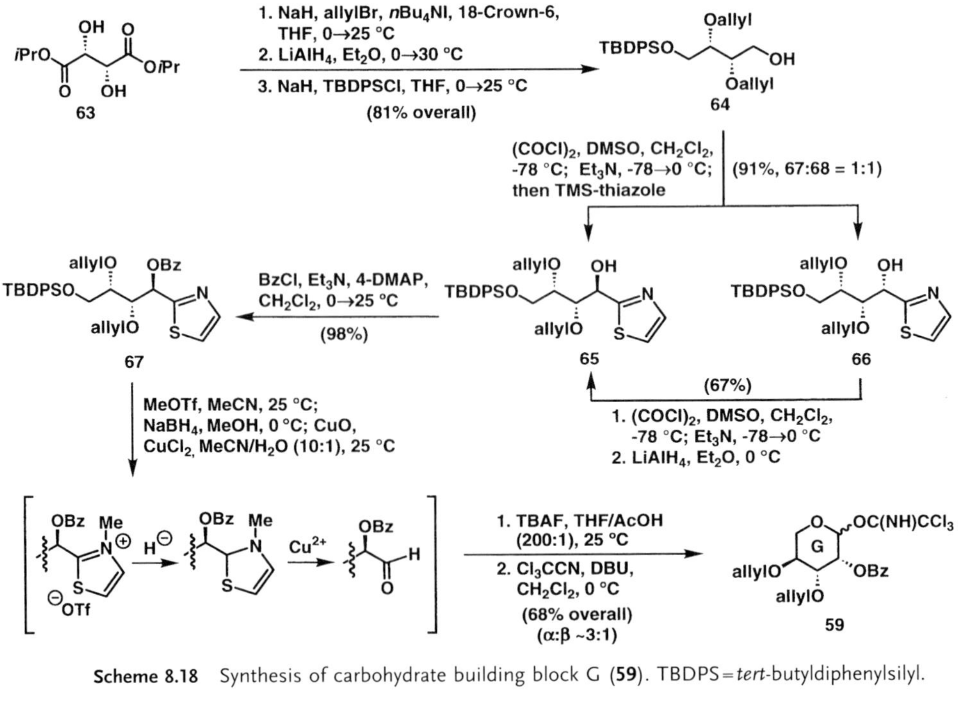

Scheme 8.18 Synthesis of carbohydrate building block G (**59**). TBDPS = *tert*-butyldiphenylsilyl.

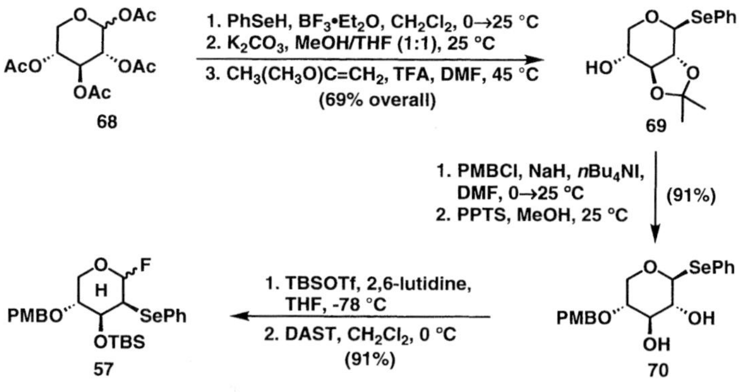

Scheme 8.19 Synthesis of carbohydrate building block H (**57**). PPTS = pyridinium p-toluenesulfonate.

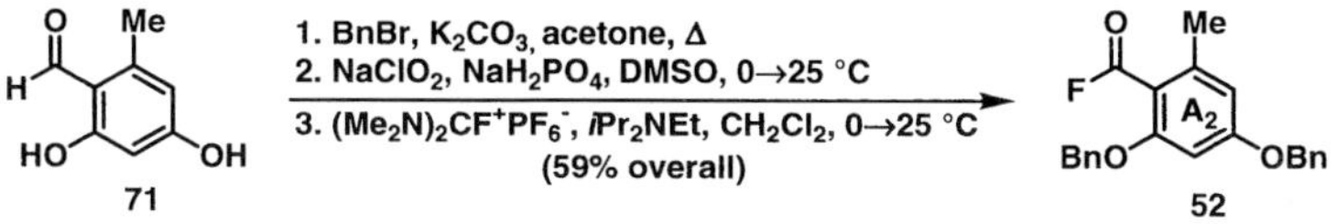

Scheme 8.20 Synthesis of acyl fluoride **52**.

these three steps. Protection of the remaining hydroxy group (PMBCl, NaH, nBu$_4$NI) furnished the PMB ether from which the acetonide was removed under acidic conditions (PPTS, MeOH), affording diol **70** in 91% yield. After considerable experimentation, it was discovered that treatment of diol **70** with TBSOTf in THF at −78 °C provided selective C-3 protection directly, whereas reaction of the same diol in CH$_2$Cl$_2$ at −78 °C gave exclusive protection of the C-2 hydroxy group. Subsequent treatment of the C-2 alcohol with DAST at 0 °C facilitated the 1,2-selenium migration, providing 2-phenylselenoglycosyl fluoride **57** in 91% yield for these final two steps.

The construction of the last component, acyl fluoride **52**, is briefly outlined in Scheme 8.20. Benzylation of bis-phenol **71** (BnBr, K$_2$CO$_3$), oxidation to the carboxylic acid (NaClO$_2$), and final reaction with (Me$_2$N)$_2$CF$^+$PF$_6^-$ in the presence of iPr$_2$NEt furnished the chromatographically stable acyl fluoride **52** in 59% overall yield.

8.3.2.3 Assembly of the FGHA$_2$ Fragment

With all four building blocks in hand, final assembly of the FGHA$_2$ fragment was initiated as shown in Scheme 8.21. Coupling between tin acetal **60** and trichloroacetimidate **59** in the presence of TMSOTf, followed by acidic workup and careful methylation of the resulting alcohol (NaH, MeI), afforded the desired β-mannoside containing the 1,1′-disaccharide of **72** in 64% overall yield. Removal of the G ring benzoate (NaOH, MeOH) was followed by benzylation (NaH, BnBr) to afford

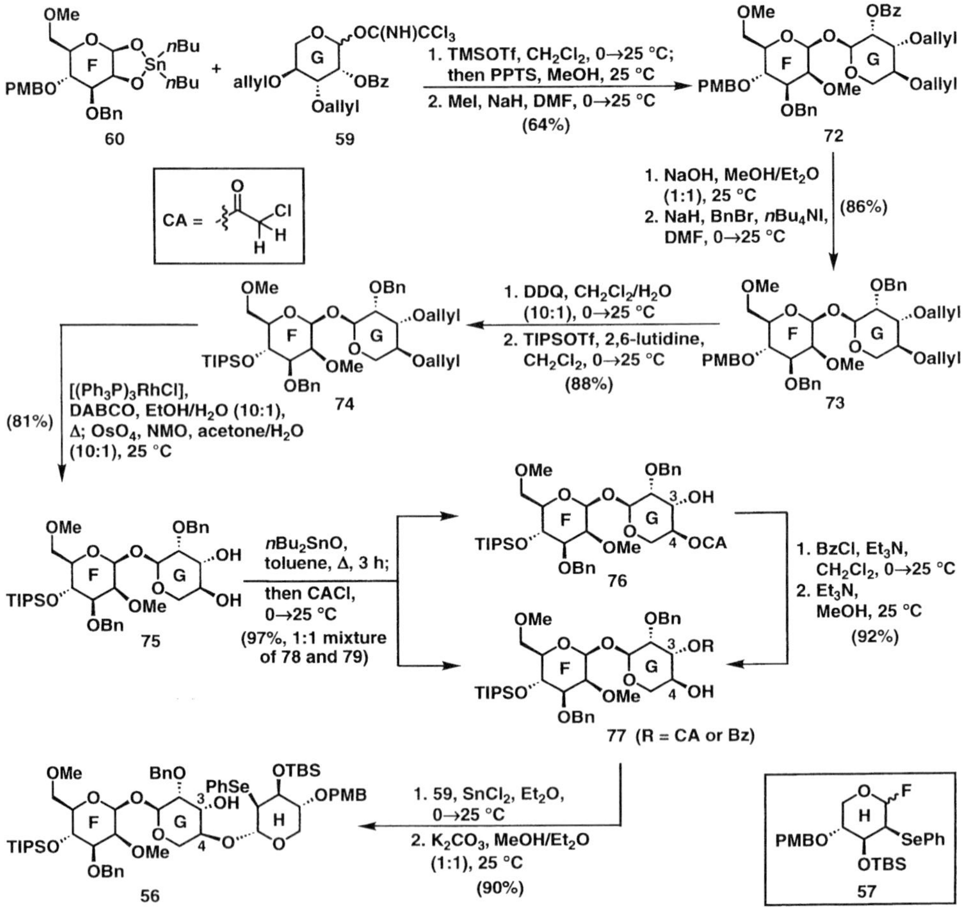

8 Synthesis of Complex Carbohydrates: Everninomicin 13,384-1

Scheme 8.21 Assembly of FGH fragment **56**. DABCO = 1,4-diazabicyclo[2.2.2]octane, CA = chloroacetyl.

benzyl ether **73** in 86% yield. The PMB ether on ring F was then exchanged for a TIPS ether by a DDQ oxidation, followed by silylation with TIPSOTf and 2,6-lutidine, furnishing **74** in 88% yield for the two steps. In preparation for coupling with ring H, the allyl groups were removed by treatment with (Ph$_3$P)$_3$RhCl/DABCO and OsO$_4$/NMO, furnishing diol **75** in 81% yield. At this stage, from experiments with a GH model system, it was believed that the C-3 hydroxy group on ring G would provide the correct stereochemistry during orthoester formation. Many attempts were therefore made to facilitate selective C-3 protection of **75** by use of tin acetal chemistry and different solvent combinations; however, in the best case identified, a 1:1 mixture of chromatographically separable regioisomers **76** and **77** (R = CA) was formed in 97% combined yield. Gratifyingly, the overall material throughput could be made more efficient by recycling the C-4 ester (**76**).

Thus, benzoylation of **76** (BzCl, Et$_3$N, catalytic 4-DMAP) followed by selective removal of the chloroacetate with Et$_3$N in MeOH furnished the C-3 benzoate **77** (R=Bz) in 92% overall yield. The correctly protected C-3 ester **77** (R=CA or Bz) was then coupled with the 2-phenylselenoglycosyl fluoride **57** in the presence of SnCl$_2$ to afford the desired trisaccharide as a single anomer. The chloroacetate or benzoate protecting group was then subsequently cleaved under mild conditions with K$_2$CO$_3$ in MeOH, affording hydroxyselenide **56** in 90% yield for the two steps.

Our initial attempt to construct the orthoester moiety of the FGH fragment is shown in Scheme 8.22. Treatment of alcohol **56** with TBAF in THF at 0 °C predominantly induced removal of the H-ring TBS group. Furthermore, when the resulting diol was subjected to the Sinaÿ orthoester formation procedure, *syn*-elimination and orthoester formation was accompanied by a Ferrier-type rearrangement [39], providing allylic orthoester **78** in 61% overall yield. This serendipitous result initially appeared extremely promising, as it shortened our proposed sequence to introduce this double bond by several steps. At the same time, we postulated that the resultant stereochemistry of orthoester formation was correct (i.e., opposite to that depicted for **78** in Scheme 8.22), although, as is often the case with assumptions, X-ray analysis of a subsequent intermediate proved this hypothesis to be in error. However, we were to advance several steps before we obtained this information. First, exposure of the newly formed olefin (**78**) to standard dihy-

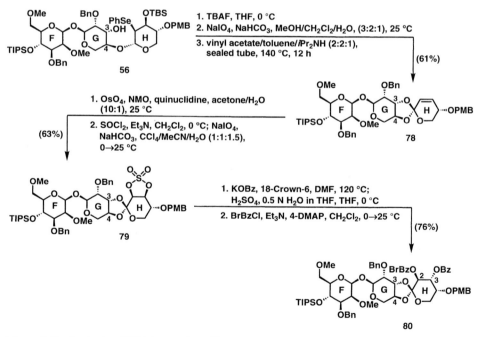

Scheme 8.22 Synthesis of FGH intermediate **80**.

droxylation conditions (OsO$_4$, NMO) gave the *cis* diol in an approximately 8:1 ratio of diastereoisomers. After unsuccessful attempts to monoprotect this diol for oxidation/reduction purposes (in order to invert the *C*-2 position of ring H to generate the stereochemistry present in the natural product), the opening of a cyclic sulfate was attempted. Thus, treatment of this diol with SOCl$_2$ and Et$_3$N, followed by oxidation (NaIO$_4$) of the sulfur atom to the corresponding sulfone, furnished cyclic sulfate **79** in 63% overall yield. Opening of this cyclic sulfate with KOBz in DMF at 120 °C provided the monobenzoate, ^{1}H NMR analysis of which indicated that opening of the cyclic sulfate had occurred with undesired diaxial opening, revealing a *trans* relationship between the ring H substituents at *C*-2 and *C*-3 and that the benzoate was at the *C*-3 position. Furthermore, transformation into the *p*-bromobenzoate (BrBzCl, Et$_3$N) yielded a crystalline derivative **80** (76% yield for the two steps) which was subjected to X-ray analysis, revealing that the orthoester center had the wrong stereochemistry and that cyclization during allylic orthoester formation had occurred from the undesired face.

In the hope of reversing this outcome, orthoester formation was then attempted from the *C*-3 coupled disaccharide (i.e., the product resulting from the coupling of *C*-4 ester **76** and donor **57**, Scheme 8.21). However, expected *syn*-elimination and cyclization under Sinaÿ conditions to afford the desired allylic orthoester occurred in only 45% yield and as an inseparable 4:1 mixture of diastereomers. These results having proven unsatisfactory, a slightly longer, but more secure, sequence was adopted, as described in Scheme 8.23. Hydroxyselenide **56** was subjected to the Sinaÿ-type orthoester formation procedure as before, furnishing 2-deoxy orthoester **81** in 81% yield. At this stage, removal of the H ring TBS group was required, but it was observed that treatment of **81** with TBAF afforded exclusive removal of the F ring TIPS group, necessitating an exchange of protecting groups. Thus, treatment of **81** with TBAF in THF and subsequent benzoylation (BzCl, Et$_3$N, catalytic 4-DMAP) followed by desilylation under buffered conditions (TBAF, AcOH, THF) furnished the correct ring H alcohol **82** in 87% overall yield for these three steps. The resulting alcohol **82** was then dehydrated by treatment with Martin's sulfurane [40], furnishing olefin **83** in 85% yield. Before the required dihydroxylation reaction was attempted, it was necessary to exchange the benzoate group on ring F for a TBS group in order to ensure the success of subsequent steps. Towards this end, compound **83** was debenzoylated with K$_2$CO$_3$ in MeOH, and the resulting alcohol was treated with TBSCl in the presence of NaH and 18-crown-6 to afford silyl ether **84** in 72% overall yield. At this stage, the stereochemistry of the orthoester was established unambiguously through comparison of the ^{1}H NMR spectra of the allylic orthoester **84** (Scheme 8.23) to **78** (Scheme 8.22). Treatment of the highly sensitive olefinic orthoester **84** with OsO$_4$/NMO under the activating influence of quinuclidine [41] provided 1,2-diol **85** as the major product in 70% yield and as an 8:1 mixture of *cis* diastereoisomers.

Because of the extreme sensitivity of olefins **83** and **84**, an alternative approach to compound **85** was also explored, as illustrated in Scheme 8.24. Olefin **83** was directly subjected to the same dihydroxylation procedure as before, affording the corresponding *cis* diol as a 10:1 mixture of diastereoisomers. The benzoate moiety

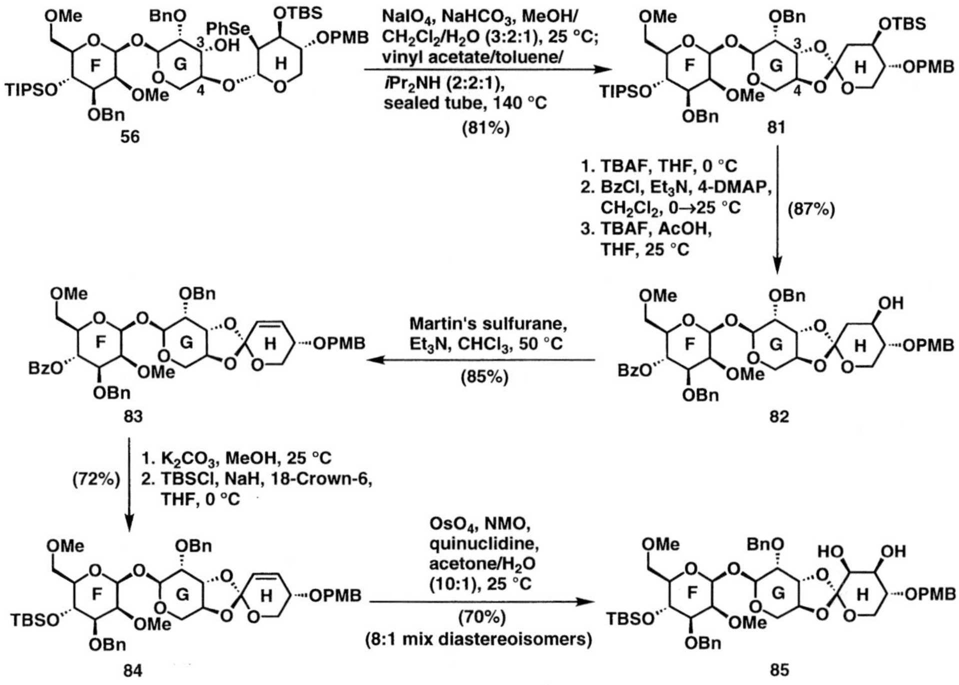

Scheme 8.23 The synthesis of the FGHA$_2$ fragment **85**.

in this compound was then removed with K$_2$CO$_3$ in MeOH to provide triol **86** in 97% overall yield. Reaction of **86** with triphosgene in pyridine installed the carbonate on the *cis* diol, and a TBS group was easily incorporated on the remaining hydroxy group to afford **87** in 89% yield. Finally, basic methanolysis of the cyclic carbonate (NaOH, MeOH) gave diol **85** (95% yield), identical to that derived as indicated in Scheme 8.23. This alternative synthesis was also useful in that the H ring C-2 and C-3 coupling constants of intermediate **87**, as well as other derivatives, confirmed the desired stereochemistry of the ring H diol for this series of compounds.

Completion of the synthesis of the FGHA$_2$ fragment (**10**) is shown in Scheme 8.25. Ring H diol **85** was regioselectively converted into the C-3 monobenzoate by treatment with *n*Bu$_2$SnO and BzCl (~5:1 ratio of the desired product together with its undesired regioisomer) [42]. At this stage, inversion of the C-2 stereochemistry was required in order to obtain the 1,2-*trans* diol system present on ring H. To this end, we investigated a nearly exhaustive set of different oxidizing agents (such as Dess–Martin periodinane, Swern, TPAP/NMO, *n*Bu$_2$SnO/Br$_2$) and reducing agents (such as Li(*t*BuO)$_3$AlH, NaBH$_4$, Na(AcO)$_3$BH, LiAlH$_4$, K- and L-selectrides, LiEt$_3$BH). The best combination identified involved oxidation with Dess-Martin periodinane [43] followed by reduction with Li(*t*BuO)$_3$AlH to afford, through the corresponding ketone, the desired alcohol **88** in 78% overall yield

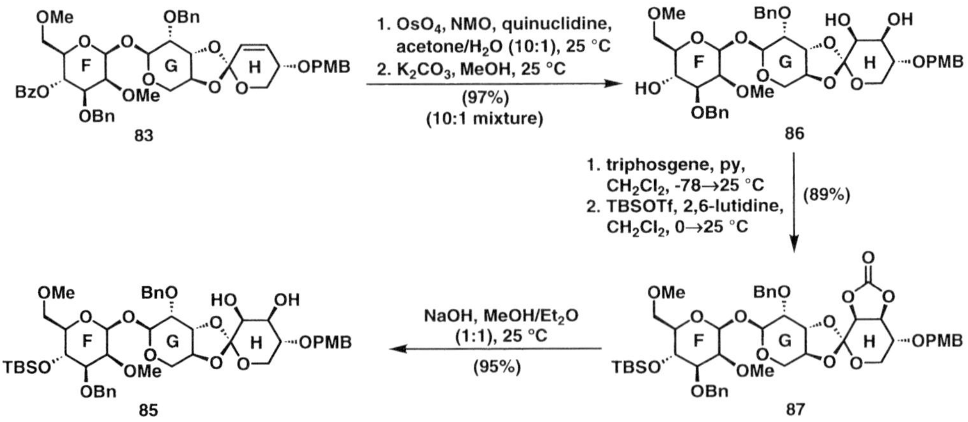

Scheme 8.24 Alternative synthesis of the FGH fragment **85**.

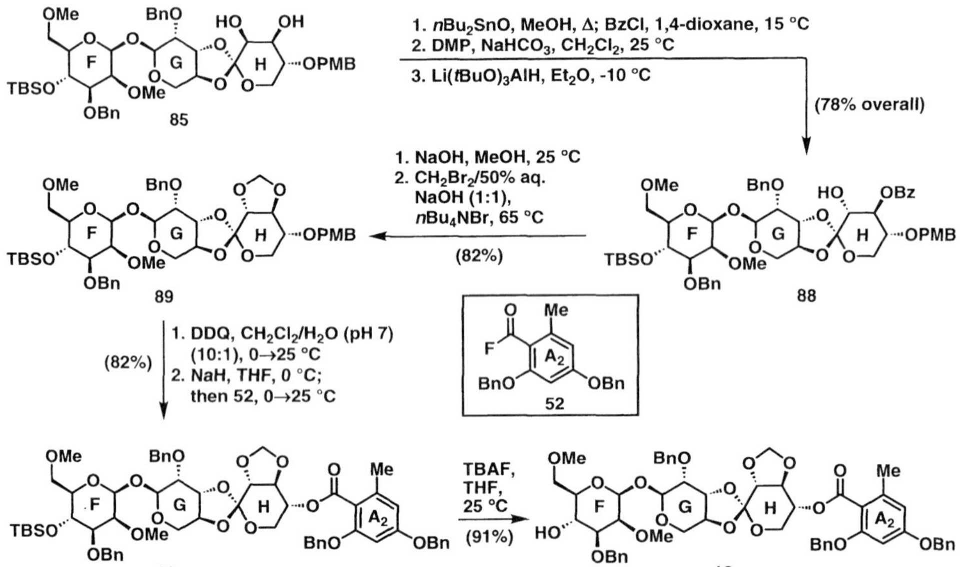

Scheme 8.25 Completion of the synthesis of the FGHA$_2$ fragment **10**. DMP = Dess-Martin periodinane.

from **85**. The benzoate group was removed from **88** (NaOH in MeOH), setting the stage for engagement of the resulting *trans*-1,2-diol system as the desired methylene acetal moiety present in the natural product. This goal was achieved by slow addition of the diol to a mixture of aqueous NaOH, CH$_2$Br$_2$, and nBu$_4$NBr at 65 °C [44] to provide acetal **89** in 82% yield for the two steps. The remaining modifications to complete the FGHA$_2$ fragment **10** involved DDQ-mediated removal of the PMB group from **89**, followed by esterification of the resulting compound with acyl fluoride **52** in the presence of NaH in THF, affording **90** in 82% overall yield. Finally, removal of the TBS group with TBAF furnished the targeted FGHA$_2$ fragment **10** in 91% yield.

8.3.3
Construction of the DE Disaccharide

8.3.3.1 Retrosynthetic Analysis and Construction of Building Blocks for the DE Fragment

Scheme 8.26 depicts the DE fragment as the appropriately functionalized key intermediate **9** and its retrosynthetic analysis to disaccharide **91**, disconnection of which, as shown, gives the building blocks **92** and **93**. The protecting groups on **91** were carefully chosen so as to be flexible enough to permit extension at either end in order to provide the larger A$_1$B(A)CDE or DEFGHA$_2$ fragments, respectively. After much experimentation, the ensemble illustrated by **9** was found to be the most viable form for incorporation into the final target **1**. Examination of **9** revealed two challenging features: (a) the β-mannoside linkage bridging rings D and E, and (b) the tertiary center on ring D. We selected the Kahne sulfoxide-based glycosidation reaction [20a] as the procedure to couple **92** and **93** in the forward direction, and, in addition, a benzylidene ring was incorporated onto **92** to ensure the stereocontrolled formation of the β-mannoside bond, as first reported by Crich [20b,c]. Both the projected reaction conditions and the use of a benzylidene ring suited the proposed protecting group strategy, so the issue of branching the synthesis to include ring D was examined next. We believed that the introduction of the methyl group at C-3 of ring D at the monosaccharide stage would be laden with difficulty [3], and so the planned nucleophilic attack was postponed until after coupling of the D and E ring precursors had been achieved.

The construction of building block D (**92**) proceeded smoothly from intermediate **94** [45], as summarized in Scheme 8.27. Thus, regioselective tin acetal-mediated protection of the C-3 hydroxy group in **94** as a PMB ether (nBu$_2$SnO;

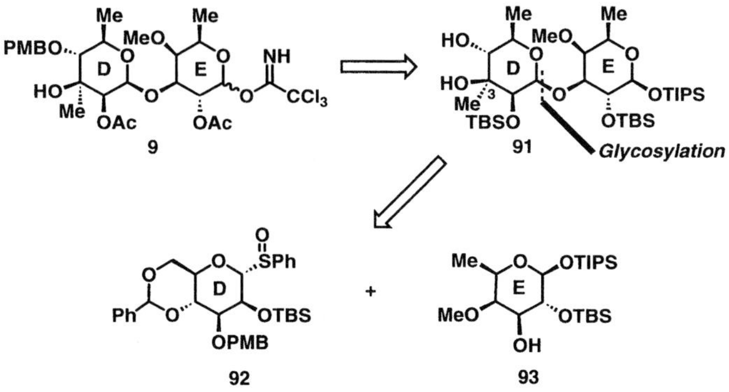

Scheme 8.26 Retrosynthetic analysis of DE fragment **9**.

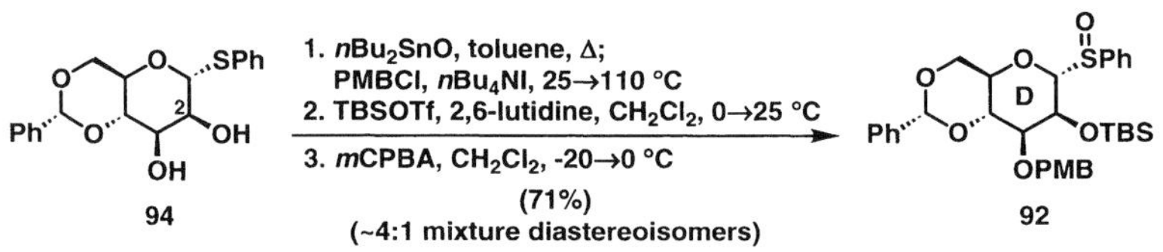

Scheme 8.27 Synthesis of carbohydrate building block D (**92**). *m*CPBA=3-chloroperoxybenzoic acid.

PMBCl/*n*Bu₄NI), followed by silylation at C-2 (TBSOTf, 2,6-lutidine) and *m*CPBA-mediated oxidation of the sulfur moiety, furnished the desired sulfoxide **94** in 71% overall yield for these three steps and as a 4:1 mixture of chromatographically separable diastereoisomers.

Scheme 8.28 illustrates the synthesis of the requisite acceptor fragment, ring E (**93**). Beginning with galactose diol **95** [46], tin acetal-mediated monoprotection (*n*Bu₂SnO, then PMBCl/*n*Bu₄NI) produced the C-3 PMB ether, and silylation of the C-2 hydroxy group (TBSOTf, 2,6-lutidine) completed intermediate **96** in 84% yield for the two steps. Deoxygenation of the C-6 position was effected by initial acidic cleavage of the benzylidene group [Zn(OTf)₂/EtSH] to provide the diol, followed by selective tosylation of the resultant primary hydroxy group to afford **97** in 75% overall yield. LiAlH₄ reduction of **97** achieved the desired deoxygenation, and was followed by methylation of the remaining C-4 hydroxy group (NaH, MeI), providing **98** in 85% yield. Oxidative cleavage of the phenylthio group from

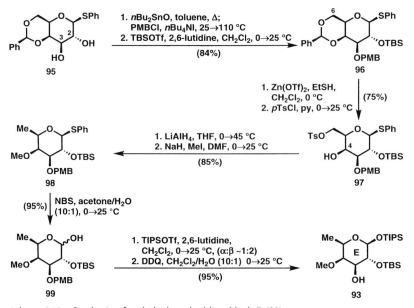

Scheme 8.28 Synthesis of carbohydrate building block E (**93**).

98 with NBS in aqueous acetone (95% yield) was followed by silylation (TIPSOTf, 2,6-lutidine), giving a 1:2 ratio of α:β anomers. Significantly, both anomers could be taken through the remainder of the sequence. Finally, the PMB group was oxidatively removed by treatment with DDQ, furnishing the targeted building block **93** in 95% yield for these final two steps.

8.3.3.2 Assembly of the DE Fragment

With syntheses of the D and E ring fragments achieved, we then pursued their coupling and elaboration. Scheme 8.29 outlines the union of building blocks **92** and **93** to form intermediates **9** and **91**. Coupling of **92** and **93** under Kahne/Crich conditions (**92**, Tf$_2$O, DTBMP, −78 °C, followed by the addition of **93**) proceeded smoothly to afford the desired β-mannoside, from which the PMB group was removed (DDQ) to afford alcohol **100** in 67% yield overall. At this juncture, it

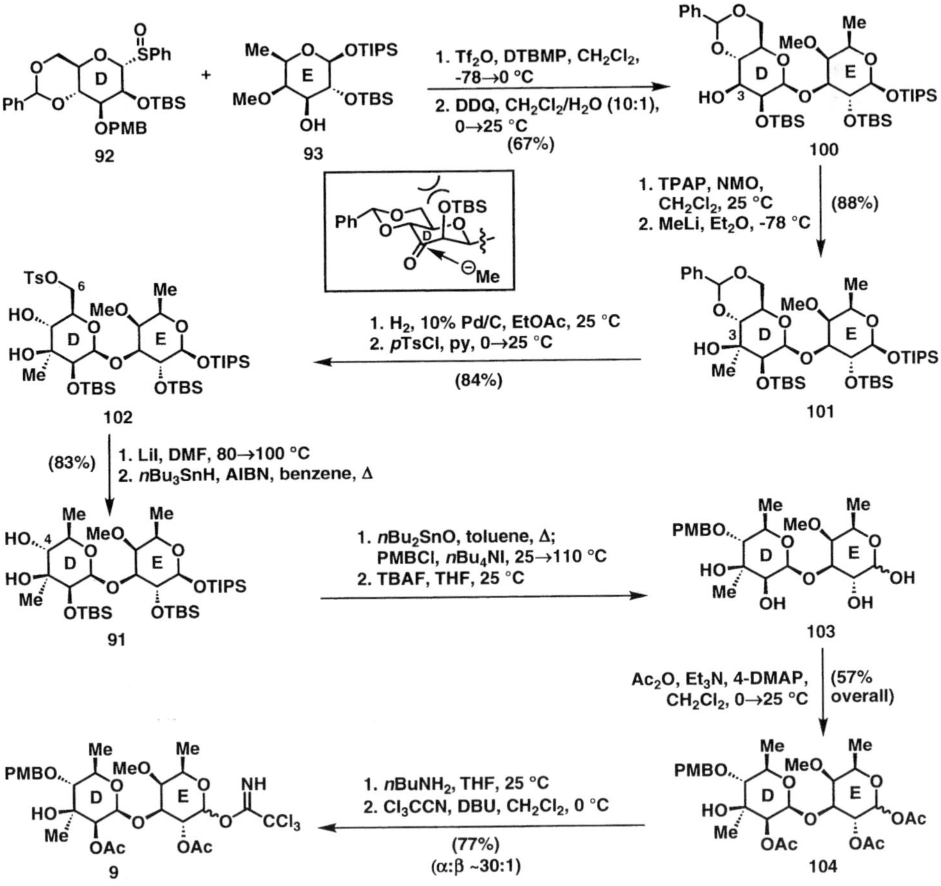

Scheme 8.29 Assembly of DE fragment **9**. AIBN = 2,2′-azobisisobutyronitrile.

was considered prudent to introduce the C-3 branching in ring D prior to deoxygenation of the C-6 position in the same ring. To this end, alcohol **100** was oxidized with TPAP/NMO, and, pleasantly, treatment of the resulting ketone with MeLi in ether at −78 °C produced the desired tertiary alcohol **101** in 88% overall yield as a single diastereoisomer. As shown in the inset box in Scheme 8.29, it is reasonable to presume that the equatorially orientated β-mannoside bond and the bulky, axially orientated TBS group at C-2 provided a decisively biased encounter for the incoming nucleophile during this reaction, with the approach as drawn. In contrast with this analysis, reagent approach towards a D-ring monosaccharide bearing a 1α substituent would have encountered opposing 1,2- and 1,3-interactions.

Nevertheless, the resulting tertiary alcohol in **101** was found to be highly hindered and difficult to protect under standard conditions, and it was therefore postulated that it could remain unprotected throughout the remainder of the sequence. Deoxygenation of the C-6 position in ring D was the next task, and as such the benzylidene group of **101** was removed by mild hydrogenolysis with catalytic Pd/C, and the resulting primary alcohol was selectively tosylated by treatment with pTsCl in pyridine, furnishing **102** in 84% overall yield. Initial attempts to reduce tosylate **102** directly resulted in decomposition, but exchange of the tosylate for an iodide (LiI), followed by reduction with nBu$_3$SnH/AIBN, furnished the desired compound **91** in 83% yield for the two steps.

8.3.3.3 Test of Strategies

At this stage, the availability of intermediate **91** provided an opportunity to test the first of two strategies to elaborate the DE fragment to the final target, as mentioned above. Scheme 8.30 summarizes the progress made by following the first approach. Coupling of A$_1$B(A)C 2-phenylseleno-1-fluoro donor **7** with the DE diol **91** in the presence of SnCl$_2$ in ether solution afforded pentasaccharide **105** in 62% yield. The next task involved a test of orthoester formation under Sinaÿ-type conditions. As such, the selenium moiety was oxidized and then heated in a sealed tube at 140 °C in toluene/vinyl acetate/diisopropylamine (2:2:1) to afford A$_1$B(A)CDE orthoester **106** as a single product in 60% overall yield. It was initially envisioned that precursors **105** or **106** might eventually be coupled with the FGHA$_2$ fragment, but all attempts to elaborate these intermediates any further resulted in decomposition. Nevertheless, the construction of model system **106** was highly informative in that it confirmed that the diol system of ring D could be used in a late-stage coupling reaction and that an orthoester-forming reaction with advanced intermediates was possible. In the hope that the second approach (DE + FGHA$_2$) would be more successful, donor **9** was proposed as the subtarget. Returning to Scheme 8.29, the remaining tasks for the completion of target **9** included protection of the C-4 hydroxy group of ring D and activation of the C-1 carbon of ring E. To this end, regioselective protection of ring D at C-4 was achieved with nBu$_2$SnO and PMBCl/nBu$_4$NI, followed by desilylation with TBAF to afford triol **103**; finally, this compound was peracetylated (Ac$_2$O, Et$_3$N, catalytic 4-DMAP) to afford triacetate **104** in 57% overall yield as a roughly equal mixture of α and β anomers. Exposure of

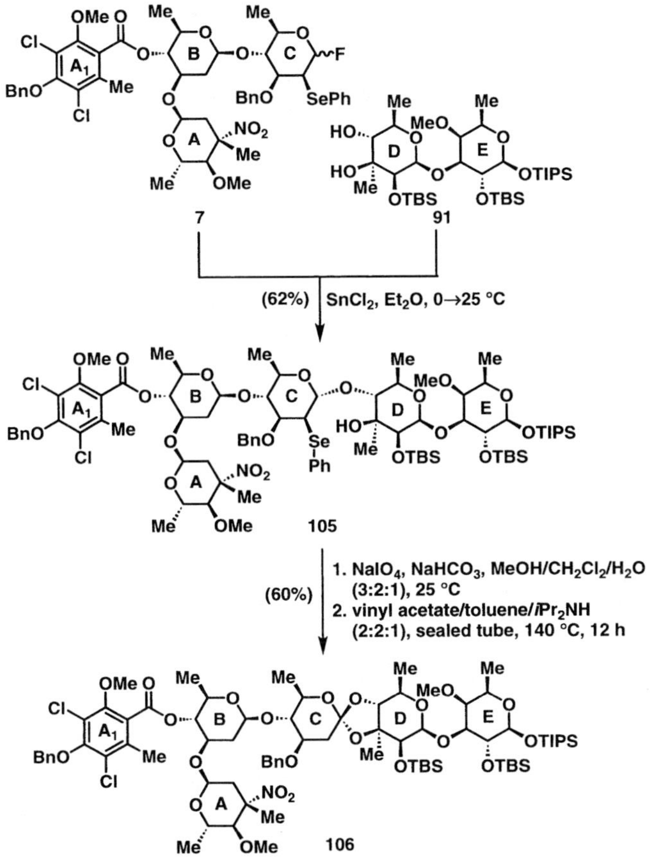

Scheme 8.30 Synthesis of the A₁B(A)CDE system **106**.

triacetate **104** to the mild action of nBuNH₂ in THF resulted in selective removal of the anomeric acetate, furnishing the lactol, conversion of which into trichloroacetimidate **9** was accomplished by treatment with Cl₃CCN and DBU in 77% yield for the two steps, yielding the α anomer with high selectivity.

8.3.4
Assembly of the DEFGHA₂ Fragment

The key coupling of DE trichloroacetimidate **9** and FGHA₂ alcohol **10** is shown in Scheme 8.31. Treatment of **9** with hydroxy fragment **10** in the presence of BF₃·Et₂O at −20 °C in CH₂Cl₂ furnished oligosaccharide **107** in 55% yield with the desired α-glycoside bond between rings E and F. In addition to **107**, this reaction also produced a small amount of the corresponding EF α-glycoside (5%), as well as another isomer possessing the EF β-glycoside bond (18% yield). The lack of complete structural information regarding this last by-product prompted a need

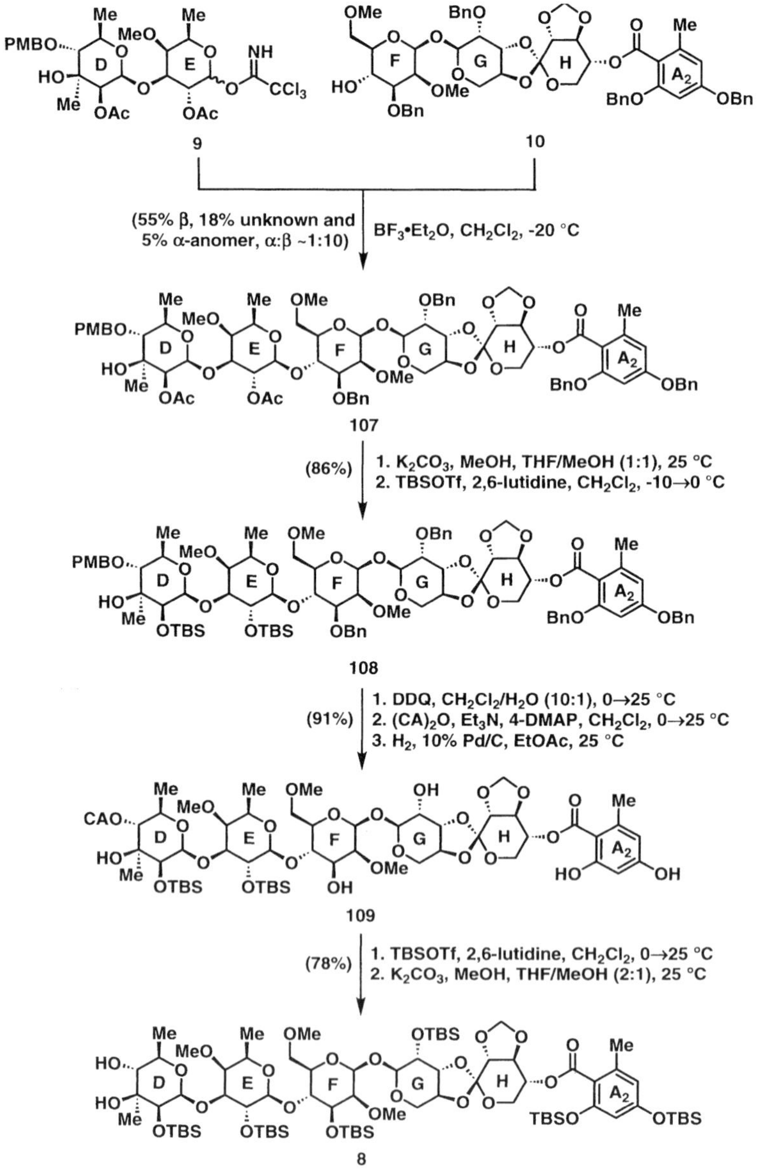

Scheme 8.31 Completion of the synthesis of the DEFGHA$_2$ fragment **8**.

to confirm the structure of the major isomer **107** unambiguously, a goal achieved through degradation studies with everninomicin 13,384-1 (**1**) as described below.

The next task was to devise a suitable protecting group strategy that would enable coupling of the A$_1$B(A)C and DEFGHA$_2$ fragments and their elaboration to the final target. In all, three sets of protecting groups on fragment DEFGHA$_2$

were tested before a successful route was found. Hexabenzyl diol **116** (for structure, see Scheme 8.33) was initially targeted as a potential partner in the projected final coupling reaction with the $A_1B(A)C$ fragment, but problems in the coupling of **116** with the $A_1B(A)C$ fragment, including low glycosidation yields and rupture of the highly sensitive CD orthoester moiety, were discouraging. During these studies, it was found that hexabenzyl diol **116** could also be obtained from degradation of **1** (Scheme 8.33) and, reassuringly, the synthetic and semisynthetic materials were identical. The hexaacetylated counterpart of **116** (not shown) was the next choice attempted, but glycosidation with the same $A_1B(A)C$ partner again gave poor coupling yields.

The third generation strategy involved the adoption of the hexa-TBS derivative **8** (Scheme 8.31) as the $DEFGHA_2$ coupling partner. Preliminary attempts to couple hexasilyl derivative **8** (derived from degradation as shown in Scheme 8.33) with the $A_1B(A)C$ partner appeared promising, so an efficient synthesis of this compound (**8**) was devised as described in Scheme 8.31. The acetate groups were removed from **107** by treatment with K_2CO_3 in MeOH, and the newly released hydroxy groups were silylated with TBSOTf and 2,6-lutidine to provide the bis-TBS intermediate **108** in 86% yield for the two steps. The PMB group was removed through the action of DDQ and replaced with a chloroacetate group [$(CA)_2O$, Et_3N, catalytic 4-DMAP], followed by removal of the benzyl ethers by hydrogenolysis to afford pentaol **109** in 91% overall yield. Four additional TBS groups were then installed (TBSOTf, 2,6-lutidine), and the chloroacetate group was carefully removed from the resulting compound with K_2CO_3 in MeOH, affording the targeted hexa-TBS diol **8** in 78% yield for these final two steps.

8.3.5
Completion of the Total Synthesis of Everninomicin 13,384-1

The final steps in the total synthesis of everninomicin 13,384-1 (**1**) are illustrated in Scheme 8.32. Coupling of the $A_1B(A)C$ glycosyl fluoride donor **7** with the $DEFGHA_2$ hexa-TBS diol acceptor **8** by use of $SnCl_2$ in ether proceeded smoothly with complete stereocontrol to afford the 2-phenylseleno glycoside **110** in 70% yield. Formation of the remaining orthoester site was then accomplished with the same facility as achieved earlier through oxidation of the selenide to selenoxide followed by heating in a sealed tube at 140 °C, affording the fully protected everninomicin 13,384-1 derivative **111** in 65% yield as a single isomer. The remaining tasks for the generation of the natural product from **111** included removal of the benzyl and silyl protecting groups. After extensive experimentation with different catalysts, solvents, and buffers, the optimal conditions to remove the benzyl ethers without damaging the chlorine, nitro, or orthoester sites proved to be hydrogenolysis in the presence of 10% Pd/C and $NaHCO_3$, using tBuOMe as solvent. Finally, the resulting hexa-TBS derivative was globally deprotected upon treatment with excess TBAF in THF, furnishing the targeted molecule (**1**) in 75% overall yield. Most gratifyingly, synthetic everninomicin 13,384-1 (**1**) was identical with an authentic sample by the usual criteria [47].

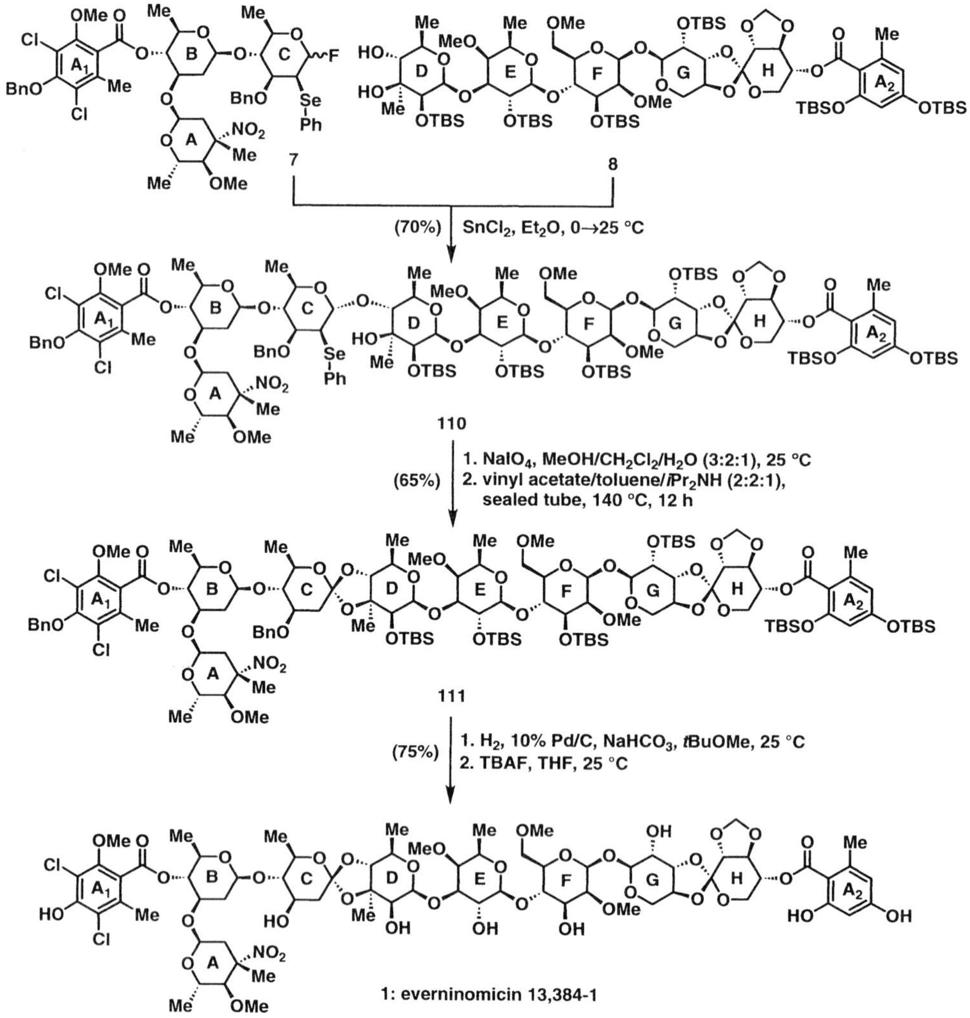

Scheme 8.32 Final stages and completion of the total synthesis of everninomicin 13,384-1 (**1**).

Finally, as mentioned briefly above, degradation of the natural product was very helpful in facilitating initial coupling studies of the $A_1B(A)C$ and $DEFGHA_2$ fragments, as well as for spectroscopic comparisons. It was found that both the hexabenzylated derivative **116** and the hexa-TBS derivative **8** could be obtained from the natural product, as shown in Scheme 8.33. Treatment of naturally derived everninomicin 13,384-1 (**1**) [47] with TBSOTf in the presence of 2,6-lutidine furnished the fully silylated derivative **113**; analogously, treatment with NaH and BnBr afforded the fully benzylated derivative **112**. Treatment of **112** or **113** with dilute aqueous HCl and subsequent addition of aqueous KOH furnished the targeted hexasilylated diol **8** in 58% overall yield or the hexabenzylated diol **116** in 84%

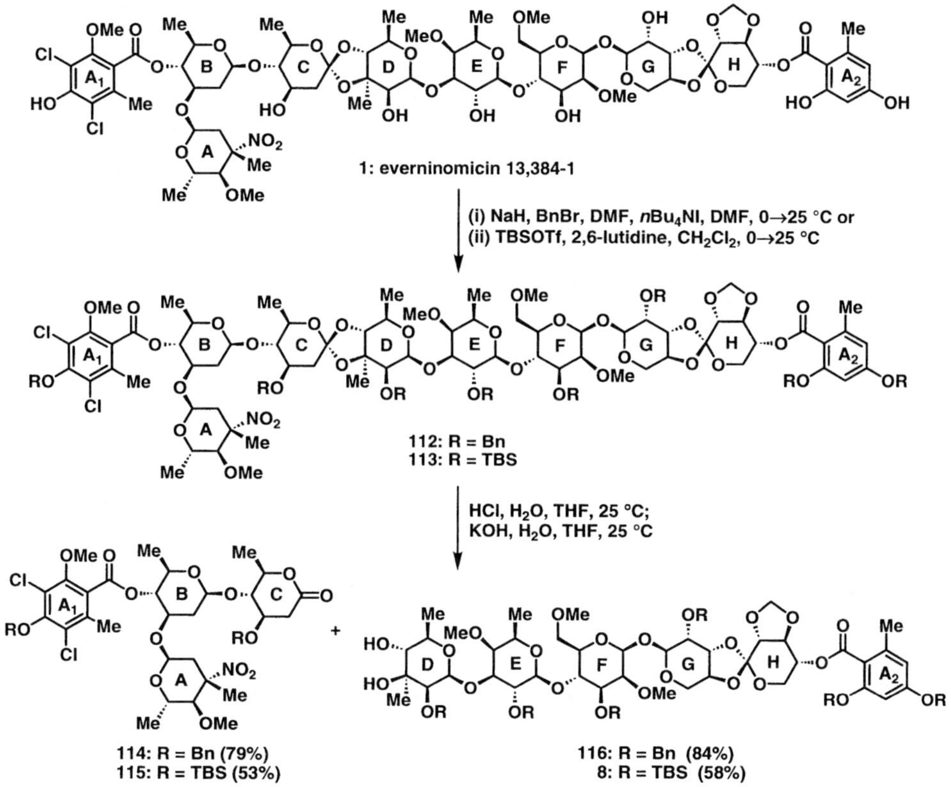

Scheme 8.33 Degradation studies with naturally derived everninomicin 13,384-1 (**1**).

overall yield, along with their corresponding protected δ-lactones (**114** and **115**). In addition to facilitating the total synthesis of everninomicin 13,384-1 (**1**), these degradative routes provided access to useful quantities of the complex DEFGHA$_2$ fragment **8**, a potentially valuable intermediate for the semisynthesis of everninomicin analogues and libraries for biological screening purposes.

8.4
Conclusion

The described research program culminating in the total synthesis of everninomicin 13,384-1 (**1**) served as an opportunity to develop a number of novel synthetic reactions and strategies, to explore their scope and generality, and to apply them in the context of complex situations. Most notable among these new methods are the stereocontrolled construction of 1,1′-disaccharides, and the 1,2-phenylseleno migrations on carbohydrate templates and their use in stereocontrolled glycosidation reactions. Other processes applied and championed in this synthesis include:

selective silylation of carbohydrate diols under unique solvent conditions, the use of acyl fluorides for the formation of sterically hindered esters, the Sinaÿ orthoester formation procedure, the Kahne-Crich sulfoxide glycosidation, the Schmidt trichloroacetimidate method, the Mukaiyama glycosyl fluoride methodology, and the tin acetal technique for differentiating 1,2-diols. In addition, considerable knowledge was acquired in the field of effective orthogonal protecting group ensembles, and much light was shed on conformational effects resulting in selective functionalization of carbohydrate substrates. Through these synthetic forays, the everninomicin field has moved within reach of further studies including semisynthesis of designed analogues of everninomicin 13,384-1 (**1**), solid-phase synthetic applications and combinatorial chemistry, and exploration of problems in chemical biology.

8.5
References

1 For such a review, see: K. C. Nicolaou, H. J. Mitchell, *Angew. Chem.* **2001**, *113*, 1624–1672; *Angew. Chem. Int. Ed.* **2001**, *40*, 1576–1624.

2 A. K. Ganguly, B. Pramanik, T. C. Chan, O. Sarre, Y.-T. Liu, J. Morton, V. M. Girijavallabhan, *Heterocycles* **1989**, *28*, 83–88; b) J. A. Maertens, *Curr. Opin. Anti-Infect. Invest. Drugs* **1999**, *1*, 49–56; c) J. A. Maertens, *IDrugs* **1999**, *2*, 446–453.

3 For a review on the orthosomicin class, see: P. Juetten, C. Zagar, H. D. Scharf, *Recent Prog. Chem. Synth. Antibiot. Relat. Microb. Prod.* **1993**, 475–549 and references cited therein.

4 D. E. Wright, *Tetrahedron* **1979**, *35*, 1207–1237.

5 a) N. Cappuccino, A. Bose, A. K. Ganguly, J. Morton, *Heterocycles* **1981**, *15*, 1621–1641; b) W. Ollis, I. Sutherland, B. Taylor, C. Smith, D. Wright, *Tetrahedron* **1979**, *35*, 993–1001.

6 a) V. M. Girijavallabhan, A. K. Saksena, F. Bennet, E. Jao, M. Patel, A. K. Ganguly, United States Patent **1997**, US 5 652 226; b) V. M. Girijavallabhan, A. K. Saksena, F. Bennet, E. Jao, M. Patel, A. K. Ganguly, United States Patent **1998**, US 5 795 874.

7 H. Wolf, *FEBS Lett.* **1973**, *36*, 181–186.

8 a) L. Belova, T. Tenson, L. Xiong, P. McNicholas, A. Mankin, *Proc. Natl. Acad. Sci. U.S.A.* **2001**, *98*, 3726–3731; b) F. Aarestrup, L. Jensen, *Antimicrob. Agents Chemother.* **2000**, *44*, 3425–3427.

9 a) A. K. Ganguly, V. M. Girijavallabhan, G. Miller, O. Sarre, *J. Antibiot.* **1982**, *35*, 561–570; b) A. K. Ganguly, O. Sarre, S. Szmulewicz, United States Patent **1975**, 3 920 629; c) A. K. Ganguly, V. M. Girijavallabhan, O. Sarre, H. Reimann, United States Patent **1978**, 4 129 720; d) A. K. Ganguly, O. Sarre, United States Patent **1975**, 3 915 956; e) A. K. Ganguly, J. L. McCormick, T. M. A. K. Saksena, P. R. Das, *Tetrahedron Lett.* **1997**, *38*, 7989–7992; e) A. K. Ganguly, J. L. McCormick, T. M. Chan, A. K. Saksena, P. R. Das, United States Patent **1998**, 5 763 600; f) A. K. Ganguly, J. L. McCormick, L. Jinping, A. K. Saksena, P. R. Das, R. Pradip, T. M. Chan, *Bioorg. Med. Chem. Lett.* **1999**, *9*, 1209–1214.

10 a) K. C. Nicolaou, H. J. Mitchell, H. Suzuki, R. M. Rodríguez, O. Baudoin, K. C. Fylaktakidou, *Angew. Chem.* **1999**, *111*, 3523–3528; *Angew. Chem. Int. Ed.* **1999**, *38*, 3334–3339; b) K. C. Nicolaou, R. M. Rodríguez, K. C. Fylaktakidou, H. Suzuki, H. J. Mitchell, *Angew. Chem.* **1999**, *111*, 3529–3534; *Angew. Chem. Int. Ed.* **1999**, *38*, 3340–3345; c) K. C. Nicolaou, H. J. Mitchell, R. M. Rodríguez, K. C. Fylaktakidou, H. Suzuki, *Angew.*

Chem. **1999**, *111*, 3535–3540; *Angew. Chem. Int. Ed.* **1999**, *38*, 3345–3350.
11 K. Toshima, K. Tatsuta, *Chem. Rev.* **1993**, *93*, 1503–1531.
12 R. Lemieux, K. Hendriks, R. Stick, K. James, *J. Am. Chem. Soc.* **1975**, *97*, 4056–4062.
13 T. Mukaiyama, Y. Murai, S. Shoda, *Chem. Lett.* **1981**, 431–432.
14 a) W. Rosenbrook, Jr., D. A. Riley, P. A. Lartey, *Tetrahedron Lett.* **1985**, *26*, 3–4; b) G. H. Posner, S. R. Haines, *Tetrahedron Lett.* **1985**, *26*, 5–8; c) G. A. Olah, J. T. Welch, Y. D. Vankar, M. Nojima, I. Kerekes, J. A. Olah, *J. Org. Chem.* **1979**, *44*, 3872–3881; d) T. Mukaiyama, Y. Hashimoto, S. Shoda, *Chem. Lett.* **1983**, 935–938; e) Y. Araki, K. Watanabe, F.-H. Kuan, K. Itoh, N. Kobayachi, Y. Ishido, *Carbohydr. Res.* **1984**, *127*, C5–C9; f) H. Kuntz, W. Sager, *Helv. Chim. Acta* **1985**, *68*, 283–287; g) M. Burkart, Z. Zhang, S.-C. Hung, C.-H. Wong, *J. Am. Chem. Soc.* **1997**, *119*, 11743–11746.
15 K. C. Nicolaou, R. E. Dolle, D. P. Papahatjis, J. L. Randall, *J. Am. Chem. Soc.* **1984**, *106*, 4189–4192.
16 S. Hashimoto, M. Hayashi, R. Noyori, *Tetrahedron Lett.* **1984**, *25*, 1379–1382.
17 R. R. Schmidt, J. Michel, *Angew. Chem.* **1980**, *92*, 763–765; *Angew. Chem. Int. Ed. Engl.* **1980**, *19*, 731–732.
18 R. R. Schmidt, *Angew. Chem.* **1986**, *98*, 213–236; *Angew. Chem. Int. Ed. Engl.* **1986**, *25*, 212–235.
19 R. Ferrier, R. Hay, N. Vethaviyasar, *Carbohydr. Res.* **1973**, *27*, 55–61.
20 a) D. Kahne, S. Walker, Y. Chang, D. Van Engen, *J. Am. Chem. Soc.* **1989**, *111*, 6881–6882; b) D. Crich, S. Sun, *J. Am. Chem. Soc.* **1998**, *120*, 435–436; c) D. Crich, H. Li, Q. G. Yao, D. J. Wink, R. D. Commer, A. L. Rheingold, *J. Am. Chem. Soc.* **2001**, *123*, 5826–5828.
21 K. C. Nicolaou, T. Ladduwahetty, J. L. Randall, A. Chucholowski, *J. Am. Chem. Soc.* **1986**, *108*, 2466–2467.
22 M. Trumtel, P. Tavecchia, A. Veyrières, P. Sinaÿ, *Carbohydr. Res.* **1990**, *202*, 257–275.
23 K. C. Nicolaou, F. L. van Delft, S. R. Conley, H. J. Mitchell, Z. Jin, R. M. Rodríguez, *J. Am. Chem. Soc.* **1997**, *119*, 9057–9058.
24 a) A. David in *Preparative Carbohydrate Chemistry* (Ed.: S. Hanessian), Marcel Dekker, Inc., **1997**, 69–83; b) T. B. Grindley, *Adv. Carb. Chem. Biochem.* **1998**, *53*, 17–142.
25 a) P. Juetten, H. D. Scharf, G. Raabe, *J. Org. Chem.* **1991**, *56*, 7144–7149; b) J. Dornhagen, H. D. Scharf, *Tetrahedron* **1985**, *41*, 173–175; c) P. Juetten, J. Dornhagen, H. D. Scharf, *Tetrahedron* **1987**, *43*, 4133–4140.
26 K. C. Nicolaou, R. M. Rodríguez, H. J. Mitchell, F. L. van Delft, *Angew. Chem.* **1998**, *110*, 1975–1977; *Angew. Chem. Int. Ed.* **1998**, *37*, 1874–1876.
27 For a recent review, see: M. Schuster, S. Blechert, *Angew. Chem.* **1997**, *109*, 2124–2145; *Angew. Chem. Int. Ed. Engl.* **1997**, *36*, 2036–2056.
28 K. C. Nicolaou, H. J. Mitchell, F. L. van Delft, F. Rübsam, R. M. Rodríguez, *Angew. Chem.* **1998**, *110*, 1972–1974; *Angew. Chem. Int. Ed.* **1998**, *37*, 1871–1874.
29 J. A. Marco, M. Cards, J. Murga, F. González, E. Falomir, *Tetrahedron Lett.* **1997**, *38*, 1841–1844.
30 M. Hirama, I. Nishizaki, T. Shigemoto, S. Ito, *J. Chem. Soc., Chem. Commun.* **1986**, 393–394.
31 a) J. E. Baldwin, G. A. Höfle, O. W. Lever, Jr., *J. Am. Chem. Soc.* **1974**, *96*, 7125–7127; b) R. K. Boeckman, Jr., K. J. Bruza, *J. Org. Chem.* **1979**, *44*, 4781–4788.
32 P. S. Bailey, J. E. Keller, *J. Org. Chem.* **1968**, *33*, 2680–2684.
33 G. Solladie, A. Rubio, M. Carreno, J. Ruano, *Tetrahedron: Asymmetry* **1990**, *1*, 187–198.
34 L. A. Carpino, A. El-Faham, *J. Am. Chem. Soc.* **1995**, *117*, 5401–5402.
35 A. Y. Chemyak, K. V. Antonov, N. K. Kochetkov, *Biorg. Khim.* **1989**, *15*, 1113–1127.
36 C. Czernecki, K. Vijayakumaran, G. Ville, *J. Org. Chem.* **1986**, *51*, 5472–5474.
37 S. Mehta, B. M. Pinto, *J. Org. Chem.* **1993**, *58*, 3269–3276.
38 a) A. Dondoni, G. Fantink, M. Fogagnolo, A. Medici, P. Pedrini, *J. Org.*

Chem. **1988**, *53*, 1748–1761; b) A. DONDONI, A. MARRA, D. PERRONE, *J. Org. Chem.* **1993**, *58*, 275–277.

39 P. BHATE, D. HORTON, W. PRIEBE, *Carbohydr. Res.* **1985**, *144*, 325–331.

40 J.C. MARTIN, R. J. ARHART, *J. Am. Chem. Soc.* **1971**, *93*, 4327–4329.

41 a) F. HE, Y. BO, J.D. ALTOM, E.J. COREY, *J. Am. Chem. Soc.* **1999**, *121*, 6771–6772; b) E.J. COREY, S. SARSHAR, M.D. AZIMIOARA, R. NEWBOLD, M.C. NOE, *J. Am. Chem. Soc.* **1996**, *118*, 7851–7852.

42 X. WU, F. KONG, *Carbohydr. Res.* **1987**, *162*, 166–169.

43 a) D.B. DESS, J.C. MARTIN, *J. Org. Chem.* **1983**, *48*, 4155–4157; b) D.B. DESS, J.C. MARTIN, *J. Am. Chem. Soc.* **1991**, *113*, 7277–7279; c) S.D. MEYER, S.L. SCHREIBER, *J. Org. Chem.* **1994**, *59*, 7549–7552.

44 K.S. KIM, W.A. SZAREK, *Synthesis* **1978**, 48–50.

45 T. OSHITAR, M. SHIBASAKI, T. YOSHIZAWA, M. TOMITA, K. TAKAO, S. KOBAYASHI, *Tetrahedron* **1997**, *53*, 10993–11006.

46 K.C. NICOLAOU, C.W. HUMMEL, Y. IWABUCHI, *J. Am. Chem. Soc.* **1992**, *114*, 3126–3128.

47 We thank Dr. Ashit Ganguly of Schering-Plough Corp. for a generous sample of natural everninomicin 13, 384-1 (**1**).

9
Chemical Synthesis of Asparagine-Linked Glycoprotein Oligosaccharides: Recent Examples

Yukishige Ito and Ichiro Matsuo

9.1
Introduction

Accumulating evidence strongly suggests that the oligosaccharide parts of glycoproteins play pivotal roles in various biological events [1]. To gain precise understanding of their functional roles and underlying molecular mechanism, access to structurally defined oligosaccharides is of critical importance [2].

Glycoprotein oligosaccharides are highly diverse. Two major groups among them are asparagine-linked (*N*-linked) and serine/threonine-linked (*O*-linked) glycans [3], typical structures of which are given in Fig. 9.1. They are further classified into subgroups, each of which has structural diversity arising from variable degrees of branching and terminal modification. Several types of minor groups, some of them attracting particular attention, have also been identified [4].

Of course, it is not surprising at all that different proteins often carry different oligosaccharides. However, the structure of a given glycoprotein varies depending upon the species, organ, and cell. Additionally, most glycoproteins are not uniform, but consist of various "glycoforms" differing in type, length, branching and terminal decoration of oligosaccharides, and number or site of glycosylation(s). The isolation of oligosaccharides from natural sources is therefore highly challenging and the range of oligosaccharides that can be obtained in pure form should be severely limited.

In contrast, chemical synthesis of oligosaccharides has an obvious advantage [5] in that, once established, various types of oligosaccharides can be produced in large amounts. As alternatives, enzymatic or chemoenzymatic approaches are certainly promising [6], but it has to be kept in mind that, of the vast array of enzymes (glycosyltransferases) involved in biosynthetic pathways of these molecules, only a small fraction has been overexpressed and far fewer even of these are commercially available. Even in this latter case, economic problems cannot be ignored, because of the costs of enzymes and sugar nucleotides. Additionally, glycosyltransferases have narrow substrate specificities, being able to catalyze the formation of single types of glycosidic linkage (one enzyme-one linkage). It is obvious that several glycosyltransferases responsible for the critical steps in the biosynthesis of asparagine-linked glycoprotein glycans are not suitable for such use. This is because: (1) these enzymes

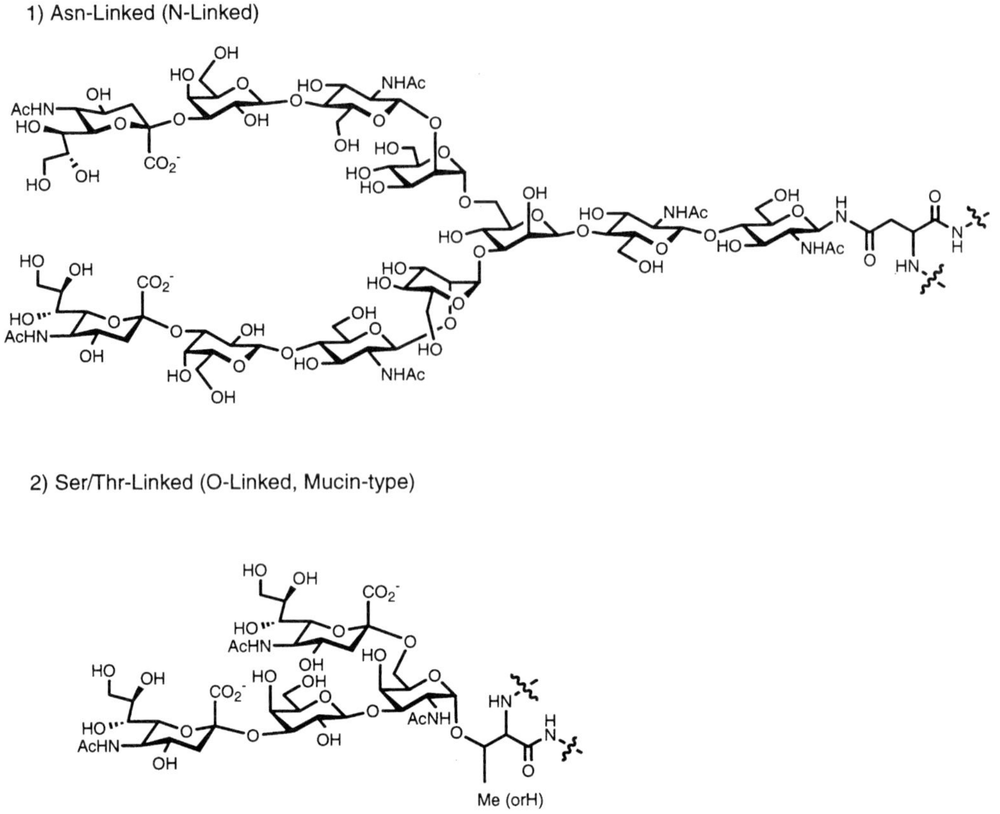

1) Asn-Linked (N-Linked)

2) Ser/Thr-Linked (O-Linked, Mucin-type)

Fig. 9.1 Typical structures of glycoprotein glycans.

are membrane-bound proteins and not amenable to large-scale isolation and/or overexpression, and (2) they require lipid-linked (dolichol) acceptor and/or donor substrates, preparation of which is itself a significant challenge [7]. Reconstitution of such enzymatic systems on a preparative scale is likely to be extremely difficult. As things stand, the applicability of enzyme-based approaches is rapidly expanding, but still severely limited. Chemical synthesis is more flexible by far. In theory, there is no limitation with respect to the range of compounds that can be targeted.

However, the chemical synthesis of complex oligosaccharides is by no means a trivial task, due to the following obstacles. Firstly, it tends to be highly time- and labor-consuming, requiring multi-step operations consisting of iterative O-glycosylation and partial deprotection (Scheme 9.1). Selection of suitable glycosyl donor (**D**), promoter, and temporary protecting group (**P**) is of pivotal importance. The last of these should be stable under various glycosylation conditions, but removable in high yield without affecting the permanent protecting group (**R**) [8].

All monosaccharide components should be properly designed, so that selective deprotection of specific hydroxy groups is possible. Preparation of these blocks

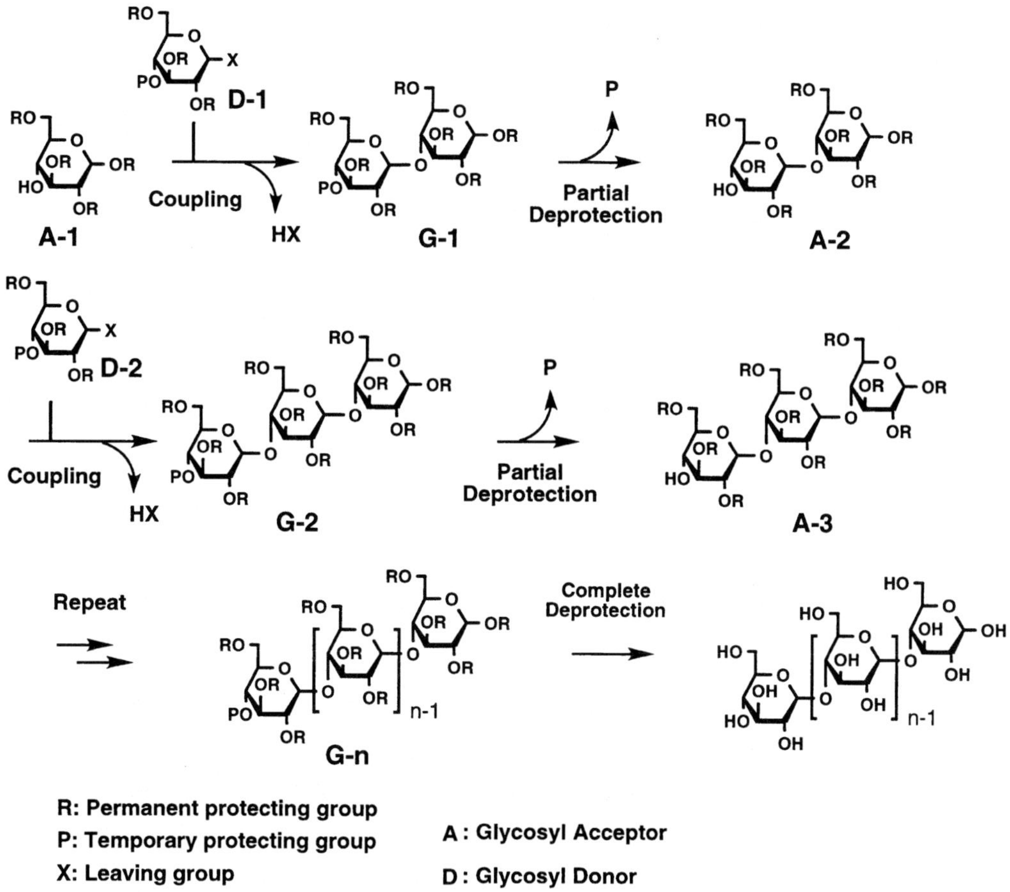

Scheme 9.1 General oligosaccharide synthesis scheme.

R: Permanent protecting group
P: Temporary protecting group
X: Leaving group
A: Glycosyl Acceptor
D: Glycosyl Donor

again requires a number of steps mainly consisting of selective protection of hydroxy groups [9]. Among them, tin oxide-mediated (either Bu_2SnO or $(Bu_3Sn)_2O$) alkylation [10], and reductive cleavage of benzylidene acetals [11] are of exceptional utility (Scheme 9.2).

Besides regioisomeric issues, more fundamental are stereochemical problems [12]. Each coupling step potentially generates two stereoisomers: α- and β-glycosides (Scheme 9.3), and the stereochemical outcome of a glycosylation is not always predictable, so has to be regarded as a case-by-case matter. Seemingly trivial structural changes in either donor or acceptor may perturb or even reverse the stereoselectivities found in very similar systems.

The biological properties of stereo- and regioisomers are often drastically different, while their separation and structure assignment are usually difficult. Therefore, any structural ambiguity in terms of isomerism should be eliminated. Because the synthesis of large oligosaccharides inevitably requires multi-step trans-

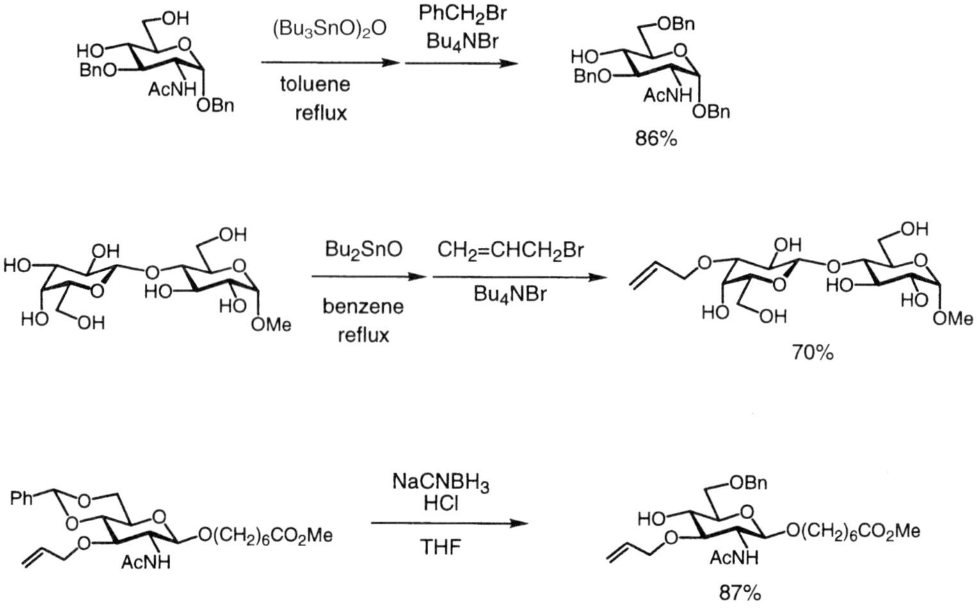

Scheme 9.2 Tin-mediated alkylation and reductive cleavage of benzylidene acetals.

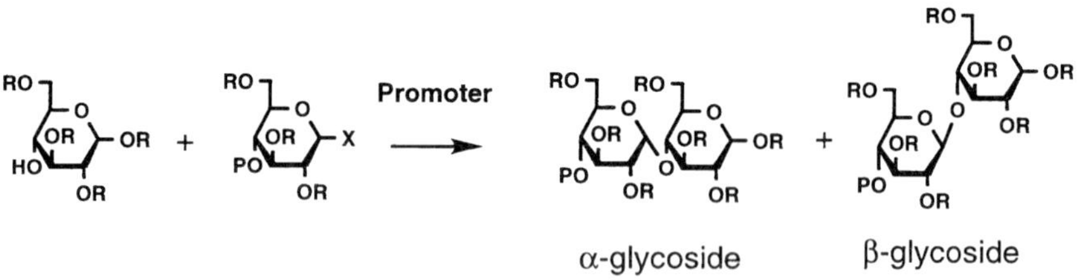

Scheme 9.3 Formation of stereoisomeric products in glycosylations.

formations and isomer separation is not always possible after each step, it is highly desirable that all reactions should proceed in a controlled and unambiguous manner. A poorly designed synthesis is likely to be troublesome and low-yielding, and may even result in an intractable mixture from which isolation of the target molecule is desperately difficult.

In this chapter, recent reports on the synthesis of oligosaccharides derived from asparagine (Asn)-linked glycoproteins are taken as examples. Of the various types of glycoprotein glycans, Asn-linked oligosaccharides are the most diverse, complex, and biologically important. In particular, their recently delineated role in protein quality control is highly intriguing [13], though it is not our aim to provide an extensive review of this subject. Rather, we would like to summarize some important examples to provide readers with the basic concepts and recent trends. For earlier studies, some excellent reviews are provided as reference [14].

9.2
Synthesis of Asn-Linked Oligosaccharides: Basic Principles

Asparagine (Asn)-linked oligosaccharides can be classified into three subgroups: high-mannose type, complex-type, and hybrid type (Fig. 9.2). It may be noted that

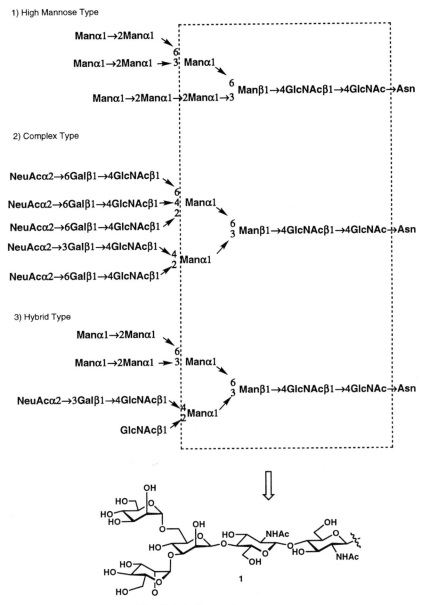

Fig. 9.2 Typical structures of Asn-linked oligosaccharides.

all of these contain a "core" pentasaccharide **1**, consisting of two α-linked mannose (Man), two β-linked N-acetylglucosamine (GlcNAc) units, and one β-linked Man unit. These oligosaccharides are linked to protein through N-glycosidic linkages formed between the reducing end terminal GlcNAc and the side chain carboxamide of Asn.

Chemical synthesis of complex oligosaccharides such as Asn-linked glycans requires repetitive glycosylation reactions, in which case – as already mentioned – stereochemical control is the most significant issue. The major monosaccharide components of Asn-linked oligosaccharides and their possible anomeric configurations are summarized in Fig. 9.3. On the basis of glycosidic linkage configurations and the relative orientations of the 1- and 2-positions, they can be divided into 1,2-*trans*-β (type I), 1,2-*trans*-α (type II), 1,2-*cis*-α (type III), 1,2-*cis*-β (type IV), and 2-acetamido-β (type-V).

For type-I and -II linkages, neighboring group participation from ester groups situated at the 2-positions can be used (Scheme 9.4). In such cases, glycosylation is believed to proceed by way of cyclic acyloxonium ion intermediates **2**, which in most cases precludes the formation of 1,2-*cis* isomers. Generally speaking, this is a very reliable strategy, and complete stereochemical control can be expected in most cases. However, formation of orthoesters is often a serious side reaction. Type-V (2-acetamido-β-) glycosides are most typically synthesized from N-phthaloyl-protected (Phth-protected) donors **3**. Participation by the carbonyl groups of Phth is generally accepted as an explanation for the excellent β-selectivity.

When the 2-OH of a glycosyl donor is protected with an ether-type group (benzyl or allyl, for instance), predominant formation of α isomers (both type-II and -III) is usually the case. In the absence of neighboring group participation, formation of axially oriented (α) glycosides is generally favored (Scheme 9.5). Once activated, the glycosyl donor **4** is transformed into the rapidly interconverting ion-pairs **5a** and **5b**. Because of the anomeric effect, the β-oriented ion-pair **5b** has much higher reactivity, which produces the α-glycoside. While rather rare in Asn-linked glycoproteins, α-linked N-acetylgalactosamine (GalNAc) is widespread as a constituent of Ser/Thr-linked glycan chains [15]. It is usually prepared in a manner similar to that used in the case of type-II linkage, most often by use of 2-azido substituted donor **6**.

Type-IV glycosides are not included in the above discussion and need special consideration. Their formation is essential for the synthesis of Asn-linked glycan chains, which incorporate a core pentasaccharide containing β-Man(1 → 4)GlcNAc. Formation of this linkage is difficult, because neither neighboring group participation, nor stereoelectronic control can be utilized for this purpose. Various approaches have been investigated to remove this difficulty [16]. Among recent endeavors, intramolecular aglycon delivery [17] and the glycosyl sulfoxide approach [18] seem the most successful (Scheme 9.6). The utility of the former technique in the synthesis of Asn-linked glycoproteins has been clarified [19], while evaluation of the latter in this respect awaits future progress. Some aspects of β-mannosylation are discussed in some depth in the following sections.

Similar considerations apply to sialic acids. Sialic acids, most typically N-acetylneuraminic acid (NeuAc) are important constituent of N- and O-linked glycopro-

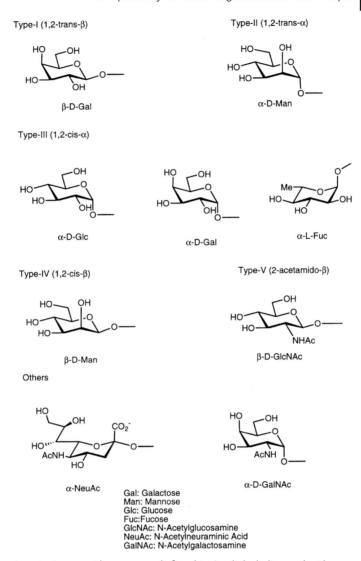

Fig. 9.3 Sugar residues commonly found in Asn-linked oligosaccharides.

teins as well as glycosphingolipids and reside at the non-reducing ends of glycan chains with α configurations. (In this particular case, "α-glycoside" refers by definition to an equatorially oriented glycoside, in contrast to commonly encountered hexopyranosides.) Among various methods used to construct the challenging α-NeuAc linkage [20], activation of thioglycoside **7** in acetonitrile is acquiring widespread acceptance [21].

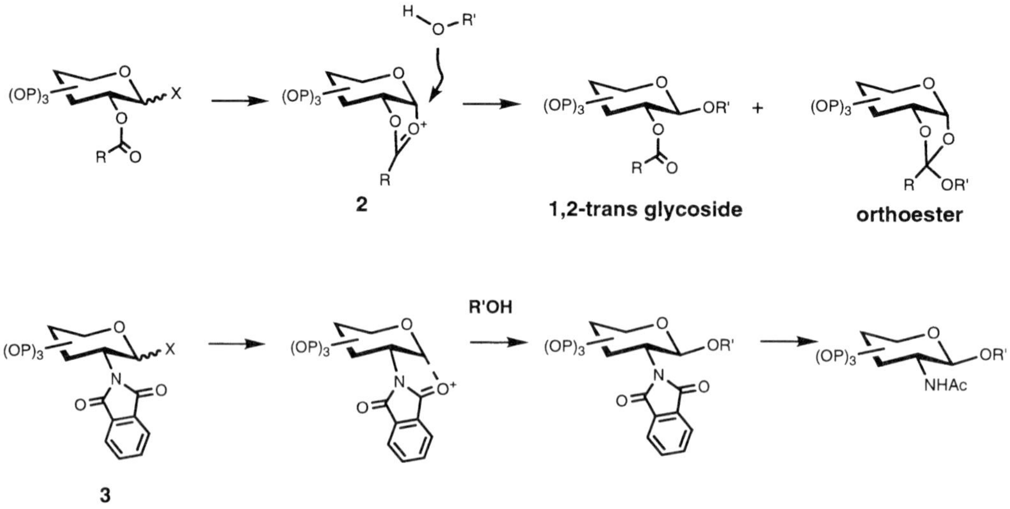

Scheme 9.4 Formation of 1,2-*trans* glycosides: neighboring group participation.

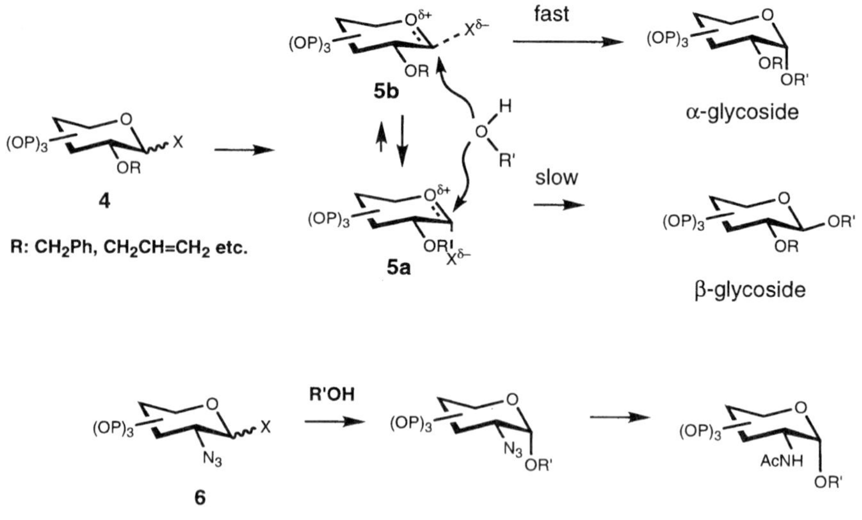

Scheme 9.5 Formation of α-glycosides: stereoelectronic control.

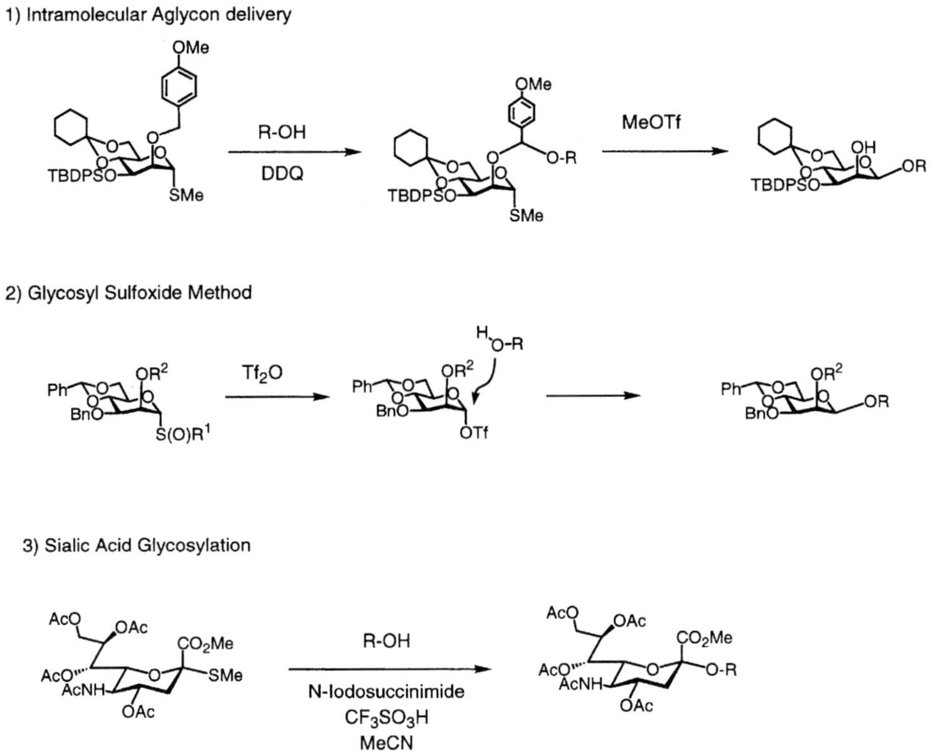

1) Intramolecular Aglycon delivery

2) Glycosyl Sulfoxide Method

3) Sialic Acid Glycosylation

Scheme 9.6 Formation of difficult linkages.

9.3
Chemical Synthesis of Complex Oligosaccharides

In order to show how chemical synthesis of complex oligosaccharide can be achieved, some examples are selected from recent synthesis of Asn-linked glycans.

9.3.1
Classical Examples

Before the mid-1980s, oligosaccharide synthesis was performed with glycosyl halides as donors. Despite the technical difficulties involved in preparing and handling unstable glycosyl halides, some impressively complex oligosaccharides were synthesized successfully by this classical approach. Such examples are given below.

Paulsen exploited "insoluble" Ag salt (silver silicate) for the problematic β-manno glycoside formation (Scheme 9.7) [22]. 3,6-O-Allyl-2,4-O-benzyl-mannosyl bromide **8** was treated with the acceptor **9** to provide β-linked product **10** [23].

Scheme 9.7 Synthesis of core pentasaccharide.

The stereoselectivity of this type of reaction can be interpreted as follows (Fig. 9.4). Thus, the glycosyl halide **8**, activated by the insoluble salt, has its counter-ion absorbed on the solid surface, so α to β anomerization is suppressed. Glycosylation therefore predominantly proceeds in an S_N2-like manner to give the anomerically inverted (β) product.

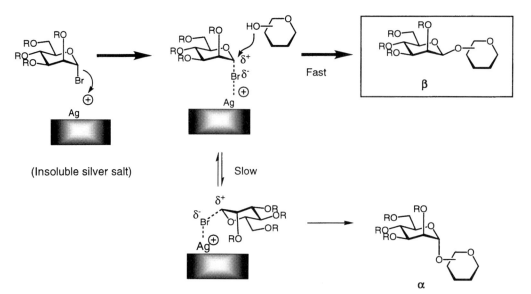

Fig. 9.4 Insoluble silver salt-mediated glycosylation.

It should be noted that the stereochemical outcome of insoluble Ag salt-mediated glycosylation is highly dependent on the reactivity of substrates; it requires a reactive donor and acceptor. Less reactive substrates may cause anomerization of the reactive intermediate to give significant amounts of α-glycosides. In this particular example, the steric hindrance of the 4-OH group was relieved by the creation of the 1,6-anhydro form, forcing the hexopyranoside ring into the $^1C_4$ conformation. It was also found that the nature of donor protection is also important. In order to attain optimum β-selectivity, 2-, 3-, and 6-OH should be protected by ether-type protecting groups, while electron-withdrawing acyl protection at the 4-position was reported to be beneficial [24].

The allyl groups of **10** were removed and the resultant diol was doubly glycosylated with mannosyl chloride **11** to give **12**, deprotection of which completed the synthesis of the tetrasaccharide core **13** of Asn-linked glycans. Masked tetrasaccharide **12** was used as an intermediate for the synthesis of the complex-type octasaccharide **16** [25] (Scheme 9.8). To that end, it was deacetylated to liberate the 2-OH groups, which were glycosylated with lactosamine bromide **14** to give the octasaccharide **15**. This was transformed into **16** after ring-opening of the 1,6-anhydro sugar and complete deprotection.

To reach a similar goal, Ogawa and co-workers took a substantially different approach (Scheme 9.9) [26]. In this case, chitobiose derivative **17** was used directly as the acceptor to react with mannosyl bromide **8**. Since the reactivity of **17** is low, this glycosylation required excess donor and proceeded with poor selectivity. However, trisaccharide **18** was still isolated in a synthetically useful yield (40%), together with the corresponding α isomer (36%). Deallylation gave the correspond-

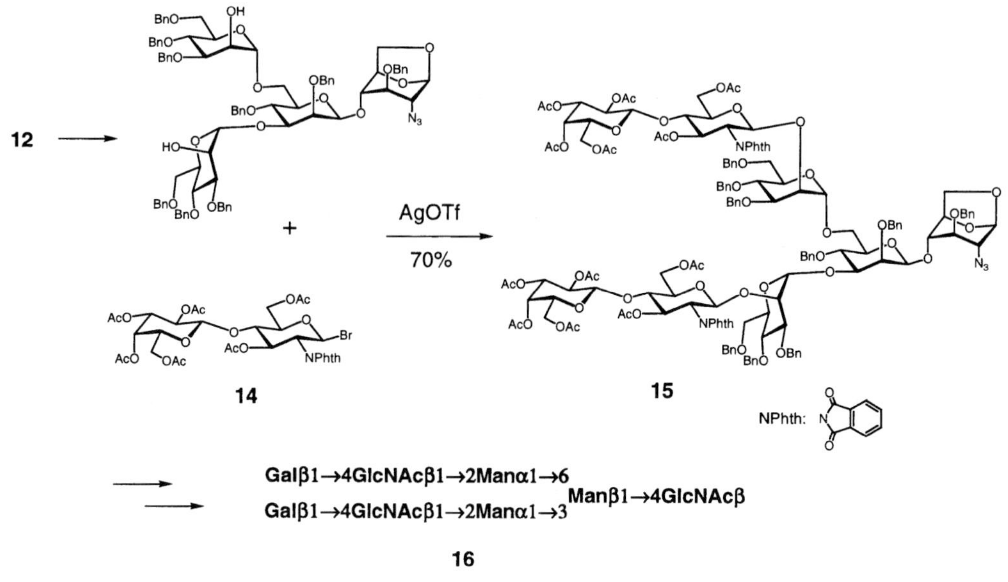

Scheme 9.8 Synthesis of complex-type oligosaccharides – I.

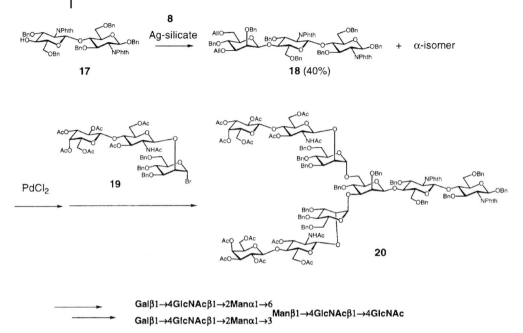

Scheme 9.9 Synthesis of complex-type oligosaccharides – II.

ing diol, which was in turn doubly glycosylated with trisaccharide bromide **19** by use of silver triflate as a promoter to give the nonasaccharide **20** [27].

The same group also subsequently synthesized the α-2,6-linked sialic acid containing the undecasaccharide **21**, by use of the trichloroacetimidate **22** in place of **19** [28].

9.3.2
Trichloroacetimidate Approach to Complex-Type Glycan Chains

Trichloroacetimidate is currently one of the most widely used glycosyl donors, very often regarded as the first choice in various aspects of oligosaccharide synthesis. It is easy to prepare (with trichloroacetonitrile and base), reasonably stable (can be isolated by silica gel column chromatography), and is strongly activated by exposure to catalytic amounts of Lewis acid (such as $BF_3 \cdot OEt_2$ or TMSOTf) at low temperature (typically $<0\,^\circ C$) [29]. It has been used for the synthesis of numerous oligosaccharides, glycoconjugates, and natural products.

Schmidt et al. originally developed trichloroacetimidate-based glycosylation, relying on it entirely in a versatile strategy for the synthesis of complex-type oligosaccharides [30] (Scheme 9.10). The synthesis starts from the coupling between the glucose-derived trichloroacetimidate **23** and the masked GlcNAc **24**. After deacetylation, treatment with mannosyl donor **25** proceeded regioselectively to afford **26**. The remaining hydroxy group was transformed into a triflate, which was treated with Bu_4NNO_2 to give **27** after benzylation and acidic removal of the benzylidene group. It should be noted that the difficulty inherent in β-manno glycosylation was avoided by the introduction of β-linked glucose, which was eventually converted into mannose through S_N2-type inversion. The primary hydroxy group in **27** was then again regioselectively glycosylated with **25** to afford a tetrasaccharide that was further glycosylated with **28** and deacetylated to give **29**. This in turn was doubly glycosylated with the lactosamine (Galβ1 → 4GlcNAc) donor **30** and converted into the octasaccharide trichloroacetimidate **31**. Final coupling with GlcNAc equivalent **24** took advantage of the solvent effect of acetonitrile [31] and proceeded in a β-selective manner to give **32**. Final deprotection completed the synthesis of the target molecule **33**.

9.3.3
n-Pentenyl Glycosides as Glycosyl Donors

Fraser-Reid and co-workers developed the usefulness of n-pentenyl glycoside **34** as a glycosyl donor [32]. Although possessing excellent stability under various conditions, it can readily be activated with iodonium ion, generated from N-iodosuccinimide (NIS) and silyl triflate via **35** (Scheme 9.11). The donor reactivity can be masked in the form of a bromine adduct **36** and can be regenerated by reductive debromination as shown below.

Its versatility was nicely demonstrated in the synthesis of nonamannoside component of a high-mannose type glycan chain (Scheme 9.12) [33]. Pentamannosyl donor **42** was prepared from dibromopentyl glycoside **37**. Glycosylation with n-pentenyl glycoside **38** and partial deprotection gave **39**. Second glycosylation and deacetylation gave **40**, which was double-glycosylated with the same donor to provide **41**. Final treatment with Zn afforded n-pentenyl glycoside **42**, which was used as the pentasaccharide donor.

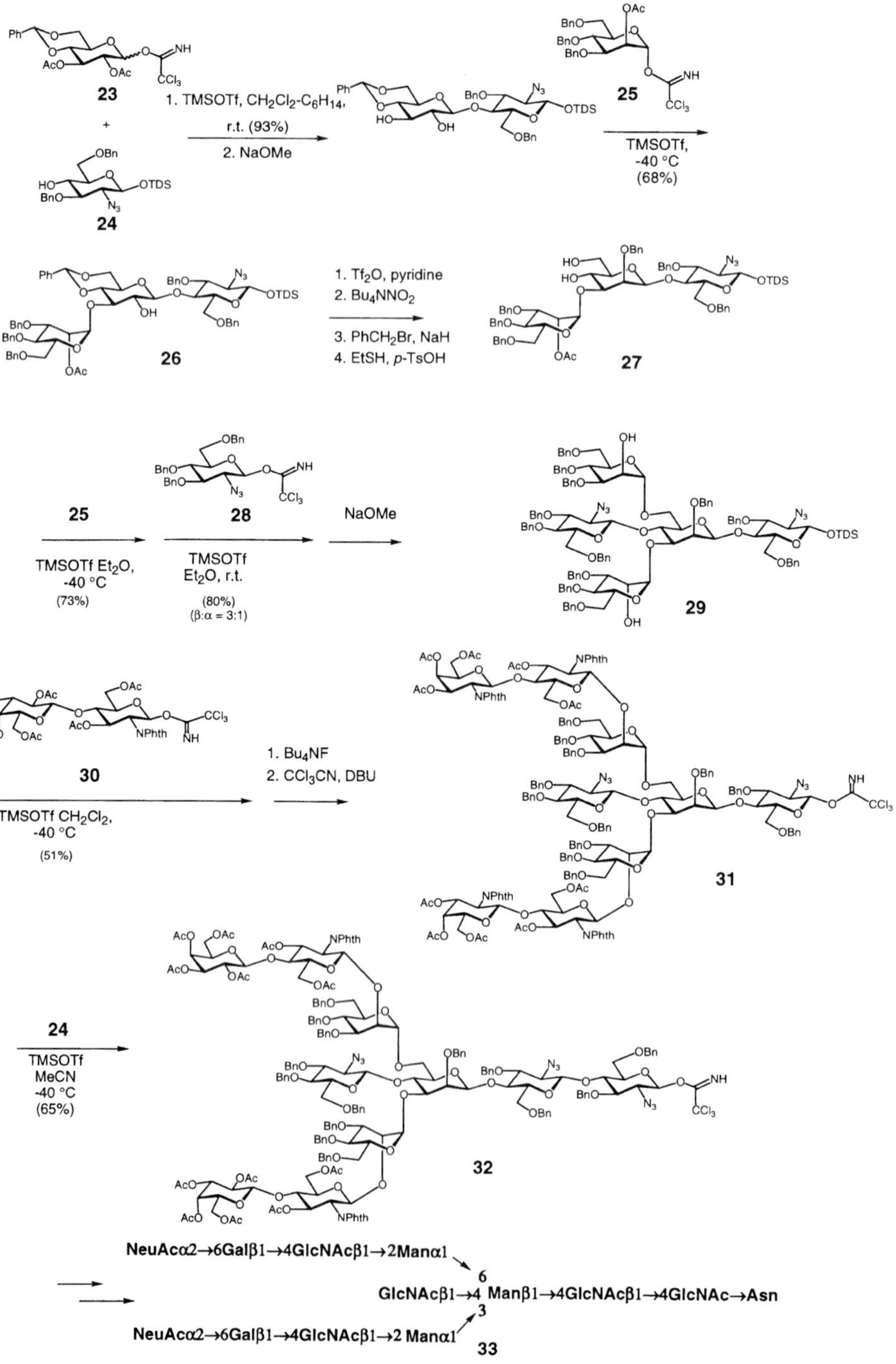

Scheme 9.10 Trichloroacetimidate approach to complex oligosaccharides.

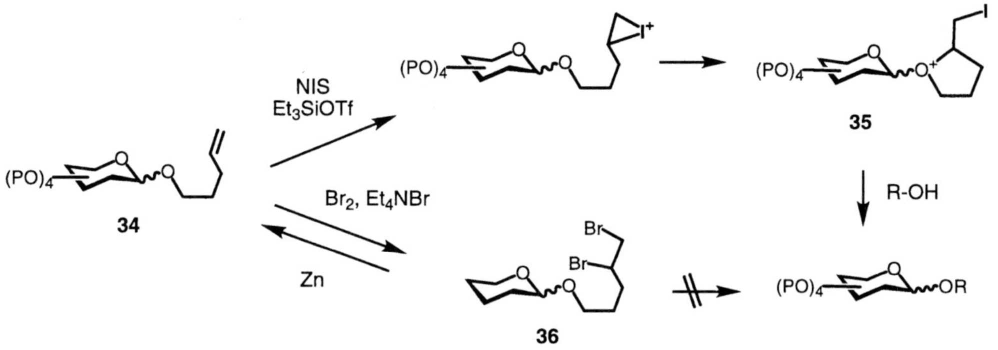

Scheme 9.11 *n*-Pentenyl glycoside as glycosyl donor.

Linear trisaccharide **44** (Scheme 9.12) was prepared from **43** in a similar manner, and was then used as the donor to couple with **37**. The tetrasaccharide product was selectively deprotected to give **45** and then subjected to fragment condensation, again with pentasaccharide **42**, to give nonasaccharide **46**.

9.3.4
Glycal Approach to Complex Oligosaccharides

Glycals, obtainable from glycosyl bromides by reduction with Zn, have proven versatile intermediates in complex oligosaccharide synthesis. Extensive works by Danishefsky and co-workers have convincingly demonstrated the versatility of glycal-based strategy [34]. One very impressive example is their recent synthesis of blood group H type 2 glycopeptide [35]. This synthesis is extremely rich in terms of its chemistry and clearly demonstrates the power of the glycal strategy. Important keys in glycal chemistry are: (1) the stereoselective oxidation of glycal **47** to epoxide **48** (1,2-anhydro sugar) [36], which can serve as a glycosyl donor through activation by Lewis acid ($ZnCl_2$, for example) to give the 1,2-*trans* glycoside **49**, and (2) the rearrangement of iodosulfonamide **50** to the 2-aminoglycoside **51** (Scheme 9.13).

Since the epoxide derived from trisaccharide glycal **52** was not satisfactory as a donor, it was first transformed into thioglycoside **53**. This was then coupled with disaccharide **54** with the aid of MeOTf to give **55**. The *C*-2 position of Glc was inverted by oxidation-reduction, and the product, now with a *β*-manno glycoside, was desilylated to give diol **56** (Scheme 9.14).

The selectively protected lactosamine **61** was prepared from lactal **57** in a highly efficient manner [37], in a procedure contrasting with the conventional azidonitration approach [38], which results in a mixture of stereoisomers. Thus, deacetylation, regioselective allylation by a Bu_2SnO-mediated process, and benzylation gave **58**. Successive treatment with IDCP (iodonium dicollidine perchlorate)/ $TMSCH_2CH_2SO_2NH_2$ gave 2-iodo-sulfonamide glycoside **59**, which was treated with thiolate to provide **60**. It was subsequently converted to phthalimide **61** to provide *β*-selective glycosylation (see below).

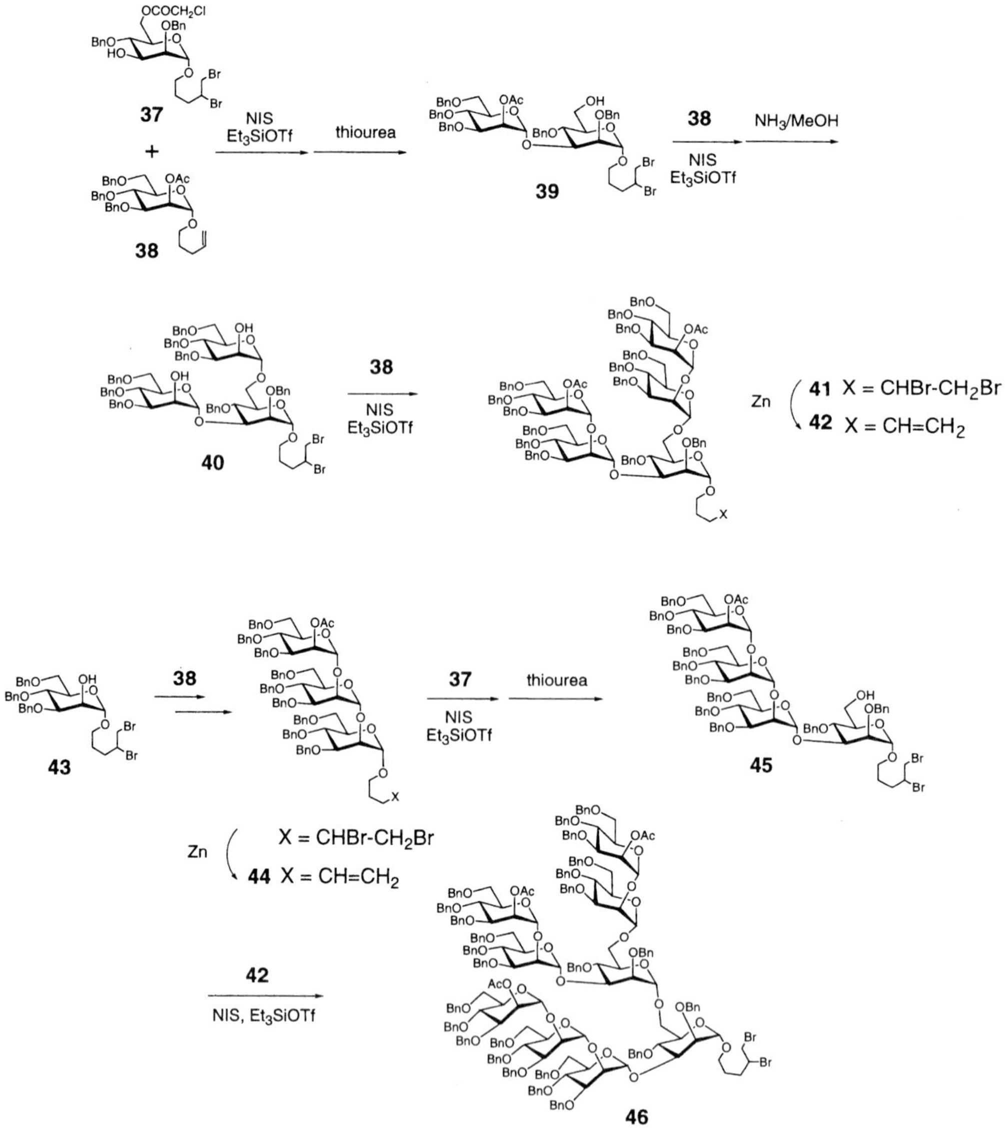

Scheme 9.12 *n*-Pentenyl glycoside approach to high-mannose-type oligosaccharides.

On the other hand, construction of the H-type 2 trisaccharide block took full advantage of the glycal assembly technology. Conversion of galactal **62** into the α-epoxide **63** by use of DMDO (dimethyldioxirane) proceeded in a stereoselective manner. Compound **63** was then coupled with glucal **64** in the presence of ZnCl$_2$ as an activator and the now liberated C-2 OH was glycosylated with fucosyl donor **65**. The product **66** was transformed into **67** in a manner similar to that described for **61**.

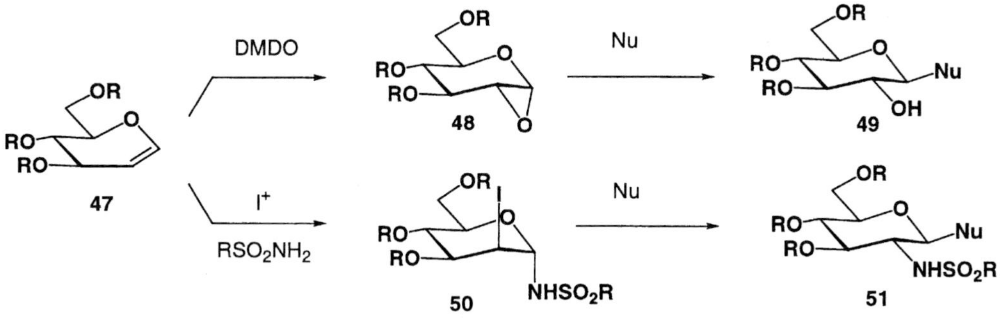

Scheme 9.13 Glycals as versatile intermediates.

Coupling of the first block **56** with **61** then gave a 9-mer (Scheme 9.15), which was regioselectively deprotected to afford **68**, and final coupling with **67** then gave the 15-mer **69**. The subsequent transformations are marvelously elaborate. Namely, conversion of Phth into NHAc was followed by iodosulfonamidation-hydrolytic rearrangement, global deprotection (with Na/NH$_3$), and N-acetylation to provide a free 15 mer. It was converted into glycosylamine **70**, which was in turn coupled with the aspartic acid-containing peptide by use of HOBt-HBTU as coupling agent to complete the synthesis of the glycopeptide.

9.3.5
Intramolecular Aglycon Delivery Approach

In the above approaches in this chapter, β-gluco derivatives are used as a precursor of β-manno glycosides. Intramolecular aglycon delivery (IAD) was first developed by Hindsgaul [39] and subsequently investigated by Stork [40] and by our group [41]. These methods utilize mixed acetonide **71**, dimethylsilyl ketal **72**, and p-methoxybenzylidene acetal **73** as the tethered intermediate (Scheme 9.16). Since the C-2 oxygen of Man is axially oriented, glycosyl transfer from these intermediates is possible only from the β-face, provided that intramolecularity of the reaction is assumed. It is a highly attractive approach for β-mannosylation, because it guarantees the exclusive formation of the correct stereoisomer [42]. In particular, p-methoxybenzyl-assisted (PMB-assisted) IAD seems to be the optimum, successfully providing Manβ1 → 4GlcNAc derivatives in ∼80% yields and in a strictly stereoselective manner [43]. This method was applied as the key transformation for the synthesis of a complex-type undecasaccharide bearing a terminal α2 → 3-linked sialic acid.

The trisaccharide core (Manβ1 → 4GlcNAcβ1 → 4GlcNAc) was synthesized from thioglycoside **74**, a highly optimized donor for β-mannosylation (Scheme 9.17). It was first treated with acceptor **75** in the presence of DDQ to afford mixed acetal **76**, which serves as the tethered intermediate for IAD. It was subsequently treated with MeOTf and 2,6-di-*tert*-butyl-4-methylpyridine (DTBMP) in 1,2-dichloroethane to afford an 78% yield (based on **75**) of the β-mannoside **77**, which was transformed into **78**.

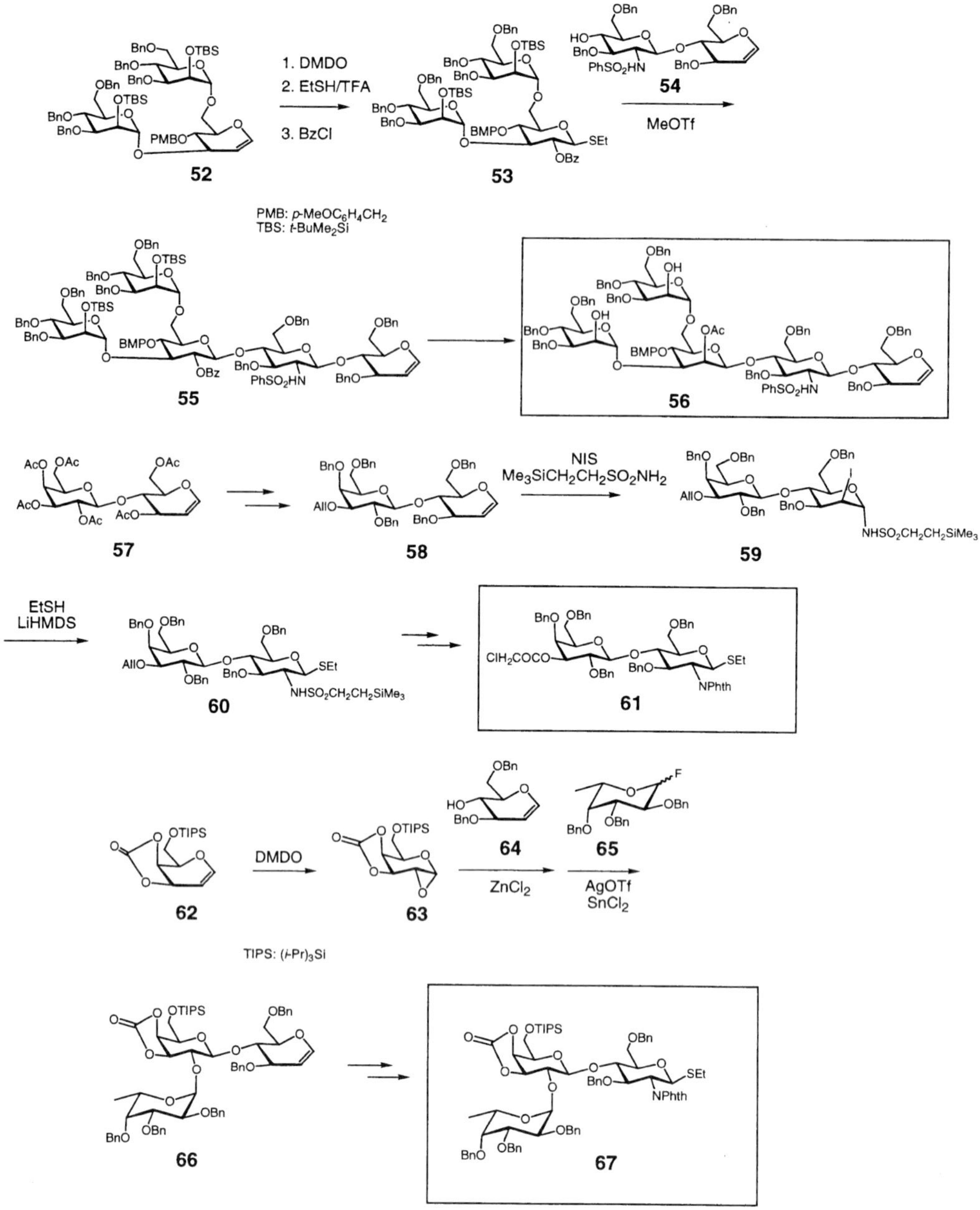

Scheme 9.14 Glycal approach to blood group H type 1 glycopeptide – I.

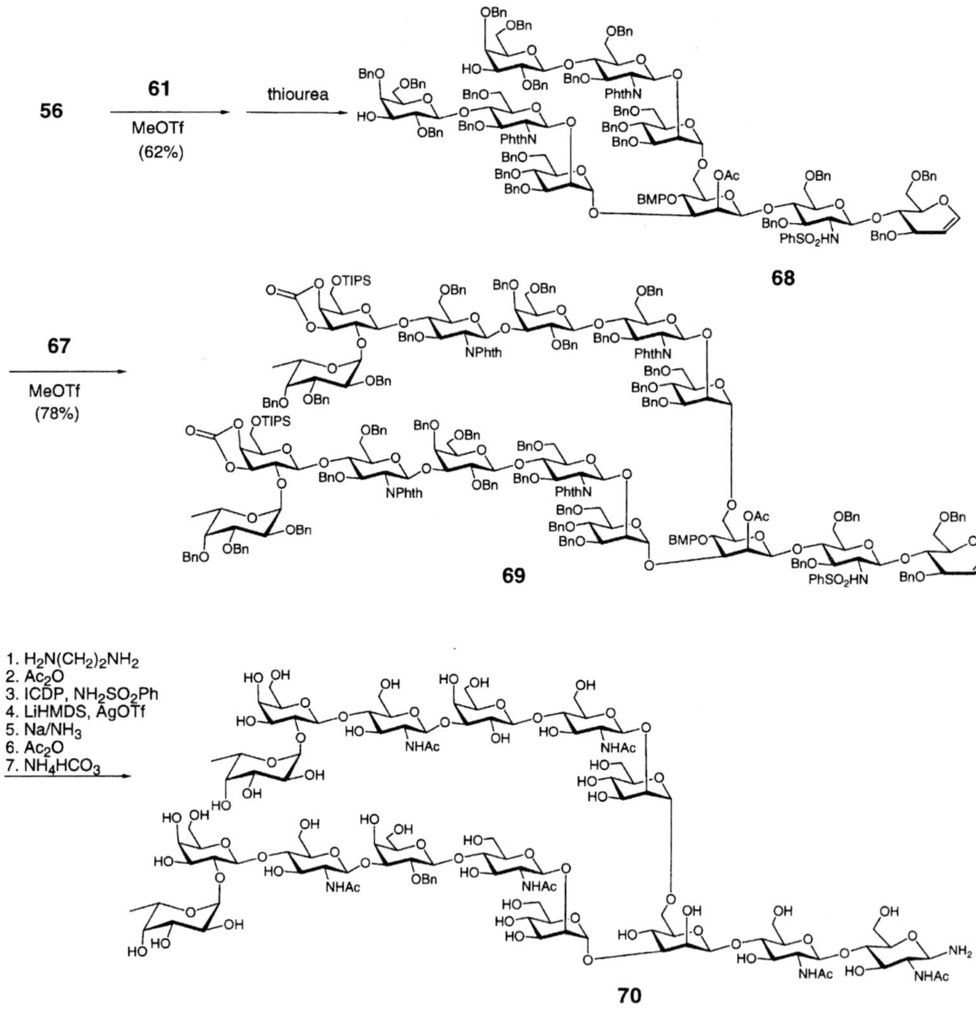

Scheme 9.15 Glycal approach to blood group H type 1 glycopeptide – II.

The sialic acid-containing tetrasaccharide (NeuAcα2 → 3Galβ1 → 4GlcNAcβ1 → 2Man) portion was prepared from disaccharides **79** and **80** (Scheme 9.18). The former component was synthesized by glycosylation with sialic acid donor **7**, based on the technique developed by Hasegawa, taking advantage of the solvent effect of acetonitrile, which is particularly useful for controlling the stereochemistry of sialylation. The sialic acid thioglycoside **7** was coupled with the galactose-derived acceptor **81** to give an α-glycoside, which was isolated as **79** in 57% yield after acetylation. The anomeric silyl group was removed and the liberated hemiacetal was converted into the trichloroacetimidate **82**.

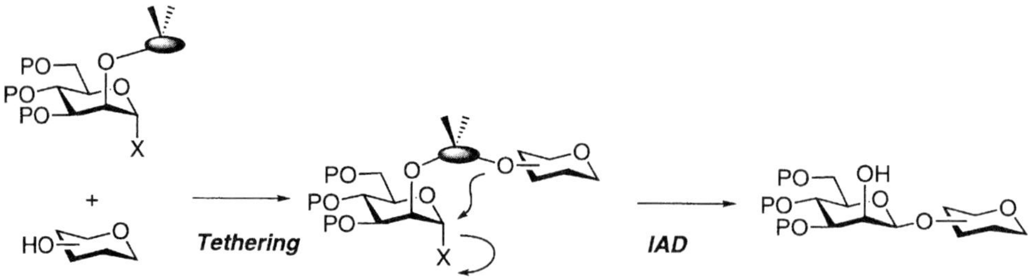

71~73

71	Me_2C
72	Me_2Si
73	p-C_6H_4CH

Scheme 9.16 Intramolecular aglycon delivery approach to β-manno glycosides.

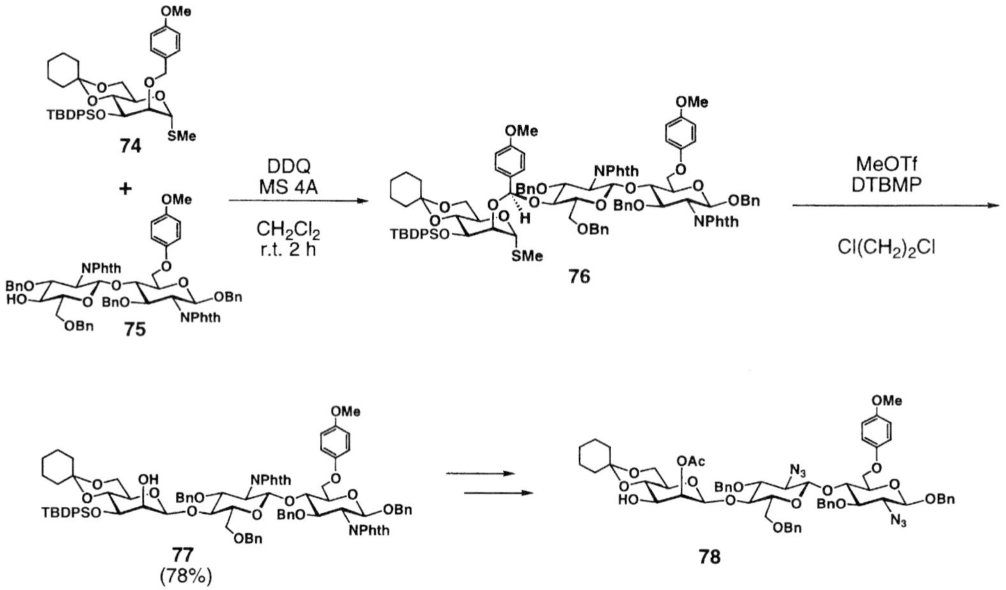

Scheme 9.17 Synthesis of disialylated undecasaccharide – I.

The preparation of the other disaccharide component **80** commenced with glycosylation of the mannose derivative **83** with the glucosamine donor **84** to afford **85**, which was converted to azide **80**. Coupling between **82** with **80** was performed with TMSOTf as an activator, and the product was transformed into tetrasaccharide donor **86** [44].

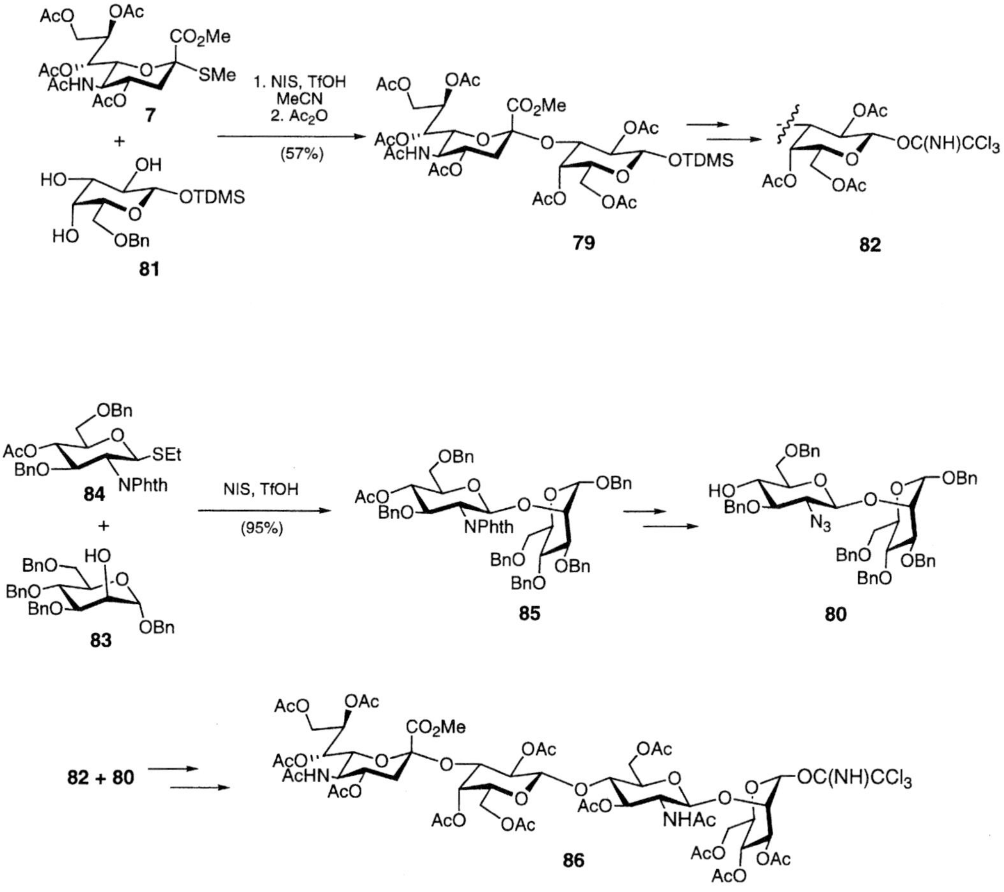

Scheme 9.18 Synthesis of disialylated undecasaccharide – II.

Glycosylation of core trisaccharide **78** with **86** with the aid of $BF_3 \cdot OEt_2$ (Scheme 9.19) was followed by acetylation and selective deprotection (to generate the Man 4,6-diol) to give **87**, which was coupled again with further **86**. Undecasaccharide **88**, obtained in 50% yield, was deprotected in three steps to give the free undecasaccharide **89** [19].

9.3.6
New Protecting Group Strategy

Ley and co-workers have developed a very concise strategy to produce the high-mannose-type glycan chain through the use of: (1) a novel hydroxy protecting group [45], and (2) chemoselective activation of a selenoglycoside [46]. Specifically, they exploited cyclohexane-1,2-diacetal, which is able to block *trans*-1,2-diols (3- and 4-OH of mannose, for example) selectively. The synthesis commenced with

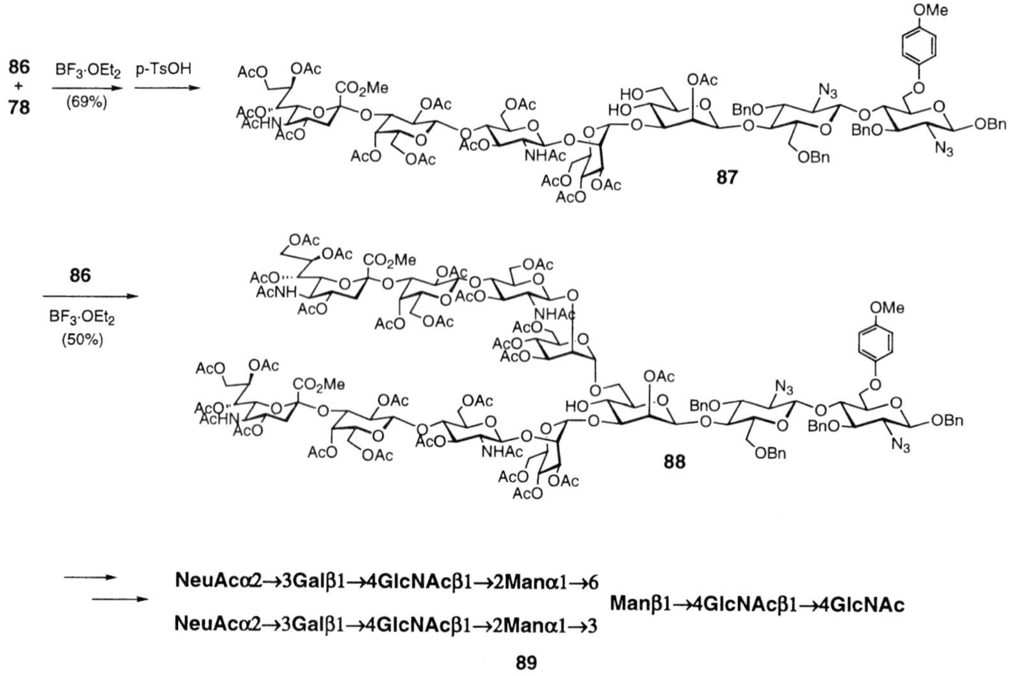

Scheme 9.19 Synthesis of disialylated undecasaccharide – III.

the coupling of selenoglycosides **90** and **91**. The former component can be selectively activated, because the cyclohexane-1,2-diacetal-protected compound has reduced donor reactivity. The product **92** was then used as the donor in reaction with thiomannoside diol **93** to give the branched pentasaccharide **94**. In this case, chemoselective activation of **92** was possible because of the higher reactivity of selenoglycoside in relation to thioglycoside. It was then used as the donor for combination with the linear tetramannoside fragment to afford **95** (Scheme 9.20).

9.3.7
Linear Synthesis of Branched Oligosaccharide

In general, a convergent strategy is advantageous over a stepwise one for the synthesis of complex molecules. This is very true for oligosaccharides, as shown in the sections above. In contrast, Seeberger's synthesis of high-mannose-type oligosaccharides, depicted in Scheme 9.21, used a linear strategy [47]. Although seemingly unusual, this synthetic strategy is in line with their aim of establishing a synthetic route adaptable for automated solid-phase synthesis [48], for which linear strategy is required.

The target nonasaccharide was retrosynthetically disconnected to three monosaccharide blocks **98**, **99**, and **101**. n-Pentenyl β-mannoside **98** was prepared from glycal

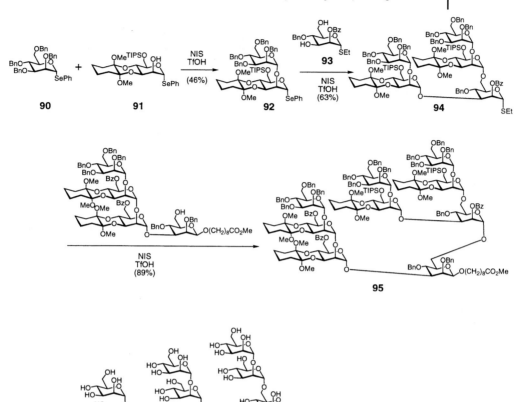

Scheme 9.20 Cyclohexane diacetal as a novel protecting group in oligosaccharide synthesis.

96. It was first converted into an epoxide and then treated with 5-pentenol and ZnCl$_2$. The resulting β-glucoside was converted into β-mannoside 97, after oxidation-reduction, and the *p*-bromobenzyl (PBB) group was selectively removed to give 98. This was glycosylated with trichloroacetimidate 99 and desilylated to give 100. Incorporation of 3,6-di-*O*-benzoylated mannosyl donor 101 was followed by two deacylation-glycosylation steps to give 102, which was fully deprotected to give 103.

9.3.8
Chemoenzymatic Approach to Complex-type Glycans

A majority of complex-type glycans carry terminal lactosamine (Galβ1 → 4GlcNAc), further decorated with α2 → 3- or α2 → 6-linked sialic acid (NeuAc). Since galactosyltransferase and α2 → 3 and α2 → 6 sialyltransferases are commercially available, their

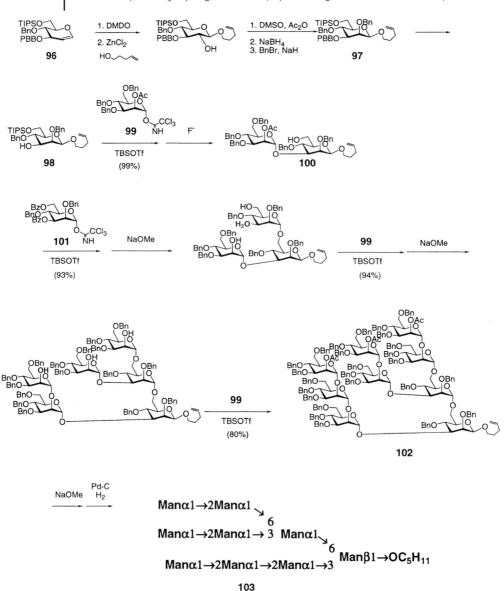

Scheme 9.21 Linear approach to a high-mannose-type oligosaccharide.

use in complex-type glycan synthesis is a logical choice [49]. In their systematic studies, Unverzagt et al. have extensively exploited such chemoenzymatic approaches [50] for the successful production of various types of Asn-linked glycans, as exemplified in Scheme 9.22 [51]. For β-mannoside construction, they adapted an intramolecular S_N2-type inversion developed by Kunz [52]. The core trisaccharide component

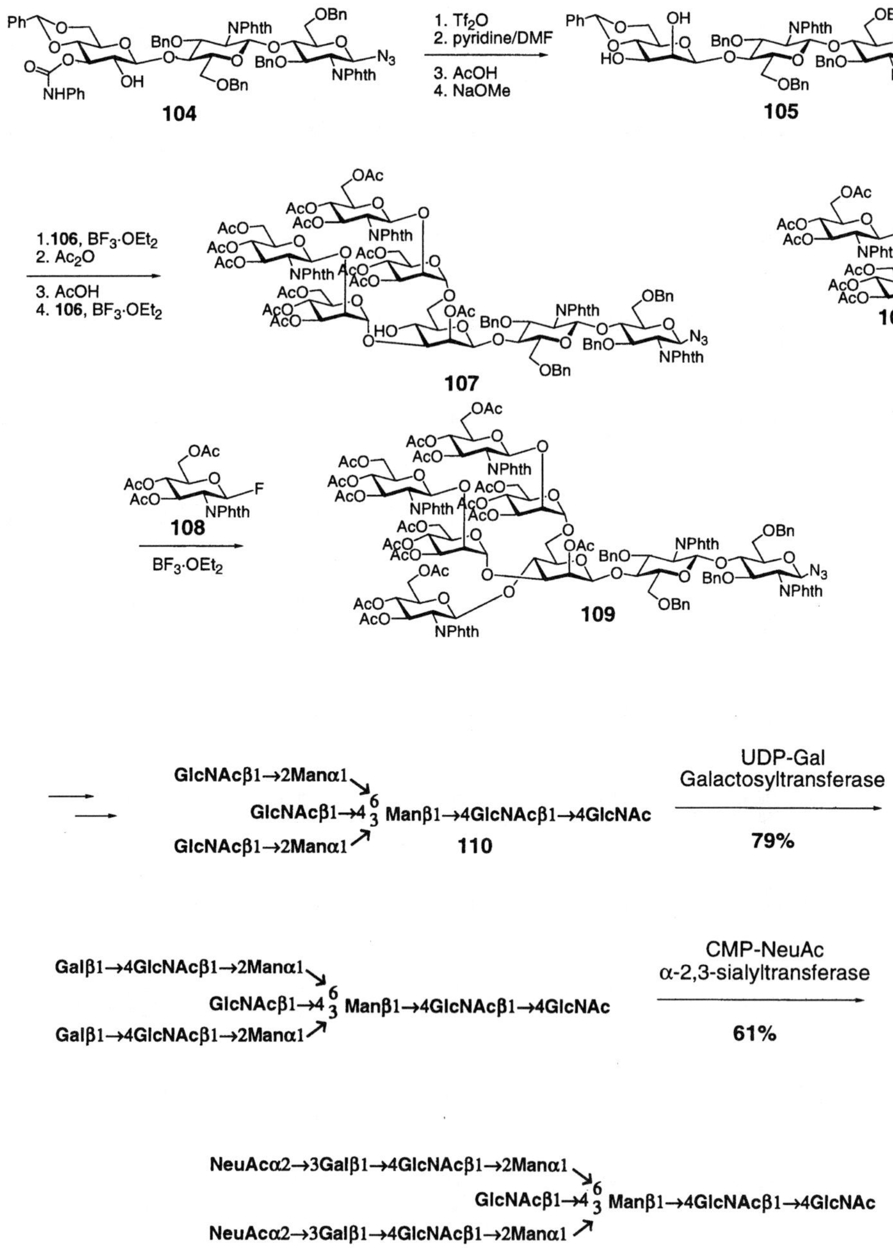

9.3 Chemical Synthesis of Complex Oligosaccharides | **277**

Scheme 9.22 Chemoenzymatic approach to a complex-type oligosaccharide.

was first constructed as β-glucoside **104**, with a carbamate moiety at C-3. The C-2 position was activated as a triflate, causing cyclization to give a carbonate that was cleaved under basic conditions to give **105**, now with the manno configuration.

A glycosylation-deprotection-glycosylation series with **106** as a donor afforded **107**. Incorporation of the central GlcNAc at the highly hindered C-4 position of Man was successfully accomplished by use of fluoride **108**. The resulting octasaccharide **109** was completely deprotected to afford **110**, which was subjected to enzymatic incorporation of galactose and sialic acid with UDP-galactose and CMP-sialic acid as glycosyl donors.

The costs of enzymes and sugar nucleotides may seem to be the bottleneck for such a strategy. However, it is still extremely useful for the terminal modification (for instance, incorporation of Gal, NeuAc, and Fuc) of glycan chains, at least on laboratory scales. It must also be stressed that advances in overexpression of glycosyltransferases and the development of sugar nucleotide recycling in the last decade are likely to allow the scaling up of the chemoenzymatic route [53].

9.4
References

1   R. A. Dwek, *Chem. Rev.* **1996**, *96*, 683–720.
2   C. R. Bertozzi, L. L. Kiessling, *Science.* **2001**, *291*, 2357–2364.
3   B. O. Fraser-Reid, K. Tatsuta, J. Thiem, *Glycoscience: Chemistry and Chemical Biology I-III*, Springer, **2001**.
4   J. F. G. Vliegenthart, F. Casset, *Curr. Opin. Struct. Biol.* **1998**, *8*, 565–571; L. Wells, K. Vosseller, G. Hart, *Science.* **2001**, *291*, 2376–2378, D. J. Moloney, V. M. Panin, S. H. Johnston, J. Chen, L. Shao, R. Wilson, Y. Wang, P. Stanley, K. Irvine, R. S. Haltiwagner, T. F. Vogt, *Nature.* **2001**, *406*, 369–375.
5   B. G. Davis, *J. Chem. Soc. Perkin Trans. 1*, **2000**, 2137–2160
6   N. Wymer, E. J. Toone, *Curr. Opin. Chem. Biol.* **2000**, *4*, 110–119, M. M. Palcic: *Curr. Opin. Biotech.* **1999**, *10*, 616–624, G. M. Watt, P. A. S. Lowden, S. L. Flitsch, *Curr. Opin. Struct. Biol.* **1997**, *7*, 652–660, Y. Ichikawa, G. C. Look, C.-H. Wong, *Anal. Biochem.* **1992**, *202*, 215–238; S. David, C. Augé, C. Gautheron, *Adv. Carbohydr. Chem. Biochem.* **1991**, *49*, 175–237.
7   G. M. Watt, L. Revers, M. C. Webberley, I. B. H. Wilson, S. L. Flitsch, *Angew. Chem. Int. Ed. Engl.* **1997**, *21*, 2354–2356.
8   T. W. Greene, P. G. M. Wuts, *Protective Groups in Organic Synthesis 3rd Ed.*, John Wiley and Sons, New York, **1999**.
9   G.-J. Boons, K. Hale, *Organic Synthesis with Carbohydrate* p. 26–55, Sheffield Academic, Sheffield, **2000**.
10 S. David, S. Hanessian, *Tetrahedron.* **1985**, *41*, 643–663
11 P. J. Garegg, H. Hultberg, S. Wallin, *Carbohydr. Res.* **1982**, *1108*, 97–101, M. P. Denino, J. B. Etienne, K. C. Duplantier, *Tetrahedron Lett.* **1995**, *36*, 669–672, M. Ek, P. J. Garegg, H. Hultberg, S. Oscarson, *J. Carbohydr. Chem.* **1983**, *2*, 305–311, M. Oikawa, W. C. Liu, Y. Nakai, S. Koshida, K. Fukase, K. Kusumoto, *Synlett.* **1996**, 1179–1180, L. Jiang, T.-H. Chan, *Tetrahedron Lett.* **1998**, *39*, 355–358.
12 H. Paulsen, *Angew. Chem. Int. Ed. Engl.* **1982**, *21*, 155–173.
13 N. B. Nicita, A. J. Petrescu, G. Negroiu, R. A. Dwek, S. M. Petrescu, *Chem. Rev.* **2000**, *100*, 4697–4711.
14 H. Paulsen, *Angew. Chem. Int. Ed. Engl.* **1990**, *29*, 823–839, G. Arsequell, G. Valencia, *Tetrahedron Asymmetry.* **1999**, *10*, 3045–3094.

15 H. G. Garg, K. von dem Brauch, H. Kunz, *Adv. Carbohydr. Chem. Biochem.* **1994**, *50*, 277–310.

16 J. J. Gridley, H. M. I. Osborn, *J. Chem. Soc. Perkin Trans I.* **2000**, 1472–1491.

17 Y. Ito, Y. Ohnishi, T. Ogawa, Y. Nakahara, *Synlett.* **1998**, 1102–1104.

18 D. Crich, S. Sun, *J. Org. Chem.* **1997**, *62*, 1198–1199, D. Crich, S. Sun, *Tetrahedron.* **1998**, *54*, 8321–8348.

19 J. Seifert, M. Lergenmüller, Y. Ito, *Angew. Chem. Int. Ed.* **2000**, *39*, 531–534.

20 G.-J. Boons, A. V. Demchenko, *Chem. Rev.* **2000**, *100*, 4539–4565, Y. Ito, M. Numata, M. Sugimoto, T. Ogawa, *J. Am. Chem. Soc.* **1989**, *111*, 8508–8510, C. De Meo, A. V. Demchenko, G.-J. Boons, *J. Org. Chem.* **2001**, *66*, 5490–5497.

21 A. Hasegawa, H. Ohki, T. Nagahama, H. Ishida, *Carbohydr. Res.* **1991**, *212*, 277–281.

22 H. Paulsen, R. Lebuhn, O. Lockhoff, *Carbohydr. Res.* **1982**, *103*, C7–C11.

23 H. Paulsen, R. Lebuhn, *Liebigs Ann. Chem.* **1983**, 1047–1072.

24 C. A. A. van Boeckel, T. Beetz, S. F. Aelst, *Tetrahedron.* **1984**, *40*, 4097–4107, F. Yamasaki, T. Nukada, Y. Ito, S. Sato, T. Ogawa, *Tetrahedron Lett.* **1989**, *30*, 4417–4420.

25 H. Paulsen, R. Lebuhn, *Carbohydr. Res.* **1984**, *130*, 85–101.

26 T. Ogawa, T. Kitajima, T. Nukada, *Carbohydr. Res.* **1983**, *123*, C5–C7.

27 T. Ogawa, T. Kitajima, T. Nukada, *Carbohydr. Res.* **1983**, *123*, C8–C11.

28 T. Ogawa, M. Sugimoto, T. Kitajima, K. K. Sadozai, T. Nukada, *Tetrahedron Lett.* **1986**, *27*, 5739–5742.

29 R. R. Schmidt, *Adv. Carbohydr. Chem. Biochem.* **1994**, *50*, 21–123.

30 S. Weiler, R. R. Schmidt, *Tetrahedron Lett.* **1998**, *39*, 2299–2302.

31 R. R. Schmidt, M. Behrendt, A. Toepfer, *Synlett.* **1990**, 694–696.

32 B. Fraser-Reid, U. Udodong, Z. Wu, H. Ottosen, J. R. Merritt, C. S. Rao, C. Roberts, R. Madsen, *Synlett.* **1992**, 927–942.

33 J. R. Merritt, E. Naisang, B. Fraser-Reid, *J. Org. Chem.* **1994**, *59*, 4443.

34 M. T. Bilodeau and S. J. Danishefsky, in *Modern Methods in Carbohydrate Synthesis*, S. H. Khan, R. A. O'Neill Eds. pp. 171–193, Harwood Academic, Amsterdam, **1996**.

35 Z.-G. Wang, X. Zhang, M. Visser, D. Live, A. Zatorski, U. Iserloh, K. O. Lloyd, S. J. Danishefsky, *Angew. Chem. Int. Ed. Engl.* **2001**, *40*, 1728–1732.

36 R. L. Halcomb, S. J. Danishefsky, *J. Am. Chem. Soc.* **1989**, *111*, 6661–6666.

37 S. J. Danishefsky, K. Koseki, D. A, Griffith, J. Gervay, J. M. Peterson, F. E. McDonald, T. Oriyama, *J. Am. Chem. Soc.* **1992**, *114*, 8331–8333.

38 R. U. Lemieux, R. M. Ratcliffe, *Can. J. Chem.* **1979**, *57*, 1244–1251.

39 F. Barresi, O. Hindsgaul, *J. Am. Chem. Soc.* **1991**, *113*, 9376–9377, *Synlett.* **1992**, 759–761, *Can J. Chem.* **1994**, *72*, 1447–1465.

40 G. Stork, G. Kim, *J. Am. Chem. Soc.* **1992**, *114*, 1087–1088, G. Stork, J. J. La Clair, *J. Am. Chem. Soc.* **1996**, *118*, 247–248.

41 Y. Ito, T. Ogawa, *Angew. Chem. Int. Ed. Engl.* **1994**, *33*, 1765–1767, A. Dan, Y. Ito, T. Ogawa, *J. Org. Chem.* **1995**, *60*, 4680–4681, Y. Ito, T. Ogawa, *J. Am. Chem. Soc.* **1997**, *119*, 5562–5566.

42 M. Lergenmüller, T. Nukada, K. Kuromachi, A. Dan, T. Ogawa, Y. Ito, *Eur. J. Org. Chem.* **1999**, 1367–1376.

43 Y. Ohnishi, H. Ando, T. Kawai, Y. Nakahara, Y. Ito, *Carbohydr. Res.* **2000**. *328*, 263–276.

44 J. Seifert, T. Ogawa, S. Kurono, Y. Ito, *Glycoconjugate J.* **2000**, *17*. 407–423.

45 P. Grice, S. V. Ley, J. Pietruszka, H. M. I. Osborn, H. W. M. Priepke, S. L. Warriner, *Chem. Eur. J.* **1997**, *3*, 431–440, S. V. Ley, D. K. Baeschlin, D. J. Dixon, A. C. Foster, S. J. Ince, H. W. M. Prieoke, D. J. Reynolds, *Chem. Rev.* **2001**, *101*, 53–80.

46 S. Mehta, B. M. Pinto, *Tetrahedron Lett.* **1991**, *32*, 4435–4438.

47 D. M. Ratner, O. J. Plante, P. H. Seeberger, *Eur. J. Org. Chem.* **2002**, 826–833.

48 P. H. Seeberger, W.-C. Haase, *Chem. Rev.* **2000**, *100*, 4349–4393.

49 O. Blixt, J. Brown, M. Schur, W. Wakarchuk, J. C. Paulson, *J. Org. Chem.* **2001**, *66*, 2422–2448.

50 C. Unverzagt, *Carbohydr. Res.* **1998**, *305*, 423–431, *Angew. Chem. Int. Ed. Engl.* **1996**, *35*, 2350–2352.
51 C. Unverzagt, J. Seifert, *Tetrahedron Lett.* **2000**, *41*, 4549–4553.
52 H. Kunz, W. Günther: Angew. Chem. **1988**, 100, 1118–1119, Angew. Chem. Int. Ed. Engl. **1988**, *27*, 1086–1087, C. Unverzagt, *Angew. Chem. Int. Ed. Engl.* **1994**, *33*, 1102–1104.
53 W. Fitz, C.-H. Wong, Preparative Carbohydrate Chemistry, S. Hanessian Ed. pp. 485–504, Marcel Dekker, New York, 1997.

10
Chemistry and Biochemistry of Asparagine-Linked Protein Glycosylation

BARBARA IMPERIALI and VINCENT W.-F. TAI

10.1
Protein Glycosylation

10.1.1
Introduction

Glycosylation is potentially the most complex category of protein modification reactions in eukaryotic systems. A remarkable degree of diversity is introduced, due to the wide array of available monosaccharide building blocks, as well as the potential for different chemical linkages between each pair of carbohydrates. The size of each oligosaccharide chain can also vary greatly, ranging from simple monosaccharides to complex branched structures composed of as many as 40 saccharide units. In addition, many proteins are glycosylated at multiple sites with different carbohydrate groups, generating further diversity. In many cases, enzyme-catalyzed protein glycosylation affects both the structural framework [1, 2] and the functional capabilities [3–5] of the modified protein. Glycoproteins have been implicated in processes as varied as the immune response, proper intracellular targeting, intercellular recognition and protein folding, stability, and solubility. The major carbohydrate modifications of proteins fall into three general categories: N-linked modification of asparagine [6, 7], O-linked modification of serine or threonine [8], and glycosylphosphatidyl inositol derivatization of the C-terminus carboxyl group [9]. One or more enzymes that demonstrate different peptide sequence requirements and reaction specificities catalyze each of these transformations. Other novel carbohydrate modifications, including the C-mannosylation of tryptophan [10], have also recently been documented.

10.1.2
Asparagine-Linked Glycosylation and Oligosaccharyl Transferase

This chapter focuses on the chemistry and biochemistry of asparagine-linked protein glycosylation, the most common eukaryotic glycosylation reaction. A single membrane-associated enzyme, oligosaccharyl transferase (OT), catalyzes N-linked glycosylation. The transformation involves the co-translational transfer of a saccharide from a dolichol-linked pyrophosphate donor to an asparagine side chain (in the consensus sequence Asn-Xaa-Ser/Thr) within a nascent polypeptide. The reaction is illustrated in Fig. 10.1. The preferred saccharide in eukaryotes, both in vitro and in vivo, is the triantennary branched structure -GlcNAc$_2$-Man$_9$-Glc$_3$ [7] (except in trypanosomatid protozoa, which lack Dol-P-Glc synthase and therefore transfer the -GlcNAc$_2$-Man$_9$ saccharide [11, 12]). The central amino acid of the polypeptide substrate, Xaa, can be any of the encoded residues except proline [13]. With regard to the hydroxyamino acid, in vivo, threonine-containing sequences are almost three times more likely than the corresponding serine-containing peptides to be glycosylated [14]. However, the efficiency of Asn-Xaa-Thr glycosylation in vitro exceeds that of the Asn-Xaa-Ser sequence by as much as 40-fold [15].

The cellular location of this co-translational modification is on the lumenal face of endoplasmic reticulum (ER) membrane, and the polypeptide is glycosylated while being biosynthesized on membrane-associated ribosomes docked on the cytoplasmic face of the ER membrane (Fig. 10.2). Approximately 14 residues of the nascent peptide must clear the lumenal surface of the ER membrane before oligosaccharyl transferase-mediated glycosylation can occur [16]. This suggests that the OT active site is positioned 30–40 Å away from the surface of the ER membrane. These studies further suggest that the active site of the enzyme resides in the large soluble domains of the enzyme subunits. The subsequent diversification of the protein glycoconjugates formed in the OT-catalyzed step arises from the collective action of a series of glycosyl hydrolase and glycosyltransferase processing steps that occur in the ER and Golgi apparatus *after* the addition of the initial triantennary tetradecasaccharide. The transferase enzymes catalyze the incorporation of a broader diversity of carbohydrate units, including fucose, sialic acid, and galactose. These enzymes therefore generate the structural diversity associated with mature *N*-linked glycoproteins in eukaryotic cells [17].

To date, oligosaccharyl transferases from a number of species have been biochemically characterized. In general, the oligosaccharyl transferase enzymes are multimeric, membrane-associated enzymes in which the functional domains of the enzyme are localized in the lumen of the ER. Evidence now confirms that this

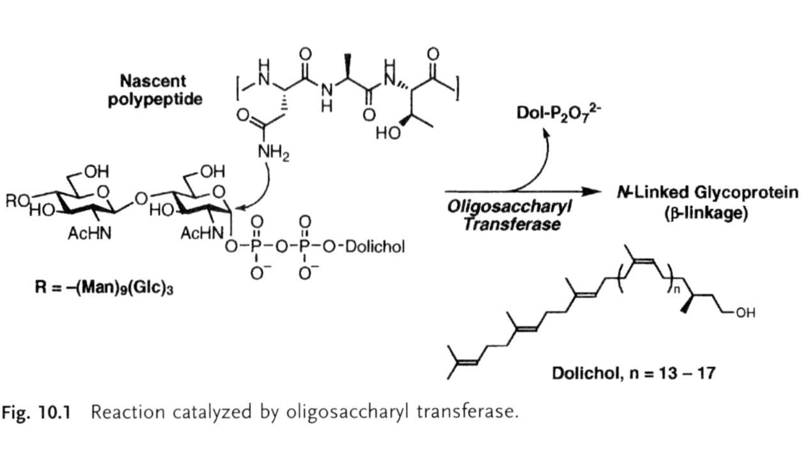

Fig. 10.1 Reaction catalyzed by oligosaccharyl transferase.

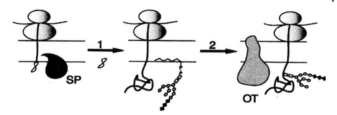

Fig. 10.2 Cellular location of asparagine-linked glycosylation. SP=signal peptidase, OT=oligosaccharyl transferase, 1=protein synthesis, 2=oligosaccharyl transferase-catalyzed protein glycosylation.

enzymatic step appears to be conserved throughout eukaryotic evolution from yeast to mammals [7, 18].

The system that has been studied in the most detail is that from the yeast *Saccharomyces cerevisiae* [7, 18, 19]. A total of nine distinct subunits (Ost1p, Ost5p, Swp1p, Wbp1p, Ost2p, Stt3p, Ost4p, Ost3p, and Ost6p) have been associated with the activity of the yeast enzyme. Each subunit includes at least one trans-membrane hydrophobic domain, and several of the subunits are glycosylated (Ost1p, Wbp1p, and Stt3p). Additionally, many of the subunits include amino terminus signal sequences that are cleaved prior to complete maturation of the protein. Several of the subunits are glycosylated with high-mannose N-linked oligosaccharides, indicating that mature OT is a self-processing enzyme. Genetic knockout experiments have revealed that five of these subunits (Ost1p, Swp1p, Wbp1p, Ost2p, and Stt3p) are essential for cell viability and absolutely required for in vivo OT activity, while the remaining subunits appear to influence the glycosylation efficiency, but are not absolutely essential for catalytic activity. A schematic of the current understanding of the subunit composition of yeast OT is presented in Fig. 10.3. Biochemical studies to date have also revealed that the OT complex is organized into three sub-complexes [20], comprising Ost1p-Ost5p, Swp1p-Wbp1p-Ost2p, and Stt3p-Ost4p-Ost3p(or Ost6p). Current efforts in OT biochemistry are focusing on defining the functional roles of each of the subunits of the enzyme complex and the significance of each of the sub-complexes.

10.2
Small-Molecule Probes of the Biochemistry of Oligosaccharyl Transferase

The complexity of the oligosaccharyl transferase enzyme and the limited availability of pure protein, due to low endogenous expression levels, has placed a heavy reliance on the application of small-molecule probes to gain insight into the function of this central and essential eukaryotic process. These studies include the development of synthetic probes to label essential subunits covalently, as well as substrate analogues and inhibitors targeted at both the peptide and carbohydrate binding sites of the enzyme.

284 | 10 Chemistry and Biochemistry of Asparagine-Linked Protein Glycosylation

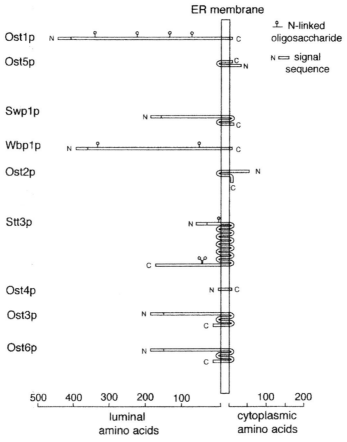

Fig. 10.3 Subunit composition of yeast (*S. cerevisiae*) oligosaccharyl transferase complex illustrating predicted membrane orientation. Taken from ref. 7, with permission.

10.2.1
Photoaffinity and Affinity Labeling of Oligosaccharyl Transferase

The goal in affinity labeling studies is to develop probes that show structural similarity to the native substrates for an enzyme, but also include reactive moieties that can covalently label functionality once bound to the cognate protein. The reactive functionality is in general represented by an electrophilic moiety that can label a nucleophilic functionality, or a photolabile group that, once activated by photolysis, reveals a reactive group that can target any nearby residues. The advantage of the latter strategy is that it is not so selective, so the likelihood of a covalent modification event is greater. Inevitably, the disadvantage of this strategy is that the labeling may be less selective.

10.2 Small-Molecule Probes of the Biochemistry of Oligosaccharyl Transferase

While the native peptide substrate for *N*-linked glycosylation is a nascent polypeptide, studies on OT have been enabled by the fact that even simple *N*- and *C*-terminal-capped tripeptides can serve as substrates for the enzyme [13]. It has therefore been possible, through chemical synthesis, to prepare a wide variety of substrate analogues as conformational and mechanistic probes of OT.

Early attempts to label oligosaccharyl transferase with photoaffinity probes based on the peptide substrate were carried out on crude enzyme preparations with a peptide that included a *p*-azidobenzoyl-modified lysine residue as the Xaa residue in the substrate peptide N^α-[^{3}H]Ac-Asn-Xaa-Thr-CONH$_2$ [21] (see **1**, Fig. 10.4). The precursor aryl azide is converted into a reactive aryl nitrene species upon irradiation at 254 nm. These studies were carried out in an attempt to identify the oligosaccharyl transferase subunit that bound to the peptide substrate. While labeling of a 60 kDa protein, together with a modest reduction in OT activity ($\sim 15\%$), was observed in these studies, Lennarz and co-workers ultimately established that the loss of activity was not actually associated with OT inactivation but rather that the protein that was modified was the peptide-binding ER protein – protein disulfide isomerase (PDI) [22, 23]. These studies were complicated by the use of a crude enzyme preparation, together with poor affinity of the aryl azide for OT. More recently, new photoaffinity labeling studies have been carried out with a modified probe including a *p*-benzoylphenyl alanine as the Xaa residue in the OT substrate peptide (see **2**, Fig. 10.4) [24]. The design of this probe is better than that of **1** because there is a shorter tether between the peptide-binding fragment and the photochemically activated unit. Other advantages of the benzophenone moiety are that it is chemically more stable than the aryl azide precursors, and it is activated by light of longer wave-

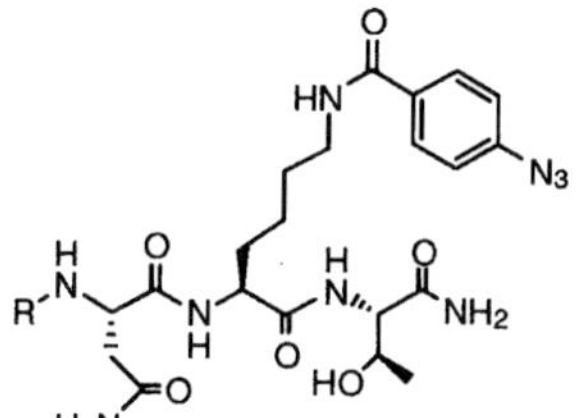

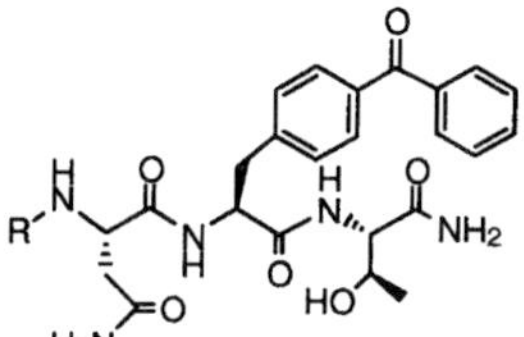

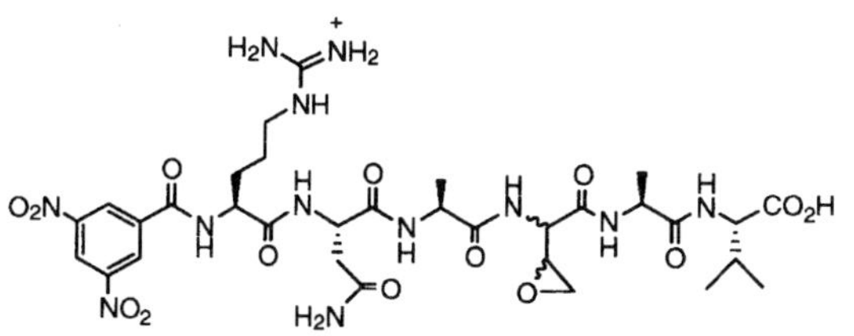

Fig. 10.4 Photoaffinity and affinity labeling agents for oligosaccharyl transferase.

length (350 nm), which is less damaging to proteins than the wavelength used for the aryl azide activation and less likely to cause nonspecific inactivation. Finally, the activated ketyl radical intermediate reacts preferentially with unreactive C–H bonds even in the presence of aqueous solvent and bulk nucleophiles [25]. Photoinactivation studies with **2b** in the presence of crude yeast microsomes showed specific labeling of two proteins of molecular mass 62 kDa and 64 kDa, which correspond to two glycoforms of the Ost1p subunit of OT. The identity of the labeled protein was confirmed by use of an Ost1p knockout yeast strain that included an introduced epitope-tagged analogue of Ost1p. These studies therefore establish that the large type I membrane protein subunit of OT, Ost1p, includes the peptide-binding site for asparagine-linked glycosylation activity.

The development of affinity-labeling agents with electrophilic functionality has been more limited, due to the stringent requirements for peptide binding to OT. Specifically, replacement of key functionality either in the asparagine or in the hydroxyamino acid residues commonly abolishes binding to the enzyme. One exception to this is the study of epoxyethylglycyl peptides as inactivators of porcine liver OT [26]. Bause and co-workers have found that peptide **3** inactivates OT in a time-dependent and concentration-dependent manner. Additionally, the inactivation depends on the presence of the dolichylpyrophosphate glycosyl donor. In this case, a dinitrobenzoyl group was introduced into the peptide as a hapten to enable identification of covalently modified subunits. Bause and co-workers found that the asparagine-containing epoxyethylglycine peptide covalently labeled both a 48 kDa and a

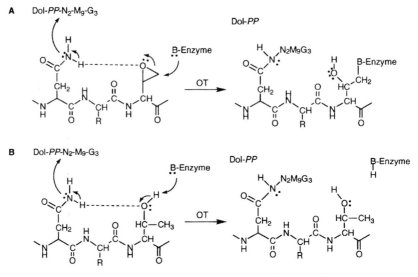

Fig. 10.5 Proposed mechanism for inactivation of oligosaccharyl transferase by epoxyethylglycine-containing peptides (reproduced from ref. 26, with permission). A = Suicide inactivation resulting in subunit double labeling. B = Catalytic mechanism of OT according to Bause.

66 kDa subunit of OT. These porcine liver OT subunits correspond to ribophorin I and OST48, which are homologous to the yeast subunits Ost1p and Wbp1p. When inactivation was carried out in the presence of radiolabeled glycosyl donor, it was found that the inactivated subunits included radioactivity corresponding both to the radiolabeled sugars as well as to the dinitrobenzoyl marker. This result indicates that the inactivating peptide may be glycosylated through the native catalytic machinery as it is reacting with a nucleophilic active site residue. This is illustrated in the mechanism proposed by Bause and co-workers [26] (Fig. 10.5). While one of the two subunits modified in these experiments corresponds to the analogous subunit that was labeled in the photoaffinity studies, it is not clear at present why these electrophilic probes label more than one subunit of OT.

10.2.2
Investigation of Peptide-Based Substrate Analogues as Inhibitors of Oligosaccharyl Transferase

10.2.2.1 Inhibitors of N-Linked Glycosylation and Glycoprotein Processing

Inhibitors targeting oligosaccharyl transferase or the substrates or products of this enzyme can disrupt the biosynthesis of N-linked glycoproteins. As shown in Fig. 10.6, the glycosyl donor, Dol-PP-GlcNAc$_2$Man$_9$Glc$_3$, is biosynthesized from do-

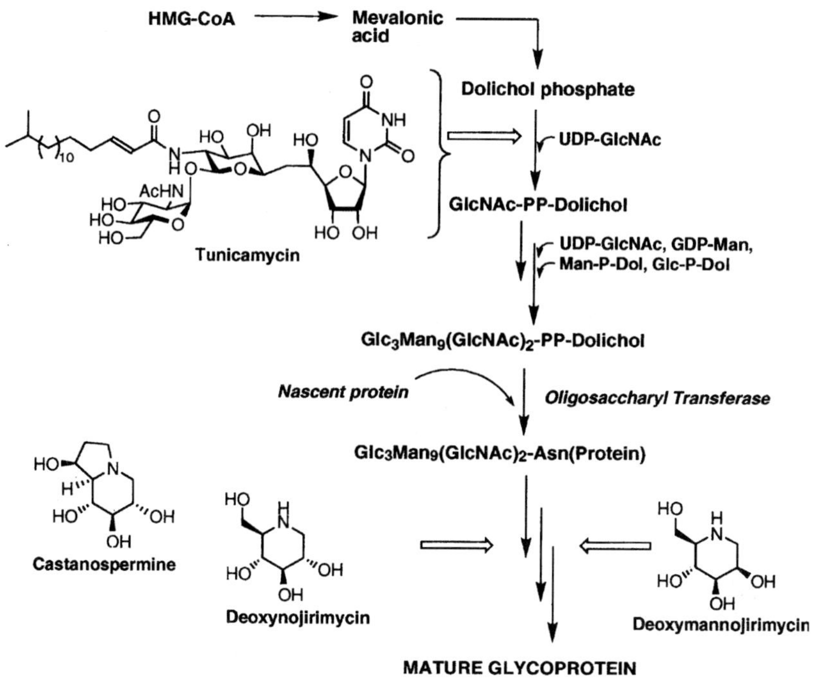

Fig. 10.6 Survey of N-linked glycosylation inhibitors.

lichol phosphate by the dolichol pathway. Inhibition of any of the enzymes along this pathway will diminish the supply of the saccharide substrate available for N-linked glycosylation. Currently, the microbial antibiotic tunicamycin is the most widely used inhibitor for N-linked glycosylation [27]. Tunicamycin acts by inhibiting GlcNAc phosphotransferase, the first step of the dolichol pathway, which transfers GlcNAc monophosphate from UDP-GlcNAc to dolichol monophosphate. However, use of tunicamycin requires several cell cycles before depletion of the saccharide donor to arrest protein glycosylation. This inhibition is neither direct nor specific for protein glycosylation; moreover, tunicamycin is both too toxic to be useful as a drug and too structurally complex for facile synthetic manipulation. Additionally, there are a number of natural products that also target processing steps in the N-linked glycosylation pathway. Since neither of the strategies for inhibiting N-linked glycosylation directly targets OT, there has been considerable interest in the development of inhibitors that specifically target the biochemical step that results in glycoconjugate formation.

10.2.2.2 Peptide-Based Analogues and Inhibitors

The development of mechanistic probes and inhibitors of OT has taken advantage of the simple peptide substrate requirements for the enzyme. For example, the design, synthesis, and evaluation of conformationally constrained oligopeptides that include the asparagine and a hydroxyamino acid have resulted in the proposal that OT recognizes peptide substrates in a specific Asx-turn conformational motif [28, 29]. This motif is characterized by a 10-membered ring hydrogen-bonding network involving the asparagine side chain and the backbone amide hydrogen of the hydroxyamino acid.

Peptides including a number of synthetic, non-encoded amino acids in the tripeptide recognition motif for OT have been investigated. Bause and co-workers have demonstrated that OT is highly specific for the hydroxyamino acid in the Asn-Xaa-Thr/Ser sequence. Studies with porcine liver OT and a series of peptide analogues show that replacement of Yaa in the tripeptide N-benzoyl-Asn-Gly-Yaa-NHCH$_3$ with virtually any residue other than serine or threonine greatly compromises acceptor activity in OT-catalyzed glycosylation [30, 31]. For example, incorporation of cysteine and L-*threo*-hydroxynorleucine as Yaa results in 375-fold and 750-fold reductions in rate, respectively. Similarly, the enzyme is extremely sensitive to the stereochemistry at both the α- and the β-carbon of the hydroxyamino acid. Integration of any other diastereomers of L-threonine as Yaa effects a precipitous drop in glycosylation rate, suggesting that OT has a highly stereospecific hydrophobic binding pocket for the threonine side chain in the glycosylation tripeptide sequon.

A number of peptide analogues in which the critical asparagine has been replaced with an alternative residue have also been prepared [30, 32, 33]. A summary of the results from these studies is presented in Fig. 10.7. Many of the peptides examined (**4–10**) fail to demonstrate any binding to OT even at elevated concentrations (10 mM). Substitution of the native amide functionality with a thio-

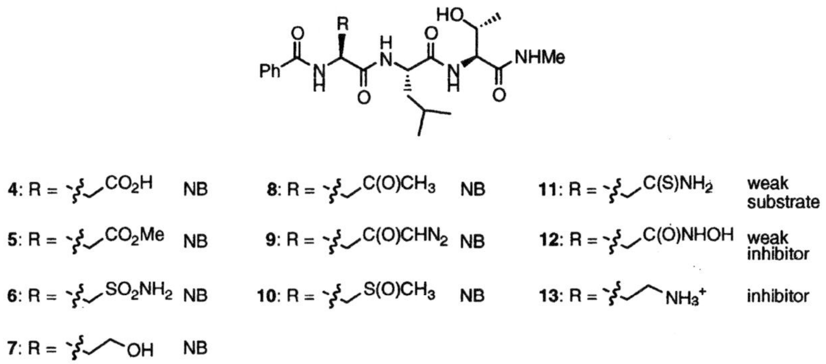

10.2 Small-Molecule Probes of the Biochemistry of Oligosaccharyl Transferase | 289

4: R = CO₂H   NB
5: R = CO₂Me   NB
6: R = SO₂NH₂   NB
7: R = OH   NB

8: R = C(O)CH₃   NB
9: R = C(O)CHN₂ NB
10: R = S(O)CH₃ NB

11: R = C(S)NH₂   weak substrate
12: R = C(O)NHOH   weak inhibitor
13: R = NH₃⁺   inhibitor

Fig. 10.7 Summary of peptide analogue studies in which asparagine is replaced with alternative synthetic, non-encoded amino acids (NB = no binding) [30, 32, 33].

amide afforded a peptide with comparable K_m but reduced turnover [32]. This compound was later examined in detail in an effort to gain insight into the metal cation dependency of N-linked glycosylation [34]. Integration of a hydroxylamine moiety in place of the asparagine amide afforded a compound with weak inhibitory activity [30].

By far the most interesting tripeptide analogue included replacement of the carboxamide functionality of asparagine with the corresponding reduced amine species from the residue 1,4-diaminobutanoic acid (Dab). Peptides including this residue showed modest inhibitory properties [32]. For example, the linear tripeptide N-benzoyl-Dab-Leu-Thr-NHMe (**13**) shows competitive inhibition of yeast OT with a K_i of approximately 1 mM. Constraint of the backbone conformation, to lock the peptide into an Asx-turn, improved binding, affording a tripeptide (**14**) with a 100 µM K_i (Fig. 10.8) [35].

Further improvement of peptide inhibitory properties was made possible by considering the fact that glycosylation substrates in vivo are not simple tripeptides but, rather, extended sequences with numerous residues flanking the tripeptide sequon in both C- and N-terminal directions. As the structure of oligosaccharyl transferase remains unknown, it was not possible to use a rational approach to design the extended binding site. Statistical studies of the sequences of N-linked glycoproteins were therefore used as a guide for defining the ideal residues adjacent

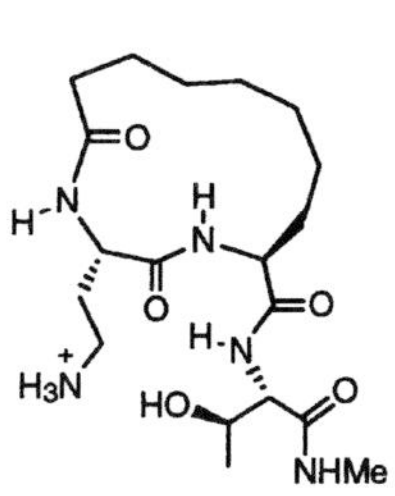

Fig. 10.8 Prototype inhibitor of oligosaccharyl transferase.   **14**

to the threonine of the tripeptide sequon [14]. The next phase of inhibitor development thus included elongation of the peptidyl structures to provide extended binding determinants for interaction with oligosaccharyl transferase. Modification of compound **14** to include flanking *N*-terminal residues was not synthetically straightforward with the macrocyclic structure, but extension in the *C*-terminal direction was feasible. In order to facilitate these studies, an efficient solid-phase synthesis of this class of inhibitors was developed [35, 36]. Through use of this chemistry, a family of sixteen pseudohexapeptides was prepared for evaluation as inhibitors of oligosaccharyl transferase. All the inhibitors included a key cyclic tripeptide core with the Dab amino acid as well as a nitrophenylalanine residue at the *C*-terminus to allow facile and accurate quantification of inhibitor concentrations. The principal variation amongst the inhibitor structures was the identity of the flanking amino acid residues. Basic (lysine), acidic (glutamic acid), neutral (valine), and polar, uncharged (threonine) residues at each of the variable positions were evaluated (Fig. 10.9). Additionally, the peptides were all examined with both a fungal (yeast) and mammalian (porcine liver) oligosaccharyl transferase.

The results of inhibition studies on the family of sixteen peptides were quite dramatic. The estimated K_i values for each of the inhibitors for oligosaccharyl transferase from both yeast and porcine liver are reported in Tab. 1. The K_i values range from high µM to low nM, with optimum binding observed when the residues at positions 4 and 5 are valine and threonine (K_i = 25 nM for yeast OT and 30 nM for porcine liver OT). In all cases, placement of the basic residue, lysine, at either position 4 or 5 significantly worsens binding. The results from these studies are in excellent agreement with the statistical analyses of Gavel and von Heijne [14] in their systematic study of peptides sequences of known glycopro-

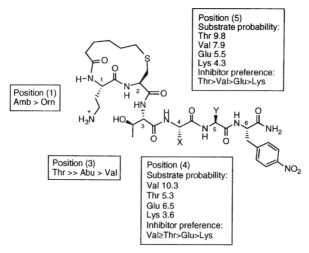

Fig. 10.9 Summary of oligosaccharyl transferase inhibitors, illustrating amino acid preferences in positions 1, 3, 4, and 5, with variants of peptide **14** (taken from ref. 36, with permission).

Tab. 10.1 Equilibrium dissociation constants (K_i) for yeast and porcine liver oligosaccharyl transferase inhibition by cyclo(hex-Amb-Cys)-Thr-Xaa-Yaa-Nph-NH$_2$ variants (taken from ref. 36, with permission).

Group	Xaa-Yaa	K_i (μM) yeast OT	K_i (μM) porcine OT
a	Lys-Lys	8.5 ±1.5	14 ±4
a	Glu-Lys	4.3 ±0.5	4.5 ±1.2
a	Glu-Glu	4.3 ±0.8	4.0 ±0.8
a	Lys-Glu	3.2 ±0.6	5.5 ±1.5
a	Thr-Glu	1.6 ±0.1	2.0 ±0.2
a	Val-Glu	0.95±0.15	1.1 ±0.1
b	Thr-Lys	0.75±0.07	2.7 ±0.3
b	Val-Lys	0.60±0.20	2.6 ±0.4
c	Glu-Val	0.60±0.08	0.68±0.075
c	Lys-Val	0.55±0.14	1.40±0.10
c	Lys-Thr	0.43±0.07	0.31±0.06
c	Glu-Thr	0.31±0.05	0.31±0.06
c	Thr-Thr	0.14±0.02	0.26±0.02
c	Thr-Val	0.10±0.03	0.41±0.09
c	Val-Val	0.09±0.015	0.22±0.04
c	Val-Thr	0.025±0.008	0.03±0.006

a variants with $K_i \geq 1$ μM. *b* variants showing the highest selectivity between yeast and porcine OT, *c*. variants with $K_i \leq 1$ μM.

teins in the protein literature. At position 4, for example, the most frequently observed residue is valine, and indeed the same residue was found to be optimum in the inhibitor study. Similarly, the least frequently observed residue is lysine, and it is this residue that is found in this position in the poorest of the inhibitors.

An important observation in these studies concerns the efficacy of the inhibitors with the oligosaccharyl transferase preparations from mammalian and fungal sources. Detailed genetic analysis of oligosaccharyl transferase from a number of different species reveals a complex oligomeric architecture that appears to be highly conserved throughout eukaryotic evolution [18]. It would therefore be anticipated that the cyclic tripeptide core should exhibit little species selectivity. However, this situation may be different with the extended binding determinants, because there is less evolutionary pressure for the residues that interact with these determinants to be conserved. It was observed in some cases (for example, residues 4 and 5=Thr-Lys and Val-Lys) that there was about a fourfold difference in the inhibition potency. These data suggest that it may be possible to design species-specific inhibitors of N-linked glycosylation by exploiting the extended binding determinants. Recently, more potent derivatives of the inhibitors discussed have been reported [37]; the most potent compound **15** shows a K_i of 10 nM against *S. cerevisiae* OT (Fig. 10.10).

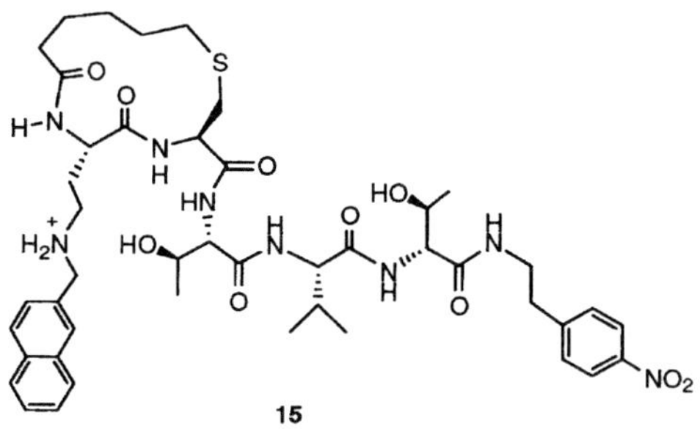

Fig. 10.10 A potent inhibitor of *S. cerevisiae* oligosaccharyl transferase [37].

10.2.2.3 Interim Summary

While the peptidyl inhibitors that have been described are very potent for in vitro analyses of oligosaccharyl transferase activity, studies with these compounds in whole-cell assays of N-linked glycosylation have shown the compounds to be ineffective. This is presumably because the peptides are too polar and strongly solvated to passively cross the plasma membrane and the ER membrane to gain access to the enzyme in its native cellular location in the lumen of the ER. The development of bioavailable OT inhibitors remains an important objective in the study of asparagine-linked glycosylation.

10.2.3
Investigation of Carbohydrate-Based Substrate Analogues as Probes of Oligosaccharyl Transferase Function

The dolichol-linked tetradecasaccharide (Dol-PP-GlcNAc$_2$Man$_9$Glc$_3$), biosynthesized by the dolichol pathway, is the preferred substrate for OT both in vivo and in vitro. As a result of the poor availability of the full-length substrate, most in vitro studies have been carried out with truncated derivatives such as the chitobiosyl analogue (Dol-PP-GlcNAc$_2$). Sharma and co-workers showed that Dol-PP-GlcNAc$_2$Man was also a substrate, while no transfer was observed for Dol-PP-GlcNAc$_2$Man$_9$ and Dol-PP-GlcNAc [38]. In contrast, Bause and co-workers reported that Dol-PP-GlcNAc was in fact a poor substrate for OT [30]. This latter result has been confirmed independently [39]. The K_m values for the full-length and the truncated substrate were determined to be 0.5 µM and 1.2 µM respectively, by use of Tyr-Asn-Leu-Thr-Ser-Val as the peptide acceptor with a yeast OT preparation. In an independent study, Gibbs and Coward found the K_m values for the full substrate and the truncated substrate to be 33 µM and 65 µM respectively. In this case the tripeptide Bz-Asn-Leu-Thr-CONH$_2$ was used as the glycosyl acceptor [40]. The differences between these kinetic parameters may be due either to the low concentrations of sugar substrates used in the early studies or to the size of the peptide substrates, which has been shown to induce an effect on the binding of the sugar substrates. Using yeast genetic techniques, Burda and Aebi also showed that various truncated dolichol-linked sugars were substrates for OT in vivo [41].

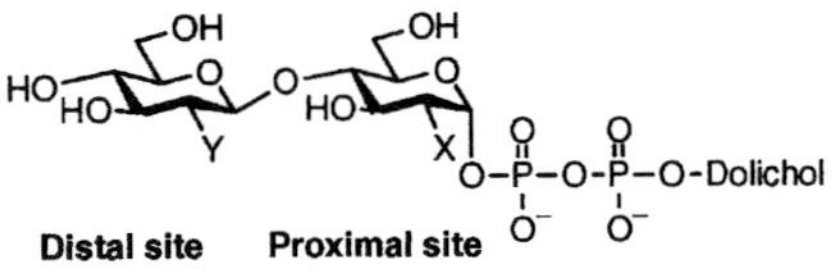

10.2 Small-Molecule Probes of the Biochemistry of Oligosaccharyl Transferase

Fig. 10.11 Structures of dolichol pyrophosphate-linked saccharides.

Compound	Y	X
16	NHCOCH$_3$	NHCOCH$_3$
17	NHCOCH$_3$	NHCOCF$_3$
18	NHCOCH$_3$	F
19	OH	NHCOCH$_3$

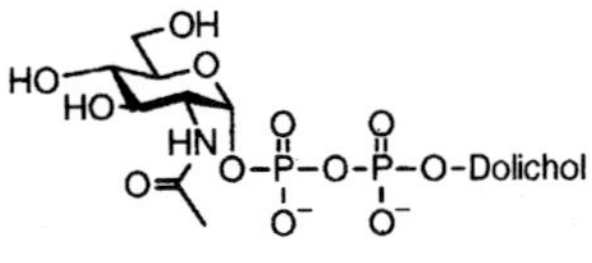

Tab. 10.2 Kinetic constants for compounds **16–19**.

Compound	K_m (µM)	V_{max} (pmol min^{-1})	V_{max}/K_m (rel)	K_i (µM)
16	64.3	10.3	1	
19	26.0	3.2	0.76	
17				154 ± 14
18				252 ± 71

A substrate analogue approach has been employed to explore the specificity of the glycosyl donor for N-linked glycosylation. In particular, the intriguing fact that the glycosyl donor comprises two N-acetylglucosamine saccharides proximal to the dolichol pyrophosphate leaving group suggested a functional role for the C-2 acetamido groups. To address this issue, unnatural dolichol-linked disaccharides (**17–19**), with replacement of the acetamido group in the saccharide units, were designed, synthesized, and evaluated as substrates or inhibitors for OT (Fig. 10.11) [39].

The unnatural substrate analogue Dol-PP-GlcNAc-Glc (**19**), with substitution of a glucose unit in the distal saccharide site, was found to be a substrate for OT. As shown in Tab. 2, the K_m values for **16** and **19** are 64 µM and 26 µM, respectively. Overall, the unnatural substrate **19** is 76% as effective (V_{maxapp}/K_{mapp}) as **16** under the in vitro analysis conditions. Interestingly, the monosaccharide analogue Dol-PP-GlcNAc (**20**) also showed product formation (11% relative to **16**). On the other hand, disaccharide analogues **17** and **18**, with replacement of N-acetylglucosamine in the proximal saccharide site, were inhibitors for OT with K_i values of 154 µM and 252 µM, respectively. The minimum glycosyl donor for N-linked glycosylation is Dol-PP-GlcNAc (Fig. 10.12). The efficiency of this process can be improved by the addition of an extra saccharide unit. The requirement of N-acetylglucosamine in the distal site is not absolute, since the unnatural analogue **19** displayed comparable transferase activity. The addition of a distal saccharide unit may simply assist the binding of the glycosyl donor. However, this acetamido group may be important in modulating the conformation of the glycosylated prod-

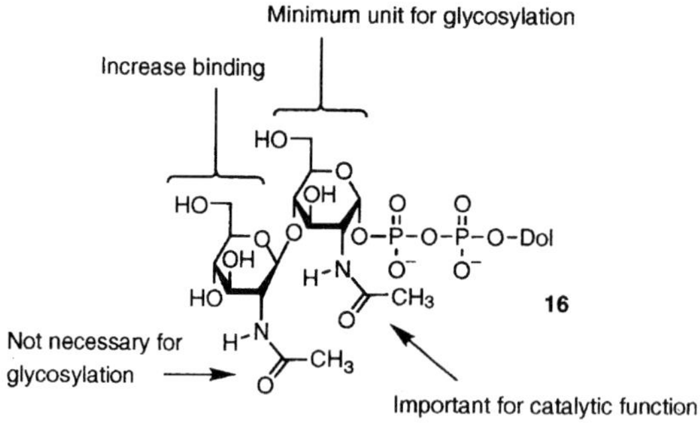

Fig. 10.12 Substrate specificity of glycosyl donors for oligosaccharyl transferase (taken from ref. 39, with permission).

uct [42]. On the other hand, the acetamido group in the proximal sugar plays a significant role in OT catalysis, as shown by the unnatural analogues **17** and **18**. The loss of transferase activities for **17** and **18** cannot simply be explained by the reduced binding (2.4 to 3.9 times less than the in vitro substrate) of the unnatural analogues, the acetamido group may be involved directly in the catalytic machinery, or alternatively, may interact with the enzyme through critical interactions in the catalytic site.

In a study on the chain length effect of the polyisoprene donor, Coward and co-workers prepared four new lipid-linked chitobiose pyrophosphates and evaluated these as substrates for yeast OT [43]. Replacement of dolichol in the disaccharide substrate by phytanyl, dihydrofarnesyl, dodecanyl, or citronellyl has detrimental effects on the transferase activity. The analogues with shorter isoprenoid chains or saturated alkyl chains attached to chitobiosyl diphosphate were poor substrates for OT, thus indicating that dolichol may be necessary for anchoring the sugar substrate into the membrane in order to ensure favorable interactions with OT. The significance of the terminal, saturated isoprenoid unit, adjacent to the pyrophosphate moiety, as well as the presence of the *cis*-isoprenoid units in dolichol still remains to be addressed.

10.2.3.1 Possible Mechanisms for Glycosyl Transfer

Despite significant discussion relating to the activation of the carboxamide of asparagine for N-linked glycosylation [32, 44–46], there has not yet been a report on studies of the mechanism of carbohydrate donor activation and glycosyl transfer. Study of the role of the glycosyl donor and its leaving group should ultimately help better understanding of the complete N-linked glycosylation process. Three possible mechanisms for the nucleophilic substitution and subsequent glycosylation are shown in Fig. 10.13 [47]. The first mechanism, (a), involves a direct S_N2 displacement reaction with the activated asparagine attacking the C-1 position of the first GlcNAc with the concomitant release of dolichol pyrophosphate (Dol-PP). This type of mechanism has been postulated in systems such as N-acetylglucosaminyl transferase V [48] and chitin synthases [49]. The second mechanism, (b), involves an S_N1-type displacement with the initial release of Dol-PP to generate

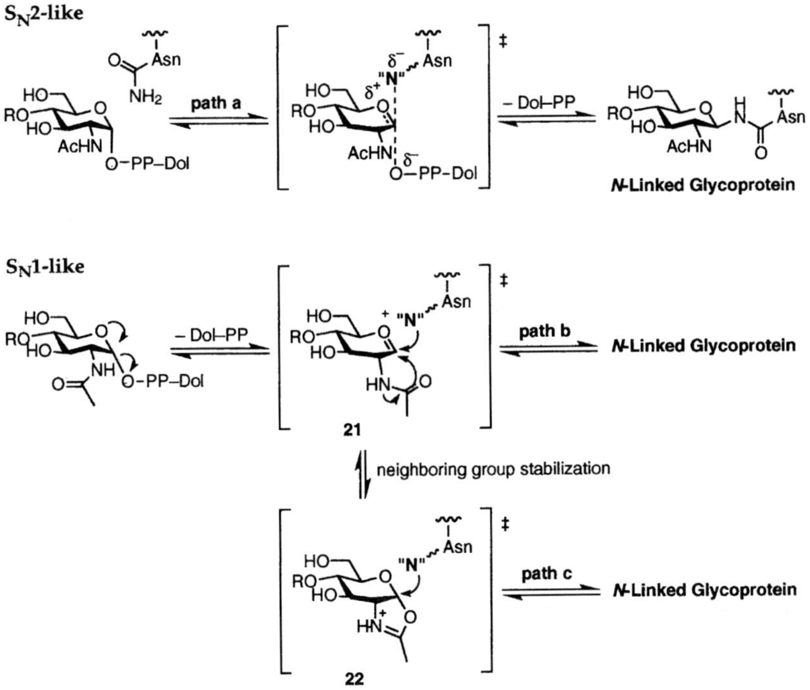

Fig. 10.13 Possible mechanisms for N-linked glycosylation (taken from ref. 47, with permission).

an oxocarbenium ion **21**, followed by the attack of the activated asparagine. The third mechanism, (c), is a modification of (b) in which the high-energy oxocarbenium ion **21** is further stabilized by the delocalization of positive charges through the participation of the C-2 acetamido group, forming an oxazolinium species **22**. Quantum mechanical calculations by Brameld and co-workers have provided a quantitative picture of the relative energies of the potential intermediates **21** and **22** (R = H) [50]. The oxazolinium species **22** was found to be more stable than the oxocarbenium species **21** by 16.9 kcal mol^{-1} by use of the HF/6-31G force field in a solvation model. It is difficult to distinguish between mechanisms that involve a discrete oxocarbenium ion (path b) or an oxocarbenium ion-like transition state (path a) [51]. On the other hand, the oxazolinium species **22** presents a intriguing target for investigation of the mechanism of glycosyl transfer with the use of small-molecule inhibitors or substrates.

The availability of mechanistic information from other enzymes that utilize N-acetylhexopyranose as part of their substrates has exploded in the past decade. Enzymes such as lysozyme [52], chitinase [53–56], chitobiase [57, 58], hevamine [59–61], N-acetyl β-hexoaminidase [62–64], lytic transglycosylase [65], and MurG [66] are some that fall into this category. Both the crystal structures of these enzymes and kinetic studies with substrate analogues and inhibitor analogues have provided insights into understanding of OT catalysis. Studies on chitinases and related enzymes from glycosyl hydrolase families 18 and 20 have provided supporting evidence for the existence of the oxazolinium species **22** in these systems. These enzymes act with overall retention of configuration but their structures showed only a single glutamic acid in the enzymes' active sites. This raises the question of whether a second carboxylate is necessary to form the classical glycosyl-enzyme intermediate. Crystal structure of *Serratia marcescens* chitobiase complex with the substrate chitobiose revealed a distortion of the nonreducing GlcNAc unit to a sofa conformation in which the acetamido oxygen atom was located 3 Å from the C-1 position of the nonreducing GlcNAc [57]. This observation strongly suggests that the C-2 acetamido group acts as a participating, intermediate nucleophile. Further evidence was provided by the structure of plant hevamine complex with the inhibitor allosamidin [60]. The (dimethylamino)-oxazoline unit of the aglycon, allosamizoline, was positioned at the active site, mimicking the oxazolinium species of the proposed mechanism.

10.2.3.2 Probing of the Mechanism of Oligosaccharyl Transferase with Potential Inhibitors

Imperiali and co-workers have examined a number of small molecules as potential inhibitors for oligosaccharyl transferase. Simple glycosidase inhibitors, such as 1-deoxynojirimycin, 1-deoxymannojirimycin, and castanospermine, targeting the oxocarbenium species (see Fig. 10.6) showed no inhibitory effect on OT even at elevated concentrations (5 mM), and neither did the substrate-like disaccharides chitobiose **23** and chitobiosylamine **24** (1 mM) (Fig. 10.14). The chitobionhydroximolactone **25** and the corresponding *N*-phenyl carbamate derivative **26** [67] developed by Vasella also showed no inhibitory activity with OT [68].

The glucoallosamidin A pseudodisaccharide **27** is composed of an *N*-acetylglucosamine linked to the aglycon, allosamizoline (**28**). This compound, first isolated from acid degradation of glucoallosamidin A, is a member of the allosamidin family of chitinase inhibitors. Glucoallosamidin A pseudodisaccharide was synthesized and evaluated as an inhibitor for OT that might mimic the oxazolinium intermediate species **22**. The additional GlcNAc unit was incorporated to provide ad-

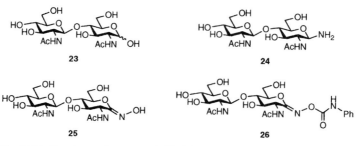

Fig. 10.14 Structures of compounds **23–26**.

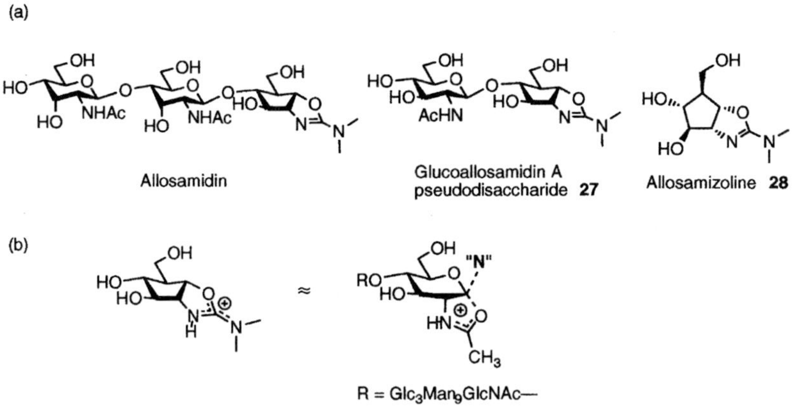

Fig. 10.15 (a) Structures of allosamidin and derivatives; (b) comparison of protonated allosamizoline (**28**) and oxazolium (**22**) species.

ditional binding, based on understanding of the substrate glycosyl donors. A comparison of protonated allosamizoline and the postulated oxazolinium species **22** in OT catalysis is shown in Fig. 10.15. The positive charge on the oxazoline nitrogen of allosamizoline can be delocalized onto the oxygen atom and the dimethylamino group, while the cyclopentane ring is in a half-chair or sofa conformation. This compound inhibits *Candida albicans* chitinase with an IC_{50} of 1.3 µg ml^{-1} [69], and insect *Chironomus tentans* chitinase with a K_i of 16.9 µM [70].

The key step in the synthesis of glucoallosamidin A pseudodisaccharide (**27**) involved glycosylation of protected allosamizoline **30** with oxazoline **29** in the presence of camphorsulfonic acid (CSA) to afford protected **31**, which upon subsequent deprotection gave **27** (Scheme 10.1). However, the glucoallosamidin A pseudodisaccharide showed no inhibitory activity for OT even at 5 mM. The synergistic effect of dolichol monophosphate was also examined, but compound **27** (1 mM) did not show allosteric inhibition in the presence of dolichol monophosphate (3 µM). Interestingly, dolichol monophosphate alone inhibited OT by up to 35% (10 µM) in a concentration-dependent manner. However, an increase in dolichol phosphate concentration above 10 µM was not accompanied by further inhibition. This effect may be explained by the formation of micelles, affecting the actual amount of free Dol-P available for inhibition. The mode of inhibition of dolichol

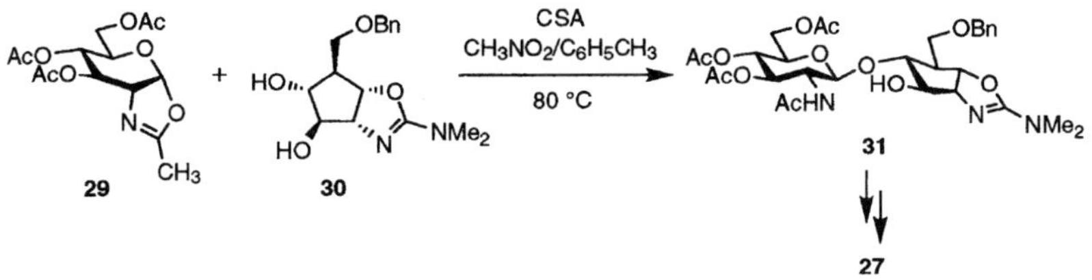

Scheme 10.1 Synthesis of glucoallosamidin A pseudodisaccharide **27**.

monophosphate was not determined. Other than competing for the same binding site as the glycosyl donor, dolichol monophosphate has been shown to perturb the fluidity of the lipid bilayer [71], which in turn may affect the activity of OT.

Kobayashi and co-workers have shown that chitobiose oxazoline **32** is a substrate for *Bacillus* chitinase, which gives chitin upon enzymatic polymerization, while GlcNAc oxazoline **33** can be employed to prepare chitobiose (Scheme 10.2) [72, 73]. Knapp and co-workers have pointed out that GlcNAc oxazoline **33** is easily hydrolyzed under acidic conditions. The alternative, more stable GlcNAc thiazoline **34** was prepared by these authors and was shown to be a competitive inhibitor of jack bean *N*-acetylhexoaminidase with a K_i of 280 nM [63]. Chitobiose oxazoline **32** and thiazoline derivative **46** (see Scheme 10.7) were synthesized and evaluated as substrates or inhibitors for OT.

Chitobiose oxazoline **32** can be readily prepared in two steps from commercially available chitobiose octaacetate **35** (Scheme 10.3). By Davis' approach [74], treatment of **35** with trimethylsilyl bromide (TMSBr), boron trifluoride etherate ($BF_3 \cdot Et_2O$), and collidine in dichloroethane afforded the known oxazoline **36**, which upon ammonolysis gave chitobiose oxazoline **32**.

Knapp and co-workers have devised an elegant route for the preparation of GlcNAc thiazoline **34**. Upon treatment of D-glucosamine pentaacetate (**37**) with Lawesson's reagent **38** (0.66 equiv.) [75], protected GlcNAc thiazoline **40** was obtained via the intermediate thioamide **39** (Scheme 10.4). It should be noted that the in situ cyclization of **39** requires that the leaving group at *C*-1 be in the β con-

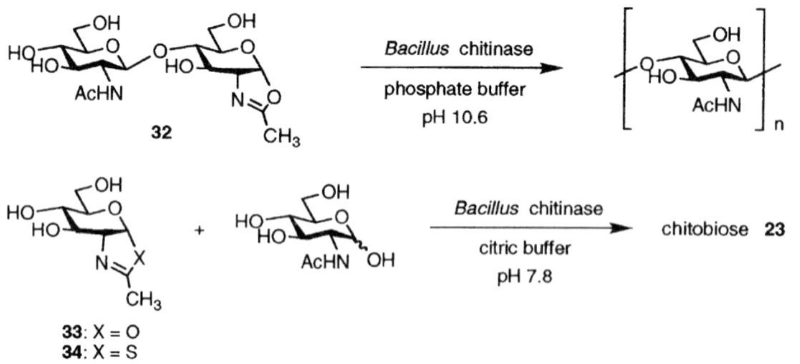

Scheme 10.2 Enzymatic synthesis involving oxazolines **32** and **33**.

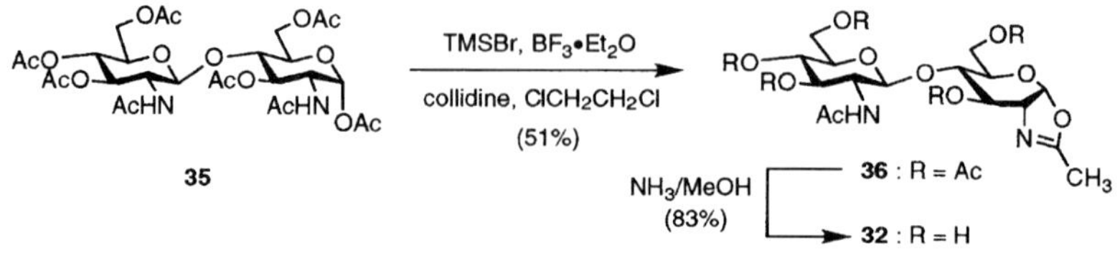

Scheme 10.3 Synthesis of chitobiose oxazoline **32** (taken from ref. 47, with permission).

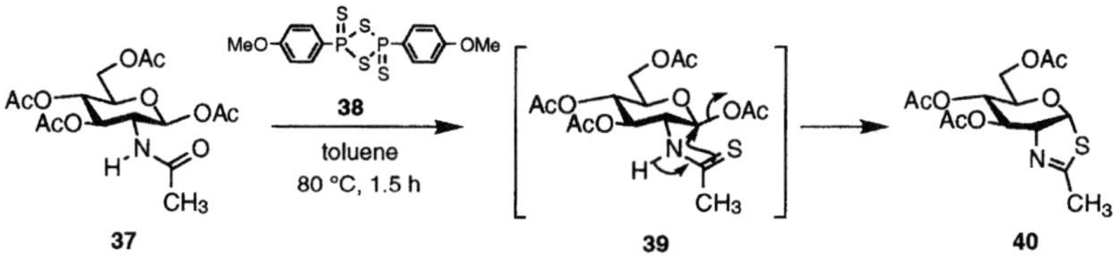

Scheme 10.4 Knapp's synthesis of thiazoline **40**.

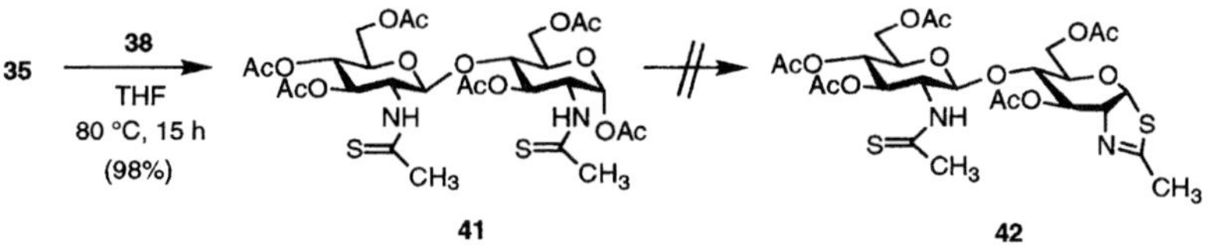

Scheme 10.5 Attempted synthesis of thiazoline **42**.

figuration. Since the C-1 acetyl group of chitobiose octaacetate **35** is in an α linkage, thionation with Lawesson's reagent in THF did not give the cyclized product **42** but the expected dithioacetamide **41**, in high yield (Scheme 10.5). Attempts to convert dithioacetamide **41** into thiazoline **42** under conditions similar to those used for the preparation of oxazoline **36** were unsuccessful.

After several unsuccessful attempts, thiazoline formation was finally achieved by one-pot thionation and cyclodehydration of the reducing sugar **43** with Lawesson's reagent. Upon treatment of reducing sugar **43** with 1.2 equivalents of **38** in toluene at 80 °C, the desired GlcNAc thiazoline **40** was isolated in moderate yield (Scheme 10.6). The reaction is believed to proceed through the intermediates **44** and **45**, followed by cyclodehydration. The dehydrating agent was previously proposed to be the mixed S,O-phosphine ylide formed from S,O exchange of Lawesson's reagent **38** [76]. It should be noted that treatment of 1-hydroxyl-2-acetamido glycosides with Lawesson's reagent is a versatile means of preparing glycosyl thia-

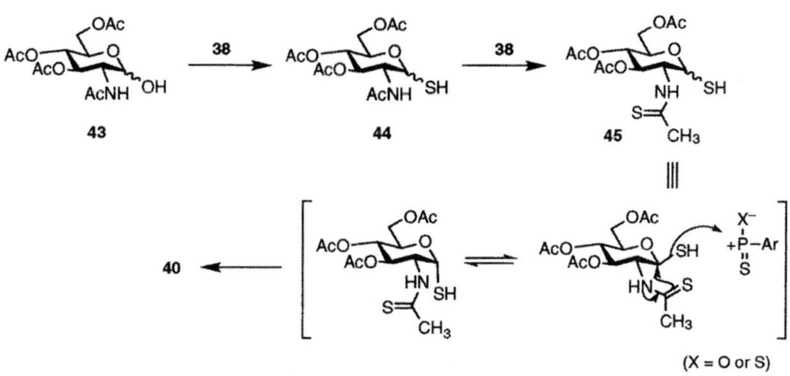

Scheme 10.6 Synthesis of thiazoline **40**.

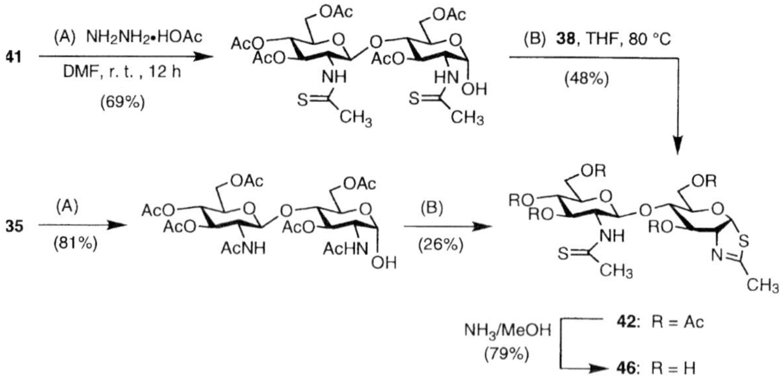

Scheme 10.7 Synthesis of N'-thioacetamido chitobiosyl thiazoline **46**.

zolines since this reaction does not depend on the stereochemistry of the leaving group at C-1, due to the equilibration of α and β anomers.

By application of the methodology developed for the monosaccharide, N'-thioacetamido chitobiosyl thiazoline **46** was readily prepared by treatment of reducing sugars, derived from **41** or **35**, with Lawesson's reagent and subsequent deprotection (Scheme 10.7). Thiazoline **46** displayed significant acid stability relative to the chitobiose oxazoline **32**. At pH 4.8, oxazoline **32** was hydrolyzed immediately, while the thiazoline **46** took 2 hours to hydrolyze completely at pH 2.3. Both compounds are sufficiently stable at neutral pH for inhibition study of OT. However, these compounds showed no inhibition activity against OT. Furthermore, chitobiose oxazoline **32** was not a substrate for OT either in the absence or presence of dolichol monophosphate.

10.2.3.3 Interim Summary

Compounds targeting oxocarbenium species, such as 1-deoxynojirimycin, 1-deoxymannojirimycin, castanospermine, swainsonine, chitobionhydroximolactone, and the corresponding N-phenyl carbamate, show no inhibition with OT. Additionally, compounds targeting oxazolinium species, such as glucoallosamidin A pseudodisaccharide, chitobiose oxazoline, and the thiazoline derivative also show no effects on OT. The failure of these compounds to bind to OT may be due to insufficient binding relative to the native substrate, which is a tetradecasaccharide. Alternatively, the polyisoprenyl pyrophosphate (Dol-PP) may experience significant binding to OT. Both Sharma and Coward have independently shown that the K_m for the disaccharide substrate **16** ($K_m = 65$ μM) is only twice that of the tetradecasaccharide substrate ($K_m = 33$ μM) [14, 38, 40], suggesting that the remaining saccharide units do not contribute significantly to the binding of OT. Coward and coworkers have shown that replacement of dolichol in **16** with various short-chain isoprenoids could only restore about 10% of the OT activity [43]. Moreover, if dolichol phosphate inhibits OT ($IC_{30} = 5$ μM) by competing for the same binding site as the glycosyl donor, this suggests a second role of the polyisoprene, beyond anchoring the sugar substrate to the membrane bilayer. Breuer and Bause have proposed a dolichol recognition sequence [77] in Ribophorin I (corresponding to Ost1 in yeast) in pig liver OT [78], but no specific dolichol recognition site has yet been identified in yeast OT complex.

10.3
Conclusions

The basic principles of asparagine-linked glycosylation have been recognized for a considerable time. In particular, progress has been made in defining the native substrates for the process, and in understanding of the minimum recognition elements necessary for these substrates to bind to oligosaccharyl transferase. This insight has resulted in the design, synthesis, and evaluation of numerous peptide and carbohydrate substrates and inhibitors as probes for investigating the mechanistic details of the glycosylation process. In recent years, significant progress has also been made in understanding the molecular architecture of oligosaccharyl transferase. OTs from a wide range of eukaryotic organisms have now been characterized; the enzymes are typically large, hetero-oligomeric complexes with four or more membrane-associated subunits. The next phase of investigations into N-linked glycosylation will be focused on the generation of a complete molecular description of the enzyme. To this end, the complementary application of both bioorganic and biophysical methods, together with the tools of contemporary molecular biology, should hopefully result in the ultimate goal of defining the detailed structure and molecular mechanism of the OT enzyme complex.

10.4
References

1 S. E. O'Connor, B. Imperiali, Chem. Biol. 1996, 3, 803–812.
2 S. Trombetta, A. J. Parodi, Adv. Protein Chem. 2002, 59, 303–344.
3 A. Varki, Glycobiology 1993, 3, 97–130.
4 R. A. Dwek, Biochem. Soc. Trans. 1994, 23, 1–25.
5 J. Roth, Chem. Rev. 2002, 102, 285–303.
6 B. Imperiali, Acc. Chem. Res. 1997, 30, 452–459.
7 R. Knauer, L. Lehle, Biochim. Biophys. Acta 1999, 1426, 259–273.
8 P. Vandensteen, P. M. Rudd, R. A. Dwek, G. Opdenakker, Crit. Rev. Biochem. Mol. Biol. 1998, 33, 151–208.
9 T. Kinoshita, N. Inoue, Curr. Opin. Chem. Biol. 2000, 4, 632–638.
10 A. Loeffler, M.-A. Doucey, A. M. Jansson, Biochemistry 1996, 35, 12005–12014.
11 A. J. Parodi, Glycobiology 1993, 3, 193–199.
12 M. Bosch, S. Trombetta, U. Engstrom, A. J. Parodi, J. Biol. Chem. 1988, 263, 17360–17365.
13 R. D. Marshall, Biochem. Soc. Symposia 1974, 40, 17–26.
14 Y. Gavel, G. von Heijne, Prot. Eng. 1990, 3, 433–442.
15 E. Bause, Biochem. Soc. J. 1984, 12, 514–517.
16 I. Nilsson, G. von Heijne, J. Biol. Chem. 1993, 268, 5798–5801.

17 R. Kornfeld, S. Kornfeld, Ann. Rev. Biochem. 1985, 54, 631–664.
18 S. Silberstein, R. Gilmore, FASEB J. 1996, 10, 849–858.
19 D. J. Kelleher, R. Gilmore, J. Biol. Chem. 1994, 269, 12908–12917.
20 D. Karaoglu, D. J. Kelleher, R. Gilmore, J. Biol. Chem. 1997, 272, 32513–32520.
21 J. K. Welply, P. Shenbagamurthi, F. Naider, H. R. Park, W. J. Lennarz, J. Biol. Chem. 1985, 260, 6459–6465.
22 R. Noiva, H. Kimura, J. Roos, W. J. Lennarz, J. Biol. Chem. 1991, 266, 19645–19649.
23 M. LaMantia, T. Miura, H. Tachikawa, H. A. Kaplan, W. J. Lennarz, T. Mizunaga, Proc. Natl. Acad. Sci. USA 1991, 88, 4435–4457.
24 Q. Yan, G. P. Prestwich, W. J. Lennarz, J. Biol. Chem. 1999, 274, 5021–5025.
25 G. Dorman, G. D. Prestwich, Biochemistry 1994, 33, 5661–5673.
26 E. Bause, M. Wesemann, A. Bartoschek, W. Breuer, Biochem. J. 1997, 322, 95–102.
27 A. D. Elbein, Trends Biochem. Sci. 1981, 6, 291–293.
28 B. Imperiali, K. L. Shannon, K. W. Rickert, J. Am. Chem. Soc. 1992, 114, 7942–7943.
29 B. Imperiali, J. R. Spencer, M. D. Struthers, J. Am. Chem. Soc. 1994, 116, 8424–8425.
30 E. Bause, W. Breuer, S. Peters, Biochem. J. 1995, 312, 979–985.
31 W. Breuer, R. A. Klein, B. Hardt, A. Bartoschek, E. Bause, FEBS Lett. 2001, 501, 106–110.
32 B. Imperiali, K. L. Shannon, M. Unno, K. Rickert, J. Am. Chem. Soc. 1992, 114, 7944–7945.
33 T. Xu, R. M. Werner, K.–C. Lee, J. C. Fettinger, J. T. David, J. K. Coward, J. Org. Chem. 1998, 63, 4767–4778.
34 T. L. Hendrickson, B. Imperiali, Biochemistry 1995, 34, 9444–9450.
35 T. L. Hendrickson, J. R. Spencer, M. Kato, B. Imperiali, J. Am. Chem. Soc. 1996, 118, 7636–7637.
36 C. Kellenberger, T. L. Hendrickson, B. Imperiali, Biochemistry 1997, 36, 12554–12559.
37 M. D. L. Ufret, B. Imperiali, Bioorg. Med. Chem. Lett. 2000, 10, 281–284.
38 C. B. Sharma, L. Lehle, W. Tanner, Eur. J. Biochem. 1981, 116, 101–108.
39 V. W.-F. Tai, B. Imperiali, J. Org. Chem. 2001, 66, 6217–6228.
40 B. S. Gibbs, J. K. Coward, Bioorg. Med. Chem. 1999, 7, 441–447.
41 P. Burda, M. Aebi, Biochim. Biophys. Acta 1999, 1426, 239–257.
42 S. E. O'Connor, B. Imperiali, Chem. Biol. 1998, 5, 427–437.
43 X. G. Fang, B. S. Gibbs, J. K. Coward, Bioorg. Med. Chem. Lett. 1995, 5, 2701–2706.
44 E. Bause, L. Gunter, Biochem. J. 1981, 195, 639–644.
45 J. Lee, J. K. Coward, Biochemistry 1993, 32, 6794–6801.
46 R. S. Clark, S. Bannerjee, J. K. Coward, J. Org. Chem. 1990, 55, 6275–6285.
47 V. W.-F. Tai. Mechanistic Studies of Oligosaccharyl Transferase Employing Synthetic Inhibitor and Substrate Analogues; Massachusetts Institute of Technology: Cambridge, 2001.
48 M. Kaneko, O. Kanie, T. Kajimoto, C.-H. Wong, Bioorg. Med. Chem. Lett. 1997, 7, 2809–2812.
49 J. Grugier, J. Xie, I. Duarte, J.-M. Valery, J. Org. Chem. 2000, 65, 979–984.
50 K. A. Brameld, W. D. Shrader, B. Imperiali, W. A. Goddard III, J. Mol. Biol. 1998, 280, 913–923.
51 G. Davies, M. L. Sinnott, S. G. Withers, Glycosyl Transfer; Sinnott, M. L., Ed.; Academic Press: San Diego, 1997; Vol. I, pp. 119–209.
52 K. Harata, M. Muraki, Acta Crystallograph. 1995, D51, 718–724.
53 J. P. Hart, A. F. Monzingo, M. P. Ready, S. R. Ernst, J. D. Robertus, J. Mol. Biol. 1993, 229.
54 A. Perrakis, I. Tews, Z. Dauter, A. B. Oppenheim, I. Chet, K. S. Wilson, C. E. Vorgias, Structure 1994, 2, 1169–1180.
55 I. Tews, A. C. Terwisscha van Scheltinga, A. Perrakis, K. S. Wilson, D. W. Dijkstra, J. Am. Chem. Soc. 1997, 119, 7954–7959.
56 T. Hollis, A. F. Monzingo, K. Bortone, S. Ernst, R. Cox, J. D. Robertus, Prot. Sci. 2000, 9, 544–551.

57 I. Tews, A. Perrakis, A. Oppenheim, Z. Dauter, K. S. Wilson, C. E. Vorgias, Nature Struct. Biol. 1996, 3, 638–648.
58 S. Drouillard, S. Armand, G. J. Davies, C. E. Vorgias, B. Henrissat, Biochem. J. 1997, 328, 945–949.
59 A. C. Terwisscha van Scheltinga, K. H. Kalk, J. J. Beintema, B. W. Dijkstra, Structure 1994, 2, 1181–1189.
60 A. C. Terwisscha van Scheltinga, S. Armand, K. H. Kalk, A. Isogai, B. Henrissat, B. W. Dijkstra, Biochemistry 1995, 34, 15619–15623.
61 A. C. Terwisscha van Scheltinga, M. Hennig, B. W. Dijkstra, J. Mol. Biol. 1996, 262, 243–257.
62 V. Rao, C. Guan, P. Van Roey, Structure 1995, 3, 449–457.
63 S. Knapp, D. Vocadlo, Z. Gao, B. Kirk, J. Lou, S. G. Withers, J. Am. Chem. Soc. 1996, 118, 6804–6805.
64 D. J. Vocadlo, C. Mayer, S. He, S. G. Withers, Biochemistry 2000, 39, 117–126.
65 E. J. van Asselt, K. H. Kalk, B. W. Dijkstra, Biochemistry 2000, 39, 1924–1934.
66 S. Ha, D. Walker, Y. G. Shi, S. Walker, Prot. Sci. 2000, 9, 1045–1052.
67 D. Beer, J.-L. Maloisel, D. M. Rast, A. Vasella, Helv. Chim. Acta. 1990, 73, 1918–1922.
68 J. Pohlmann, B. Imperiali, unpublished results.
69 Y. Nishimoto, S. Sakuda, S. Takayama, Y. Yamada, J. Antibiot. 1991, 44, 716–722.
70 M. Spindler-Barth, R. Blattner, C. E. Vorgias, K. D. Spindler, Pesticide Science 1998, 52, 47–52.
71 C. Valtersson, G. van Düyn, A. J. Verkleij, T. Chojnack, B. de Kruijff, G. Dallner, J. Biol. Chem. 1985, 260, 2742–2751.
72 S. Kobayashi, T. Kiyosada, S.-I. Shoda, Tetrahedron Lett. 1997, 38, 2111–2112.
73 S. Kobayashi, T. Kiyosada, S.-I. Shoda, J. Am. Chem. Soc. 1996, 118, 13113–13114.
74 M. Colon, M. M. Staveski, J. T. Davis, Tetrahedron Lett. 1991, 32, 4447–4450.
75 M. P. Cava, M. I. Levinson, Tetrahedron 1985, 41, 5061–5087.
76 T. Nishio, J. Chem. Soc. Perkin Trans I 1993, 1113–1117.
77 C. F. Albright, P. Orlean, P. W. Robbins, Proc. Natl. Acad. Sci. USA 1989, 86, 7366–7369.
78 W. Breuer, E. Bause, Eur. J. Biochem. 1995, 228, 689–696.

11
Conformational Analysis of C-Glycosides and Related Compounds: Programming Conformational Profiles of C- and O-Glycosides

Peter G. Goekjian, Alexander Wei, and Yoshito Kishi

11.1
Introduction

An increasing appreciation of the roles of carbohydrates, glycolipids, and glycoproteins in biological function has uncovered new opportunities in carbohydrate-based drug design [1]. C-Glycosides are carbohydrate analogues in which the anomeric oxygen has been substituted with a carbon atom, with intriguing prospects as carbohydrate isosteres or as non-hydrolyzable recognition elements. Several comprehensive reviews are available on the synthesis [2], conformational analysis [3], and biological activities [4] of C-glycosides. In this chapter, we focus on diamond lattice-assisted conformational analysis of C-glycosides, and present a set of simple rules that allow us to program conformational profiles into a molecular architecture design based on carbohydrates. The objective of this approach is not to define the precise conformation of a given C-glycoside, but rather to introduce a useful analytical device with which: (1) to identify strategic steric interactions that have a significant influence on its local conformational preferences, and (2) to suggest rational structural modifications for modulating its global conformational profile [5].

For the sake of simplicity, and to illustrate the relationship between C- and O-glycosides, we shall adopt the nomenclature commonly used for O-glycosides. Thus, the carbon analogue of lactose will be called C-lactose, or will be represented by the shorthand structural notation Gal-β(1' → 4)-C-Glc. The numbering system for C- and O-disaccharides will be the same, namely 1 through 6 for the reducing sugar and 1' through 6' for the terminal, non-reducing sugar (Fig. 11.1). The exocyclic methylene unit linked to C1' will be referred to as the a carbon.

The conformational analysis of C- and O-disaccharides can be divided into questions concerning: (1) the pyranose rings, and (2) the interglycosidic linkage (i.e., the glycosidic and aglyconic bonds) [6, 7]. As the pyranose rings nearly always adopt the $^4C_1$ chair conformation, discussion in this chapter focuses on the conformational preferences of the glycosidic and aglyconic bonds, which can be defined quantitatively by the dihedral angles ϕ and φ, respectively (Fig. 11.2). However, we will deliberately use qualitative descriptors, such as *exo/anti* or *C-anti-C* conformations. In this manner, we avoid over-interpretation of the experimental

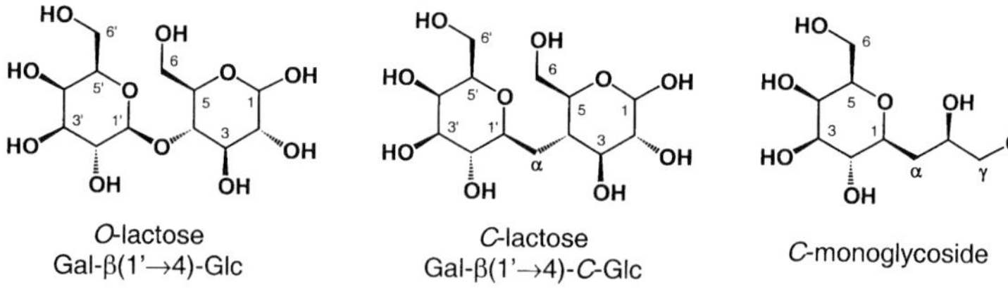

Fig. 11.1 Nomenclature and numbering used in this chapter for C-glycosides and their parent O-glycosides.

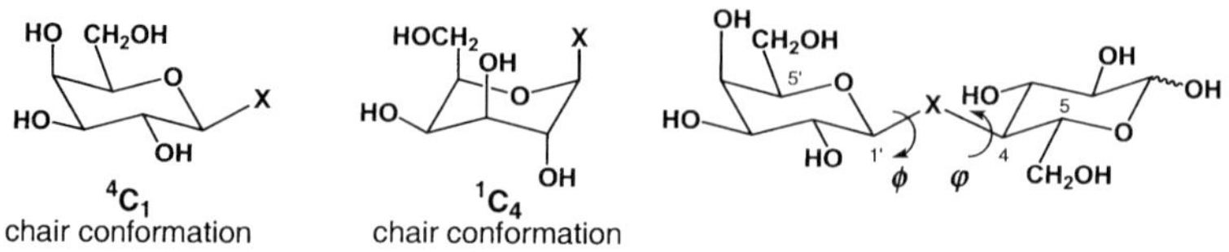

Fig. 11.2 Conformational definitions for C-glycosides (X=CH$_2$) and O-glycosides (X=O). ϕ and φ represent the dihedral angles formed by O5'-C1'-X-C4 and C1'-X-C4-C5, respectively. The positive sense of rotation is shown with the ring held in place and the exocyclic bond rotating.

data and, at the same time, focus sharply on the unique value of this approach for carbohydrate-based molecular architecture design.

11.2
Stereoelectronic Effects and the *exo*-Anomeric Conformation

The modern era of conformational analysis of carbohydrates began with the recognition of the anomeric and *exo*-anomeric effects. The anomeric effect, first described by Lemieux [8], refers to the tendency of a C1 substituent on an O-pyranoside to favor the axial (a) over the equatorial (β) configuration, in spite of unfavorable steric interactions. For example, methyl a-glucoside is thermodynamically preferred over methyl β-glucoside under equilibrating conditions by a ratio of roughly 2:1 (Fig. 11.3). The molecular orbital interpretation of this effect invokes a stabilizing overlap between a nonbonding (lone pair) orbital of ring oxygen O5 and the σ^* antibonding orbital of the polarized C1–O1 bond. The degree of stabilization depends on the efficiency of orbital mixing; maximum overlap is achieved when the orbitals are aligned in an antiperiplanar fashion. This form of stabilization is therefore generally referred to as a stereoelectronic effect [9]. The concept of orbital overlap provides a rational basis to explain the experimental outcome: the C1–O1 bond in methyl a-glucoside is antiperiplanar to a lone pair orbital of O5, whereas the C1–O1 bond in methyl β-glucoside is not (Fig. 11.3). The anomeric effect has been estimated to provide stabilization energies as high as 1.5 kcal mol^{-1}.

The term "*exo*-anomeric effect" was also introduced by Lemieux to describe the distinctive conformational preference observed in the glycosidic bonds of many py-

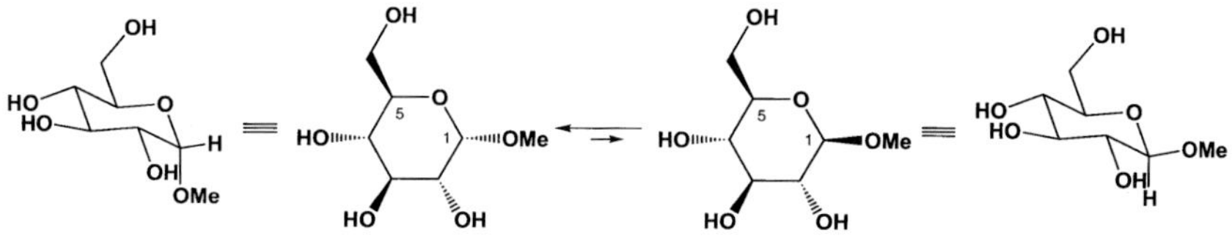

HCl / MeOH / 25 °C
α:β = 2:1

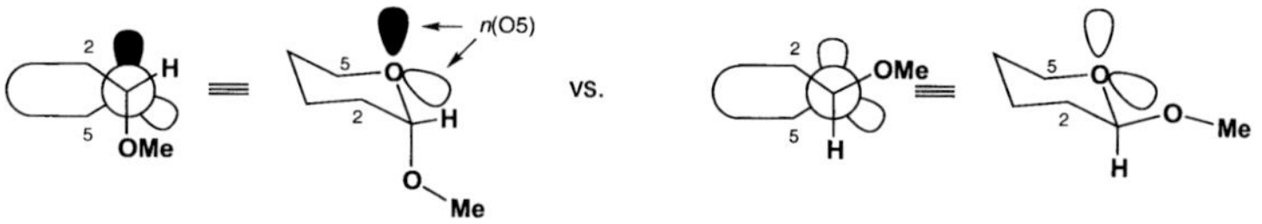

Fig. 11.3 The stereoelectronic basis of the anomeric effect. Methyl α-glycoside (*left*) is stabilized by electronic overlap between one of the nonbonding lone pair (*n*) orbitals of O5 (filled in black) and the antibonding C1–O1 σ* orbital (not shown), which are aligned in antiperiplanar fashion. This stabilization is absent in methyl β-glycoside (*right*), which has minimal n → σ* orbital overlap.

ranosides [8]. This effect is clearly illustrated by comparison of the three staggered rotamers of an α-glycosidic bond, in which the *exo/anti* conformation with the aglyconic carbon antiperiplanar to the ring C2 carbon is preferred over the *nonexo/gauche* and *exo/gauche* conformations (Fig. 11.4). This conformational preference has proven to be valid for oligosaccharides as well as for simple O-alkyl glycosides.

The strong preference for the *exo*-anomeric conformation in O-glycosides has been attributed to a combination of steric and stereoelectronic effects. The application of the stereoelectronic effect to the *exo*-anomeric conformation is analogous to that used to describe the anomeric effect: maximum electronic stabilization is achieved by the existence of an antiperiplanar relationship between the nonbonding orbital of the glycosidic oxygen O1 and the σ* antibonding orbital of the polarized C1–O5 bond in the pyranose ring (Fig. 11.4). This interaction should stabilize the *exo/anti* and the *exo/gauche* conformations, but not the *nonexo/gauche* conformer. With respect to steric interactions, a gradient exists in which the *exo/anti* conformation experiences the least destabilization (a C–O gauche interaction between O5 and R) and the *exo/gauche* conformation the most (a minimum of two gauche interactions, plus 1,3-diaxial-like interactions between the pyranose ring and R in the case of axial C-glycosides). The distinct preference for the *exo/anti* conformer is thus supported by either line of reasoning.

While both steric and stereoelectronic effects had been appreciated in carbohydrate conformational analysis for some time, ambiguity remained as to their relative importance in the conformational behavior of O-glycosidic and O-aglyconic bonds, particularly in their native environments. For example, the anomeric effect has been shown to be strong in nonpolar solvents, but weak in aqueous solutions [9a]. It was therefore not immediately clear what level of impact stereoelectronic effects should have on the *exo*-anomeric conformation, as logical arguments could be made in either direction. Direct *experimental* evidence was critically needed to

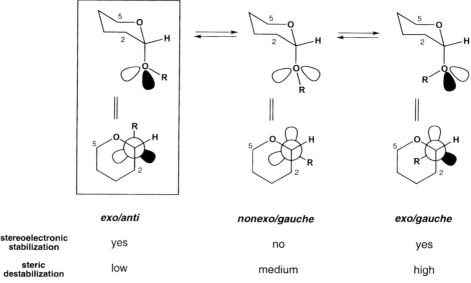

	exo/anti	nonexo/gauche	exo/gauche
stereoelectronic stabilization	yes	no	yes
steric destabilization	low	medium	high

Fig. 11.4 The *exo*-anomeric effect as illustrated in axial O-glycosides. Stereoelectronic stabilization is provided by electronic overlap between one of the nonbonding (*n*) orbitals of O1 (filled in black) and the antibonding C1–O5 σ* orbital (not shown), aligned in antiperiplanar fashion. The relative stereoelectronic and steric effects of the three staggered rotamers about the glycosidic bond are listed.

resolve this fundamental question. However, to the best of our knowledge, there was no experiment that could separately address the relevance of steric versus electronic effects in the conformational preference of O-glycosides.

The use of C-glycosides as conformational models for O-glycosides was expected to bring valuable insights into the relevance of stereoelectronic stabilization. The replacement of the glycosidic oxygen with a methylene carbon effectively eliminates the stereoelectronic component from the *exo*-anomeric effect, which can consequently be evaluated strictly in terms of steric destabilization. Perturbations in secondary structure caused by the $O \rightarrow C$ replacement are expected to be minor, because discrepancies in bond lengths (C–C: 1.54 Å vs. C–O: 1.43 Å) are compensated by the corresponding difference in glycosidic bond angle (C–C–C: 109° vs. C–O–C: 116°). This results in less than a 2% change in distance across the interglycosidic linkage, negligible in comparison with the errors associated with the experimental measurements and their theoretical interpretation.

A very important advantage of using C-glycosides as model systems is the ease with which their solution conformations can be evaluated. Time-averaged conformations of the C-glycosidic bond can be determined by ^{1}H NMR spectroscopy from $^3J_{H,H}$ coupling constants and the well established Karplus relationship [10]. The procedure is equally useful for determining the conformational tendencies of the C-aglyconic bond in C-disaccharides. These spectroscopic handles are essential for the development and validation of conformational hypotheses, such as on the importance of local steric interactions. The preferred conformations of C-glycosides can also be compared to those of the corresponding parent O-glycosides, to determine the relative influence of electronic interactions from their conformational similarities or dissimilarities.

11.3
Conformational Analysis of C-Glycosides: C-Monoglycosides

The classic Karplus equation relates $^3J_{H,H}$ coupling constants to dihedral angles by a simple cosine function [10]. However, this relationship is affected by the orientation of electronegative substituents around the protons of interest and requires correction factors for accurate interpretations. A parameterized Karplus relationship was developed by Altona [11]. With respect to C-glycosides, the $^3J_{H,H}$ coupling constants between the C1 proton and the Cα pro-(R) and pro-(S) protons are obtained from NMR experiments, then correlated with values derived from modified Karplus relationships corresponding to ideally staggered conformations or from semiempirical energy minimizations (Fig. 11.5). These values can be fitted to the experimental coupling constants $J_{1,R}$ and $J_{1,S}$ to yield an approximate, weighted conformational distribution [5f]. In this manner, one can determine local solution conformations directly from experimental data, and investigate the role of specific interactions on conformational behavior to a first degree of approximation.

It should be noted that a similar Karplus relationship between $^3J_{H,C}$ coupling constants and dihedral angle is known for ^{1}H–C–O–^{13}C systems. In principle, this relationship can be applied to conformational analysis of O-glycosidic and O-

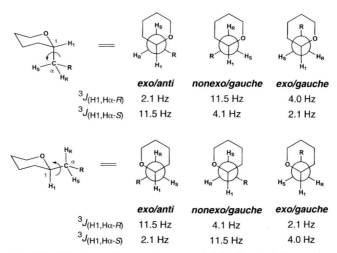

Fig. 11.5 Vicinal coupling constants expected for the ideal staggered conformers around the C-glycosidic bond of α (axial) and (equatorial) C-glycosides [5a, f]. Newman projections are along the C1–Cα bond.

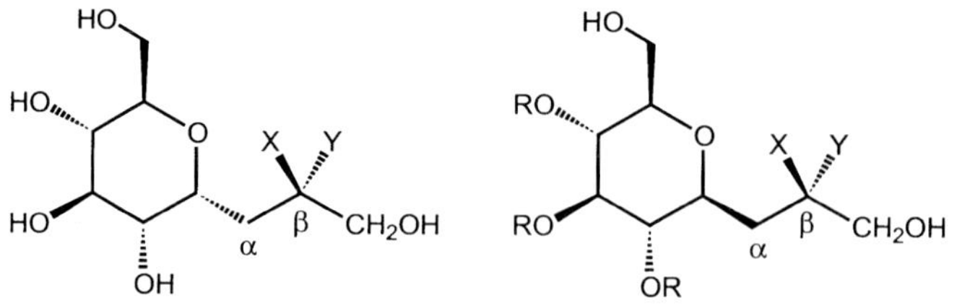

11 Conformational Analysis of C-Glycosides and Related Compounds

1 : X=OH; Y=H
2 : X=H; Y=OH

3 : X=OH; Y=H; R=H
4 : X=H; Y=OH; R=H

Fig. 11.6 The structures of axial and equatorial C-monosaccharides 1–4.

Tab. 11.1 Vicinal coupling constants (CD$_3$OD, 500 MHz) and estimated conformational distributions for C-monosaccharides 1~4 [5 a, f]. Distributions were estimated from a three-state model of the staggered conformers derived from theoretical calculations [3 c]. Errors in calculated populations are estimated at ±10%.

	C-glycosidic (C1–C) bond					C-aglyconic (Cα–Cβ) bond				
	$J_{1,R}$	$J_{1,S}$	exo/anti	nonexo/gauche	exo/gauche	$J_{R,\beta}$	$J_{S,\beta}$	C-anti-C	C-anti-O	C-anti-H
1	3.3	11.3	95	5	0	10.1	2.9	80	10	10
2	4.1	10.1	85	15	0	6.0	6.4	40	35	25
3	8.9	2.8	75	5	20	6.9	5.3	45	25	30
4	9.7	2.5	80	10	10	3.1	9.7	75	10	15

aglyconic bonds, and several applications have demonstrated its potential [12]. While dihedral angle analysis of the C1–O1 bond might appear to be the most direct method for addressing conformational issues in O-glycosides, this approach has not yet been widely adopted, primarily due to the technical difficulties encountered, including ^{13}C enrichment of the carbon of interest.

Returning to the subject of $^3J_{H,H}$ coupling constant analysis, a large number of C-monosaccharides have been examined and have universally demonstrated a strong preference for the *exo*-anomeric conformation [5a,f]. Four C-monosaccharides 1–4 are chosen for illustration (Fig. 11.6 and Tab. 11.1).

A pattern is immediately apparent in the coupling constants measured between the C1 and Cα protons for compounds 1–4. For α(axial)-C-glycosides 1 and 2, a small coupling constant observed between the C1 and Cα-pro-(R) protons ($J_{1,R}$) corresponds to a gauche relationship, and a large coupling constant observed between the C1 and Cα-pro-(S) protons ($J_{1,S}$) corresponds to an antiperiplanar relationship. On the other hand, a small $J_{1,S}$ coupling and a large $J_{1,R}$ coupling are observed for β(equatorial)-C-glycosides 3 and 4. Altogether, these data convincingly demonstrate that the *exo/anti* conformer is preferred over the two other conformers.

Interpretation of $^3J_{H,H}$ coupling constants in terms of discrete conformational states is in fact well justified. For the sake of argument, the modified Karplus equation can translate coupling constant information into a dihedral angle ϕ, yielding a value of +55° for the axial C-glycoside **1** and –80° for the equatorial C-glycoside **4**. These are tantalizingly similar to the values of +55° reported for methyl α-D-glucopyranoside and –70° for methyl β-D-glucopyranoside, respectively [13]. The data could therefore be interpreted in terms of a single dominant conformer slightly distorted from the ideally staggered state. However, variable-temperature NMR experiments have shown that the conformational behavior of C-glycosides **1–4** is better described as a mixture of staggered conformers rather than as a single, non-staggered conformer. Conformer populations were derived by fitting the observed coupling constants to the theoretical values reported for individual staggered conformers [5f, 3c]. The conformational distributions thus estimated for **1–4** (Tab. 11.1) validate the general conclusion that C-glycosidic bonds have a strong preference for the *exo/anti* conformation over the remaining conformers.

The conformational distributions estimated in Tab. 11.1 are also informative for describing the behavior of the aglyconic $C\alpha$–$C\beta$ bond. The $C\alpha$–$C\beta$ bond of axial C-glucoside **1** predominantly adopts the extended *C-anti-C* conformation, whereas that of epimer **2** does not exhibit a well defined conformational preference. This pattern is also observed in the equatorial C-glucosides **3** and **4**; the $C\alpha$–$C\beta$ bond of **3** shows a nearly statistical distribution of the three conformers, whereas that of epimer **4** predominantly adopts the extended *C-anti-C* conformation. The conformational differences due to the $C\beta$ configuration can be understood clearly by analyzing the unfavorable *gauche* and *syn*-pentane interactions across the $C\alpha$–$C\beta$ bond for the three ideal staggered conformers depicted in Fig. 11.7. For the axial C-glucoside **1**, the *C-anti-C* conformer is uniquely free both of unfavorable *syn*-pentane and of carbon-carbon *gauche* interactions, but the remaining *C-anti-O* and *C-anti-H* conformers suffer from these unfavorable steric interactions. For the axial C-glucoside **2**, none of the three ideal staggered conformers is free of such steric interactions; the *C-anti-C* conformer suffers from a *syn*-pentane interaction between the C1–O5 bond and the $C\beta$ hydroxyl group, the *C-anti-O* conformer from a *gauche* interaction involving the C1–$C\gamma$ carbon backbone, and the *C-anti-H* conformer suffers from both [14]. The same analysis can be performed on the equatorial C-glucosides **3** and **4**.

Two additional conformational features are worth mentioning. Firstly, it is remarkable that the preferences for the *exo/anti* conformation in C-glycosides **2** and **3** are only slightly compromised by the conformational instability of the aglyconic $C\alpha$–$C\beta$ bond. The reason for this is easily revealed upon examination of a molecular model: the adoption of other staggered conformations about the C-glycosidic bond serves only to increase torsional strain, rather than to relieve it. Thus, in situations in which unfavorable steric interactions exist across the C-glycosidic linkage, conformational distortion takes place predominantly around the C-aglyconic bond. Interestingly, this discrimination can also be found in the conformational behavior of O-glycosides [13].

A second, more subtle effect is noticeable on comparison of the conformational stabilities of α- and β-C-glycosides. Axial C-glycosides **1** and **2** appear to have slightly stronger preferences for the *exo/anti* conformation than equatorial C-glycosides **3** α(axial)-*C*-glucosides **1** (X=OH; Y=H) and **2** (X=H; Y=OH)

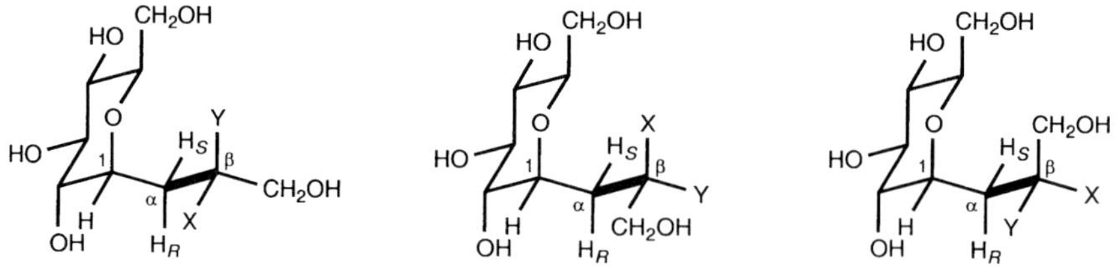

1 : C-anti-C
2 : C-anti-C

1 : C-anti-H
2 : C-anti-O

1 : C-anti-O
2 : C-anti-H

β(equatorial)-*C*-glucosides **3** (X=OH; Y=H) and **4** (X=H; Y=OH)

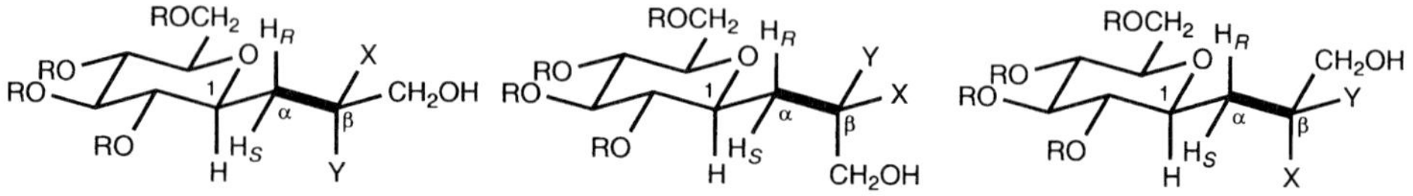

3 : C-anti-C
4 : C-anti-C

3 : C-anti-O
4 : C-anti-H

3 : C-anti-H
4 : C-anti-O

Fig. 11.7 Conformational analysis of the aglyconic Cα–Cβ bond (marked in bold) of *C*-monosaccharides **1** (X=OH, Y=H), **2** (X=H, Y=OH), **3** (X=OH, Y=H, R=H), and **4** (X=H, Y=OH, R=H). For the conformational analysis of the *C*-glycosidic bond, see Fig. 11.5.

and **4**. The differences are barely large enough to be significant, but this trend has been observed in all subsequent studies and appears to be quite general. A number of such interesting and intriguing features have been recognized in the conformations of *C*-glycosides. However, we will not go into further detail here, but instead call attention to several pertinent references in the literature [5 f, g, l, 3 c].

The effects of hydrogen bonding and solvation on *C*-glycoside conformations are summarized in Tab. 11.2 and Tab. 11.3 [5 a, f]. Overall, one sees very little difference in coupling constants between the polyols **1** and **3** in D_2O, their peracetates in $CDCl_3$, or their permethyl ethers in $CDCl_3$, thus indicating only a minor redistribution of conformers (Tab. 11.2) [15]. Similarly, NMR studies on the polyols in D_2O, MeOH-d_4, DMSO-d_6, and pyridine-d_5 show little effect of solvent polarity on the conformations (Tab. 11.3). Interestingly, a similar lack of sensitivity has been observed in conformational studies on parent *O*-glycosides [16a]. This trend may seem somewhat surprising, given the dramatic differences in hydrogen bonding, electronegativity, and solvent polarity. However, as modest variations in conformational equilibria often belie the complexity of the compensatory effects accompanying changes in solvation energy or solvent reorganization, one needs to be cautious in interpreting the implications of such observations [16].

From the results as a whole, we can summarize certain conformational characteristics of *C*-glycosides:

Tab. 11.2 Vicinal coupling constants (Hz) for the polyols **1** and **3** and their protonated forms (solvent as indicated, 500 MHz). Compound **1** is an a(axial)-C-glycoside with one major aglyconic conformation, and **3** is a β(equatorial)-C-glycoside with a distribution of aglyconic conformations [5 f]. *Exo/anti* and **C-anti-C** populations were estimated from a three-conformer model.

	-OR	Solvent	$J_{1,R}$	$J_{1,S}$	exo/anti	$J_{R,\beta}$	$J_{S,\beta}$	C-anti-C
1	-OH	D$_2$O	2.9	11.9	>95	10.5	2.6	85
	-OAc	CDCl$_3$	2.9	11.7	>95	9.7	3.9	75
	-OMe	CDCl$_3$	2.8	11.8	>95	9.7	3.0	75
3	-OH	D$_2$O	9.2	2.8	75	6.5	6.2	40
	-OAc	CDCl$_3$	9.0	2.9	75	6.0	6.0	35
	-OMe	CDCl$_3$	9.2	2.7	75	4.4	7.6	20

Tab. 11.3 Solvent effects observed for the vicinal coupling constants (Hz) of **1** and **3** (500 MHz) *Exo/anti* and **C-anti-C** populations were estimated using a three-conformer model.

	Solvent	$J_{1,R}$	$J_{1,S}$	exo/anti	$J_{R,\beta}$	$J_{S,\beta}$	C-anti-C
1	D$_2$O	2.9	11.9	>95	10.5	2.6	85
	CD$_3$OD	3.3	11.3	95	10.1	2.7	80
	DMSO	2.5	11.5	>95	10.1	2.6	80
	C$_5$D$_5$N	3.8	10.7	90	9.6	3.5	75
3	D$_2$O	9.2	2.8	75	6.5	6.2	40
	CD$_3$OD	8.9	2.8	75	6.9	5.3	45
	DMSO	8.9	2.8	75	6.6	5.8	45
	C$_5$D$_5$N	9.2	2.8	75	7.1	5.2	45

1. a strong preference for the *exo/anti* conformation about the C-glycosidic bond appears to be universal,
2. in cases in which there are unfavorable steric interactions about the interglycosidic linkage, conformational distortion will occur preferentially around the C-aglyconic bond, and
3. conformational perturbations due to hydrogen bonding or solvent effects are generally small or negligible.

Importantly, all of these characteristics have been recognized for the parent O-glycosides, which makes a strong case for the notion that the conformational behavior of C-glycosides is comparable to that of O-glycosides, at least to a first degree of approximation. Of course, the experimental results discussed above do not preclude a stereoelectronic contribution to the conformational tendencies of O-glycosides, nor should they be taken to mean that C-glycosides are identical to their parent O-glycosides in every respect. However, they do demonstrate that steric effects are sufficient to reproduce the general conformational features of O-glycosides.

11.4
1,4-Linked C-Disaccharides: the Importance of syn-Pentane Interactions

The carbon analogues of the 1,4-linked disaccharides C-maltose (Glc-α(1' → 4)-C-Glc; **5**), C-cellobiose (Glc-β(1' → 4)-C-Glc; **7**), and their C2' epimers **6** and **8**, respectively, provide the next level of insight into the conformational profiles of C-glycosides [5c, h, i]. The observed $^3J_{H,H}$ coupling constants establish that the C-glycosidic bonds of C-disaccharides **5–8** preferentially adopt the *exo/anti* conformation, whereas the C-aglyconic bonds do not preferentially adopt a single, distinct staggered conformation (Fig. 11.8). As in the C-monoglycoside series, these data demonstrate that the C-disaccharides minimize steric destabilization primarily through rotation of the C-aglyconic bonds and, once again, the equivalent conformational behavior has been observed for the corresponding O-disaccharides [13]. In this section we take a closer look at some of the factors affecting the conformational behavior of the C-aglyconic bond. This provides not only a better understanding of the local steric interactions involved, but also some direction as to how such interactions can be rationally modified.

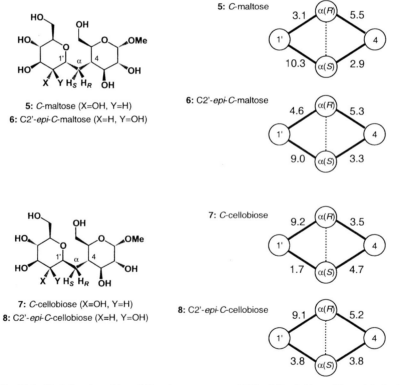

Fig. 11.8 Methyl α-glycosides of C-maltose (**5**), C2'-epi-C-maltose (**6**), C-cellobiose (**7**), and C2'-epi-C-cellobiose (**8**), and their observed coupling constants [5c, i].

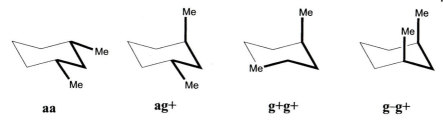

Fig. 11.9 Analysis of *n*-pentane conformers superimposed on a chair conformation of cyclohexane.

The most influential steric effect in the C-aglyconic conformations of **5–8** is the *syn*-pentane interaction. This is historically derived from the hydrocarbon *n*-pentane, which can adopt four different staggered conformations: **aa**, **ag⁺**, **g⁺g⁺**, and **g⁻g⁺**, with + and − indicating the relative sense of rotation. These are ranked in ascending order of torsional strain energy; the **g⁻g⁺** (*syn*-pentane) conformer is estimated to be higher in strain energy than the **aa** conformer by more than 3.3 kcal mol⁻¹ [6]. By superimposing these conformers on the chair conformation of cyclohexane (Fig. 11.9), one can appreciate that the Me/Me-interaction in the

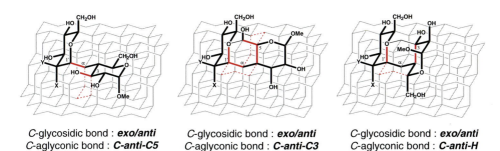

Fig. 11.10 Diamond lattice-assisted conformational analysis: *C*-maltoside (**5**: X=OH, Y=H) and C2′-*epi*-*C*-maltoside (**6**: X=H, Y=OH) [5 c, i]. 1,3-Diaxial-like relationships across the interglycosidic linkage are highlighted with bold red bonds [14]. The corresponding cyclohexane chairs are shown by dashed red lines.

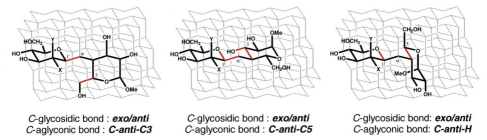

Fig. 11.11 Diamond lattice-assisted conformational analysis: *C*-cellobioside (**7**: X=OH, Y=H) and C2′-*epi*-*C*-cellobioside (**8**: X=H, Y=OH) [5 c, i]. 1,3-Diaxial-like relationships across the interglycosidic linkage are highlighted with bold red bonds.

g⁻g⁺ conformer resembles the Me/Me-interaction in the 1,3-diaxial conformer of *syn*-1,3-dimethylcyclohexane. Similarly, the **ag⁺** conformer of *n*-pentane resembles the case of *anti*-1,3-dimethylcyclohexane.

In the case of *C*-disaccharides, the cyclohexane framework can be extended into a three-dimensional diamond lattice to facilitate a priori conformational analysis about the interglycosidic linkage (Figs. 11.10 and 11.11, see p. 315) [5c]. Diamond lattice-assisted conformational analysis is especially useful for identifying *syn*-pentane relationships in staggered conformations. For example, it can be easily recognized that the C3–OH and the Ca–C1' bonds have a 1,3-diaxial-like relationship in the *C-anti-C5* conformer of 1,4-linked *C*-disaccharide **5** (cf., the cyclohexane chair-conformation marked in red in Fig. 11.10) [14]. In the diamond lattice-assisted conformational analysis, *syn*-pentane interactions are envisioned as 1,3-diaxial-like interactions, and are therefore referred to as 1,3-diaxial-like interactions throughout this chapter.

As determined from the experimentally measured $^3J_{H,H}$ coupling constants (Fig. 11.8), the *C*-glycosidic bonds of *C*-maltose (**5**) and its C2'-epimer (**6**) predominantly adopt the *exo/anti* conformation [5c, i]. Therefore, conformational analysis around only the *C*-aglyconic bond is sufficient to predict the secondary structure profiles for the *C*-disaccharides, at least to a first degree of approximation. The three ideally staggered conformers of **5** and **6** around the *C*-aglyconic bond (i.e., the *C-anti-C5*, *C-anti-C3*, and *C-anti-H* conformers) are superimposed on a diamond lattice, keeping the *C*-glycosidic bond in the *exo/anti* conformation (Fig. 11.10). This diamond lattice-assisted conformational analysis reveals at least one destabilizing 1,3-diaxial-like interaction in each conformer. This would suggest that none of these conformers should be favored over the others. From this analysis, it is also evident that the functional group at C2' has no direct involvement with the *C*-aglyconic conformations in question. Therefore, one would anticipate the overall conformational profiles of **5** and **6** to be similar.

These predictions are indeed supported by the experimentally measured $^3J_{H,H}$ coupling constants summarized in Fig. 11.8. Firstly, the coupling constants about the Ca–C4 bonds were found to be small (≤ 5.5 Hz), thereby showing that the *C*-aglyconic bond does not exist predominantly in either the *C-anti-C5* or *C-anti-C3* conformations. However, from these experimentally observed coupling constants alone, it is not straightforward to refine their conformational properties further. In this regard, Jimenez-Barbero and others have provided interesting insights into the conformational behavior of the *C*-aglyconic bond through extensive NOESY studies coupled with theoretical modeling. Secondly, the coupling constants about the *C*-glycosidic and *C*-aglyconic bonds of **5** and **6** are very similar, which strongly indicates their overall conformational similarity. It is interesting to note that the preference of the *C*-glycosidic bond for the *exo/anti* conformer is slightly compromised in **6** by the inversion of the C2' configuration; this is probably due to the removal of a destabilizing 1,3-diaxial-like interaction from the competing *nonexo/gauche* conformer (cf. Fig. 11.4). However, these interactions do not play as dominant a role in the *C*-glycosidic conformation as they do in the *C*-aglyconic conformation.

A priori conformational analysis of *C*-cellobiose (**7**) and its C2'-epimer **8** can be performed in parallel fashion to that of **5** and **6** above (Fig. 11.11) [5c, i]. As would be expected, the observed $^3J_{H,H}$ coupling constants once again support the predicted conformational profiles:

1. the C-glycosidic bonds predominantly adopt the *exo/anti* conformation,
2. the C-aglyconic bonds do not predominantly adopt either the *C-anti-C3* or the *C-anti-C5* conformations, and
3. the two C-disaccharides **7** and **8** share similar global conformational profiles.

With respect to the conformational behavior of the C-aglyconic bond, extensive conformational studies on C-lactose (Gal-β(1' → 4)-C-Glc) have been carried out by several groups, providing an instructive complement to our first-approximation analysis [5q, 17]. These authors have suggested that the C-aglyconic bond of this molecule exists as a mixture of the *C-anti-H* conformer and a non-staggered conformation between the *C-anti-C3* and *C-anti-C5* conformers, in which the C1' carbon nearly eclipses the C4–H bond [17a]. In a global sense, this observation is in line with the conclusion obtained from diamond lattice-assisted conformational analysis: in the absence of a low-energy staggered conformation with minimal torsional strain, the C-aglyconic bond is likely to adopt several conformations, the geometries and strain energies of which are determined by complex, secondary effects. Such conformations cannot be determined intuitively, but require semiempirical methods such as those used in the studies above.

Although there is little doubt that a detailed study of conformational characteristics is interesting and valuable for various purposes, one should also keep in mind that there is great value in models that predict approximate conformational behavior from simple and well established principles. In this respect, the diamond lattice-assisted conformational analysis provides two unique qualities for carbohydrate conformational analysis: (1) it identifies steric interactions that have a significant influence on the conformational profile of a given substrate, and (2) it suggests simple yet rational structural modifications for inducing changes in conformational profile.

One more aspect of C-aglyconic conformation deserves mention here. Conformational studies on Gal-α(1' → 3)-C-Man (**9**) by Vogel revealed an interesting trend in steric destabilization by 1,3-diaxial-like interactions [18]. In spite of unavoidable 1,3-diaxial-like relations between the Gal and Man units in all three staggered conformers of **9**, the observed $^3J_{H,H}$ coupling constants across the C-aglyconic linkage do not show the characteristic pattern observed in C-disaccharides **5–8**. Instead, the large values observed for $J_{1',S}$ and $J_{3,S}$ indicate a preference for the *C-anti-C4* conformer, which suffers from a 1,3-diaxial-like interaction between two ring bonds (Fig. 11.12).

This is an intriguing observation, particularly when compared with the diamond lattice-assisted conformational analysis of C-maltose (**5**) (Fig. 11.10). At first glance, the steric environment for the *C-anti-C4* conformer of **9** might appear to be similar to that of the *C-anti-C3* conformer of **5**, because each exhibits a 1,3-diaxial-like interaction between the C1'–O5 bond of the pyranose ring and a C–C bond of the opposing ring. Upon closer inspection, it is apparent that the *C-anti-C3* conformer of **5** is further destabilized by 1,3-diaxial-like interactions between the C1'–Cα and C5–C6 bonds, as well as by incidental interactions between the C5'–C6' and C6–OH bonds

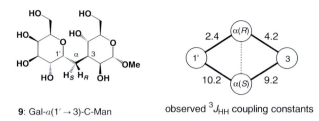

9: Gal-α(1′→3)-C-Man

observed $^3J_{HH}$ coupling constants

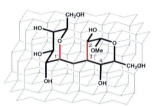
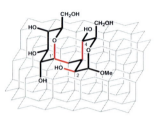

C-glycosidic bond : **exo/anti**
C-aglyconic bond : **C-anti-C2**

C-glycosidic bond : **exo/anti**
C-aglyconic bond : **C-anti-C4**

C-glycosidic bond : **exo/anti**
C-aglyconic bond : **C-anti-H**

Fig. 11.12 Diamond lattice-assisted conformational analysis of Gal-α(1′→3)-C-Man (**9**) [18]. 1,3-Diaxial-like relationships across the interglycosidic linkage are highlighted with bold red bonds [14].

[14]. In contrast, the C-anti-C4 conformer of **9** has no corresponding 1,3-diaxial interaction. Furthermore, the incidental interactions not only appear to be sterically less severe, but may even provide some stabilization because of the antiferroelectric alignment of the polarized C2–OH and O5′–C5′ bonds.

Similar trends have also been observed for Gal-α(1′→3)-C-Glc [5c], L-Fuc-α(1′→3)-C-GalNAc [19], and Man-α(1′→2)-C-Man [20]. Overall, these results suggest that steric destabilization resulting from ring/ring 1,3-diaxial-like interactions (e.g., the C-anti-C4 confomrer of **9**) is not as severe as that involving at least one exocyclic bond (e.g., the C-anti-C2 and C-anti-H conformers of **9**) [21].

11.5
Prediction of Conformational Preference and Experimental Validation

The examples discussed demonstrate that, by examining local steric interactions, one can qualitatively predict the effect of a structural modification on the conformational preference for a given C-glycoside, and presumably for its parent O-glycoside as well. This presents an exciting opportunity for applying diamond lattice-assisted conformational analysis as a simple but rational device to program carbohydrate conformational profiles: the removal, or the addition, of specific 1,3-diaxial-like destabilizations can produce a dramatic change in local conformational bias across the C-aglyconic bond, consequently with a profound effect on global conformation. Very importantly, the projected changes in C-glycoside conformation can be validated by ^{1}H NMR coupling constant analysis, as illustrated in the examples below.

11.5 Prediction of Conformational Preference and Experimental Validation

Example 1: The pyranose ring of glucose exists in a $^4C_1$ chair conformation, whereas that of 1,6-anhydroglucose is locked in a $^1C_4$ chair conformation. Inversion of the reducing-end pyranose ring in C-cellobiose (**7**) and C2'-*epi*-C-cellobiose (**8**) is anticipated to have a dramatic effect on conformational preferences across the interglycosidic linkage, due to the removal of 1,3-diaxial-like destabilizations from certain conformers. This prediction can be evaluated experimentally with 1,6-anhydro-C-cellobiose (**10**) and C2'-*epi*-1,6-anhydro-C-cellobiose (**11**), which bear 1,6-anhydroglucose residue at the reducing end (Fig. 11.13). A diamond lattice-assisted conformational analysis shows that the *exo/anti-C-anti-C3* conformer is uniquely free of 1,3-diaxial-like steric destabilization, and therefore one would predict that **10** and **11** should exist predominantly in this global conformation. The $^3J_{H,H}$ coupling constants observed for C-disaccharides **10** and **11** clearly show a predominantly antiperiplanar arrangement between the C*apro*-S and the C4 protons, which is indeed consistent with this prediction [5c, i].

A similar a priori analysis can be applied to the conformational effect of inverting the reducing-end pyranose ring of C-maltose (**5**) and C2'-*epi*-C-maltose (**6**). This prediction was also verified by $^3J_{H,H}$ coupling constant analysis with 1,6-anhydro-C-maltose (**12**) and C2'-*epi*-1,6-anhydro-C-maltose (**13**) (Fig. 11.14) [5c, i].

Example 2: Diamond lattice-assisted conformational analysis of C-maltose (**5**) reveals that the *C-anti-C5* conformer is destabilized by a 1,3-diaxial-like interaction

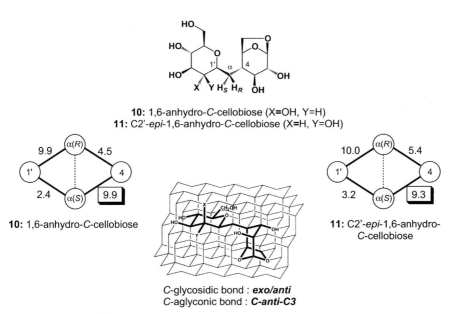

Fig. 11.13 Observed $^3J_{H,H}$ coupling constants and diamond lattice-assisted conformational analysis: 1,6-anhydro-C-cellobioside (**10**: X=OH, Y=H) and its C2' epimer (**11**: X=H, Y=OH) [5c, i]. The most notable changes in the $^3J_{H,H}$ coupling constants, relative to **7** and **8**, respectively, are highlighted.

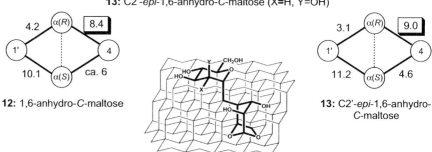

Fig. 11.14 Observed $^3J_{H,H}$ coupling constants and diamond lattice-assisted conformational analysis: 1,6-anhydro-C-maltose (**12**: X=OH, Y=H) and its C2′ epimer (**13**: X=H, Y=OH) [5c, i]. The most notable changes in the $^3J_{H,H}$ coupling constants, relative to **5** and **6**, respectively, are highlighted.

between the C1′–Cα bond and the exocyclic C3–OH bond (Fig. 11.15) [5c, d, e, i]. Similarly, the *C-anti-C3* conformer of *C*-cellobiose (**7**) is destabilized by a 1,3-diaxial-like interaction between the C1′–Cα bond and the exocyclic C5–C6 bond (Fig. 11.16). Removal of these groups, or inversion of their configurations, should create a local environment free of 1,3-diaxial-like destabilizations across the C-aglyconic bond. The ^{1}H NMR coupling constant analysis of 3-deoxy-*C*-maltose (**14**) and 5-dehydroxymethyl-*C*-cellobiose (**15**) unequivocally demonstrates this to be the case: the coupling constants about the C-aglyconic bond of **14** are dramatically different from those of **5**, and indicate an antiperiplanar arrangement between the C1′ and Cα pro-*S* protons and between the Cα pro-*R* and C4 protons, equivalent to a single, predominant conformation consisting of the staggered *exo/anti-C-anti-C5* conformers [5d, e, i]. The corresponding NMR analysis for **15** indicates that it too exists predominantly in a single conformation, this time made up of the *exo/anti-C-anti-C3* conformers.

Example 3: On the other hand, one can identify a structural modification that should not have a significant impact on the global conformational profile of a given substrate. The conformational similarities mentioned for *C*-maltose and *C*-cellobiose and their respective C2′ epimers represent such a case (Fig. 11.8). This is because the C2′ position is not directly involved in the 1,3-diaxial-like steric interactions in question, and therefore structural modification at that position would be expected and shown to result in minimal conformational change [5c, i].

Fig. 11.15 Observed $^3J_{H,H}$ coupling constants and diamond lattice-assisted conformational analysis: C-maltose (**5**: X=OH) and 3-deoxy-C-maltose (**14**: X=H) [5 d, e, i]. 1,3-Diaxial-like relationships across the interglycosidic linkage are highlighted with bold red bonds. The most notable change observed in the $^3J_{H,H}$ coupling constants is highlighted.

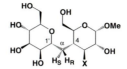

5: C-maltose (X=OH)
14: 3-deoxy-C-maltose (X=H)

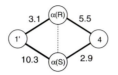 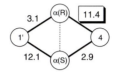

5: C-maltose   **14:** 3-deoxy-C-maltose

C-glycosidic bond : *exo/anti*
C-aglyconic bond : *C-anti-C5*

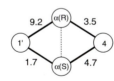 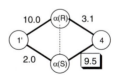

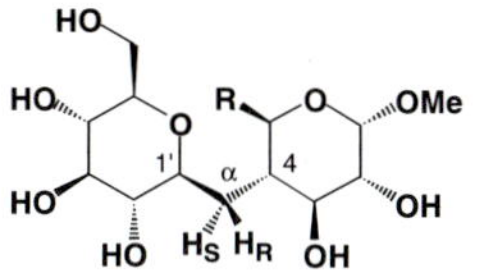

7: C-cellobiose   **15:** 5-deshydroxymethyl-C-cellobiose

Fig. 11.16 Observed $^3J_{H,H}$ coupling constants and diamond lattice-assisted conformational analysis: C-cellobiose (**7:** R=CH$_2$OH) and 5-dehydroxymethyl-C-cellobiose (**15:** X=H) [5 d, e, i]. 1,3-Diaxial-like relationships across the interglycosidic linkage are highlighted with bold red bonds. The most notable change observed in the $^3J_{H,H}$ coupling constants is highlighted.

7: C-cellobiose (R=CH$_2$OH)
15: 5-deshydroxymethyl-C-cellobiose (R=H)

C-glycosidic bond : *exo/anti*
C-aglyconic bond : *C-anti-C3*

11.6
Programming Oligosaccharide Conformation

We are now at a point at which it becomes possible to define a discrete set of simple rules that can be used to design *C*-glycosides with specific conformational profiles. Summarizing from the beginning, we have established the following generalities:

1. the *C*-glycosidic bond always prefers the *exo/anti* conformer,
2. the conformational preference of the *C*-glycosidic bond has priority over that of the *C*-aglyconic bond,
3. the *C*-aglyconic bond will avoid sterically destabilizing 1,3-diaxial-like interactions whenever possible, and
4. remote substituent effects seldom play an influential role on *C*-glycosidic conformations.

With respect to the conformational preference of the *C*-aglyconic bond in *C*-disaccharides, we add the following observations:

5. staggered conformers without 1,3-diaxial-like interactions are strongly preferred,
6. staggered conformers with 1,3-diaxial-like interactions between C1' and an exocyclic substituent are severely destabilized, and
7. conformers with 1,3-diaxial-like interactions between O5' and a ring carbon (ring/ring interactions) are relatively stable, unless accompanied by severe 1,3-diaxial-like or incidental interactions.

A remarkable example of prioritization in conformational profiles can be seen in the case of the a-(1' → 3)-*C*-linked disaccharides (Fig. 11.17) [5d, 18]. Three a-*C*-galactosides linked in 1,3-fashion to galactose, mannose, and glucose have been synthesized and subjected to conformational studies, which revealed that each *C*-disaccharide exhibits a dramatically different conformational profile from the others: Gal-a(1' → 3)-*C*-Gal (**16**) predominantly adopts the *C-anti-C2* conformation [5d], Gal-a(1' → 3)-*C*-Man (**9**) exists predominantly as the *C-anti-C4* conformer [18], and Gal-a(1' → 3)-*C*-Glc (**17**) does not exhibit a distinct preference for either [5d]. The difference between the *C-anti-C4* conformer of **9** and the *C-anti-H* conformer of **17** might seem subtle, but the latter also suffers from a *gauche* interaction involving the C2–C3 bond. Thus, three fundamentally different conformational profiles are accessible simply by varying the pattern of 1,3-diaxial-like relationships.

11.7
Conformational Design of C-Trisaccharides based on a Human Blood Group Antigen

The rules outlined above suggest a unique strategy for programming the conformational profiles of both *C*- and *O*-glycosides through manipulation of 1,3-diaxial-like steric interactions. We have chosen to demonstrate this concept with the H-

11.7 Conformational Design of C-Trisaccharides based on a Human Blood Group Antigen

Type II blood group trisaccharide [5j,o,p]. The human blood group Type II antigens are membrane-bound glycosphingolipids commonly found on the surfaces of red blood cells, and consist of a variable oligosaccharide group linked through a β-lactosyl spacer to the terminal hydroxyl group of ceramide (Fig. 11.18). The fine structure of the blood group determinant is critical for recognition; the Type I antigens found on the surface of epithelial cells, for example, typically do not elicit a response from biological receptors primed for recognition of the isomeric Type II antigens, even though they differ by only a single glycosidic linkage [22].

The H-Type II trisaccharide is an ideal substrate for demonstrating rational control over carbohydrate secondary structure, for two reasons. Firstly, with diamond lattice-assisted conformational analysis (Fig. 11.19), one can immediately recognize that the C3' hydroxy group (Y=OH) and C5 hydroxymethyl group (X=CH$_2$OH) are in strategic positions for modulation of the conformational behavior of their respective C-aglyconic bonds. Removal of either of these groups can be expected to result in significant conformational changes both locally and globally; flexibility would be replaced with a strong preference for well defined conformers. Secondly, the biological role of the blood group trisaccharide provides us with the enticing possibility of correlating recognition with the conformational behavior of the carbohydrate ligands. While a number of complex issues exist for protein-carbohydrate recognition, our ultimate purpose is to show that a priori conformational analysis can provide, through rational and strategic modifications, the critical link between the secondary structure of a given carbohydrate ligand and its biological activity.

The conformational analysis of the carbon analogue of the H-type II trisaccharide can be divided into two parts: (1) analysis of the C1'–Ca'–C4 linkage between the D-galactose and D-glucosamine moieties, and (2) analysis of the C1''–Ca''–C2' linkage between the L-fucose and D-galactose moieties. From the examples given in the previous sections, we can expect little or no interaction between the two disaccharide units: the 2'-fucosylmethyl functionality should have little bearing on the conformation of the C1'–Ca'–C4 bridge, and the glucosamine ring should not affect the conformational behavior across the C1''–Ca''–C2' bridge. We can also expect the last remaining C-glycosidic (C1–Ca) bond to show the same conformational behavior as observed in simple C-monoglycosides, again with minimal effect on the other linkages. Therefore, the conformational analysis of the C-trisaccharide can be regarded as the sum of three independent C-glycoside systems, and provide an approximate conformational model of the parent O-trisaccharide.

With respect to the C-trisaccharide **18**, the conformational behavior of the D-Gal(1' → 4)-D-GlcNAc linkage should be similar to that of C-cellobiose (**7**), meaning that the C-glycosidic (C1'–Ca') bond can be expected to adopt the *exo/anti* conformation, while the C-aglyconic (Ca'–C4) bond is subject to severe steric destabilization in each of the three staggered conformers. The conformation across the L-Fuc(1'' → 2')-D-Gal linkage of **18** can be analyzed in the same manner: the C1''–Ca'' bond should exist in the predicted *exo/anti* conformation, while the Ca''–C2' bond should suffer from unavoidable 1,3-diaxial-like destabilizations in all of the staggered conformations. As discussed for the C-monoglycosides and the 1,4-linked C-disaccharides, conformational distortion would be expected to occur pri-

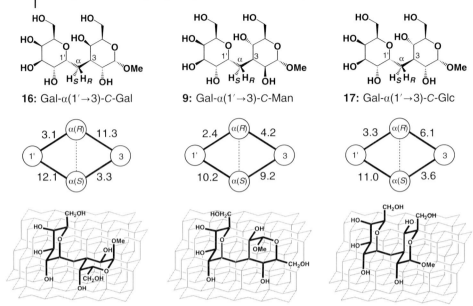

Fig. 11.17 Observed $^3J_{H,H}$ coupling constants (Hz) and diamond lattice-assisted conformational analysis: Gal-α(1′→3)-C-Gal (**16**), Gal-α(1′→3)-C-Man (**9**), Gal-α(1′→3)-C-Glc (**16**) [5d, 18a]. 1,3-Diaxial-like relationships across the interglycosidic linkage are highlighted with bold red bonds [14].

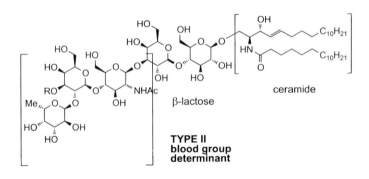

Fig. 11.18 The structures of Type I and Type II human blood group antigens.

11.7 Conformational Design of C-Trisaccharides based on a Human Blood Group Antigen

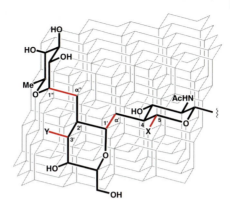

	Fuc-α(1″→2′)-C-Gal	Gal-β(1′→4)-C-GlcNAc
C-glycosidic bond :	exo/anti	exo/anti
C-aglyconic bond :	C-anti-C1′	C-anti-C3

Fig. 11.19 Diamond lattice-assisted conformational analysis of the carbon analogue of the H-type II blood group trisaccharide **18** (X=CH$_2$OH) [5j, o, p]. 1,3-Diaxial-like relationships across the interglycosidic linkage are highlighted with bold red bonds.

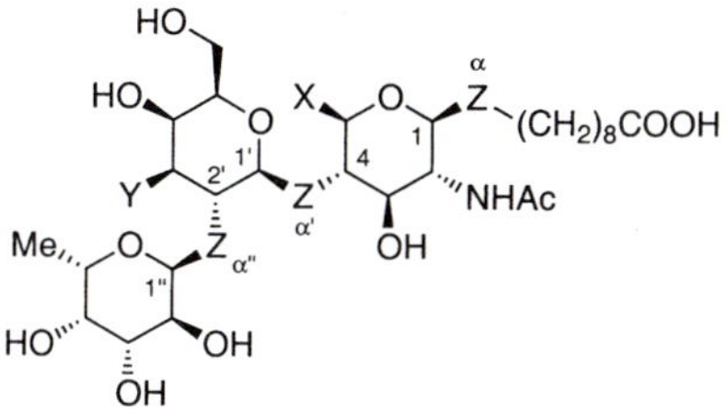

18: X=CH$_2$OH, Y=OH, Z=CH$_2$
19: X=CH$_2$OH, Y=H, Z=CH$_2$
20: X=H, Y=OH, Z=CH$_2$
21: X=Y=H, Z=CH$_2$

22: X=CH$_2$OH, Y=OH, Z=O
23: X=CH$_2$OH, Y=H, Z=O
24: X=H, Y=OH, Z=O
25: X=Y=H, Z=O

Fig. 11.20 The structures of the H-Type II blood group determinant O- and C- trisaccharides and their analogues.

marily around the C-aglyconic bond. The C-trisaccharide **18** is therefore predicted to prefer conformations in which: (1) all three C-glycosidic bonds adopt the *exo/anti* conformation, and (2) neither of the two interglycosidic C-aglyconic bonds have a strong preference for any single staggered conformer.

This sets the stage for conformational programming of structurally modified analogues of the H-Type II blood group trisaccharide (Fig. 11.20). With the fully functional carbon analogue **19** as a point of reference, we can design conformationally modified H-Type II analogues by systematically eliminating key 1,3-diaxial-like interactions:

1. removal of the C3′ hydroxyl group (i.e., **18 → 19**) should encourage the C1″–Cα″–C2′ bridge to adopt a stable conformation made up of the staggered *exo/anti*-C1″-*anti*-C1′ conformers, while leaving the conformation across the galactose/glucosamine linkage unaffected,
2. removal of the C5 hydroxymethyl group (i.e., **18 → 20**) should encourage the C1′–Cα′–C4 bridge to adopt a stable conformation made up of the *exo/anti* – C1′-*anti*-C3 conformers, without affecting the conformational behavior of the fucose/galactose system, and lastly,
3. removal of both the C3′ hydroxyl group and the C5 hydroxymethyl group (i.e., **18 → 21**) should produce an analogue that exists predominantly in the ideally staggered conformation shown in Fig. 11.19.

In this manner, four different conformational profiles for the H-Type II trisaccharide can be programmed by employing small but rational structural modifications. This analysis should be applicable both to the C-trisaccharides and to their parent oxygen congeners, so the same structural modifications should also be possible in the native H-Type II blood group *O*-trisaccharide, with comparable changes in conformational profile.

With this design plan in hand, the four C-trisaccharides **18–21** and their four corresponding O-trisaccharides **22–25** were synthesized and analyzed by ^{1}H NMR spectroscopy (Fig. 11.21) [5j,n,o]. As expected, the C-glycosidic bonds in all four analogues exist predominantly in the *exo/anti* conformation. More importantly, each of the four C-trisaccharides **18–21** proved to have a unique conformational profile, exactly as programmed:

1. the C-aglyconic bonds of **18** do not demonstrate a strong preference for any staggered conformation characterized by large $^3J_{H,H}$ coupling constants (i.e., the C1″-*anti*-C1′ and C1″-*anti*-C3′ conformations or the C1′-*anti*-C3 and C1′-*anti*-C5 conformations),
2. the 3′-deoxy analogue **19** prefers a distinct conformation across the L-Fuc(1″ → 2′)-D-Gal linkage, with the C-aglyconic Cα″–C2′ bond adopting a predominantly extended C′-*anti*-C1′ conformation, characterized by the large coupling constant between the C′(S) and the C2′ protons,
3. the 5-deshydroxymethyl analogue **20** prefers a distinct conformation across the D-Gal(1′ → 4)-D-GlcNAc linkage, with the C-aglyconic Cα′–C4 bond predominantly adopting an extended C1′-*anti*-C3 conformation, characterized by the large coupling constant between the Cα(S) and the C4 protons, and
4. the doubly modified analogue **21** exhibits a well defined conformational preference across both linkages and exists predominantly as a single conformer (cf. the conformation shown in Fig. 11.19).

11.7 Conformational Design of C-Trisaccharides based on a Human Blood Group Antigen | 327

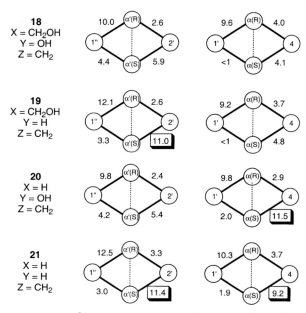

Fig. 11.21 The $^3J_{H,H}$ coupling constants observed for the H-Type II blood group determinant C-trisaccharide (**18**) and its analogues **19–21**. The most notable change observed in the $^3J_{H,H}$ coupling constants is highlighted.

The differences in secondary structure in these C-trisaccharides are striking. The predominant time-averaged solution conformations for **18–21** were generated from the observed $^3J_{H,H}$ coupling constants, complemented by experimental data obtained from 2D-NOESY NMR spectroscopy (Fig. 11.22). The distance-dependent nuclear Overhauser effects (NOEs) yielded additional insights into the conformational changes induced upon removal of 1,3-diaxial-like destabilizations. For example, a strong NOE across the Gal(1′ → 4)-GlcNAc linkage of compound **19** was observed between the methine protons at C1′ and C3, which corresponds to the C-anti-H conformer, but was replaced in the C5-dehydroxymethyl analogue **20** by an exchange between the C1′ and the equatorial C5 protons, characteristic of the C-anti-C3 conformation. A similarly large difference in NOE interactions across the Fuc(1″ → 2′)-Gal linkage was noted, an exchange between the C1″ and the C1′ protons in compound **20** (C-anti-H) being replaced by one between the C1″ and equatorial C3′ protons in C3′-deoxy C-trisaccharide **19** (C-anti-C1′). Overall, these NOEs have reinforced our position that the four C-trisaccharides adopt the ground-state conformational profiles as programmed by the manipulation of steric interactions.

In addition to the examples cited here, a variety of C-glycosides have been synthesized and subjected to conformational analysis, which has provided ample evidence of the many interesting structural characteristics in this class of compounds [23–28]. In this chapter we have deliberately focused on a limited number of C-glycosides in order to illustrate the logical progression toward our analytical

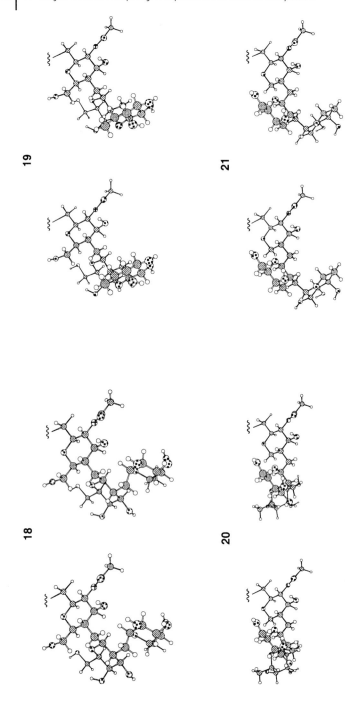

Fig. 11.22 Stereoviews of C-trisaccharides 18–21. In each compound the C1 hydrocarbon side chain has been truncated and the relative orientation of the N-acetylglucosamine ring is fixed [5j, o, p].

approach based on diamond lattice-assisted conformational analysis. In this manner, we hope to highlight the unique value of this approach toward carbohydrate-based molecular-architecture design. The mechanism of our analysis can be summarized as follows: (1) to identify specific steric interactions with significant influences on local conformational profiles, and (2) to suggest simple but rational structural modifications for modulating of global conformational profiles.

The conformational analysis outlined in this chapter can be applied to molecules and/or structural moieties other than carbohydrate analogues. A number of natural products are known to contain C-glycosides as structural elements, and are represented here by palytoxin [3a, 29, 30] and maitotoxin [31, 32]. Interestingly, all C-glycosidic and C-aglyconic bonds present in both palytoxin and maitotoxin exhibit, at least to a first approximation, the conformational properties predicted by the principles discussed in this chapter. The following two examples suggest how this approach might be extended to a new level of molecular-architecture design based on carbohydrates and related compounds.

On the basis of the conformational analysis discussed, it was predicted, and confirmed by NMR studies, that the C56–C75 segment of maitotoxin would predominantly adopt the conformation depicted in Fig. 11.23 (*top*) [31]. This preferred conformation yields an extended secondary structure in the backbone of maitotoxin, as indicated by the direction of the arrows. Similarly, it was predicted, and proven, that the C31–C43 segment would exist predominantly in the conformation shown in Fig. 11.23 (*bottom*). Unlike in the C56–C75 case, however, this pre-

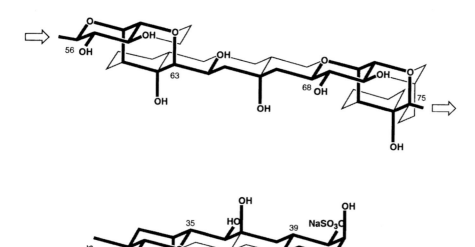

Fig. 11.23 The preferred conformations observed for two partial structures of the marine natural product maitotoxin. The arrows indicate the direction in which the backbone of the molecule extends [31a].

ferred conformation yields a hairpin-like secondary structure in the backbone of maitotoxin. Thus, by positioning C-glycosides with predictable conformational preferences at strategic positions, one can design and assemble larger molecular architecture arrays.

11.8
Conformational Design: Relationship to Biological Activity

We now turn our attention toward the glycomimetic function of C-glycosides in a biological setting. The stoichiometric binding of carbohydrate ligands to proteins has been characterized mostly in terms of: (1) polar interactions and key hydrogen bonds, (2) complementary hydrophilic or hydrophobic domains on the ligand and receptor site, and (3) enthalpic and entropic parameters optimizing hydrogen bond networks between carbohydrate, protein, and interstitial water molecules [16b]. Less well understood is the role of shape selectivity in protein-carbohydrate interactions; the problem here is somehow to relate the free solution conformation of the ligand to its receptor-bound conformation. The interglycosidic linkage of the latter is often distorted into conformations disfavored by local steric interactions, as exemplified in a number of X-ray crystal structures of protein-carbohydrate complexes [33].

In this regard, biological evaluation of conformationally programmed oligosaccharide ligands should be valuable for addressing the importance of carbohydrate secondary structure to receptor binding. Specifically, ligands favoring stable but unnatural solution conformations should display greater free energy differences between their solution and receptor-bound states, reflected in lower binding affinities. This argument is valid under the assumption that changes in binding energy due to ligand modification are greater than any changes in solvation energy (i.e., the number of solute-associated water molecules) due to differences in conformation, a condition generally true when variations in molecular volumes or surface areas are small [34]. It must also be assumed that decreases in binding energy are not due to the loss of key polar interactions. With these assumptions, our goal will be to show that large changes in conformational preference correlate strongly with binding affinity.

The capacity with which C-glycosides can serve as glycomimetic ligands will also justify their use as conformational models of O-glycosides, which up to this point has been based largely on spectroscopic data. While there is strong evidence that the native O-glycosides are subject to the same steric influences as observed in the C-glycosides, the somewhat subjective evaluation of their ground-state conformations leaves room for various interpretations. In fact, the central issue determining whether C-glycosides are suitable isosteres of O-glycosides is *not* the precision with which their solution secondary structures can be compared, but the accuracy with which C-glycosides can mimic the function of O-glycosides in a biological setting. Strong correlation would validate all of the transitive assumptions made with respect to the conformational similarities of C- and O-glycosides, in-

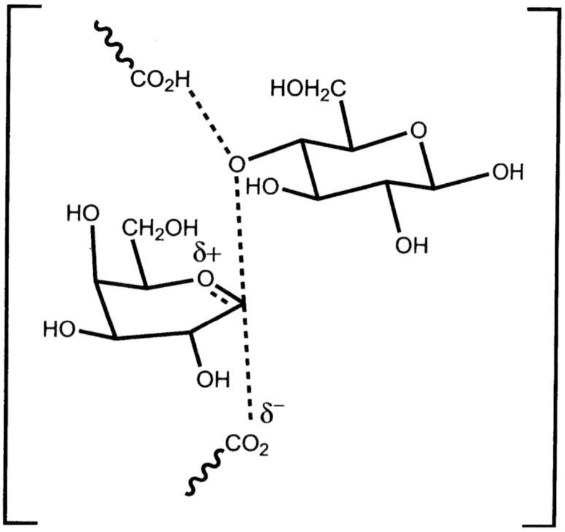

Fig. 11.24 A proposed transition state for enzymatic glycosidic cleavage [4a].

cluding the use of a priori, diamond lattice-assisted conformational analysis for programming conformational profiles of carbohydrate ligands.

It is important to keep in mind that proteins such as lectins and carbohydrate transporters recognize carbohydrates in their ground-state conformations, and should therefore bind to C- and O-glycosides with almost equal capacity. On the other hand, enzymes that cleave glycosidic bonds (glycosidases) provide special environments for stabilizing the transition states of these transformations (Fig. 11.24). In these cases, C-glycoside isosteres cannot properly mimic either the developing electrostatic charge or the three-dimensional geometry of the transition state derived from the parent O-glycoside. Thus, unless an additional device is incorporated in its structure, C-glycoside substrate analogues should not be expected to serve as glycomimetic inhibitors of these enzymes [4].

We provide two specific examples to illustrate that C- and O-glycosides can be recognized by carbohydrate-binding proteins in nearly identical fashion, and that large changes in their conformational profiles correlate with significant differences in biological activity. These examples provide solid experimental evidence that C-glycosides are indeed excellent mimics of O-glycosides.

11.8.1
C-Lactose vs. O-Lactose

Because the structural difference between C-lactose (**26**) and C-cellobiose (**7**) is determined solely by the configuration of the C4′ hydroxyl group, they would be expected to have virtually identical conformational preferences about the interglycosidic linkage (cf., Fig. 11.11). This has been verified by the extensive NMR conformational studies carried out by several groups (Fig. 11.25) [5q, 17]. The observed $^3J_{H,H}$ coupling constants demonstrate that the C-glycosidic bond predominantly adopts the *exo/anti* conformation, whereas the C-aglyconic bond does not strongly prefer either the *C-anti-C3* or the *C-anti-C5* conformation. However, the observed

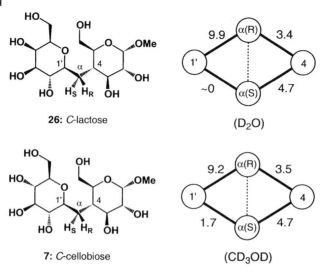

Fig. 11.25 Methyl C-lactoside (**26**) and C-cellobioside (**7**) and their observed $^3J_{H,H}$ coupling constants [5 c, i, q].

$^3J_{H,H}$ coupling constants could not directly yield more detailed information on the conformational behavior of the C-aglyconic bond. Additional NMR studies, including NOESY and off-resonance ROESY experiments, were conducted in order to gain a more defined conformational picture of C-lactose; a comprehensive computational molecular modeling study was also performed [35]. In spite of all these efforts, some disagreements still remain over the precise conformational preferences of the C-aglyconic bond in solution. Nevertheless, there appears to be a consensus on its general conformational properties: it is fairly flexible, with a relatively flat energy surface from the C-anti-C3 conformation through the C-anti-C5 conformation, and a significant contribution from the C-anti-H conformation.

With respect to glycomimetic function, studies by Sinaÿ have demonstrated that C-lactose binds with the same potency as the parent O-lactose to a series of monoclonal antibodies [36]. The most direct evidence of the conformational similarity between O- and C-lactose was ultimately provided by an X-ray crystal structure of C-lactose bound to peanut lectin (Fig. 11.26) [5 q]. The crystal structure reveals that the C-glycosidic bond is nearly in the *exo/anti* conformation, whereas the C-aglyconic bond exists in the near-eclipsed conformation midway between the C-anti-C3 and C-anti-C5 conformations. Most importantly, the bound conformation of C-lactose is practically identical to that of the parent O-lactose, including all the major protein-sugar interactions (Fig. 11.27). Interestingly, the lectin-bound conformation of C-lactose is remarkably similar to one of the two major conformations in the free solution state, as suggested by Berthault [17 a]. It is also worthwhile to add that, according to recent work by Kawagishi, the association and dissociation rates of C-lactose to peanut lectin are practically identical to those of O-lactose [3 b].

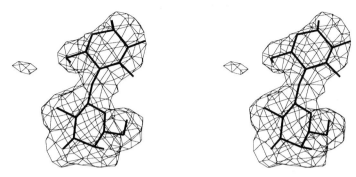

Fig. 11.26 Stereoview of the electron density corresponding to C-lactose bound to peanut lectin [5q].

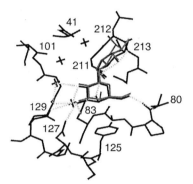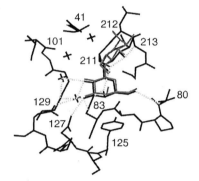

Fig. 11.27 Stereoview of the C-lactose/peanut lectin interactions. An X-ray crystal structure of an O-lactose molecule complexed to lectin (light lines) is superimposed over C-lactose for visual comparison. Water molecules and hydrogen bonds are represented as crosses and as broken lines, respectively. Hydrogen bonds observed between peanut lectin and C-lactose include: Asp83/Gal-O3, Gly104/Gal-O3, Asn127/Gal-O3, Asp83/Gal-O4, Ser211/Gal-O4, Ser211/Gal-O5, Asp80/Gal-O6, Ser 211/Glu-O3, and Gly213/Glu-O3. The amino acid residues found within 4 Å of C-lactose include: Asp80, Ala82, Asp83, Gly103, Gly104, Tyr125, Asn127, Ser211, Leu212, and Gly214 [5q].

The results discussed above make a convincing case that C-lactose is an excellent isostere for O-lactose. However, it should be recalled that proteins that recognize the same molecule often use different structural binding motifs. The binding of C-lactose with galactose-binding lectins such as ricin-B and galectin-1 may be one such example [37]. Finally, it is worth adding that Jimenez-Barbero has suggested that the C-lactose conformation recognized by *E. coli* β-galactosidase does not correspond either with its solution conformation or with that of O-lactose, and that this enzyme selects the *exo/gauche-C-anti-H* conformation of C-lactose, which represents less than 10% of the solution population. Later, it was suggested that the bound conformation of O-lactose is different from that of C-lactose in the enzyme-active site [38].

11.8.2
Human Blood Group Trisaccharides

The trisaccharide portion of the H-type II human blood group determinant possesses the basic structural requirements for binding with high specificity and micromolar affinity to several leguminous lectins, including lectin I of *Ulex europaeus* (UEA-I). As discussed above, we can program specific conformational profiles for the trisaccharide by strategic and rational structural modifications, as identified by diamond lattice-assisted conformational analysis. The consequently large changes in secondary structure can then be correlated with their binding affinities. We selected and synthesized four structurally analogous C-trisaccharides **18–21** and demonstrated that each one exhibits the predicted conformational characteristics (Figs. 11.20–11.22). The same strategic structural modifications were also applied toward the corresponding O-trisaccharides **22–25**, resulting in a parallel set of O- and C-trisaccharides with programmed conformational profiles [5 o, p].

If it is assumed that differences in lectin-trisaccharide binding are due primarily to changes in conformational energies, one would expect:

1. isosteric C-trisaccharides and O-trisaccharides to exhibit similar binding affinities,
2. the four C-trisaccharides **18–21** to exhibit decreases in binding affinity as a function of their relative changes in conformational behavior, and
3. the four corresponding O-trisaccharides **22–25** to exhibit parallel affinity profiles, a consequence of integration of the first two predictions.

These projections are based on the premise that the receptor-binding site is fixed and therefore highly selective, rather than participating in an "induced fit" with the substrate. The exclusive selectivity of the lectin UEA-I for the H-type II trisaccharide (L-Fuc-a(1″ → 2)-D-Gal-β(1′ → 4)-D-β-GlcNAc) over the H-type I trisaccharide (L-Fuc-a(1″ → 2)-D-Gal-β(1′ → 3)-D-β-GlcNAc) is a typical example of specificity observed in oligosaccharide-binding proteins. This simplifying argument purposely disregards the complex interplay between entropic and enthalpic parameters. We therefore assume that UEA-I defines a specific bound conformation for the trisaccharide epitopes **18–25**, and that differences in the free energy of binding are derived essentially from changes associated with ligand structure [16 b].

With respect to the decrease in binding affinity, the predominantly staggered conformations adopted by the Ca″–C2′ or the Ca′–C4 aglyconic bonds of **19** or **20** in their unbound solution states have less torsional strain than those of **18**, resulting in larger free energy differences between the free and bound states of these ligands. The C-trisaccharide with the lowest conformational energy, doubly modified ligand **21**, would therefore be expected to have the weakest binding affinity of all. The same argument is of course equally applicable to the O-trisaccharide series **22–25**. It should be emphasized again that a general and unambiguous method for the conformational analysis of O-glycosides has yet to be established, and that there is room for ambiguity even with C-glycosides, as we have seen in the case of C-lactose. The great advantage of using C-glycosides as models, however, is the reliability with which conformational behavior and structural modifications can be correlated.

The relative binding affinities of the trisaccharide epitopes **18–25** were determined from the IC_{50} values from an enzyme-linked biotin-avidin assay (Tab. 11.4) [5 p]. The differences in binding affinity between any given C-trisaccharide and its isosteric O-trisaccharides are negligible ($\Delta\Delta G° \leq 0.2$ kcal mol^{-1}), firmly establishing the glycomimetic function of the C-glycoside series. This dispels any concerns that the glycosidic acetal oxygens have an important role in binding, at least for this particular system; indeed, the similarities in affinity are even better than expected. The retention of binding observed for the C-trisaccharide **18** is particularly remarkable in the light of the accumulation of structural deviations due to the replacement of three glycosidic oxygens by three methylene groups.

Comparison of the IC_{50} values for the structurally modified C-trisaccharides **19–21** reveals a substantial decrease in their binding affinities relative to that of **18** [5 p]. As predicted, the absence of either the C3' hydroxyl or the C5 hydroxymethyl substituent results in a loss of biological activity for **19** and **20**, respectively. Binding is weakened to an even greater extent in the case of **21**, demonstrating that the effect of removal of both substituents is cumulative. The binding profile for compounds **18–21** correlates directly with the changes in conformational preferences determined for these ligands. A gradient in binding affinity is also observed for O-trisaccharide **22** and the structurally modified analogues **23–25**, whose losses in biological activity exactly parallel those observed for the C-trisaccharides **19–21**. From this, one can infer that the decreased affinities of the defunctionalized O-trisaccharides are also due to changes in conformational profile.

An exhaustive survey of carbohydrate epitopes based on the H-type II human blood group determinant was undertaken by Lemieux, and is of particular relevance to the work presented here [39]. These workers observed that the removal of certain hydroxyl groups from O-trisaccharide **22** methyl ester resulted in a sharp decrease in its binding affinity to various protein receptors, including lectin UEA-I. The binding profiles of the epitopes of **22** methyl ester are different for each receptor, but they very interestingly all share a common sensitivity to the loss of the

Tab. 11.4 Competitive binding assay data for C-trisaccharides **18–21** and their corresponding O-trisaccharides **22–25** to the lectin lectin UEA-1 [5 p]. Free energy differences for C- and O-trisaccharides are relative to **18** and **22**, respectively.

	C-trisaccharides			O-trisaccharides	
	IC_{50} (µM)	$\Delta\Delta G°$ (kcal/mol)		IC_{50} (µM)	$\Delta\Delta G°$ (kcal/mol)
18	2.6±0.6	+0 [a]	22	1.9±0.3	+0.0
19	55.5±1.4	+1.8	23	59.0±18.4	+2.0
20	13.7±2.1	+1.0	24	14.1±2.4	+1.2
21	97±29	+2.2	25	107±2	+2.4

a) The free energy difference between **18** and **22** is ~0.2 kcal mol^{-1}, within experimental error.

C3′ hydroxyl group. Furthermore, the absence of the C6 hydroxyl group from **22** methyl ester did not cause a decrease in binding for the receptors studied, but rather resulted in a slight increase. Although one cannot rigorously exclude the possibility that the C3′ hydroxyl group participates in a key hydrogen bond with the protein in any of the four receptor–ligand systems, its relevance in defining ligand conformation correlates strongly with its invariable importance in binding affinity. Moreover, it is clear that the C6 hydroxyl group is not an important polar substituent, and therefore a loss in binding due to the removal of the C6 carbon is even more convincingly related to a change in shape. Steric influences in the H-type II trisaccharide epitopes were not explicitly cited by Lemieux as a relevant element in binding. Nevertheless, his data support our argument that ground-state conformational energies are an important factor in protein-carbohydrate recognition.

11.9
Concluding Remarks

Conformational analysis of *C*-glycosides has proven to be a fruitful and valuable endeavor toward programming conformational profiles and designing molecular architectures based on carbohydrates. An awareness of the impact of local steric interactions on secondary structure has allowed us to establish several qualitative rules of thumb regarding the conformational behavior of *C*-oligosaccharides, which should also apply to that of their parent *O*-glycosides. The *C*-glycosidic bond has a strong preference for the *exo*-anomeric (*exo/anti*) conformation, determined essentially by the avoidance of destabilizing *gauche* interactions. The *C*-aglyconic bond is also sensitive to steric destabilization, particularly due to 1,3-diaxial-like relationships, and favors staggered conformations free of severe forms of this destabilization. In cases in which such interactions are unavoidable, the *C*-aglyconic bond can be regarded as conformationally labile, and will always defer to the conformational preferences of the *C*-glycosidic bond. Other physical parameters such as hydrogen bonding, solvation, and remote stereoelectronic effects have relatively minor impacts on the conformational profiles of *C*-glycosides.

With diamond lattice-assisted conformational analysis, 1,3-diaxial-like steric interactions can first be evaluated and then verified experimentally by $^3J_{H,H}$ coupling constants. This provides us with a mechanism for identifying functional groups in conformationally strategic positions, and for programming oligosaccharide ligands with modified secondary structures. Importantly, the lessons learned from the conformational analysis of structurally modified *C*-glycosides can be translated directly into the rational design of conformationally modified *O*-glycosides.

Finally, the recognition of *C*-oligosaccharides by different classes of carbohydrate-binding proteins demonstrates several key issues. Firstly, *C*-glycoside isosteres have proven to be excellent glycomimetic ligands for receptors recognizing the ground-state conformations of their corresponding *O*-glycosides. Secondly, *C*-oligosaccharides with modified conformational profiles have significantly different binding affinities, correlating with the free energy difference of their solution and bound conformational states. Thirdly, application of the same structural modifications to the corresponding *O*-oligosaccharides produces nearly identical changes in binding affinity. Thus, by applying *C*-glycosides as conformational models, we have defined a logical relationship between carbohydrate architecture and biological activity. The relevance of nonbonded interactions in oligosaccharide conformation opens the possibility of designing conformational profiles toward glycomimetic ligands with enhanced biological activities.

11.10
Acknowledgements

Financial support from the National Institutes of Health (NS 12108) is gratefully acknowledged.

11.11
References

1 For recent comprehensive reviews on the chemistry and biology of glycosides, see for example: (a) FRASER-REID, B.O.; TATSUTA, K.; THIEM, J.; Eds. *Glycoscience: Chemistry and Chemical Biology*, Springer-Verlag: Heidelberg, **2001**. (b) ERNST, B.; HART, G.W.; SINAY, P.; Eds. *Carbohydrates in Chemistry and Biology*, Wiley-VCH: Weinheim, **2000**.

2 For reviews on the synthesis of *C*-glycosides, see for example: (a) GYORGYDEAK, Z.; PELYVAS, I.F. "*C*-Glycosylation," in *Glycoscience*; FRASER-REID, B.O., TATSUTA, K., THIEM, J., Eds.; Springer-Verlag: Heidelberg, **2001**; Vol. 1, 691. (b) BEAU, J.-M.; VAUZEILLES, B.; SKRYDSTRUP, T. "*C*-Glycosyl analogs of oligosaccharides and glycosyl amino acids," in *Glycoscience*; FRASER-REID, B.O., TATSUTA, K., THIEM, J., Eds.; Springer-Verlag: Heidelberg, **2001**; Vol. 3, 2679. (c) LIU, L.; MCKEE, M.; POSTEMA, M.H.D. "Synthesis of *C*-saccharides and higher congeners," *Cur. Org. Chem.* **2001**, *5*, 1133. (d) SKRYDSTRUP, T.; VAUZEILLES, B.; BEAU, J.-M. "Synthesis of *C*-oligosaccharides," in *Carbohydrates in Chemistry and Biology*, ERNST, B., HART, G.W., SINAY, P., Eds.; Wiley-VCH: Weinheim, **2000**; Vol. 1, 495. (e) POSTEMA, M.H.D. and CALIMENTE, D. "*C*-Glycoside synthesis: Recent developments and current trends," in *Glycochemistry: Principles, Synthesis and Applications*: WANG, P.G. and BERTOZZI, C. Eds.; Marcel Dekker: New York, **2000**. Chapter 4, 77. (f) DU, Y.; LINHARDT, R.J. "Recent advances in stereoselective *C*-glycoside synthesis." *Tetrahedron* **1998**, *54*, 9913. (g) TOGO, H.; HE, W.; WAKI, Y.; YOKOYAMA, M. "*C*-Glycosidation technology with free radical reactions." *Synlett.* **1998**, 700. (h) NICOTRA, F. "Synthesis of *C*-glycosides of biological interest." *Top. Cur. Chem.* **1997**, *187*, 55. (i) BEAU, J.-M.; GALLAGHER, T. "Nucleophilic *C*-glycosyl donors for *C*-glycoside synthesis." *Top. Cur. Chem.* **1997**, *187*, 1. (j) SINAŸ, P. "Synthesis of oligosaccharide mimetics." *Pure Appl. Chem.* **1997**, *69*, 459. (k) POSTEMA, M.H.D., Ed. *C-Glycoside Synthesis*; CRC Press: Boca Raton, FL, **1995**. (l) LEVY, D.E.; TANG, C. *The Chemistry of C-Glycosides*; Elsevier Science: Oxford, **1995**.

3 For reviews on the conformation of *C*-glycosides, see for example: (a) KISHI, Y. *Pure Appl. Chem.* **1989**, *61*, 313 and **1993**, *65*, 771. (b) KISHI, Y. *Tetrahedron* **2002**, *58*, 6239. (c) JIMENEZ-BARBERO, J.; ESPINOSA, J.F.; ASENSIO, J.L.; CANADA, F.J.; POVEDA, A. *Adv. Carbohydr. Chem. Biochem.* **2001**, *56*, 235. (d) HIDEYA, Y.; HIRONOBU, H. *Trends in Glycoscience and Glycotechnology* **2001**, *13*, 31.

4 For recent reviews on the biological activity of glycomimetics, see for example: (a) Sears, P.; Wong, C.-H. *Angew. Chem. Int. Ed.* **1999**, *38*, 2301. (b) Koeller, K. M.; Wong, C.-H. *Nature Biotechnology* **2000**, *18*, 835. (c) Compain, P.; Martin, O. R. *Bioorg. Med. Chem.* **2001**, *9*, 3077.

5 For the work on the conformational analysis of C-glycosides from this laboratory, see: (a) Wu, T.-C.; Goekjian, P. G.; Kishi, Y. *J. Org. Chem.* **1987**, *52*, 4819. (b) Goekjian, P. G.; Wu, T.-C.; Kang, H.-Y.; Kishi, Y. *J. Org. Chem.* **1987**, *52*, 4823. (c) Babirad, S. A.; Wang, Y.; Goekjian, P. G.; Kishi, Y. *J. Org. Chem.* **1987**, *52*, 4825. (d) Wang, Y.; Goekjian, P. G.; Ryckman, D. M.; Kishi, Y. *J. Org. Chem.* **1988**, *53*, 4151. (e) Miller, W. H.; Ryckman, D. M.; Goekjian, P. G.; Wang, Y.; Kishi, Y. *J. Org. Chem.* **1988**, *53*, 5580. (f) Goekjian, P. G.; Wu, T.-C.; Kishi, Y. *J. Org. Chem.* **1991**, *56*, 6412. (g) Goekjian, P. G.; Wu, T.-C.; Kang, H. Y.; Kishi, Y. *J. Org. Chem.* **1991**, *56*, 6422. (h) Wang, Y.; Babirad, S. A.; Kishi, Y. *J. Org. Chem.* **1992**, *57*, 468. (i) Wang, Y.; Goekjian, P. G.; Ryckman, D. M.; Miller, W. H.; Babirad, S. A.; Kishi, Y. *J. Org. Chem.* **1992**, *57*, 482. (j) Haneda, T.; Goekjian, P. G.; Kim, S. H.; Kishi, Y. *J. Org. Chem.* **1992**, *57*, 490. (k) O'Leary, D. J.; Kishi, Y. *J. Org. Chem.* **1993**, *58*, 304. (l) Wei, A.; Kishi, Y. *J. Org. Chem.* **1994**, *59*, 88. (m) O'Leary, D. J.; Kishi, Y. *J. Org. Chem.* **1994**, *59*, 6629. (n) O'Leary, D. J.; Kishi, Y. *Tetrahedron Lett.* **1994**, *35*, 5591. (o) Wei, A.; Haudrechy, A.; Audin, C.; Jun, H.-S.; Haudrechy-Bretel, N.; Kishi, Y. *J. Org. Chem.* **1995**, *60*, 2160. (p) Wei, A.; Boy, K. M.; Kishi, Y. *J. Am. Chem. Soc.* **1995**, *117*, 9432. (q) Ravishankar, R.; Surolia, A.; Vijayan, M.; Lim, S.; Kishi, Y. *J. Am. Chem. Soc.* **1998**, *120*, 11297.

6 For an authentic monograph on the conformational analysis of organic compounds in general, see: Eliel, E. L.; Wilen, S. H. *Stereochemistry of Organic Compounds*; John Wiley and Sons: New York, **1994**.

7 For comprehensive monographs on carbohydrate conformation, see for example: (a) Rao, V. S. R.; Qasba, P. Q.; Balaji, P. V.; Chandrasekaran, R. *Conformation of Carbohydrates*; Harwood Academic Publishers: London, **1998**. (b) Hounsell, E. F.; Ragazzi, M.; Editors, *Conformational Studies of Carbohydrates*. Special issue of *Carbohydr. Res.*, **1997**, *300*.

8 Lemieux, R. U.; Koto, S.; Voisin, D. "The exo-anomeric effect." *ACS Symp. Ser.* **1979**, *87* (Anomeric Eff.: Origin Consequences), 17–29 and references cited therein.

9 For authentic monographs on stereoelectronic effects, see for example: (a) Kirby, A. J. *The Anomeric Effect and Related Stereoelectronic Effects at Oxygen*; Springer-Verlag: Berlin, **1983**; Vol. 15. (b) Deslongchamps, P. *Stereoelectronic Effects in Organic Chemistry*, Pergamon Press: Oxford, **1983**.

10 Karplus, M. *J. Am. Chem. Soc.* **1963**, *85*, 2870.

11 Haasnoot, C. A. G.; De Leeuw, F. A. A. M.; Altona, C. *Tetrahedron* **1980**, *36*, 2783.

12 For example, see: (a) Spoormaker, T.; De Bie, M. J. A. *Rec. Trav. Chim. Pays-Bas* **1980**, *99*, 154. (b) Bose, B.; Zhao, S.; Stenutz, R.; Cloran, F.; Bondo, P. B.; Bondo, G.; Hertz, B.; Carmichael, I.; Serianni, A. S. *J. Am. Chem. Soc.* **1998**, *120*, 11,158. (c) Mulloy, B.; Frenkiel, T. A.; Davies, D. B. *Carbohydr. Res.* **1988**, *184*, 39. (d) Rundlof, T.; Kjellberg, A.; Damberg, C.; Nishida, T.; Widmalm, G. *Magn. Reson. Chem.* **1998**, *36*, 839.

13 Lemieux, R. U.; Koto, S. *Tetrahedron* **1974**, *30*, 1933.

14 The destabilization energy due to 1,3-diaxial-like interactions depends on the groups involved (see general discussion on A values in Reference 6). 1,3-Diaxial-like interactions across the interglycosidic linkage may also be accompanied by additional diaxial-like relationships between remote groups, referred to as "incidental" interactions. For example, incidental diaxial-like relationships exist between the C5'–C6' and C6–OH bonds in the C-anti-C3 conformer of **5**, and the O5'–C5' and C3–OH bonds in the C-anti-H conformer (Fig. 11.10). These may contribute further to steric destabilization, albeit as secondary interactions.

15 The lack of influence of hydrogen bonding in water was further shown by the following NMR studies. In the diol **3** with R=Bn, X=OH, and Y=H, the coupling constants observed in CDCl$_3$ were found to correspond almost perfectly to the *C-anti-C* conformation, indicating that the hydroxyl group at Cβ may form an intramolecular hydrogen bond with the pyranose ring oxygen in CDCl$_3$. In contrast, the coupling constants measured for the polyol of **3** in D$_2$O were found to be dramatically different from those of the diol in CDCl$_3$, but similar to those observed for the permethyl derivative of **3** in CDCl$_3$. This suggests that the hydrogen bond between the Cβ hydroxyl and C5 ring oxygen has a strong effect on the conformation of **3** in D$_2$O.

16 (a) LEMIEUX, R. U.; PAVIA, A. A.; MARTIN, J. C.; WATANABE, K. A. *Can. J. Chem.* **1969**, *47*, 4427. (b) LEMIEUX, R. U. *Acc. Chem. Res.* **1996**, *29*, 373. (c) PRALY, J. P.; LEMIEUX, R. U. *Can. J. Chem.* **1987**, *65*, 213. (d) GREGOIRE, F.; WEI, S. H.; STREED, E. W.; BRAMELD, K. A.; FORT, D.; HANELY, L. J.; WALLS, J. D.; GODDARD, W. A.; ROBERTS, J. D. *J. Am. Chem. Soc.* **1998**, *120*, 7537.

17 (a) RUBINSTEIN, G.; SINAŸ, P.; BERTHAULT, P. *J. Phys. Chem. A* **1997**, *101*, 2536. (b) ESPINOSA, J.-F.; CANADA, F. J.; ASENSIO, J. L.; MARTIN-PASTOR, M.; DIETRICH, H.; MARTIN-LOMAS, M.; SCHMIDT, R. R.; JIMENEZ-BARBERO, J. *J. Am. Chem. Soc.* **1996**, *118*, 10862.

18 (a) FERRITTO, R.; VOGEL, P. *Tetrahedron: Asymmetry* **1994**, *5*, 2077. (b) PASQUARELLO, C.; DEMANGE, R.; VOGEL, P. *Bioorg. Med. Chem. Lett.* **1999**, *9*, 793.

19 VIODE, C.; VOGEL, P. *J. Carbohydr. Chem.* **2001**, *20*, 733.

20 ESPINOSA, J.-F.; BRUIX, M.; JARRETON, O.; SKRYDSTRUP, T.; BEAU, J.-M.; JIMENEZ-BARBERO, J. *Chem.-Eur. J.* **1999**, *5*, 442.

21 Alternative explanations are possible: VOGEL invoked a hydrogen bond between the C2 hydroxyl and the pyranose C5' ring oxygen [18]. However, the corresponding hydrogen bond is not possible in the *a-C*-mannobioside case [20].

22 PEREIRA, M. E.; KISAILUS, E. C.; GRUEZO, F.; KABAT, E. A. *Arch. Biochem. Biophys.* **1978**, *185*, 108.

23 For conformational studies on (1,1')-linked *C*-disaccharides, see: (a) Glc-α-(1' → 1)-α-*C*-Glc, Glc-α(1' → 1)-β-*C*-Glc, Glc-β(1' → 1)-β-*C*-Glc, Glc-α(1' → 1)-α-*C*-Man, Glc-α(1' → 1)-β-*C*-Man, Man-α(1' → 1)-β-*C*-Glc, Glc-β(1' → 1)-β-*C*-Man (trehaloses and 2-epitrehaloses): Ref. 5l. (b) Glc-β(1' → 1)-β-*C*-Glc (β,β-trehalose) and Glc-β(1' → 1)-β-CHNO$_2$-Glc: MARTIN, O. R.; LAI, W. *J. Org. Chem.* **1993**, *58*, 176. (c) Glc-β(1' → 1)-β-*C*-Glc (β,β-trehalose): DUDA, C. A.; STEVENS, E. S. *J. Am. Chem. Soc.* **1993**, *115*, 8487. (d): Man–(1' → 1)-α-*C*-Gal: ASENSIO, J. L.; CANADA, F. J.; CHENG, X.; KHAN, N.; MOOTOO, D. R.; JIMENEZ-BARBERO, J. *Chem. Eur. J.* **2000**, *6*, 1035. (e) 5'-AzaMan-α-(1' → 1)-β-*C*-Gal and 5'AzaMan-α-(1' → 1)-β-*C*-Gal: ASENSIO, J. L.; CANADA, F. J.; GARCIA-HERRERO, A.; MURILLO, M. T.; FERNANDEZ-MAYORALAS, A.; JOHNS, B. A.; KOZAK, J.; ZHU, Z.; JOHNSON, C. R.; JIMENEZ-BARBERO, J. *J. Am. Chem. Soc.* **1999**, *121*, 11,318.

24 For conformational studies on (1',2)-linked *C*-disaccharides, see: (a) Man-α-(1 → 2)-*C*-Man and Man-α(1' → 2)-CHOH-Man (*a-C*-mannobioside): ESPINOSA, J.-F.; BRUIX, M.; JARRETON, O.; SKRYDSTRUP, T.; BEAU, J.-M.; JIMENEZ-BARBERO, J. *Chem. Eur. J.* **1999**, *5*, 442. (c) L-Fuc-α(1'' → 2')-*C*-D-Gal-β(1' → 4)-*C*- D-GlcNAc (H-Type II blood group determinant trisaccharides): Ref. 5 j, o, p.

25 For conformational studies on (1',3)-linked *C*-disaccharides, please see: (a) L-Fuc-(1' → 3)-*C*-D-GlcNAc: Ref. 19. (b) Gal-α(1' → 3)-*C*-Man: Ref. 18. (c) Gal-α(1' → 3)-*C*-Gal and Gal-α(1' → 3)-*C*-Gal: Ref. 5 d, e. (d) Gal-α(1' → 3)-*C*-GalNAc, Gal-α(1' → 3)-*C*-TalNAc: PASQUARELLO, C.; PICASSO, S.; DEMANGE, R.; MALISSARD, M.; BERGER, E. G.; VOGEL, P. *J. Org. Chem.* **2000**, *65*, 4251. (e) AzaLyx-β(1' → 3)-*C*-Man: MARQUIS, C.; PICASSO, S.; VOGEL, P. *Synthesis* **1999**, 1441. (f) Sial-α(2' → 3)-CHOH-Gal (Sialyl Lewis X analogue): POVEDA, A.; ASENSIO, J. L.; POLAT, T.; BAZIN, H.; LINHARDT, R. J.; JIMENEZ-BARBERO, J. *Eur. J. Org. Chem.* **2000**, 1805. (g) 5'-AzaGal-β(1' → 3)-CHOH-Altf: BAUDAT, A.; VOGEL, P. *J. Org. Chem.* **1997**, *62*, 6252. (h) L-Fuc-β(1' → 3)-*C*-[D-Gal-β(1'' → 4)-*O*]-D-Glc (Lewis X *C*/*O*-Tri-

saccharide): Berthault, P.; Birlirakis, N.; Rubinstein, G.; Sinaÿ, P.; Desvaux, H. *J. Biomol. NMR* **1996**, *8*, 23.

26 For conformational studies on (1′,4)-linked C-disaccharides, see: (a) C-maltose, C-cellobiose, 2′-epimers and 1,6-anhydro analogues: Ref. 5 c, d, e, h, i. (b) Gal-β(1′→4)-C-Glc (C-lactose): Ref. 17. (c) L-Fuc-α(1″→2′)-C-D-Gal-β(1′→4)-C-D-GlcNAc (H-Type II blood group determinant trisaccharides): Ref. 5 j, o, p. (d) 2-deoxy-Glc-β(1′→4C)-Glc (nor-C-cellobiose): Armstrong, R. W.; Teegarden, B. R. *J. Org. Chem.* **1992**, *57*, 915.

27 For conformational studies on (1′,6)-linked C-disaccharides, see: (a) Glc-α(1→6)-C-Glc and Glc-β(1→6)-C-Glc (C-isomaltose and C-gentiobiose): Ref. 5 b, g. (b) Glc-β(1→6)-C-Glc (C-gentiobiose): Neuman, A.; Longchambon, F.; Abbes, O.; Gillier-Pandraud, H.; Perez, S.; Rouzaud, D.; Sinaÿ, P. *Carbohydr. Res.* **1990**, *195*, 187. (c) Glc-β(1→6)-C-Gal: Ref. 3 c. (d) Glc-β(1→6)-C-Gal: Ref. 3 c. (e) 5′-AzaGlc-β-(1→6)-C-Glc and 5′-Aza-Man-β-(1→6)-C-Man: Ref. 23 e.

28 For conformational studies on C-disaccharides containing a furanose ring, see: (a) Glc-α(1′→2)-C-β-Fru (C-sucrose): Ref. 5 k. (b) Glc-α(1′→1)-C-α-Ara, Glc-α(1′→1)-C-β-Ara: Ref. 5 m. (c) Glc-β(1′→1)-C-Glcf: Ref. 23 b. (d) Glc-β(1′→3)-CHF-2,3-anhGulf: Jimenez-Barbero, J.; Demange, R.; Schenk, K.; Vogel, P. *J. Org. Chem.* **2001**, *66*, 5132. (e) 5′-AzaGal-β(1′→3)-CHOH-Altf: Baudat, A.; Vogel, P. *J. Org. Chem.* **1997**, *62*, 6252.

29 For a review on palytoxin, see: Moore, R. E. *Prog. Chem. Org. Nat. Prod.* **1985**, *48*, 81.

30 (a) Cha, J. K.; Christ, W. J.; Finan, J. M.; Fujioka, H.; Kishi, Y.; Klein. L. L.; Ko, S. S.; Leder, J.; McWhorter, W. W., Jr.; Pfaff, K.-P.; Yonaga, M.; Uemura, D.; Hirata, Y. *J. Am. Chem. Soc.* **1982**, *104*, 7369. (b) Armstrong, R. W.; Beau J.-M.; Cheon, S. H.; Christ, W. J.; Fujioka, H.; Ham, W.-H.; Hawkins, L. D.; Jin, H.; Kang, S. H.; Kishi, Y.; Martinelli, M. J.; McWhorter, W. W., Jr.; Mizuno, M.; Nakata, M.; Stutz, A. E.; Talamas, F. X.; Taniguchi, M.; Tino, J. A.; Ueda, K.; Uenishi, J.-I.; White, J. B.; Yonaga, M. *J. Am. Chem. Soc.* **1989**, *111*, 7530. (c) Suh, E. M.; Kishi, Y. *J. Am. Chem. Soc.* **1994**, *116*, 11205.

31 (a) Zheng, W.; DeMattei, J. A.; Wu, J.-P.; Duan, J. J.-W.; Cook, L. R.; Oinuma, H.; Kishi, Y. *J. Am. Chem. Soc.* **1996**, *118*, 7946. (b) Cook, L. R.; Oinuma, H.; Semones, M. A.; Kishi, Y. *J. Am. Chem. Soc.* **1997**, *119*, 7928. (c) Kishi, Y. *Pure Appl. Chem.* (**1998**) *70*, 339

32 Nonomura, T.; Sasaki, M.; Matsumori, N.; Murata, M.; Tachibana, K.; Yasumoto, T. *Angew. Chem., Int. Ed.* **1996**, *35*, 1675.

33 For recent reviews, see for example: (a) Loris, R. *Biochim. Biophys. Acta* **2002**, *1572*, 198. (b) Mulloy, B.; Linhardt, R. J. *Curr. Opin. Struct. Biol.* **2001**, *11*, 623.

34 Gallicchio, E.; Kubo, M. M.; Levy, R. M. *J. Phys. Chem. B* **2000**, *104*, 6271.

35 Martin-Pastor, M.; Espinosa, J. F.; Asensio, J. L.; Jimenez-Barbero, J. *Carbohydr. Res.* **1997**, *298*, 15.

36 Wang, J.; Kovac, P.; Sinaÿ, P.; Glaudemans, C. P. J. *Carbohydr. Res.* **1998**, *308*, 191.

37 (a) Espinosa, J.-F.; Canada, F. J.; Asensio, J. L.; Dietrich, H.; Martin-Lomas, M.; Schmidt, R. R.; Jimenez-Barbero, J. *Angew. Chem., Int. Ed.* **1996**, *35*, 303. (b) Asensio, J. L.; Espinosa, J. F.; Dietrich, H.; Canada, F. J.; Schmidt, R. R.; Martin-Lomas, M.; Andre, S.; Gabius, H. J.; Jimenez-Barbero, J. *J. Am. Chem. Soc.* **1999**, *121*, 8995.

38 (a) Garcia-Herrero, A.; Montero, E.; Munoz, J. L.; Espinosa, J. F.; Vian, A.; Garcia, J. L.; Asensio, J. L.; Canada, F. J.; Jimenez-Barbero, J. *J. Am. Chem. Soc.* **2002**, *124*, 4804. (b) Espinosa, J. F.; Montero, E.; Vian, A.; Garcia, J. L.; Dietrich, H.; Schmidt, R. R.; Martin-Lomas, M.; Imberty, A.; Canada, F. J.; Jimenez-Barbero, J. *J. Am. Chem. Soc.* **1998**, *120*, 1309.

39 (a) Cromer, R.; Spohr, U.; Khare, D. P.; LePendu, J.; Lemieux, R. U. *Can. J. Chem.* **1992**, *70*, 1511. (b) Spohr, U.; Paszkiewicz-Hantiw, E.; Morishima, N.; Lemieux, R. U. *Can. J. Chem.* **1992**, *70*, 254. (c) Hindsgaul, O.; Khare, D. P.; Bach, M.; Lemieux, R. U. *Can. J. Chem.* **1985**, *63*, 2653.

12
Synthetic Lipid A Antagonists for Sepsis Treatment
WILLIAM J. CHRIST, LYNN D. HAWKINS, MICHAEL D. LEWIS, and YOSHITO KISHI

12.1
Background

Shock due to Gram-negative sepsis remains a life-threatening syndrome for which no treatment is available other than supportive therapy in an intensive care unit setting [1]. Gram-negative sepsis is a consequence of a strong and acute inflammatory response to endotoxin or lipopolysaccharide (LPS) released from the bacterial outer membrane after treatment with cytolytic antibiotics and/or action initiated by the host's defense mechanisms [2]. Triggered by detection of endotoxin, the host initiates a rapid and vigorous immune response that includes the generation of a complex cascade of cellular mediators including tumor necrosis factor-α (TNF-α), interleukin-1 (IL-1), IL-6, leukotrienes, and thromboxane A2 from monocytes and macrophages [3]. Unfortunately, the continued presence of endotoxin in the blood can result in a pathophysiological overreaction by the host, leading to the release of toxic quantities of these cellular mediators [4]. In addition, recent research has shown that LPS plays a major role in other inflammation-related diseases such as inflammatory bowel disease, some liver diseases, organ rejection, and asthma [5].

According to the current model [3 d], LPS binding protein (LBP) sequesters and catalyzes the transfer of monomerized LPS from its aggregated form, or in some cases from intact Gram-negative bacteria, to CD-14. CD-14 is a GPI-bound receptor on immune cells derived from monocytes that does not by itself have the ability to transduce the LPS binding signal directly [6]. The actual transmembrane signaling occurs after CD-14 has transferred the LPS to the toll-like receptor-4 (TLR-4), which is in turn complexed to an accessory protein MD-2. Since the internal signal transduction process, culminating in the release of a wide spectrum of cellular mediators, only occurs after the TLR-4*MD-2*LPS complex has been formed [7], a clear opportunity to inhibit this signaling exists, through replacement of LPS in the complex with an antagonistic molecule.

LPS is a large and complex molecule consisting of O-antigen, core, and lipid A regions (Fig. 12.1). Lipid A, obtained by acid hydrolysis of *E. coli* endotoxin, was first isolated by Westphal and Luderitz [8] and was found to have lethal toxicity, pyrogenicity, TNF-α and other cytokine-activating properties equal to its native LPS. Lipid A is thus regarded as the toxic principle of LPS [9].

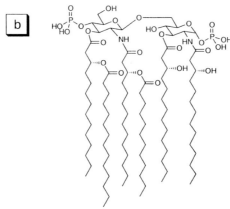

Fig. 12.1 a) A typical lipopolysaccharide from a Gram-negative bacterial cell wall. b) Structure of *E. coli* lipid A.

Shiba and co-workers achieved the chemical synthesis of *E. coli* lipid A, verified its proposed structure, and confirmed the biological activities reported for the lipid A preparation obtained from *E. coli* endotoxin [10].

12.2
Hypothesis and Approach

The lipid As of several Gram-negative bacteria were established as the principle toxicophore inducing sepsis and septic-like syndromes in humans. With this knowledge, we hypothesized that an *antagonist* specifically competitive for the lipid A receptor site might eliminate the disease-related biological events induced by endotoxins. For the design of an antagonist, we planned to use structural information relating to the naturally occurring lipid As and related compounds. In particular, we were interested in two classes of natural products: lipid X and the so-called nontoxic lipid As.

12.2.1
Monosaccharide Antagonists: Lipid X Analogues

Lipid X, a monosaccharide and a biosynthetic precursor of *E. coli* lipid A, was first isolated from a variant of *E. coli* and reported to antagonize LPS activity in vitro and in vivo (Fig. 12.2) [11]. As lipid X is structurally significantly simpler than lipid A, it seemed logical for us first to explore the potential of lipid X as our lead structure. The proposed structure of lipid X was confirmed by chemical synthesis by us and others, followed by confirmation of the reported biological activities for lipid X [12]. Encouraged by these results, we conducted a wide range of structural modifications on lipid X. Unfortunately, no successful drug candidate emerged directly from these efforts. However, we accumulated indispensable chemical and biological knowledge in this general area, which led us to the ultimate creation of lipid A antagonists.

The monosaccharide analogue ERI-1, for example, was demonstrated to show modest but definite in vitro antagonistic activity (Fig. 12.2). When this compound was tested in the mouse in vivo endotoxin challenge model, its activity was attenuated. We soon realized that, on standing, synthetic lipid X and analogues gradually became contaminated with a minute amount of disaccharide(s), which presumably formed through an intermolecular self-condensation between the C1 phosphate and the C6 hydroxyl groups. Biologically, this disaccharide contaminant(s) exhibited very potent agonistic activity, which might explain the inefficacy of the monosaccharide preparations in the in vivo biological evaluations. We later learned that Stuetz and co-workers at Sandoz had also made similar observations [13]. Through these attempts, we realized that our efforts should be focused on searching not only for a potent antagonist without agonistic activity, but also for a strategy by which to stabilize labile functional groups present in the lipid A/lipid

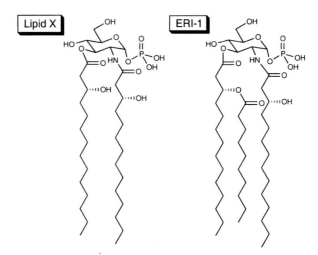

Fig. 12.2 Structures of lipid X and ERI-1.

X class of compounds. With our accumulated chemical and biological knowledge, we had an instinct that the so-called nontoxic lipid As would provide us with more valuable insights into our drug-discovery efforts than lipid X.

12.2.2
Disaccharide Antagonist of Lipid A: First Generation

The knowledge that we had gained through our exciting and learning experience with the synthetic monosaccharide antagonists directed our efforts towards the so-called nontoxic lipid As. Two different lipid A preparations, obtained from *Rhodobacter capsulatus* and *Rhodobacter sphaeroides* (previously incorrectly identified as *Rhodopseudomonas sphaeroides*), were reported to antagonize the activity of *E. coli* lipid A potently (Fig. 12.3) [14]. Undoubtedly, this was *the* biological profile we were searching for. However, before adopting these nontoxic lipid As as our lead compounds, we obviously had to resolve several issues, including unambiguously proving the proposed structure of the nontoxic lipid A and, more critically, confirming that the reported antagonistic activity was associated with chemically homogeneous lipid A-like structures. In our view, contemporary synthetic chemistry has the capacity to best deal with these types of demands.

Clearly, the first step in this effort was to establish a synthetic route to *R. capsulatus* and *R. sphaeroides* lipids As, to provide chemically homogeneous material to prove or disprove the proposed structures and to confirm the reported biological activity. Additionally, we kept in mind that our intended synthetic route should be general and sufficiently flexible to prepare not only the targeted natural products,

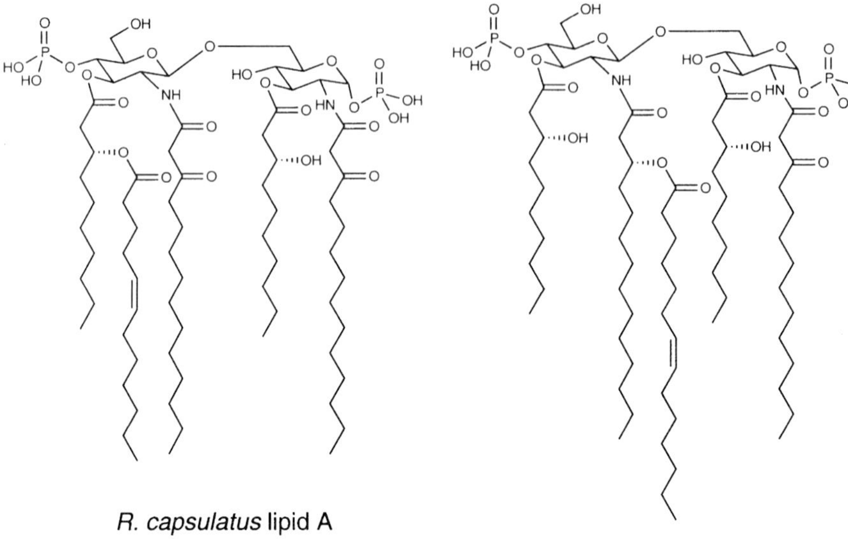

Fig. 12.3 The proposed structures of the natural nontoxic lipid As.

but their analogues for conducting a wide range of structure-activity studies, ultimately to identify a drug candidate.

A review on Shiba's seminal synthesis of E. coli lipid A provided insight into the proposed efforts [10]. Shiba used a convergent approach, in which two appropriately functionalized and protected monosaccharides were coupled to form the β-disaccharide at a late stage of the synthesis (Fig. 12.4). The key to this strategy was the use of a suitable protection/deprotection paradigm to facilitate the incorporation of various functionalities in both the glycosyl acceptor and donor at the appropriate times, to provide the desired β-disaccharide in a high yield from the Koenigs-Knorr condensation, and to allow the desired functionalities to remain intact after final deprotection.

The differences between the proposed structures of R. capsulatus and R. sphaeroides lipid As and the structure of E. coli lipid A are limited to only: (1) the presence or absence of an unsaturated fatty acid side chain, (2) the presence or absence of at least one 3-keto fatty acid, (3) the fatty acid side chain lengths, and (4) the number of fatty acid side chains. Thus, R. capsulatus, R. sphaeroides, and E. coli lipid As might look structurally similar to each other, but from the synthetic point of view these structural differences called for a strategic modification of the synthetic route. At least, the protection/deprotection paradigm used by Shiba in the synthesis of E. coli lipid A should be modified in such a way that the protecting groups adopted could be removed compatibly in the presence of the olefinic and the 1,3-ketoamido groups.

By incorporating allyl and allyloxycarbonyl protecting groups in place of the benzyl and phenoxy protecting groups used by Shiba, we successfully established a general, flexible, and effective route for the synthesis not only of the proposed

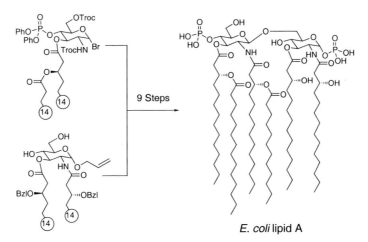

Fig. 12.4 A summary of the total synthesis of E. coli lipid A by Shiba and co-workers [10]. (1) Hg(CN)$_2$/CaSO$_4$, (2) Zn/HOAc, (3) RCO$_2$H/DCC, (4) BzlOCH$_2$Cl/iPr$_2$EtN, (5) Ir(COD)(PCH$_3$C$_6$H$_5$)$_2$)$_2$]PF$_6$, (6) I$_2$/H$_2$O, (7) (BzlO)$_2$POCl/BuLi, (8) H$_2$/Pd-black, (9) H$_2$/PdO$_2$.
Abbreviations: Bzl = benzyl, Ph = phenyl, Troc = trichloroethoxycarbonyl.

structures of both *R. capsulatus* and *R. sphaeroides* lipid As but also of their analogues [15]. The synthetic *R. capsulatus* lipid A was found to exhibit spectroscopic properties exactly matching with those reported for *R. capsulatus* lipid A derived from the natural *R. capsulatus* endotoxin, thereby confirming the proposed structure for *R. capsulatus* lipid A (Fig. 12.3). However, the synthetic *R. sphaeroides* lipid A was found to exhibit chromatographic (TLC and HPLC) and spectroscopic (^{1}H, ^{13}C, and 2D-COSY NMR and HR-MS) properties very similar to, but definitely different from, those of the lipid A preparation obtained from *R. sphaeroides* endotoxin [16], thereby indicating that some revision of the proposed structure of *R. sphaeroides* lipid A was required.

To our relief, the synthetic disaccharides, representing the proposed structures of *R. capsulatus* and *R. sphaeroides* lipid As, potently antagonized the release of the cellular mediators induced by LPS both in vitro and in vivo without any detectable agonistic properties. The stage was thus set for us to conduct a wide range of structure/activity studies based on these lipid A antagonists.

Potentially, either of the *R. capsulatus* and *R. sphaeroides* lipid As could have served equally well for us as a lead compound. However, assessing the overall efficiency projected for the structure/activity studies based on these two lead structures, we opted to focus our efforts first on the synthetic *R. capsulatus* lipid A, to improve its chemical and biological properties. During these efforts, we were continually aware of our previous experience with lipid X, and made efforts to ensure that all the observed biological activities were attributable to the parent compounds and not to minute contaminants that might have been formed from them. These efforts can be summarized as follows:

- Partial hydrolysis was observed for both the C3 and C3' acyl groups during the synthesis as well as during the biological evaluation of the synthetic lipid A. The observed instabilities were eliminated by replacing the C3 and C3' acyl groups with the corresponding ether groups.
- An impurity containing the α,β-unsaturated acyl group at C3' was detected in the preparation of the synthetic *R. capsulatus* lipid A. Apparently, this contaminant was formed through β-elimination of the β-acyloxy chain at some stage. We assumed that the structural modifications described above should also solve this problem, and indeed this was the case.
- A contaminant, paralleling a disaccharide contaminant found in the lipid X series, was detected in the current series as well. As discussed, this contaminant appeared to arise through an intermolecular self-condensation between the C1 phosphate and C6' hydroxyl groups. Indeed, formation of this contaminant was completely eliminated by conversion of the C6'-OH group into the C6'-OMe group.
- An additional contaminant, containing the C4',C6'-cyclic phosphate, was also detected in the preparation of the synthetic *R. capsulatus* lipid A. As expected, this contaminant was also completely eliminated by conversion of the C6'-OH group into the C6'-OMe group.

Although these structural modifications were driven primarily by chemical reasons, they fortunately did not affect the overall biological profile of the lead com-

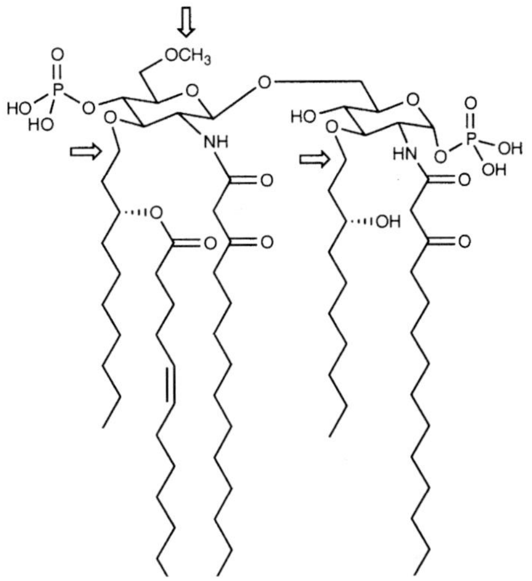

Fig. 12.5 Structure of E5531. The arrows indicate the sites of structural modifications made to the proposed *R. capsulatus* lipid A.

Tab. 12.1 Antagonism of TNF-α induced by LPS from different species of bacteria [17].

Source of LPS	TNF-α released (pg/mL)	Antagonism by E5531 IC$_{50}$ (nM)
Escherichia coli	657 ± 144	1.22 ± 0.66
Klebsiella pneumonia	615 ± 185	1.82 ± 0.79
Pseudomonas aeruginosa	475 ± 119	2.73 ± 1.79
Salmonella minnesota	898 ± 141	0.14 ± 0.06

pound. Thus, a fully stabilized, extraordinary potent endotoxin antagonist E5531, with no detectable agonistic activity, was successfully created (Fig. 12.5) [17].

E5531 potently antagonized the release of the cytokine TNF-α induced by LPSs obtained from a variety of Gram-negative bacteria in an in vitro assay (Tab. 12.1). On the other hand, E5531 was completely devoid of agonistic activity in the human and murine systems evaluated, even at concentrations 10,000 times greater than required for effective antagonism. However, when induced by non-LPS activators, E5531 was found to be ineffective in inhibiting the production of the same inflammatory mediators, thereby suggesting that E5531 specifically antagonizes the action of LPS [18].

To provide a clearer picture of E5531's mechanism of action, competitive studies were performed to test the ability of E5531 to inhibit LPS binding to its receptor. LPS binding to monocyte-derived cells activates and translocates nuclear factor κ-B (NF-κB) to the nucleus. E5531 blocked LPS's ability to stimulate NF-κB activation and translocation. Most importantly, E5531 was shown to inhibit the signaling processes of LPS and *E. coli* lipid A through the toll-like receptor-4 (TLR-4) recently shown to be the signaling cell surface receptor for LPS [7d]. These results

Tab. 12.2 Inhibitory effect of E5531 on LPS-induced increases in plasma TNF-α and lethality in BCG-primed mice [17].

EE5531 (μg/mous)	Plasma TNF-α		% Mortality
	ng/mL	% Inhibition	
0 (control)	582 ± 20	0	100
1	547 ± 45	6	80
3	432 ± 40 [a]	26	20
10	259 ± 28 [b]	55	0 [a]
30	198 ± 24 [b]	66	0 [a]
100	71 ± 18 [b]	88	0 [a]

[a] P<0.01 versus control
[b] P<0.001 versus control.

conclusively demonstrate that E5531 antagonizes LPS activity at its cell surface receptor, resulting in inhibition of transmembrane signal transduction.

The in vivo evaluation of E5531 closely mimicked the in vitro observations (Tab. 12.2). Thus, intravenous injection of 3 μg *E. coli* LPS into mice induced a rapid increase in plasma levels of TNF-α, maximal levels being reached within one hour after injection, with death as the final outcome in all animals injected. When E5531 was co-injected with LPS, the increase in TNF-α levels and subsequent mortality rate decreased with increasing concentration of E5531.

The most compelling evidence of the validity of our hypothesis and approach was demonstrated in an in vivo mouse model that closely mimics the clinical situation for sepsis (Fig. 12.6). Mice were infected with lethal doses of *E. coli*. Treatment with the antibiotic latamoxef alone gave noticeable but only transient protection from death. On treatment with the antibiotic, the bacterial load in the animal dramatically decreased (Panel b), but the endotoxin levels in these mice surged to lethal quantities post-antibiotic treatment (Panel c). On the other hand, treatment with E5531 alone provided a longer survival time for the infected animals, because the initial increase in endotoxin levels caused by the animals' immune reactions to the infection was blocked. However, the ultimate survival rate was similar to that of the non-treated mice. Most importantly, co-treatment of E5531 with the antibiotic gave sustained protection from mortality because the endotoxin released from the bacteria killed by the antibiotic was effectively antagonized by E5531 (Panel a). Overall, the results from this study provided the most compelling evidence for our approach.

12.2.3
Disaccharide Antagonist of Lipid A: Second Generation

Although E5531 fulfilled many of the criteria discussed in the original strategy, several issues led us to search for a second-generation antagonist. In particular, two improvements on E5531 appeared to be critically desirable for us to proceed

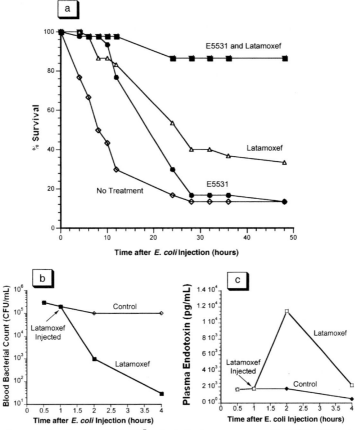

Fig. 12.6 Panel a: *E. coli* (2×10^7 colonies) plus treatment with E5531 (100 µg) alone, antibiotic (600 µg) alone, or co-treatment in a mouse, peritoneum infection model with survival over time as an endpoint. After the interperitoneum bacterial injection, *E. coli* (Panel b) and endotoxin (Panel c) levels over time in mouse plasma in the presence and in the absence of the antibiotic (600 µg) [17].

to the next stage of development, and we initiated the search for a second-generation drug candidate, with the following two specific goals:

- The time that E5531 remains active in human whole blood was found to be unexpectedly short. Obviously a drug with a longer duration of activity is more appealing for clinical application [19].
- The synthesis of E5531 is convergent and highly efficient. Nevertheless, we desired to improve the overall cost of drug synthesis further. One obvious solution is to discover a structurally simpler analogue, the production of which would require fewer steps for synthesis and purification, thereby improving the overall manufacturing costs of the drug substance.

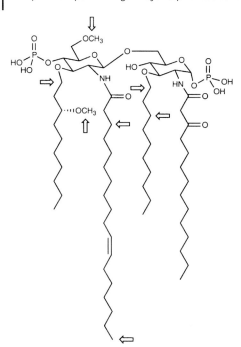

Fig. 12.7 Structure of E5564. The arrows indicate the sites of structural modifications made to the proposed *R. sphaeroides* lipid A.

To achieve these goals, we shifted our drug discovery efforts from that based on *R. capsulatus* lipid A to the one based on *R. sphaeroides* lipid A (Fig. 12.3). Like the synthetic *R. capsulatus* lipid A that inspired us to prepare E5531, the synthetic *R. sphaeroides* lipid A exhibited very potent antagonistic activities both in human systems in vitro and in murine systems in vitro and in vivo. Thus, its biological profile met with our requirements for a lead compound. Starting with the *R. sphaeroides* lipid A structure, we once again made extensive structure-activity studies, which ultimately led us to the creation of the fully stabilized, extraordinarily potent lipid A antagonist E5564, with no detectable agonistic activity (Fig. 12.7).

Overall, E5564 exhibits a biological profile exactly parallel to that found for E5531. However, E5564 has several attractive features over E5531. Firstly, the duration of action for E5564 is significantly longer than that of E5531. For example, E5564 is six times more potent after 6 hours in an ex vivo assay with human whole blood [20]. In mice, E5564 maintains its antagonistic activity up to 72 hours. Secondly, because of its simpler structure, the manufacturing cost of E5564 is expected to be significantly more economical than that of E5531. Indeed, the current synthetic route meets well with our need to produce the drug substance on a multi-kilogram scale in a cost-effective manner (Fig. 12.8). Thirdly, surprisingly and pleasantly, E5564 exhibits much better physical and chemical properties; in particular, it is five times more soluble in water than E5531, which has a significant impact on drug formulation and other challenges. Lastly, E5564

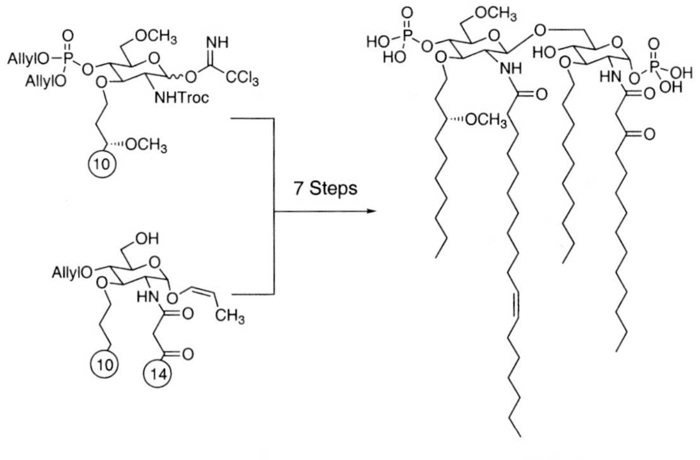

Fig. 12.8 A summary of the synthesis of E5564. *Reagents*: (1) AgOTf, (2) Zn/HOAc, (3) vaccenoyl chloride/NaHCO$_3$, (4) dil HCl, (5) (AllylO)$_2$PN*i*Pr$_2$/pyridinium triflate, (6) H$_2$O$_2$, (7) Pd(PPh$_3$)$_4$/PPh$_3$. *Abbreviations*: Troc = trichloroethoxycarbonyl.

is chemically more stable than E5531, because it does not contain an ester group, which is potentially hydrolyzed under acidic or basic conditions.

12.3
Conclusion

Our work was initiated on the basis of the hypothesis that, by using the structural information given by lipid X and the so-called nontoxic lipid As, we might be able to create an *antagonist* specifically competitive for the lipid A receptor, in the hope that such an *antagonist* may eliminate the disease-related biological events induced by endotoxin and therefore be efficacious for the treatment of this life-threatening disease associated with endotoxin. As reviewed, our efforts have successfully yielded the two extraordinarily potent endotoxin antagonists E5531 and E5564.

E5531 and E5564 represent new types of chemical entities either for a drug or for a drug candidate. In our view, our willingness to take on these types of chemical structures has provided us with new and exciting opportunities. However, it is also true that, because of the compounds' unique chemical structures, we have faced new developmental challenges not frequently encountered in the pharmaceutical industry. With great imagination and determination, our colleagues have beautifully overcome each of these challenges as they have been encountered. Our chemistry colleagues in process research have successfully addressed issues related to the scalability and economic feasibility of the synthesis of drug substances. They have also devised an economically feasible method of purification and isolation that ensures the quality of drug substances at pharmaceutically satisfactory levels. Similarly, because of the physical and chemical properties inherent in the candidate molecule, which tend to favor aggregation, drug formulation has also presented an interesting challenge. Our colleagues in drug formulation have fearlessly attacked the problem and established procedures for drug formulation that allow for the preparation of consistent and stable dosing solutions, particularly with regard to the aggregation state of the drug.

E5531 and E5564 were submitted for clinical evaluation and found to be safe. To demonstrate the antagonistic effectiveness of E5564 in humans, healthy volunteers were subjected to low levels of LPS, with and without E5564 co-treatment. The patients' biochemistry, such as cytokine levels (TNF-α), and physiological effects including body temperature were monitored. As expected, treatment with

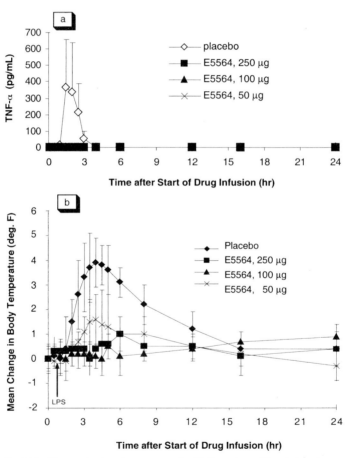

Fig. 12.9 Human in vivo time courses for TNF-α release (Panel a) or increase in body temperature (Panel b) after intravenous injection of LPS (4 ng/patient) coinjected with or without E5564 (dosages indicated) [21].

E. coli LPS alone caused a dramatic increase in TNF-α levels and a rise in mean body temperature. However, when 100 µg of E5564 were co-administered with LPS, cytokine levels, body temperature, and other symptoms were reduced to normal levels (Fig. 12.9). This controlled model confirms our initial in vivo studies conducted in mice and the potential effectiveness of E5564 in humans. To date, E5531 and E5564 are the only endotoxin antagonists of any type that have been reported to suppress the signs and symptoms of LPS in healthy humans.

More advanced clinical trials for the treatment of Gram-negative sepsis are currently in progress with the second-generation drug candidate E5564.

12.4
Acknowledgement

We would like to express our sincere thanks to all of those who have participated in this project. Their imagination, determination, and commitment have made it possible for us to have this extraordinarily challenging, exciting, and rewarding drug discovery adventure.

12.5
References

1 a) Bone, R.C. *Chest* **1991**, *100*, 802–808. b) Parrillo, J.E. *N. Engl. J. Med.* **1993**, *328*, 1471–1477.

2 a) Endo, S., Inada, K., Inoue, Y., Kuwata, Y., Suzuki, M., Yamashita, H., Hoshi, S., Yoshida, M. *Circ. Shock* **1992**, *38*, 264–274. b) Crosby, H.A., Bion, J.F., Penn, C.W., Elliott, T.S. *J. Med. Microbiol.* **1994**, *40*, 23–30. c) Dofferhoff, A.S., Nijland, J.H., de Vries-Hospers, H.G., Mulder, P.O., Weits, J., Bom, V.J. *Scand. J. Infect. Dis.* **1991**, *23*, 745–754. d) Prins, J.M., van Deventer, S.J., Kuijper, E.J., Speelman, P. *Antimicrob. Agents Chemother.* **1994**, *38*, 1211–1218.

3 a) Galanos, C., Luderitz, O., Rietschel, E.T., Westphal, O., Brade, H., Brade, L., Freudenberg, M., Schade, U., Imoto, M., Yoshimura, H. *Eur. J. Biochem.* **1985**, *148*, 1–5. b) Billiau, A., Vandekerckhove, F. *Eur. J. Clin. Invest.* **1991**, *21*, 559–573. c) Glauser, M.P., Heumann, D., Baumgartner, J.D., Cohen, J. *Clin. Infect. Dis.* **1994**, *18*, S205–216. d) Alexander, C., Rietschel, E.T. *J. Endotoxin Res.* **2001**, *7*, 167–202.

4 a) Welbourn, C.R., Young, Y. *Br. J. Surg.* **1992**, *79*, 998–1003. b) Bone, R.C. *Clin. Microbiol. Rev.* **1993**, *6*, 57–68.

5 a) Burrell, R. *Circ. Shock* **1994**, *43*, 137–153. b) Gardiner, K.R., Halliday, M.I., Barclay, G.R., Milne, L., Brown, D., Stephens, S., Maxwell, R.J., Rowlands, B.J. *Gut* **1995**, *36*, 897–901. c) Roumen, R.M., Frieling, J.T., van Tits, H.W., van der Vliet, J.A., Goris, R.J. *J. Vasc. Surg.* **1993**, *18*, 853–857. d) Suffredini, A.F., O'Grady, N.P. in *Endotoxin in Health and Disease* Brade, H. Opal, S., Vogel, S. Morrison, D. eds., Marcel Dekker Inc., New York, **1999**, 817–830. e) Yin, M., Bradford, B.U., Wheeler, M.D., Uesugi, T., Froh, M., Goyert, S.M., Thurman, R.G. *J. Immunol.* **2001**, *166*, 4737–4742.

6 a) Tobias, P.S., Soldau, K., Ulevitch, R.J. *J. Exp. Med.* **1986**, *164*, 777–793. b) Wright, S.D., Tobias, P.S., Ulevitch, R.J., Ramos, R.A. *J. Exp. Med.* **1989**, *170*, 1231–1241. c) Kitchens, R.L., Munford, R.S. *J. Immunol.* **1998**, *160*, 1920–1928.

7 a) MEDZHITOV, R., PRESTON-HURLBURT, P., JANEWAY, C. A. Jr. *Nature* **1997**, *388*, 394–397. b) YANG, R. B., MARK, M. R., GRAY, A., HUANG, A., XIE, M. H., ZHANG, M., GODDARD, A., WOOD, W. I., GURNEY, A. L., GODOWSKI, P. J. *Nature* **1998**, *395*, 217–219. c) QURESHI, S. T., GROS, P., MALO, D. *Trends Genet.* **1999**, *15*, 291–294. d) CHOW, J. C., YOUNG, D. W., GOLENBOCK, D. T., CHRIST, W. J., GUSOVSKY, F. *J. Biol. Chem.* **1999**, *274*, 10689–10692. e) IRIE, T., MUTA, T., TAKESHIGE, K. *FEBS Lett.* **2000**, *467*, 160–164. f) ANDERSON, K. V. *Curr. Opin. Immunol.* **2000**, *12*, 13–19. g) BEUTLER, B., POLTORAK, A. *Eur. Cytokine Netw.* **2000**, *11*, 143–152. h) ADEREM, A., ULEVITCH, R. J. *Nature* **2000**, *406*, 782–787. i) WASSERMAN, S. A. *Curr. Opin. Genet. Dev.* **2000**, *10*, 497–502. j) BRIGHTBILL, H. D., MODLIN, R. L. *Immunology* **2000**, *101*, 1–10. k) AKASHI, S., SHIMAZU, R., OGATA, H., NAGAI, Y., TAKEDA, K., KIMOTO, M., MIYAKE, K. *J. Immunol.* **2000**, *164*, 3471–3475. l) SHIMAZU, R., AKASHI, S., OGATA, H., NAGAI, Y., FUKUDOME, K., MIYAKE, K., KIMOTO, M. *J. Exp. Med.* **1999**, *189*, 1777–1782. m) AKASHI, S., OGATA, H., KIRIKAE, F., KIRIKAE, T., KAWASAKI, K., NISHIJIMA, M. SHIMAZU, R., NAGAI, Y., FUKUDOME, K., KIMOTO, M., MIYAKE, K. *Biochem. Biophys. Res. Commun.* **2000**, *268*, 172–177. n) YANG, H., YOUNG, D. W., GUSOVSKY, F., CHOW, J. C. *J. Biol. Chem.* **2000**, *275*, 20861–20866. o) DA SILVA, C. J., SOLDAU, K., CHRISTEN, U., TOBIAS, P. S., ULEVITCH, R. J. *J. Biol. Chem.* **2001**, *276*, 21129–21135.

8 WESTPHAL, O., LUDERITZ, O. *Angew. Chem.* **1954**, *66*, 407–417.

9 a) RIETSCHEL, E. T., BRADE, H., HOLST, O., BRADE, L., MULLER-LOENNIES, S., MAMAT, U., ZAHRINGER, U., BECKMANN, F., SEYDEL, U., BRANDENBURG, K., ULMER, A. J., MATTERN, T., HEINE, H., SCHLETTER, J., LOPPNOW, H., SCHONBECK, U., FLAD, H. D., HAUSCHILDT, S., SCHADE, U. F., DI PADOVA, F., KUSUMOTO, S., SCHUMANN, R. R. *Curr. Top. Microbiol. Immunol.* **1996**, *216*, 39–81. b) HOLST, O., ULMER, A. J., BRADE, H., FLAD, H. D., RIETSCHEL, E. T. *Immunol. Med. Microbiol.* **1996**, *16*, 83–104.

10 a) IMOTO, M., KUSUMOTO, S., SHIBA, T., RIETSCHEL, E. T., GALANOS, C., LUEDERITZ, O. *Tetrahedron Lett.* **1985**, *26*, 907–908. b) IMOTO, M., YOSHIMURA, H., YAMAMOTO, M., SHIMAMOTO, T., KUSUMOTO, S., SHIBA, T. *Tetrahedron Lett.* **1984**, *25*, 2667–2670. c) TAKADA, H., KOTANI, S., TSUJIMOTO, M., OGAWA, T., TAKAHASHI, I., HARADA, K., KATSUKAWA, C., TANAKA, S., SHIBA, T., KUSUMOTO, S., IMOTO, M., YOSHIMURA, H., YAMAMOTO, M., SHIMAMOTO, T. *Infect. Immun.* **1985**, *48*, 219–227. d) HOMMA, J. Y., MATSUURA, M., KANEGASAKI, S., KAWAKUBO, Y., KOJIMA, Y., SHIBUKAWA, N., KUMAZAWA, Y., YAMAMOTO, A., TANAMOTO, K., YASUDA, T., IMOTO, M., YOSHIMURA, H., KUSUMOTO, S., SHIBA, T. *J. Biochem.* (Tokyo) **1985**, *98*, 395–406. e) IMOTO, M., YOSHIMURA, H., SHIMAMOTO, T., SAKAGUCHI, N., KUSUMOTO, S., SHIBA, T. *Bull. Chem. Soc. Jpn.* **1987**, *60*, 2205–2214.

11 DANNER, R. L., EICHACKER, P. Q., DOERFLER, M. E., HOFFMAN, W. D., REILLY, J. M., WILSON, J., MACVITTIE, T. J., STUETZ, P., PARRILLO, J. E., NATANSON, C. *J. Infect. Dis.* **1993**, *167*, 378–384.

12 a) TAKAYAMA, K., QURESHI, N., MASCAGNI, P., NASHED, M. A., ANDERSON, L., RAETZ, C. R. *J. Biol. Chem.* **1983**, *258*, 7379–7385. b) RAY, B. L., PAINTER, G., RAETZ, C. R. *J. Biol. Chem.* **1984**, *259*, 4852–4859. c) MACHER, I. *Carbohydrate Res.* **1987**, *162*, 79–84. d) IKEDA, K., TAKAHASHI, T., SHIMIZU, C. NAKOMOTO S., ACHIWA, K. *Chem. Pharm. Bull. Jpn.* **1987**, *35*, 1383–1387. e) IKEDA, K., TAKAHASHI, T., SHIMIZU, C., NAKAMOTO, S., ACHIWA, K. *Chem. Pharm. Bull.* (Tokyo) **1987**, *35*, 1383–1387.

13 ASCHAUER, H., GROB, A., HILDEBRANDT, J., SCHUETZE, E., STUETZ, P. *J. Biol. Chem.* **1990**, *265*, 9159–9164.

14 a) LOPPNOW, H., LIBBY, P., FREUDENBERG, M., KRAUSS, J. H., WECKESSER, J., MAYER, H. *Infect. Immun.* **1990**, *58*, 3743–3750. b) TAKAYAMA, K., QURESHI, N., BEUTLER, B., KIRKLAND, T. N. *Infect. Immun.* **1989**, *57*, 1336–1338.

15 CHRIST, W. J., MCGUINNESS, P. D.; ASANO, O.; WANG, Y., MULLARKEY, M. A., PEREZ, M., HAWKINS, L. D., BLYTHE, T. A.,

Dubuc, G. R., Robidoux, A. L. *J. Amer. Chem. Soc.* **1994**, *116*, 3637–3638.

16 Purchased from the laboratories of Professor N. Qureshi.

17 Christ, W. J., Asano, O., Robidoux, A. L., Perez, M., Wang, Y., Dubuc, G. R., Gavin, W. E., Hawkins, L. D., McGuinness, P. D., Mullarkey, M. A., Lewis, M. D., Kishi, Y., Kawata, T., Bristol, J. R., Rose, J. R., Rossignol, D. P., Kobayashi, S., Hishinuma, I., Kimura, A., Asaskawa, N., Katayama, K., Yamatsu, I. *Science* **1995**, *268*, 80–83.

18 Kawata, T., Bristol, J. R., Rossignol, D. P., Rose, J. R., Kobayashi, S., Yokohama, H., Ishibashi, A., Christ, W. J., Katayama, K., Yamatsu, I., Kishi, Y. *Br. J. Pharmacol.* **1999**, *127*, 853–862.

19 Rose, J. R., Mullarkey, M. A., Christ, W. J., Hawkins, L. D., Lynn, M., Kishi, Y., Wasan, K. M., Peteherych, K., Rossignol, D. P. *Antimicrob. Agents Chemother.*, **2000**, *44*, 504–510.

20 Mullarkey, M., Rose, J. R., Bristol, J., Kawata, T., Kimura, A., Kobayashi, S., Przetak, M., Chow, J., Gusovsky, F., Christ, W. J., Rossignol, D. P. *J. Pharmacol. Exper. Ther.* **2003**, *304*, 1093–1102.

21 Lynn, M. 20th Int. Sympos. Intens. Care & Emerg. Med., Brussels, March **2000**.

13
Polysialic Acid Vaccines
HAROLD J. JENNINGS

13.1
Introduction

Polysialic acids are found on the surfaces of bacteria, where they function as virulence factors [1, 2]. Because of their accessibility and prolificity on bacterial surface, polysialic acids are also targets for bactericidal antibodies, which makes them potential vaccines. The basic structure of polysialic acids consists of contiguous sialic acid residues linked by $a2$–8-, $a2$–9-, and alternating $a2$-8- and $a2$–9 linkages. The latter pair of structures are exclusively found in bacteria. The capsular polysaccharide of group C *Neisseria meningitidis* consists of $a2$–9-linkaged polysialic acid [3] and that of *Escherichia coli* K92 consists of alternating $a2$–8- and $a2$–9-linkaged sialic acid residues [4]. The group C meningococcal polysaccharide (GCMP) is immunogenic and has been used as a vaccine against meningitis caused by group C meningococci for about 30 years [2, 5]. The fact that the GCMP is not immunogenic in infants, the most susceptible segment of the population, and that its immunological performance is suboptimal even in adults, has resulted in the development of group C meningococcal conjugate vaccines [6, 7].

The structure of the group B meningococcal polysaccharide (GBMP) consists of $a2$–8-linked polysialic acid [3]. Because $a2$–8-polysialic acid is the major topic of this review, it is referred to for the sake of brevity as PSA. Other pathogenic bacteria – including *Escherichia coli* K1 [8], *Pasteurella haemolytica* A2 [9], and *Moraxella nonliquefacians* – also have PSA capsules [10]. PSA is a unique molecule, because it not only occurs in bacteria, but is a universal mammalian developmental antigen, and is also found in other vertebrates [11, 12]. Other new forms of PSA composed of modified Neu5Ac residues have also been identified in mammalian tissues, including $a2$–8-polyNeu5Gc [13] and $a2$–8-poly KDN [14], the latter being composed of deaminated Neu5Ac residues (KDN). PSA is also expressed by many human tumors [12, 15] and has been implicated in metastasis [16]. Because it is a self-antigen, PSA is poorly immunogenic and therefore has little potential as a vaccine. However, its location on the surfaces of both bacteria and tumor cells make it a target for cytotoxic antibodies. New technologies involving the structural modification of PSA have been developed to utilize modified PSA as a vaccine against group B meningococci and *E. coli* K1 [17] and for the immunotargeting of tumor cells [18].

13.2
Group C Meningococcal Vaccines

13.2.1
Structure and Immunology of GCMP

Group C *Neisseria meningitidis* is a worldwide problem and a major contributor (~40%) to the incidence of meningococcal meningitis in developed countries. Like most other bacterial capsular polysaccharides [2], the GCMP is an attractive vaccine candidate because it is a potent virulence factor, and contains the most highly conserved and most exposed protective epitopes on its bacterial surfaces. The GCMP can be readily purified [19, 20] and is stable and nontoxic. It is also highly immunogenic in humans [21], despite the fact that it consists entirely of normally immunosuppressive sialic acid residues.

The structure of the GCMP is shown in Tab. 13.1; it consists of a linear homopolymer of sialic acid in which the individual sialic acid residues are linked by $a2$–9-ketosidic linkages [3]. Therefore, because $a2$–9-linked polysialic acid is not found in human tissues, the GCMP, with the exception of its non-reducing terminal sialic acid residues, is a good immunogen [21]. The protective epitopes span the $a2$–9-linked sialic acid residues, being situated on conventional internal short linear sequences of sialic acid [22]. While the largest proportion of group C meningococci are O-acetylated, about 15% are unacetylated [23]. In early ^{13}C NMR

Tab. 13.1 Structures of polysialic acids associated with human bacterial infections and/or human tumors [a].

Organism	Structure
Group C *Neisseria meningitidis* (O-acetylated)	-9)aNeu5Ac(2 → 8 | OAc
Group C *Neisseria meningitiditis* (unacetylated)	-9)aNeu5Ac(2 →
Group B *Neisseria meningtidis* *Escherichia coli* K1 (unacetylated) *Pasteurella haemolytica* *Moraxella liquefaciens* Human tumors	-8)aNeu5Ac(2 →
Escherichia coli K1 (O-acetylated)	-8)aNeu5Ac(2 → 7/9 | OAc
Escherichia coli K92	-8)aNeu5Ac(2 → 9) aNeu5Ac(2 →
Human tumors	-8)KDN(2 →

a) See text for references.

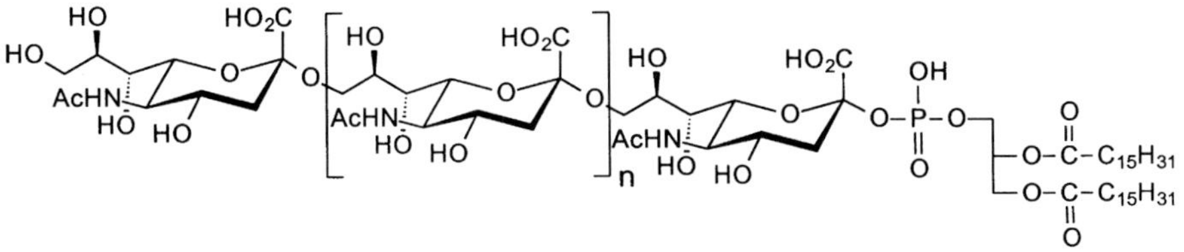

13.2 Group C Meningococcal Vaccines | 359

Fig. 13.1 Proposed structure of the lipid functional group of the GCMP.

studies it was ascertained that the O-acetyl groups of the GCMP were located on O-7 and/or O-8 of the individual sialic acid residues while a few residues remained unacetylated [3]. More recent ^{1}H NMR studies [24], however, have demonstrated that in its native state the GCMP contains a preponderance of O-8-linked acetate groups, which unfortunately have a proclivity to migrate to O-7 in solution. This increases the number of O-7-acetylated sialic acid residues to the extent that it creates structural artifacts not present in the native O-acetylated GCMP.

Another interesting and biologically active structural feature of the GCMP is that it contains a unit of diacyl glycerol linked through a phosphodiester bond [25] at the reducing end, as shown in Fig. 13.1. Although minor, this hydrophobic substituent has great immunological significance because it causes the individual polysaccharide chains to aggregate in a micellar form [25]. This maintains the GCMP in a pseudo high molecular weight form, which as in the case of all polysaccharides, has been demonstrated to be essential to its immunogenicity in humans and thus its potential as a human vaccine [26, 27].

As a result of the favorable immune properties described above, the first commercial polysaccharide vaccine against group C meningitis was employed in the late 1960s [28], and consisted of the purified O-acetylated GCMP. It is interesting to note that despite reports [29] that the unacetylated GCMP was a better immunogen in humans, it was never used as a commercial vaccine. While the GCMP has been useful in controlling outbreaks of group C meningitis, its prolonged use in humans has revealed serious deficiencies in its performance. Like most other polysaccharide vaccines, the GCMP does not prime for immunological memory [5], which is an essential property of a vaccine capable of eradicating group C meningitis. In addition, the GCMP, like most other polysaccharides, is not immunogenic in infants [30], who are the group most vulnerable to group C meningitis, and furthermore it is not even optimally immunogenic in adults. These deficiencies have now been circumvented by the development of a new generation of synthetic glycoconjugate vaccines, in which the polysaccharide fragment is covalently linked to a protein carrier [6, 7]. This simple strategy, a milestone in vaccine technology, is based on the fact that linking a normally T-cell-independent saccharide to a T-cell-dependent protein bestows the attributes of the latter (memory effect) upon the saccharide [2]. A plausible immune mechanism for this conversion has been proposed [31].

13.2.2
Group C Conjugate Vaccines

Several different approaches to the synthesis of conjugate vaccines have been described [32]. Important criteria associated with the utility of the different procedures are that the method should be generally applicable, and that the covalent linkage between the saccharide and protein should be stable. In addition, the reaction conditions should be mild enough to maintain the structural integrity of the individual components [32]. On the basis of these criteria, fragments of the O-acetylated GCMP (10–40 kD) were selectively oxidized with sodium metaperiodate to yield activated fragments with free terminal aldehyde groups [6]. The activated group C polysaccharide fragment was then covalently coupled to tetanus toxoid (TT) by reductive amination with sodium cyanoborohydride. The resultant conjugate vaccine was capable of inducing long-lasting immune responses to the polysaccharide in mice, displaying an active T-cell memory effect. Variations on the above methodology have since been extensively employed in the commercial development of conjugate vaccines, including those against group C meningitis [32]. With unique structures like that of the GCMP, this methodology – shown in Fig. 13.2 – can be used both to depolymerize the polysaccharide and simultaneously to activate the fragments with terminal aldehydes [33, 34]. The procedure [34] shown in Fig. 13.2 also depicts the removal of O-acetyl groups from the O-acetylated GCMP prior to depolymerization of the O-deacetylated GCMP and activation of the resultant fragments. The fragments were subsequently conjugated directly to TT by reductive amination. The above procedure was also used to synthesize TT conjugates of O-acetylated GCMP fragments, because the O-acetylated GCMP still contains sufficient unacetylated sialic acid residues [3, 24] for oxidative cleavage to occur. Prior removal of O-acetyl groups has certain advantages, because it prevents the formation of structural features irrelevant to the native O-acetylated GCMP, produced by unavoidable O-acetyl migration [24]. It also has other distinct immunological advantages despite the fact that ~85% of group C meningococci express the O-acetylated GCMP. In preclinical studies in mice with the tetanus toxoid conjugates of O-acetylated and O-deacetylated GCMP fragments [35], it was demonstrated that the latter conjugate was not only more immunogenic than its O-acetylated counterpart, but was also able to induce antibodies that protected against challenge by both O-acetylated and unacetylated group C meningococci [35]. Similar results had been observed in some previous comparative human studies with the O-deacetylated and O-acetylated GCMP as vaccines [29], and have also been confirmed in recent comparative human studies with their respective conjugates [7].

While the commercial development of group C meningococcal conjugate vaccines has been slow, in the United Kingdom it was recently decided, because of a rapid increase in the incidence of group C meningococcal meningitis, to immunize all children from 1–18 years with one of three commercially available group C conjugates [7, 36–38]. One of the conjugates used was made with O-deacetylated fragments [38] and the other two with O-acetylated fragments [36, 37]. All the conjugates were made with activated C polysaccharide fragments produced essentially by the procedure depicted in Fig. 13.2. All the vaccines were able to in-

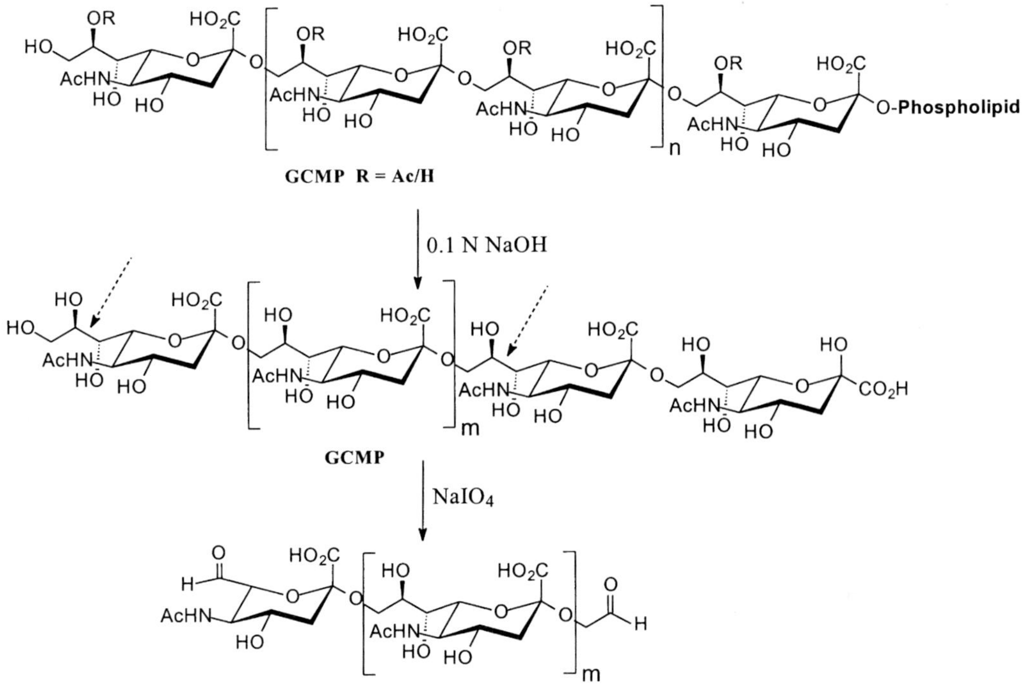

Fig. 13.2 Chemical structure of the O-acetylated GCMP and potential cleavage and activation sites in the O-acetylated GCMP.

duce immunological memory even in infants, and the vaccine program has been highly successful, the incidence of meningitis caused by group C meningococci having already dropped significantly. However, it is interesting to note that, in a recent comparative study [7] of the performance of the three conjugate vaccines in toddlers (12–18 months), the O-deacetylated saccharide conjugate was significantly more immunogenic than the others, and also induced higher bactericidal titers than its O-acetylated counterparts in the toddlers, even against O-acetylated group C meningococci. Whether these advantages will be extrapolated in the future to give more sustained levels of bactericidal activity or improved immunological memory cannot yet be determined.

Although there are other differences between the three conjugate vaccines there is strong evidence to suggest that removal of the O-acetyl groups from the O-acetylated GCMP prior to conjugation is an important factor in enhancing immunogenicity. In fact, even in adult human antisera induced by an O-acetylated GCMP vaccine the bactericidal activity could be inhibited much more efficiently by the O-deacetylated GCMP than by the O-acetylated GCMP (Fig. 13.3). This indicates that the most immunogenic epitopes are not formed by the native O-acetylated sialic acid residues but by the minor population of unacetylated sialic acid residues in the O-acetylated GCMP [35].

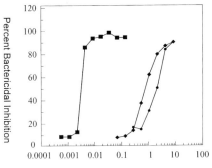

Fig. 13.3 Inhibition of the binding of O-acetylated GCMP to human O-acetylated GCMP antisera by (■) O-deacetylated GCMP, (◆) 7-O-acetylated GCMP, and (●) 8-O-acetylated GCMP.

13.3
Group B Meningococcal Vaccines

13.3.1
Structure of GBMP

The basic structure of the group B meningococcal polysaccharide (GBMP) is shown in Tab. 13.1; it consists of relatively short chains (~ 40 kD) of α2–8 polysialic acid (PSA) [3]. However, when freshly isolated from the culture medium by Cetavlon precipitation, many of these chains retain a lipoidal residue glycosidically linked to their reducing ends, as in the case of the GCMP [25]. The lipid residue from the GBMP contains the same components as that of the GCMP shown in Fig. 13.1, but its structure has been less well defined [25]. Although small, this lipid component has a profound effect on the physical and immunological properties of the GBMP. Thus, as with the GCMP, it not only causes the individual PSA chains to aggregate, but this aggregation, uniquely to the GBMP, also results in the formation of a potentially important protective epitope on the surface of group B meningococci and *E. coli* K1 (see Section 13.4.3).

Unlike the GCMP, the GBMP has not been identified in an O-acetylated form. However, *Escherichia coli* K1 produces a capsular polysaccharide structurally identical to the GBMP (Tab. 13.1), and also has a form variant that produces an O-acetylated α2–8-linked PSA capsule [8]. The O-acetyl groups are distributed on O-7 and O-9 of the α2–8-linked sialic acid residues. Interestingly, *Escherichia coli* K92 also produces a PSA capsular polysaccharide (shown in Tab. 13.1), which contains alternating sequences of α2–8- and α2–9-sialic acid residues [4].

13.3.2
Immunology of GBMP

It has been demonstrated that PSA is a potent virulence factor both in group B *Neisseria meningitidis* and in *Escherichia coli* K1 [1], and it is probably more than coincidental that, as a result of this, both organisms are important human pathogens. Group B *Neisseria meningitidis* is the most prevalent cause of meningococcal meningitis, even responsible for over 60% of cases in developed countries [28], and *E. coli* K1 is the leading cause of human neonatal meningitis [1]. However, it is unfortunate that PSA, even in its aggregated pseudo high molecular weight form, cannot be used as a vaccine, being poorly immunogenic in infants and adults [39]. While it has been reported [8] that the O-acetylated $a2$–8 PSA capsule produced by the *E. coli* K1 variant is more immunogenic than its unacetylated counterpart, and also produces antibodies specific for the latter, the increase in immunogenicity was insufficient to justify its use as a vaccine. In addition, it was also demonstrated [4] that the capsular polysaccharide of *E. coli* K92, which contains structural elements both of the GBMP and of the GCMP (Tab. 13.1), was immunogenic, but produced only a predominance of antibodies that cross-reacted with the GCMP. The immune mechanism is reluctant to produce antibodies with specificity for the $a2$–8-sialic acid linkage.

Currently there is no fully efficacious vaccine against meningitis caused by group B meningococci and the search for one remains one of the biggest scientific challenges. Because of the poor immunogenicity of the GBMP, most efforts to make a vaccine have focused on alternate surface-exposed bacterial components, such as outer membrane proteins [28] and lipooligosaccharides [40]. However, many problems associated with the development of vaccines based on these components have been identified, not the least of which is their intrinsic antigenic diversity. Therefore, because the GBMP is the only conserved antigenic structure on the surface of group B meningococci, and because the same antigenic structure is also present on *E. coli* K1, the GBMP is the vaccine of choice, provided that its poor immunogenicity can be overcome.

13.3.3
B Polysaccharide-Protein Conjugates

Although the covalent coupling of the GBMP to protein carriers results in enhanced polysaccharide-specific antibody levels, including antibodies of the IgG isotype, these levels are generally low and no bactericidal activity has been reported [6, 41]. A similar result was obtained with protein conjugates of the capsular polysaccharide of *Escherichia coli* K92, the structure which is shown in Tab. 13.1. The K92 polysaccharide is composed of alternate $a(2$–$8)$- and $a(2$–$9)$-linked sialic acid residues, which make it a potential vaccine against groups B and C *N. meningitidis* [4]. However, although conjugates of the K92 polysaccharide induce antibodies reactive towards both the GBMP and the GCMP, only the latter predominate and exhibit significant bactericidal activity [42, 43]. These results indicate that it is unlikely that protein conjugates of the GBMP will be of importance in the production of vaccines against meningitis caused by group B *N. meningitidis*. Even if this manipulation had succeeded in producing a good response, the issue of whether immune tolerance would be broken would also have to be addressed before such a vaccine could be sanctioned for human use. This is because molecular mimicry is involved in the poor immunogenicity of the GBMP, PSA having been identified in both normal and cancerous human tissues (see Section 13.5.1).

Despite its poor immunogenicity, PSA-specific antibodies can be produced in special circumstances. Hyper-immunization of a horse with group B meningococci has produced high levels of PSA-specific IgM antibodies [8], and murine monoclonal antibodies of the same specificity have been produced by similar immunization procedures [44, 45]. The potential of some of the latter antibodies to be protective against group B meningococci and E. coli K1 has been demonstrated in passive protection studies and opsonophagocytic assays. Human transformed cell lines producing protective PSA-specific antibodies have been described [46], and a human macroglobulin (IgM NOV) with the same specificity also has been reported [47]. All the above PSA-specific antibodies were of the IgM isotype with the exception of mAb 735 produced in an autoimmune New Zealand black (NZB) mouse system, which was of the IgG isotype [44]. Thus, antibodies to PSA can be produced by the use of aggressive vaccination schedules with whole organisms as the immunogen. However, such procedures cannot be used in routine human vaccination procedures and the best approach to make a vaccine based on the GBMP probably resides in its chemical manipulation (see Section 13.4.1).

13.3.4
Extended Helical Epitope of PSA

The presence of a conformational epitope in PSA was initially hypothesized to explain the inability of oligomer fragments of up to five sialic acid residues to inhibit the binding of the GBMP to a homologous horse antibody [22]. It has also been established that all PSA-specific antibodies require an unusually long segment of PSA for binding to occur [48, 49]. It has been proposed that antibodies recognize an extended helical form of this polysaccharide [50] and that a minimum of about nine residues is required in order to form this extended helical epitope [22, 48]. The validity of this hypothesis was further strengthened by the identification of a common length-dependent epitope responsible for the cross-reaction of both PSA and poly (A) with IgM NOV [51] and the known propensity of poly (A) to form helices of $n=8-10$. Because these two biopolymers share no common structure features, the cross-reaction is thought to be due to a common epitope composed of a similar helical spatial arrangement of negative charges. This is supported by the fact that $a2-8$-linked oligomers of PSA exhibit identical inhibitory properties no matter whether poly (A) or the PSA is used as the binding antigen to IgM NOV [51].

By use of potential energy calculations and nuclear magnetic resonance (NMR) it has been shown that, although mostly random coil in nature, PSA can adopt extended helical conformations in which $n \approx 9$ [50, 52]. It has also been demonstrated that the stability of this extended helical conformation is dependent on its carboxylate groups [52]. Experiments to obtain unequivocal evidence of the existence of this extended helical epitope have been carried out, with attempts being made to co-crystallize $a2-8$-linked oligomers of $n \approx 10$ with a Fab fragment of

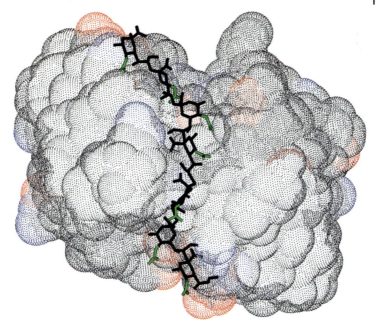

Fig. 13.4 Stereoview of the fit of the helical model of PSA to the binding surface of Fab 735. The NAc substituents are highlighted in green.

mAb 735; unfortunately these have not been successful to date. However, convincing evidence was obtained when the Fab fragment of mAb 735 was crystallized in the absence of hapten and subjected to X-ray diffraction analysis [53]. The binding site consisted of an unusually long groove, which was bimodal in that it underwent a striking reversal in shape and charge distribution along the interface between heavy and light chains. With only minor adjustments, the extended helix proposed by Brisson et al. [50], when modeled into the binding site, had a shape and charge distribution complementary to the Fab from mAb 735 (Fig. 13.4). At least eight residues are accommodated in the site, the helical twist of PSA positioning the appropriate functional groups for binding to the bimodal site.

From reported binding studies on different PSA-specific antibodies, all are specific for the extended helical epitope [22, 49, 53]. Because conformational studies indicate that this epitope is only a minor contributor to the total number of epitopes formed by PSA [50, 52], the dominance of antibodies with a specificity for this less populous epitope must be the result of immunological selection. The reluctance of the immune system to produce antibodies associated with the more populous random coil form of the polymer probably occurs because these epitopes are conformationally similar to the shorter sialyloligomers. These latter structures are also present in human tissue [54] and the production of antibody to them, rather than to the extended helical form of PSA, is even more stringently avoided.

13.4
Chemically Modified Group B Meningococcal Vaccines

13.4.1
N-Propionylated PSA Conjugate Vaccine

The failure of PSA-protein conjugates to provide satisfactory levels of protective antibody against group B meningococci prompted interest in the further chemical modification of PSA prior to its conjugation. One modification that could be made without disturbing the required helical conformation of PSA [52] was to replace the *N-acetyl* (NAc) groups of its sialic acid residues with *N-propionyl* (NPr) groups [55], as shown in Fig. 13.5. Fragments of NPr PSA (10–11 kD) previously treated with sodium metaperiodate to introduce terminal aldehyde groups, were then conjugated to TT by reductive amination [55].

The NPr PSA-TT conjugate, when administered with Freunds' complete adjuvant (FCA), was able to induce high titers of NPr PSA-specific antibodies in mice, which were highly bactericidal for group B meningococci and passively protective against both group B meningococci and *E. coli* K1 [55–57]. Despite the fact that the NPr PSA-TT conjugate was able to induce higher titers of PSA cross-reactive (IgG) antibodies than the homologous PSA-TT conjugate in mice, this PSA-specific antibody was not protective [56, 58]. Although group B meningococci were able to absorb out the protective antibodies from the antisera [56], the data in Tab. 13.2 show that PSA was unable to remove any of the bactericidal protective antibodies. Thus, it was demonstrated that NPr PSA-specific antibodies consist of two distinct populations, one of which (minor population) cross-reacts with the PSA and is not protective, whereas the other, larger population of PSA non-cross-reactive antibodies surprisingly contains all the protective antibodies [58].

This evidence indicates that the NPr PSA mimics a different epitope on the surface of group B meningococci and *E. coli* K1 than is presented by PSA alone [57, 58]. Recently this epitope has been located by electron microscopy, through the use of an NPr PSA-specific monoclonal antibody (13D9), in the capsular layers of both group B meningococci and *E. coli* K1 (see Section 13.4.3). These studies suggest that an NPr PSA-protein conjugate would be an excellent vaccine candidate

Tab. 13.2 Bactericidal titers of anti-NPrPSA serum absorbed with PSA.

Serum	Radioactive antigen binding assay		
	PSA	NPrPSA	Bactericidal titer
Anti-NPrPSA (unabsorbed)	50[a]	77	512
Anti-NPrPSA (absorbed with PSA)[b]	0	73	512
Control	0	0	<4

a) Percentage of binding to ^{3}H-labelled antigens.
b) 500 µL of serum absorbed with 250 µL of 1 mg mL^{-1} PSA (10–11 kDa) for 4 days at 4 °C.

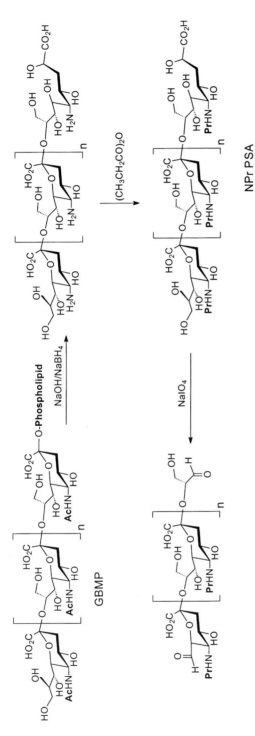

Fig. 13.5 Chemical modification of the GBMP and potential cleavage and activation sites in NPrPSA.

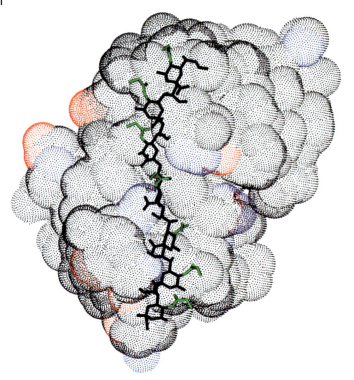

Fig. 13.6 Stereoview of the fit of the helical model of NPrPSA to the binding surface of Fab 13D9. The NPr substituents are highlighted in green.

and this has been confirmed in more recent preclinical studies with NPr PSA-protein conjugates in mice [55, 58] and primates [59].

13.4.2
Immunology of NPr PSA

Because NPr PSA mimics a protective epitope distinct from PSA on the surface of group B meningococci, it is a potential vaccine candidate. It is therefore important to define the natures both of the epitope mimic and of the mimicked epitope. To characterize these epitopes more fully, a series of NPr PSA-specific mAbs were produced by immunization of Balb/c mice with NPr PSA-TT, and screening them against a number of PSA antigens [60]. The immune properties of the majority of mAbs, as represented by mAb 13D9, are shown in Tab. 13.3, and fit the previously described profile of the PSA protective epitope [22, 49].

The mAbs bound only to extended sequences of NPr PSA, as defined by their inability to bind to (NeuNPr)$_4$-HSA. They also provided bactericidal protection against challenge in mice by group B meningococci, despite the fact that they either did not bind or bound very poorly to PSA. More recently, in an independent study [61], protective mAbs with the same specificity have been produced. Only a few mAbs that bound to (NeuNPr)$_4$-HSA were obtained [60]. These mAbs had an exclusive specificity for NPr PSA and were non-protective. Because of the paucity of mAbs, induced by NPrPSA-TT, that bound to (NeuNPr)$_4$-HSA, additional mAbs were made by use of (NeuNPSA)$_4$-TT as immunogen [60]. The immune properties of all the mAbs as represented by mAb 11G1 are shown in Tab. 13.3, and despite the fact that they all cross-reacted with PSA and the GBMP, none were protective. Recent surface plasmon resonance studies [62] have shown that these mAbs bind exclusively to the non-reducing terminal tetrasaccharide moiety of PSA, and that these epitopes occur infrequently on the surface of *E. coli* K1.

The length-dependency of the protective epitope mimic is consistent with the results of previous inhibition experiments [58] and with NMR and molecular modeling, which indicate that the epitope is in fact located on extended helical domains of NPr PSA [52]. The fact that NPr K92 polysaccharide conjugates, unlike those of NPr PSA, cannot induce bactericidal antibodies in mice is also consistent with the above hypothesis. NMR and molecular modeling of the K92 polysaccharide and its NPr analogue have shown that they cannot form extended helices, because of the innate flexibility of their alternating α2–9-sialic acid residues [43]. Confirmation of the existence of this extended helical conformation was recently obtained by X-ray crystallographic analysis of the crystallized Fab fragment of mAb 13D9 [63]. The binding site, consisting of an unusually long groove, was conformationally similar to PSA-specific mAb 735 (53) even though mAb 13D9 bound very poorly to PSA (Tab. 13.3). An extended helix of NPr PSA modeled into the binding site had a shape and charge distribution complementary to the Fab from mAb 13D9 (Fig. 13.6, see p. 368). At least eight residues are accommodated in the site, the helical twist of the NPr PSA positioning the appropriate functional groups for binding to the site.

Tab. 13.3 Biologic activity of monoclonal antibodies to NPrPSA.

Vaccine	Clone	Isotype	NPrPSA	(Neu5Pr)$_4$	PSA	GBMP[a]	Epitope size	Bactericidal activity
(Neu5Pr)$_4$-TT	11G1	IgG$_{2a}$	+	+	+	+	short	−
NPrPSA-TT	13D9	IgG$_{2a}$	+	−	−	+	extended	+
	6B9	IgG$_{2a}$	+	+	−	−	short	−
Group B meningococci	735	IgG$_{2a}$	−	−	+	+	extended	+

a) Contains aggregated PSA.

13.4.3
Protective Epitope mimicked by NPr PSA

The very close structural resemblance between NPr PSA and PSA, together with the absence of the former on the surface of group B meningococci and *E. coli* K1, would suggest that the mimicked epitope has a PSA component. This evidence is also consistent with the fact that both organisms, which have the GBMP as their only common surface antigen, were able to absorb out the bactericidal antibodies from an NPr PSA-specific mouse antiserum [58]. One can conclude from this that PSA interacts with another molecule, on the surface of group B meningococci and *E. coli* K1, to form an extended helical epitope, closely related to, but nevertheless distinct from that formed by the PSA. Both serologically distinct helical epitopes have been identified in the capsular layers of both group B meningococci and *E. coli* K1 by electron microscopy by the use of mAbs 13D9 and 735 and a gold-labeled anti-mouse IgG antibody [60]. It is plausible to propose that the lipoidal group attached to the reducing end of individual PSA chains is a factor in the formation of the mimicked epitope. This is based on the fact that mAb 13D9 only reacts with aggregated forms of the GBMP (Tab. 13.3), the formation of which is dependent on the presence of terminal lipoidal groups, which cause the individual PSA chains to aggregate [25].

13.4.4
Safety Concerns

All the evidence described above would suggest that an NPr PSA-protein conjugate is a strong vaccine candidate, which could be the missing immunogen to include in combination vaccines against bacterial meningitis. However, this type of vaccine has been shown to produce a minor population of antibodies that cross-react with PSA. Because this molecule is also found in human tissue, there is a perceived safety concern (see Section 13.5.1), and it is realistic to suppose that the acceptance of this vaccine will depend on being able either to control the induction of PSA-specific antibodies, or to prove that these antibodies are innocuous in the human system. One practical method of diminishing the production of PSA cross-reactive antibodies is to vary the adjuvant used with the NPr PSA-protein conjugate vaccine [64]. Adjuvants are required not only to enhance the overall response to the conjugates in mice, but also to achieve an isotype switch in the immune response from non-bactericidal IgG_1 to bactericidal IgG_2 and IgG_3 [64]. However, adjuvants differ markedly in their abilities to induce cross-reactive antibodies, and Freunds' complete adjuvant is probably the worst contributor to their induction [64]. In comparative studies in mice with the NPr PSA-TT vaccine in conjunction with different adjuvants, FCA induced 23% of cross-reactive PSA antibodies whereas the use of monophospholipid (MPL) resulted in a significant diminution (over 100-fold) of these cross-reactive antibodies [64]. The greater tendency of FCA to produce more cross-reactive PSA antibodies is probably also reflected in the results of two separate studies involving the production of mAbs with NPr PSA-TT conjugates with either MPL [60] or FCA [61]. Low levels of cross-reactive PSA antibodies were also observed in monkeys, immunized with a potential human vaccine (NPr PSA-RPor B), in which the outer membrane protein (RPor B) serves as both a carrier and as an adjuvant [59]. Although the amount of cross-reaction can be controlled, it is probably inevitable that the use of NPr PSA-protein conjugates will induce a subset of autoantibodies that bind to PSA and also, as has been demonstrated in some experiments, bind to polysialylated host tissues [49, 61]. On the positive side, however, there is currently no direct evidence that these cross-reactive antibodies are deleterious to host even in more meaningful in vivo studies.

In fact there is much evidence to suggest that PSA antibodies (IgG) will not be pathologic. Most adults acquire low levels of this type of antibody, presumably through asymptomatic carriage of *E. coli* K1, group B meningococci, or *M. non-liquefaciens*, and convalescence from group B meningococcal meningitis elicits a humoral response (IgM and IgG) to PSA in patients [45]. NPr PSA-protein vaccines have been used successfully without deleterious consequences as experimental human vaccines in a number of animal species [55, 57, 58, 59], and an NPr PSA-TT vaccine was even found to be safe and immunogenic in human adults [65]. However the human antisera lacked bactericidal activity, which could be due to the fact, that only a mild adjuvant (aluminum hydroxide) was used in these studies [64].

Because of the documented overexpression of PSA in vertebrates during fetal development (see Section 13.5.1), the most rigorous safety test of this type of vaccine would be to study the effects of PSA-specific antibodies on fetal development. Administration of mAb 735 (see Section 13.3.4) to pregnant rats resulted neither in lesions nor in binding to fetal brain [66]. Other even more convincing tests were carried out in female cynomologus monkeys and their offspring, following hyperimmunization with NPr PSA-TT in conjunction with FCA [67]. Prior to mating, all the female monkeys had high levels of antibodies specific for NPr PSA, and the levels of these antibodies were maintained throughout the gestation period. Antibodies of the above specificity were transmitted to the fetuses and had no harmful consequences on the development of organs and the nervous system of fetuses and sucklings up to the age of six months.

13.5
Cancer Vaccines

13.5.1
PSA on Human Cells

The structures of human tissue antigens containing contiguous α2–8 Neu5Ac linkages range from shorter fragments such as the oligomeric $(Neu5Ac)_{2-3}$ formed in mammalian gangliosides [54] to longer fragments carried by glycoprotein (NCAM) [11, 12, 15]. These latter PSA chains have been identified in lengths from eight to in excess of 55 contiguous Neu5Ac residues. Recently, shorter PSA fragments of up to seven Neu5Ac residues have also been shown to occur on a large number of brain glycoproteins, which, like the short PSA fragments found in the gangliosides, may also play a role in neural function [68].

PSA was first identified as a constituent of human cells by Finne [11], and was found to be attached to neural cell adhesion molecules (NCAM) [11, 15]. Although random coil in nature, PSA can form unique extended helical segments [52, 53], the structure which provides the recognition site for two highly specific diagnostic reagents. These are PSA-specific antibodies such as mAb 735 [44] and a PSA-specific bacteriophage-derived endoneuraminidase [69]. Used in conjunction, these reagents provide a sensitive and reliable diagnostic test for PSA, which has resulted in the identification of PSA on many other developing, neural, and cancerous tissues. PSA is widely expressed in the fetal development of all vertebrates, but postnatally becomes restricted to a few discrete regions of plasticity such as the brain and nervous system [15, 70]. It is associated with the rearrangement and migration of cells, and defects in the latter have been observed in mice genetically deficient in PSA expression [71] and mice treated with endo-N-sialidase [72]. The biological function of PSA is that it serves as a barrier to cell-cell interactions rather than having any specific affinity for a receptor [15]. That the steric properties of PSA alone are responsible for its effect on embryonic adhesion has been demonstrated by the sensitivity of the latter to small changes in ionic strength [73]. Interestingly, a related PSA, and possible cancer vaccine target, is based on a polymer of the deaminated form of neuraminic acid, 2-keto-3-deoxy-D-glycero-D-galacto-nononic acid (KDN), the structure of which is shown in Tab. 13.1. It was initially found in lower vertebrates [74], and has also been identified in various mammalian (including adult human) tissues [75, 76]. Although its function is not yet known, it is thought to be associated with embryonic development [76] and has also been identified as an oncodevelopmental antigen [14, 76]. An N-glycosylated form of PSA has also been detected in mammalian tissue [13] but has not so far been found in humans.

PSA is an oncofetal antigen [12, 15] and is expressed on a number of important tumors such as on Wilms' tumor, small cell lung cancer, neuroblastomas, and rhabdomyosarcoma [77–79]. Also, because of the established influence of PSA on cell migration, it has been closely associated with tumor metastasis [15, 80]. PSA on the surfaces of cancer cells is also attached to a glycoprotein (NCAM) but there is no evidence that NCAM causes the attached PSA to aggregate and form a complex epitope similar to that formed by the terminal lipid component of the group B meningococcal and *E. coli* K1 capsular polysaccharides (see Section 13.4.3). However, there is evidence that antibodies specific for NPr PSA bind to polysialylated tumor cells [18, 61], and the consequences of this binding have not been reported except in one case, in which mAb 13D9 proved to be non-cytotoxic (see Section 13.5.2). A plausible explanation for this binding is that PSA forms a different neoepitope on the surface of tumor cells, to which NPr PSA-specific antibodies bind, but only with weak affinity. Finne et al. [81] have hypothesized that such an epitope could originate in yet another tertiary structure of PSA, the formation of which is independent of PSA terminal attachments. By use of atomic force microscopy they found that oligomers of PSA (>12 Neu5Ac) can start to self-assemble, and are able to form filament bundle networks of increasing complexity with increasing length.

13.5.2
Potential of NPr PSA as a Cancer Vaccine

Carbohydrates are the most abundantly expressed antigens on the surfaces of most tumors and some success has been reported [82] in creating cancer vaccines based on these antigens, but the approach is not without its problems. This is due to the fact that cancer cells fail to produce saccharide markers that distinguish them from normal cells. Glycoconjugate vaccines based on the above saccharide antigens are therefore poorly immunogenic. Although the use of PSA-protein conjugate vaccines as cancer therapeutic agents has not yet been reported, it is likely that they will be no exception to the rule, since vaccines of this type have failed to produce an immune response adequate to protect against bacteria that express PSA (see Section 13.3.3). Even though NPr PSA in its conjugated form is an effective vaccine against PSA-expressing bacteria (see Section 13.4.1), there is evidence to suggest that it would not be fully effective against tumors expressing PSA. While NPr PSA-specific mAb 13D9 bound to rat (RBL-3H3) and mouse (RMA) leukemic cells, it failed to mediate the cytotoxicity of these cells [18], and in an in vivo experiment mAb 13D9 was only partially able to reduce metastasis (see below).

A novel strategy for overcoming immune tolerance to PSA and thus to facilitate the immune targeting of tumor cells is to modify its structure on the surfaces of tumor cells [18]. This was achieved by the biochemical engineering of surface PSA by exposing the cells to a chemically modified sialic acid precursor (NPr Man) as shown diagrammatically in Fig. 13.7. This is a well documented procedure that has been carried out on sialosides on the surfaces of mammalian cells in tissue culture and in vivo [83]. The permissiveness of the enzymes involved in sialic acid biosynthesis and sialoside formation allowed a wide range of different N-acylated Man precursors to be used, which resulted in the expression of unnatural N-acylated sialic acid residues on the cell surface sialooligosacchrides [83]. This procedure has been used successfully for the study of biological processes [84–86], the chemotargeting of human cells [87, 88], and immunotargeting [18] of tumor cells. Of particular interest in the biochemical engineering of PSA is that the sialotransferases responsible for its biosynthesis are less permissive, because only NPr Man [18, 89] and not its higher analogues such as NBu Man [89] can be incorporated into the sialic acid residues of PSA. In fact, NBu Man functions as a chain stopper, in which capacity it could be useful either for study of the biological role of PSA or for possible therapeutic applications.

When incubated with NPr Man in vitro, both rat and mouse leukemic cell lines (RBL-3H3 and RMA, respectively) expressed NPr PSA and its expression was both time- and dose-dependent [18]. The expression of NPr PSA and the disappearance of PSA on the bioengineered cells could be followed by flow cytometric analysis

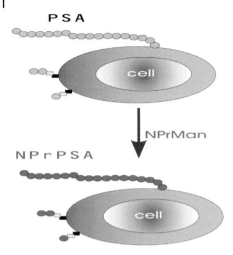

Fig. 13.7 Biosynthetic incorporation of NPrMan into cell surface PSA and other sialosides. Although, as depicted, complete incorporation is unlikely, incorporation is nevertheless probably substantial, due to the extended nature of the epitope. The proposed fit of the helical model of NPrPSA to the surface of Fab 13D9 (Fig. 13.6) suggests that four contiguous internal NPr substituents are in contact with the binding site.

by using the respective serologically distinct mAbs 13D9 (see Section 13.4.2) and 735 (see Section 13.3.4), the specificities of which are based on similar but structurally different extended helical epitopes (Fis. 13.6 and 13.4, respectively). Cancer cell cytotoxicity was achieved by use of an NPr PSA-specific antibody (mAb 13D9) in the presence of complement, and cytotoxicity was shown to be dependent on the amount of NPr PSA expressed on the cancer cells. The above procedure was also successfully applied in vivo. In a mouse solid tumor model using RMA cells, immunization of the mice and precursor (NPr Man) following installation of the solid tumor not only reduced the size of the tumors but was also able to control metastasis of the tumor effectively (Tab. 13.4) [18]. It is interesting to note that although mAb 13D9 alone had no cytotoxic effect on the above cancer cells in vitro, it did have a small effect in vivo. Whether this is due to an increased avidity of mAb 13D9 for the cancer cells due to an improved expression of a neoepitope (see Section 13.3.1) in vivo has not been established.

Although the above immunotherapeutic procedure was only partially able to inhibit the growth of tumors, its significance cannot be underestimated, because of the importance of its being able to control metastasis during cancer treatment. The procedure also has the advantage of being selective, because although NPrMan can be taken up and randomly expressed in the sialooligosaccharides of

Tab. 13.4 Antibodies against NPrPSA control tumor metastasis in vivo.

Treatment of groups of mice	Tumors in spleen[a]	Metastasis (%)
mAb 13D9 + precursor	0/5	0.0
mAb 13D9	2/6	33.3
No treatment	4/5	80.0

a) Mice inoculated with RMA tumor cells 5 days before treatment.

both normal and cancer cells, NPr PSA can be specifically immunotargeted (Fig. 13.7). In addition, while being overexpressed on cancer cells, PSA is only found in a few discrete adult tissues (see Section 13.5.1).

13.6
Acknowledgements

The author thanks Dr. Wei Zou for helpful discussions and Lynda Boucher for expert secretarial assistance.

13.7
References

1 J.B. Robbins, G.H. McCracken, E.C. Gotschlich, E. Ørskov, F. Ørskov, and L.A. Hanson. *Escherichia coli* K1 capsular polysaccharide associated with neonatal meningitis, *N. Eng. J. Med.* **1974**, *290*, 1216–1220.

2 H.J. Jennings. Capsular polysaccharides as human vaccines, *Adv. Carbohydr. Chem. Biochem.* **1993**, *41*, 155–208.

3 A.K. Bhattacharjee, H.J. Jennings, C.P. Kenny, A. Martin, and I.C.P. Smith. Structural determination of the sialic acid polysaccharide of *Neisseria meningitidis*, *J. Biol. Chem.* **1975**, *250*, 1926–1932.

4 W. Egan, T.-Y. Liu, D. Dorow, J.S. Cohen, J.D. Robbins, E,C, Gotschlich, and J.B. Robbins. Structural studies on the sialic acid polysialic acid antigen of *Escherichia coli* strain Bos-12, *Biochemistry* **1977**, *16*, 3687–3692.

5 E.C. Gotschlich. Meningococcal meningitis. In *Bacterial vaccines*, Germanier, R. (ed.). **1984**, Academic Press, New York, pp. 237–255.

6 H.J. Jennings, and C. Lugowski. Immunochemistry of groups A, B, and C meningococcal polysaccharide-tetanus toxoid conjugates, *J. Immunol.* **1981**, *127*, 1012–1018.

7 P. Richmond, R. Borrow, D. Goldblatt, J. Findlow, S. Martin, R. Morris, K. Cartwright, and E. Miller. Ability of the three different meningococcal C conjugate vaccines to induce immunologic memory after a single dose in UK toddlers, *J. Infect. Dis.* **2001**, *183*, 160–163.

8 F. Ørskov, I. Ørskov, A. Sutton, R. Schneerson, L. Wenlu, W. Egan, G.E. Hoff. Form variation in Escherichia coli K1 determined by O-acetylation of the capsular polysaccharide, *J. Exp. Med.* **1979**, *149*, 669–685.

9 C. Adlam, J.M. Knights, A. Mugridge, J.M. Williams, and J.C. Lindon. Production of colominic acid by Pasteurella haemolytica serotype A2 organisms, *FEMS Microbiol. Lett.* **1987**, *42*, 23–25.

10 S.J. Devi, R. Schneerson, W. Egan, W.F. Vann, J.B. Robbins, and J. Shiloach. Identity between polysaccharide antigens of Moraxella non-liquefacians, group B Neisseria meningitidis, and Escherichia coli K1 (non-O-acetylated), *Infect. Immun.* **1991**, *59*, 732–736.

11 J. Finne. Occurrence of unique polysialosyl carbohydrate units in glycoproteins of developing brain, *J. Biol. Chem.* **1982**, *257*, 11966–11970.

12 F.A. Troy. Polysialylation: from bacteria to brains. *Glycobiology* **1992**, *2*, 5–23.

13 C. Sato, K. Kitajima, S. Inoue, and Y. Inoue. Identification of oligo-N-glycolneuraminic acid residues in mammal-derived glycoproteins by a newly developed immunochemical reagent and biochemical methods, *J. Biol. Chem.* **1998**, *273*, 2575–2582.

14 M. ZIAK, M. MEIR, and J. ROTH. Megalin in normal tissues and carcinoma cells carries oligo/poly α2,8 deaminoneuraminic acid as a unique posttranslational modification, *Glycoconjugate J.* **1999**, *16*, 185–188.

15 U. RUTISHAUSER. Polysialic acid at the cell surface: Biophysics in service of cell interactions and tissue plasticity, *J. Cellular Biochemistry* **1998**, *70*, 304–312.

16 J. ROTH, C. ZUBER, P. KOMMINOTH, E.P. SCHEIDEGGER, M.J. WARHOL, D. BITTER-SUERMANN, P.U. HEITZ. Expression of polysialic acid in human tumors and its significance for tumor growth. In Polysialic Acid (ROTH, J., RUTISHAUSEN, U., and TROY, F.A., eds.), **1993**, pp. 335–348, Birkhauser, Basel, Switzerland.

17 H.J. JENNINGS, R. ROY, and A. GAMIAN. Induction of meningococcal group B polysaccharide-tetanus toxoid conjugate vaccine, *J. Immunol.* **1986**, *137*, 1708–1713.

18 T. LIU, Z. GUO, Q. YANG, S. SAD, and H.J. JENNINGS. Biochemical engineering of surface α2,8 polysialic acid for immunotargeting tumor cells, *J. Biol. Chem.* **2000**, *275*, 32832–32836.

19 E.C. GOTSCHLICH, T.-Y. LIU, and M.S. ARTENSTEIN. Human immunity to the meningococcus. III Preparation and immunochemical properties of the group A, group B, and group C meningococcal polysaccharides, *J. Exp. Med.* **1969**, *129*, 1349–1365.

20 C. YANG, and H.J. JENNINGS. Purification of capsular polysaccharide. In Methods in Molecular Medicine, Vol. 66: Meningococcal Vaccines: Methods and Protocols (POLLARD, A.J., and MAIDEN, M. eds.), **2000**, pp. 41–47. Human Press Inc., Totowa, N.J.

21 E.C. GOTSCHLICH, I. GOLDSCHNEIDER, and M.S. ARTENSTEIN. Human immunity to the meningococcus: IV: Immunogenicity of group A and group C meningococcal polysaccharides in human volunteers, *J. Exp. Med.* **1969**, *129*, 1367–1384.

22 H.J. JENNINGS, R. ROY, F. MICHON. Determinant specificities of the groups B and C polysaccharides of Neisseria meningitidis, *J. Immunol.* **1985**, *134*, 2651–2657.

23 G. ARAKERE, and C.E. FRASCH. Specific antibodies to O-acetyl positive and O-acetyl negative group C meningococcal polysaccharides in sera from vaccines and carriers, *Infect. Immun.* **1991**, *59*, 4349–4356.

24 X. LEMERCINIER, and C. JONES. Full 1H NMR assignment and detailed O-acetylation patterns of the capsular polysaccharides from Neisseria meningitidis used in vaccine production, *Carbohydr. Res.* **1996**, *296*, 83–86.

25 E.C. GOTSCHLICH, B.A. FRASER, O. NASHIMURA, J.B. ROBBINS, and T.-Y. LIU. Lipid on capsular polysaccharides of Gram-negative bacteria, *J. Biol. Chem.* **1981**, *256*, 8915–8921.

26 E.A. KABAT, and A.E. BEZER. The effect of variation in molecular weight on the antigenicity of dextran in man, *Arch. Biochem. Biophys.* **1958**, *28*, 306–310.

27 T.-Y. LIU, E.C. GOTSCHLICH, E.K. JONSEEN. Studies on the meningococcal polysaccharides: I. Composition and chemical properties of the group A polysaccharide, *J. Biol. Chem.* **1971**, *246*, 2849–2858.

28 H. PELTOLA. Meningococcal vaccines: Current status and future possibilities, *Drugs* **1998**, *55*, 347–361.

29 M.C. STEINOFF, E.B. LEWIN, E.C. GOTSCHLICH, and J.B. ROBBINS. Panorama Pediatric Group. Group C Neisseria meningitidis variant polysaccharide vaccines in children, *Infect. Immun.* **1981**, *34*, 144–146.

30 M.L. LEPOW, I. GOLDSCHNEIDER, M. GOLD, M. RANDOLPH, E.C. GOTSCHLICH, Persistence of antibody following immunization of children with groups A and C meningococcal polysaccharide vaccines, *Pediatrics* **1977**, *60*, 673–680.

31 G.R. SIBER. Pneumococcal disease: Prospects for a new generation of vaccines, *Science* **1994**, *265*, 1385–1387.

32 H.J. JENNINGS, and R.K. SOOD. Synthetic glycoconjugates as human vaccines. In: Neoglycoconjugates: Preparation and Applications (LEE, Y.C. and LEE, R.T. eds.). Academic Press, San Diego, **1994**, pp. 325–371.

33 E.C. BEUVERY, R. ROY, V. KANHAI, and H.J. JENNINGS. Characteristics of two types of meningococcal group C polysac-

charide conjugates using tetanus toxoid as carrier protein, *Dev. Biol. Stand.* **1986**, *65*, 197–204.

34 Z. Guo, and H. J. Jennings. Protein polysaccharide conjugation. In Methods in Molecular Medicine, Vol. 66, Meningococcal vaccines: Methods and Protocols. Polland, Humana Press Inc., Totowa, N. J., **2001**, pp. 49–54.

35 F. Michon, C.-H. Huang, E. K. Farley, L. Hronowski, and P. C. Fusco. Structure activity studies on group C meningococcal polysaccharide-protein conjugate vaccines. Effect of O-acetylation on the nature of the protective epitope. In: Physico-chemical procedures for the characterization of vaccines (Brown, F., Corbel, M., and Griffiths, E. eds.), *Dev. Biol.*, Vol. 103, Karger, Basel, **2000**, pp. 151–160.

36 C. K. Fairley, N. Begg, R. Borrow, A. J. Fox, D. M. Jones, and K. Cartright. Conjugate meningococcal serogroup A and C vaccine: Reactogenicity and immunogenicity in United Kingdom infants, *J. Infect Dis.* **1996**, *174*, 1360–1363.

37 P. Richmond, R. Borrow, E. Miller, S. Clark, F. Sadler, A. Fox, N. Begg, R. Morris, and K. Cartright. Meningococcal serogroup C conjugate vaccine is immunogenic in infancy and primer for memory, *J. Infect. Dis.* **1999**, *179*, 1569–1572.

38 P. Richmond, D. Goldblatt, P. C. Kusco, J. D. S. Fusco, I. Heron, S. Clark, R. Borrow, and F. Michon. Safety and immunogenicity of a new Neisseria meningitidis serogroup C – tetanus toxoid conjugate vaccine in healthy adults, *Vaccine* **2000**, *18*, 641–646.

39 F. A. Wyle, M. S. Artenstein, B. L. Brandt, D. L. Tramont, D. L. Kasper, P. Altieri, S. L. Berman, and J. P. Lowenthal. Immunological response of man to group B meningococcal polysaccharide antigens, *J. Infect. Dis.* **1972**, *126*, 514–521.

40 T. T. Poolman. Development of a meningococcal vaccine, *Infect. Agents Dis.* **1995**, *4*, 13–28.

41 A. Bartolini, F. Norelli, C. Ceccarini, R. Rappuoli, and P. Constantino. Immunogenicity of meningococcal B polysaccharide conjugated to tetanus toxoid or CRM197 via adipic acid dihydrazide, *Vaccine* **1995**, *13*, 463–470.

42 S. J. Devi, J. B. Robbins, and R. Schneerson Antibodies to poly[(2–8)-a-N-acetylneuraminic acid] are elicited by immunization of mice with Escherichia coli K92 conjugates: potential vaccines for groups B and C meningococci and E. coli K1, *Proc. Natl. Acad. Sci. USA* **1991**, *88*, 7175–7179.

43 R. A. Pon, N. H. Khieu, Q.-L. Yang, J. R. Brisson, and H. J. Jennings Serological and conformational properties of E. coli K92 capsular polysaccharide and its N-propionylated derivative both illustrate that induced antibody does not recognize extended epitopes of polysialic acid: Implications for a comprehensive conjugate vaccine against groups B and C. N. meningitidis, *Can. J. Chem.* **2002**, *80*, 1055–1063.

44 M. Frosch, I. Gorgen, D. Bitter-Suermann. NZB mouse system for the production of monoclonal antibodies to weak bacterial antigens: isolation of an IgG antibody to the polysaccharide capsular of Escherichia coli K1 and group B meningococci, *Proc. Natl. Acad. Sci. USA* **1985**, *82*, 1194–1198.

45 G. Rougon, C. Dubois, N. Buckley, J. L. Magnani, W. D. Zollinger. A monoclonal antibody against meningococcus group B polysaccharides distinguishes embryonic from adult N-CAM, *J. Cell Biol.* **1986**, *103*, 2429–2437.

46 H. V. Raff, D. Devereux, W. Shuford, D. Abbott-Brown, G. Maloney. Monoclonal antibody with protective activity for Escherichia coli K1 and Neisseria meningitidis group B infections, *J. Infect. Dis.* **1988**, *157*, 118–126.

47 E. A. Kabat, K. G. Nickerson, J. Liao, et al. A human monoclonal macroglobulin with specificity for $a(2 \rightarrow 8)$-linked poly-N-acetylneuraminic acid, the capsular polysaccharide of group B meningococci and Escherichia coli K1, which cross-reacts with nucleotides and with denatured DNA, *J. Exp. Med.* **1986**, *164*, 642–654.

48 J. Finne, P. H. Makela, Cleavage of the polysialosyl units of brain glycoproteins

by a bacteriophage endosialidase, *J. Biol. Chem.* **1985**, *260*, 1265–1270.

49 J. Hayrinen, H. J. Jennings, H. V. Raff. Antibodies to polysialic acid and its N-propyl derivative: binding properties and interaction with human embryonal brain glycopeptides, *J. Infect. Dis.* **1985**, *171*, 1481–1490.

50 J.-R. Brisson, H. Baumann, A. Imberty, S. Perez, H. J. Jennings. Helical epitope of the group B meningococcal $\alpha(2 \rightarrow 8)$-linked sialic acid polysaccharide, *Biochemistry* **1992**, *31*, 4996–5004.

51 E. A. Kabat, J. Liao, F. Ossermann, A. Gamian, F. Michon, H. J. Jennings. The epitope associated with the binding of the capsular polysaccharide of the group B meningococcus and of Escherichia coli K1 to a human monoclonal macroglobulin, IgM NOV, *J. Exp. Med.* **1988**, *168*, 699–711.

52 H. Baumann, J.-R. Brisson, F. Michon, R. Pon, H. J. Jennings. Comparison of the conformation of the epitope of $\alpha(2 \rightarrow 8)$-polysialic acid with its reduced and N-acyl derivatives, *Biochemistry* **1993**, *32*, 4007–4013.

53 S. V. Evans, B. W. Sigurskjold, H. J. Jennings, J.-R. Brisson, R. J. To, W. C. Tse, E. Altman, M. Frosch, C. Weisgerber, H. D. Kratzin, D. R. Rose, N. M. Young, and D. R. Bundle. Evidence for the extended helical nature of polysaccharide epitopes. The 2.8 Å resolution structure and thermodynamics of ligand polysialic acid, *Biochemsitry* **1995**, *34*, 6737–6744.

54 S. Ando, R. K. Yu. Isolation and characterization of two isomers of brain tetrasialogangliosides, *J. Biol. Chem.* **1979**, *254*, 12224–12229.

55 H. J. Jennings, R. Roy, A. Gamian Induction of meningococcal group B polysaccharide-specific IgG antibodies in mice using an N-propionylated B polysaccharide-tetanus toxoid conjugate vaccine, *J. Immunol.* **1986**, *137*, 1708–1713.

56 F. E. Ashton, J. A. Ryan, F. Michon, H. J. Jennings. Protective efficacy of mouse serum to the N-propionyl derivative of meningococcal group B polysaccharide, *Microb. Pathog.* **1989**, *6*, 455–458.

57 H. J. Jennings, A. Gamian, F. E. Ashton. N-propionylated group B meningococcal polysaccharide mimics a unique epitope on group B Neisseria meningitidis, *J. Exp. Med.* **1987**, *165*, 1207–1211.

58 H. J. Jennings, A. Gamian, F. Michon, F. E. Ashton. Unique intermolecular bactericidal epitope involving the homosialopolysaccharide capsule on the cell surface of group B Neisseria meningitidis and Escherichia coli K1, *J. Immunol.* **1989**, *142*, 3585–3591.

59 P. C. Fusco, F. Michon, J. Y. Tai, and M. S. Blake. Preclinical evaluation of a novel group B meningococcal conjugate vaccine that elicits bactericidal activity in both mice and non-human primates, *J. Infect. Dis.* **1997**, *175*, 364–372.

60 R. Pon, M. Lussier, Q. Yang, and H. J. Jennings. N-propionylated group B meningococcal polysaccharide mimics a unique bactericidal capsular epitope in group B Neisseria meningitidis, *J. Exp. Med.* **1997**, *185*, 1929–1938.

61 D. M. Granoff, A. Bartolini, S. Ricci, S. Gallo, D. Rosa, N. Ravenscroft, V. Guarnieri, R. C. Seid, A. Shan, W. R. Usinger, S. Tan, Y. E. McHugh, and G. R. Moe. Bactericidal monoclonal antibodies that define unique meningococcal B polysaccharide epitopes that do not cross-react with human polysialic acid, *J. Immunol.* **1998**, *160*, 5028–5036.

62 H. J. Jennings et al. Manuscript in preparation.

63 S. I. Patenaude, S. M. Vijay, Q.-L. Yang, H. J. Jennings, and S. V. Evans. Crystallization and preliminary X-ray diffraction analysis of antigen-binding fragments which are specific for antigenic conformations of sialic acid homopolymers, *Acta Cryst.* **1998**, *D54*, 1005–1007.

64 H. J. Jennings. N-propionylated group B meningococcal polysaccharide glycoconjugate vaccine against group B meningococcal meningitis, *Int. J. Infect. Dis.* **1997**, *1*, 158–164.

65 B. Danve, J. Bruge, N. Bouveret-Le Cam, D. Chassard, G. Rougon, and D. Schultz. Safety and immunogenicity of an N-propionylated group B meningococcal polysaccharide conjugate vaccine in volunteers. In Abstracts of the Tenth In-

ternational Pathogenic Neisseria Conference (ZOLLINGER, W., FRASCH, C., and DEAL, C. D. eds.), Baltimore, MD, **1996**, pp. 225–226.

66 K. SAUKKONEN, D. HALTIA, and M. LEINONEN. Antibodies to the capsular polysaccharide of Neisseria meningitidis group B or E. coli K1 bind to the brains of rats in vitro but not in vivo, *Microb. Pathog.* **1986**, *1*, 101–105.

67 J. BRUGE, J.-C. MOULIN, B. DANVE, N. DALLA LONGA, C. VALENTIN, C. GOLDMAN, G. ROUGON, J.-P. HERMAN, B. COQUET, M. THIONET, and D. SCHULTZ. Evaluation of the innocuity of a group B meningococcal polysaccharide conjugate in hyperimmunized pregnant cynomolgus monkeys and their offspring. In Abstracts of the Tenth International Pathogenic Neisseria Conference (ZOLLINGER, W., FRASCH, C., and DEAL, C. D. eds.), Baltimore, MD, **1986**, pp. 222–223.

68 C. SATO, H. FUKUOKA, K. OHTA, T. MATSUDA, R. KOSHINO, K. KABAYASHI, F. A. TORY, and K. KITAYIMA. Frequent occurrence of pre-existing α2–8 linked disialic and oligosialic acids with chain lengths up to 7 Sia residues in mammalian brain glycoproteins, *J. Biol. Chem.* **2000**, *275*, 15422–15431.

69 E. R. VIMR, R. D. MCCOY, H. F. VOLLGER, N. C. WILKINSON, F. A. TROY. Use of prokaryotic-derived probes to identify poly(sialic acid) in neonatal membranes, *Proc. Natl. Acad. Sci. USA* **1984**, *81*, 1971–1975.

70 G. ROUGON. Structure metabolism and cell biology of polysialic acids, *Eur. J. Cell Biol.* **1993**, *61*, 197–207.

71 H. TOMASIEWICZ, K. ONO, D. YEE, C. THOMPSON, C. GORIDIS, U. RUTISHAUSER, and T. MAGNUSON. Genetic deletion of a neural cell adhesion molecule variant (NCAM-180) produces distinct defects in the central nervous system, *Neuron* **1993**, *11*, 1163–1174.

72 K. ONO, H. TOMASIEWICZ, T. MAGNUSON, and U. RUTISHAUSER. NCAM mutation inhibits tangential neuronal migration and is phenocopied by enzymatic removal of polysialic acid, *Neuron* **1994**, *13*, 595–609.

73 P. YANG, D. MAJOR, and U. RUTISHAUSER. Role of charge and hydration in effects of polysialic acid on molecular interactions on and between cell membranes, *J. Biol. Chem.* **1994**, *269*, 23039–23044.

74 A. KANAMORI, K. KITAJIMA, S. INOUE, and Y. INOUE. Isolation and characterization of deaminated neuraminic acid-rich glycoprotein (KDN-gp-OF) in the ovarian fluid of rainbow trout, *Biochem. Biophys. Res. Commun.* **1989**, *164*, 744–749.

75 M. ZIAK, B. QU, X. ZUO, C. ZUBER, A. KANAMORI, K. KITAJIMA, S. INOUE, Y. INOUE, and J. ROTH. Occurrence of poly(α2–8-deaminoneuraminic acid) in mammalian tissues: widespread and developmentally regulated but highly selective expression on glycoproteins, *Proc. Natl. Acad. Sci. USA* **1996**, *93*, 2559–2763.

76 B. QU, M. ZIAK, C. ZUBER, and J. ROTH. Poly(α2–8-deaminoneuraminic acid) is expressed in lung on a single 150-Kda glycoprotein and is an oncodevelopmental antigen, *Proc. Natl. Acad. Sci. USA* **1996**, *93*, 8995–8998.

77 J. ROTH, C. ZUBER, P. WAGNER, D. J. TAATJES, C. WEISGERBER, P. V. HEITZ, K. RAJEWSKY, and D. BITTER-SUERMANN. Reexpression of poly(sialic acid) units of the neural cell adhesion molecule in Wilms tumor, *Proc. Natl. Acad. Sci. USA* **1988**, *85*, 2999–3003.

78 C. F. C. MOOLENAAR, F. J. MULLER, D. J. SCHOL, C. G. FIGDOR, E. BOCK, D. BITTER-SUERMANN, R. J. A. M. MICHALIDES. Expression of neural cell adhesion molecule related sialoglycoprotein in small cell lung cancer and neuroblastoma cell lines H69 and CHP-212, *Cancer Res.* **1990**, *50*, 1102–1106.

79 S. GLUER, C. SCHELP, VON SCHWEINITZ and R. GERARDY-SCHAHN. Polysialylated neural cell adhesion molecule in childhood rhabomyosarcoma, *Pediatr. Res.* **1998**, *43*, 145–147.

80 E. P. SCHEIDEGGER, P. M. LACKIE, J. PAPAY, and J. ROTH. In vitro and in vivo growth of clonal sublines of human small cell lung carcinoma is modulated by polysialic acid of the neural cell adhesion molecule, *Lab. Invest.* **1994**, *70*, 95–106.

81 J. TOIKKA, J. AALTO, L. J. HAYRINEN, PELLINIEM, and J. FINNE. The polysialic units of the neural cell adhesion molecule NCAM for filament bundle networks, *J. Biol. Chem.* **1998**, *273*, 28557–28559.

82 S. F. SLOVAN, and H. I. SCHER. Peptide and carbohydrate vaccines in relapsed prostate cancer: Immunogenicity of synthetic vaccines in man – clinical trials at Memorial Sloan-Kettering Cancer Centre, *Semin. Oncol.* **1999**, *26*, 448–454.

83 O. T. KEPPLER, R. HORSTKORTE, M. PAWLITA, C. SCHMIDT, and W. REUTTER. Biochemical engineering of the N-acyl side chain of sialic acid: biological implications, *Glycobiology* **2001**, *11*, 11R–18R.

84 O. T. KEPPLER, P. STEHLING, M. HERRMANN, H. KAYSER, D. GRUNOW, W. REUTTER, and M. PAWLITA. Biosynthetic modulation of sialic acid-dependent virus-receptor interactions of two primate polyoma viruses, *J. Biol. Chem.* **1995**, *270*, 1308–1314.

85 C. SCHMIDT, P. STEHLING, J. SCHNITZER, W. REUTTER, and R. HORSTKORTE. Biochemical engineering of neural cell surfaces by the N-propionyl-substituted neuraminic acid precursor, *J. Biol. Chem.* **1998**, *273*, 19146–19152.

86 B. E. COLLINS, T. J. FRALICH, S. ITONORI, Y. ICHIKAWA, and R. L. SCHNAAR. Conversion of cellular sialic acid expression from N-acetyl to N-glycoylneuraminic acid using a synthetic precursor, N-glycolylmannosamine pentaacetate, inhibition of myelin-associated glycoprotein binding to neural cells, *Glycobiology* **2000**, *10*, 11–20.

87 L. K. MAHAL, K. J. YAREMA, and C. R. BERTOZZI. Engineering chemical reactivity on cell surfaces through oligosaccharide biosynthesis, *Science* **1997**, *276*, 1125–1128.

88 N. W. CHARTER, N. K. MAHAL, D. E. KOSHLAND, Jr., C. R. BERTOZZI. Biosynthetic incorporation of unnatural sialic acids into polysialic acid on neural cells, *Glycobiology* **2000**, *10*, 1049–1056.

89 L. K. MAHAL, N. W. CHARTER, K. ANGATA, M. FUKUDA, D. E. KOSHLAND, Jr. and C. R. BERTOZZI. A small-molecular modulator of poly α2,8-sialic acid expression on cultured neurons and tumor cells. *Science* **2001**, *294*, 380–382.

14
Synthetic Carbohydrate-Based Vaccines
STACY J. KEDING and SAMUEL J. DANISHEFSKY

14.1
Introduction

In the late 18th century, Edward Jenner published his findings on the first successful immunization procedure designed to induce a young boy with cowpox, a bovine disease, to provide for later protection against a related human pathogen, smallpox. This process came to be known as vaccination (after *vacca*, Latin for cow). His use of vaccines against infectious diseases and subsequent investigations paved the way for the development of modern vaccines by Pasteur (rabies), von Behring (diphtheria), Salk (polio), and numerous others. Since then, investigators have attempted to find vaccines that rely on stimulation of the immune system to treat diseases such as HIV and cancer as well as to manage and reverse infections caused by bacteria, viruses and parasites.

First-generation vaccines against bacteria and viruses still used today include live-attenuated, killed, and toxoid vaccines. The preparation of vaccines based on purified subunits and the production of attenuated strains containing programmed genetic modifications have become possible thanks to improved synthetic methodology, advances in analytical methods, and the advent of molecular biology. The development of future-generation vaccines represents an important advance in immunization through recourse to fully synthetic immunogens with defined compositions, affording highly reproducible biological properties. One important direction in vaccine development is the use of conjugate vaccines. The particular focus of our laboratory is to create a cancer vaccine by using carbohydrate-based antigens. By covalently connecting the fully synthetic carbohydrate antigen to an immunogenic carrier, the immune system may be better armed to generate the response necessary for protection. Experience suggests that only through conjugation of an immunogenic carrier can an otherwise non-immunogenic carbohydrate antigen be presented to the immune system in an effective fashion. Proteins and short peptide sequences have been used successfully as immunogenic carriers, but others motifs are also being explored.

Many factors can determine the success of a carbohydrate-protein conjugate vaccine, including proper selection of the carbohydrate antigen construct, the nature of the carrier protein, the ratio of carbohydrate to protein, the length and nature of the linkage between the two moieties, as well as the homogeneity of the entire construct. Although there are different methods for generating the glycoconjugates and many more for the synthesis of the carbohydrate domains, they will not be discussed in this chapter. Instead the focus here will be on vaccines containing completely synthetic carbohydrate antigens.

Carbohydrate-based antigens proposed for inclusion in a vaccine are not readily available from natural sources. Even if one were in principle available, the measures necessary for its isolation, purification, and identification would be very tedious, low-yielding, and, in the end, impractical. The generation of sufficient quantities for useful investigations into vaccine development had hitherto been one of the main limitations to further expansion in this field. It is here that synthetic organic chemistry plays a pivotal role. Synthesis allows for the construction of complex oligosaccharide domains in a systematic fashion through the use of glycosylation methods advanced in the last half century [1–3]. The final product will be of unquestioned structural integrity and of purity appropriate for clinical application. Moreover, chemical modifications to the carbohydrate domain can be introduced during the optimization of the vaccine construct. This option is not necessarily available without a significant amount of extra effort when the antigens must be isolated from natural sources. Thus, constructs containing completely synthetic derived carbohydrate domains are now emerging as optimal for the production of new vaccines targeted to cancer, as well as bacterial and parasitic infections.

14.2
Cancer Vaccines

Many different carbohydrate epitopes are expressed on the surfaces of cells, and the architectures of glycosylation patterns on the cell surfaces of normal and cancer cells can differ. Certain carbohydrates of cancer cells frequently show aberrant glycosylation patterns, and the types of carbohydrate antigens displayed on the cell surface can be used to identify certain cancers. An enhanced level of expression of a tumor's surface antigens is often associated with disease progression and diminished prognosis for successful treatment. Vaccines based on synthetic tumor-associated carbohydrates are emerging as a promising therapy for prompting the human immune system to generate tumor-specific immune responses. The difficulty with the vaccine approach in the treatment of cancer is that the expressed candidate tumor antigens may not be sufficiently immunogenic.

Tumor antigens differentially expressed on the cancer cell surface have generally been identified and classified into four groups: gangliosides, glycophorins, blood group determinants, and the globo series [4]. The antigens have primarily been discovered through the use of monoclonal antibodies (mAb), but isolation by extraction from cancer tumors is also known.

Antibodies generated by an immune response are the primary mechanism for elimination of pathogens from the bloodstream. Vaccines based on carbohydrate antigens that produce antibodies to the tumor antigens could be ideally suited for elimination of circulating tumor cells and micrometastases [5]. There are many issues deserving of consideration in contemplating the use of carbohydrate antigens as targets for active immunotherapy. Carbohydrates are characterized as T-cell-independent antigens. They primarily produce IgM antibodies, which are effective for complement activation in the intravascular space. Despite repeated vaccinations, there is little class switching to IgG antibodies, which are the most important complement activators extravascularly. Complement activation at the cellular surface mediates inflammatory reaction, opsonization for phagocytosis, clearance of antibody complexes from the circulation, and membrane-attack-complex-mediated lysis. The use of properly conjugated vaccines is critical in surmounting this lack of T-cell help. Appropriate conjugation can induce higher-titer IgM antibodies and partial class switching to IgG antibodies. Carbohydrate-based vaccines lacking carrier protein conjugation appear to be unable to induce helper T-cell activation. This omission could well compromise the immunostimulatory process.

Still, the potential advantages of using tumor-associated carbohydrate antigens as targets for cancer immunotherapy with vaccines, especially in the adjuvant setting, are numerous. Tumor-associated carbohydrate antigens are in high abundance on the tumor cell surface, often times as high as 10^7 per cell for some antigens. With proper construction of the conjugate vaccine they also exhibit high degrees of immunogenicity. The antibodies generated are ideally suited for action in the adjuvant setting where the targets are micrometastatic and circulating tumor cells.

The preclinical goals of cancer vaccine programs can be separated into four stages:

(1) Although some of the antigens with relatively small oligosaccharide domains, particularly the gangliosides, are available from natural sources, recourse to chemical synthesis is probably necessary for antigen constructs bearing more complex oligosaccharide domains.
(2) An appropriate spacer must be introduced to the carbohydrate antigen for use in attachment to the protein domain as well as to facilitate immunological integrity of the antigens.
(3) The immunogenic carrier protein or other immunostimulant is covalently attached in the hope of generating a more competent vaccine.
(4) Studies using murine hosts are then undertaken to evaluate the immunogenicity of the construct.

On the basis of the results from the preclinical studies, petition can be made for advancing the vaccine to clinical trials in humans. An in-depth discussion of this approach used for carbohydrate cancer vaccines developed in the Danishefsky laboratory is available [6].

There are a number of different immunogenic carriers that have been explored as components of carbohydrate vaccines, each one attempting to direct the response of the immune system in a specific manner. Although most of the carbohydrate conjugate vaccines have relied on the use of carrier proteins, other moieties including lipids, dendrimers, and peptidic T-cell epitopes have also been investigated.

14.2.1
Carrier Proteins

As discussed above, the use of a suitable protein is crucial in the development of carbohydrate-based cancer vaccines. In the absence of a carrier, the immune system primarily produces IgM antibodies. The conjugation of the carbohydrate domain to an immunogenic carrier protein, however, can stimulate an increase in IgM antibody titer and class switching to IgG antibodies. Many different carrier proteins have been investigated in this regard. The list includes bovine serum albumin (BSA), keyhole limpet hemocyanin (KLH), and bacillus Calmette-Guérin (BCG). Conjugation to KLH, a very large protein (5×10^6 Da) isolated from a marine gastropod [7], has thus far proved to be superior to others in most cases [8].

The gangliosides, such as those shown in Fig. 14.1 (GM2, GM3, and GD2), have long been considered attractive targets for vaccine-based anticancer therapy, since the acidic glycosphingolipids are overexpressed in various cancer types (melanoma, glioma, seminoma, lung cancer, colon cancer, renal cancer, prostate cancer) [9–11]. GM2-reactive antibodies are cytotoxic for GM2$^+$ tumor cells and GM2 is immunogenic in humans, as indicated by the presence of naturally occurring serum antibodies to GM2. GM2 mAbs from humans are relatively easy to isolate after induction of the antibodies in melanoma patients by vaccination with GM2-containing vaccines. Melanoma-affected subjects possessing GM2 antibodies, induced either by vaccination or through natural occurrence, appear to have a more favorable prognosis. There are seemingly no deleterious effects in response to a GM2-based vaccine. Early clinical studies of GM2 with different carriers provided evidence that the GM2-KLH conjugate, with QS-21 as an adjuvant [12], was highly immunogenic, more so than GM2-BCG conjugate with Detox or GM2-Lipid A liposomes. Phase III clinical trials with GM2-KLH plus QS-21 currently in progress are not using synthetic GM2. However, a practical and total synthesis of GM2 has recently been disclosed [13], and preclinical data in mice have shown the synthetic GM2-KLH conjugate to be just as effective. A clinical trial in melanoma patients with the synthetic construct has been initiated [14].

Tumor-associated carbohydrate antigens were first discovered in epithelial cell mucins [4]. Mucins exhibit extensive serine/threonine a-linked-O-glycosylated domains in clustered form (i.e., glycosylated serine/threonine repeats). The TF (Thomsen-Friedenreich) antigen was first described as a cancer antigen in 1984 by Springer [15]. As with the Tn and sialyl-Tn (STn) antigens, the increased presence of the TF antigen on the surface of transformed cells may well be the result of changes in regulation of certain glycosyltransferases [16]. Other more complex glycophorins – 2,3-STF, 2,6-STn, and glycophorin – have not been studied as extensively, but seem to represent an interesting group of carbohydrate antigens.

More than 80% of cancers of breast, prostate, and ovarian origin express STn. In contrast, the expression levels of STn in normal tissues are much reduced and restricted to only a few epithelial tissues at secretory borders [17]. The overexpression of STn characterized in various carcinomas correlates with an aggressive phenotype and worsened prognosis [18]. Immunization with STn has been shown to induce anti-STn antibodies and to protect mice from subsequent tumor challenge with syngeneic cancer cell lines expressing STn [19]. Both active and passive immunotherapy studies have identified STn as an attractive target for antibody-mediated cancer immunotherapy. This is consistent with an expanding body of data demonstrating the ability of antibodies against defined tumor antigens to protect against circulating tumor cells and micrometastases [5, 10].

Subjects with a variety of epithelial cancers have been immunized with STn monomer conjugated to KLH plus various adjuvants. High-titer IgM and IgG antibodies against STn have resulted. A Phase III clinical trial in breast cancer patients, using Theratope® vaccine with Detox as the adjuvant, is currently underway [20]. The linkage method consists of ozonization of the STn crotyl monomer to introduce an aldehyde suitable for reductive amination with the ε-amino groups of lysine residues on KLH (see Fig. 14.2 for structures).

Studies have shown that monoclonal antibodies against STn recognize not only STn monomers, but also STn clusters (STn(c)), indicating that STn is identified in at least two distinct configurations at the tumor cell surface [21]. The STn(c) consisted of a linear tripeptide composed of serine or threonine residues with the side chain hydroxy group of each bearing the carbohydrate antigen. STn-KLH and STn(c)-KLH conjugates were prepared by use of the regular two-carbon, or the recently developed, longer, and more efficient heterobifunctional 4-(4-maleimidomethyl)cyclohexane-1-carboxylic acid hydrazide [22] (MMCCH) linker, and their resulting immunogenicities in mice were compared. Conjugation with MMCCH resulted in the highest conjugation efficiency (yield) and the highest titers against OSM and STn-positive tumor cells, and is the method of choice for the preparation of STn(c) vaccine for clinical trials [23].

Since the immune system tends to recognize clustered motifs of the glycophorin carbohydrate antigens, considerable effort has been taken to synthesize them as glycopeptides. Danishefsky and co-workers have developed a general method for the construction of glycoamino acids. Our "cassette" approach is characterized by a modular rather than a convergent approach. As shown in Scheme 14.1, the basic building block, an N-acetylgalactosamine (GalNAc) synthon, is stereospecifi-

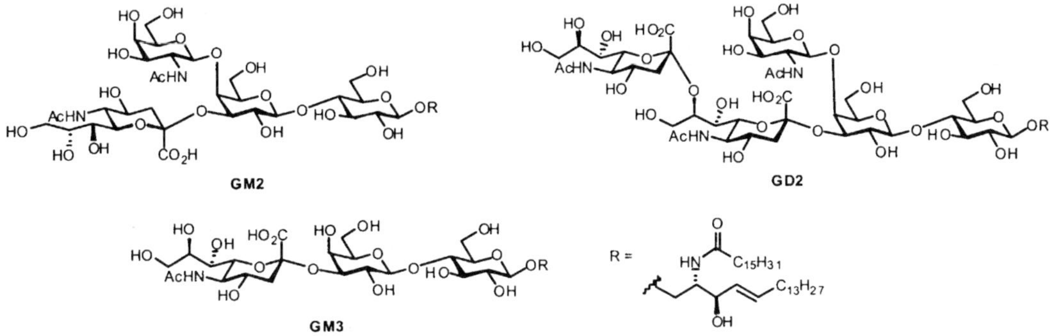

Fig. 14.1 Structures of the gangliosides used or considered for use in human tumor vaccines.

cally linked to a serine, threonine, or hydroxynorleucine, with a differentiable acceptor site on the GalNAc. This construct serves as a general insert (cassette) to be used in subsequent glycosylations with saccharides bearing suitable donor functionality [24]. The main advantage of the cassette approach is that the very difficult O-linkage step can be accomplished early in the synthesis on a simple monosaccharide. A new acceptor site is then exposed for subsequent glycosylation. The α-O-linked serine/threonine is therefore already in place as the oligosaccharide is being grown. The method has been exploited for syntheses of clustered antigen structures built on a peptide backbone [25–27].

The cassette methodology has allowed for the synthesis of a clustered epitope of Tn, shown in Fig. 14.3 [25]. The Tn(c), equipped with a suitable linker, was cross-conjugated to KLH or BSA through a heterobifunctional linker, m-maleimidobenzoyl-N-hydroxysuccinimide ester (MBS), to provide the desired vaccine constructs. Preclinical ELISA results showed the conjugates plus the adjuvant QS-21 were able to generate IgM and IgG antibodies in mice after three immunizations, with the KLH vaccine being the more immunogenic construct [25]. The cell surface reactivities of the anti-Tn(c) antibodies were evaluated by use of Tn(c) positive LS-C colon cancer cells and Tn(c) negative LS-B colon cancer cells. Sera from mice vaccinated with either of the conjugates plus adjuvant showed clear IgM reactivity and significant IgG reactivity by flow cytometry assays and complement-dependent cytotoxicity assays. A Phase I human trial in prostate cancer using the KLH conjugate has been completed. The results demonstrated positive serological results accompanied by stabilizing or declining prostate specific antigen (PSA) levels [28].

Other glycophorin clusters, including TF [25], STn [27], and 2,6-STF [29] (Fig. 14.4), as well as a Lewis^y (Le^y) cluster, have also been prepared by the cassette approach. Conjugation and immunological evaluation of the TF(c)-MBS-KLH and STn(c)-MBS-KLH have produced similar results, and analysis of other glycophorin antigens is currently being pursued.

Le^y is a blood group determinant that has been identified as an important epitope for eliciting antibodies against colon and liver cancers. It has also been implicated in prostate, breast, and ovarian tumors. The first Le^y-containing vaccine, as depicted in Fig. 14.5, incorporated the pentasaccharide Le^y allyl glycoside monomer by the use of reductive amination chemistry for linkage to a carrier protein as discussed above, while later constructs utilized MBS-derivatized KLH for conjugation. Since preclinical results in mice confirmed increased antibody formation and that the resulting antibodies recognized tumor cells expressing the antigen [30], a Phase I clinical trial in ovarian cancer patients was initiated, with the conjugate vaccine together with the immunological adjuvant QS-21 [31]. The trial was successful at the serological level in that the vaccine induced an antibody response in 75% of the subjects. Moreover, the vaccine was tolerated well, with no observable adverse effects related to autoimmunity. In an effort to promote elaboration of IgG antibodies in addition to the observed IgM antibodies, clustering of the Le^y epitope was investigated [32]. Although the Le^y(c)-peptide-MBS-KLH conjugate was capable of eliciting both IgM and IgG responses, the specificity of the re-

Fig. 14.2 Structures of STn monomer and its corresponding conjugate vaccines.

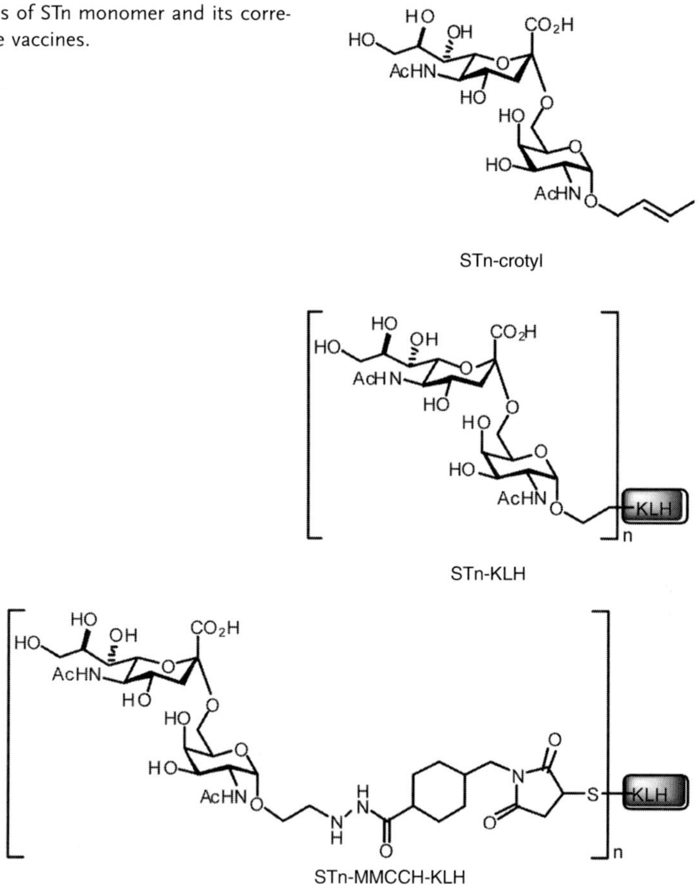

STn-crotyl

STn-KLH

STn-MMCCH-KLH

sponse was limited to the immunizing epitope (ELISA). However, FACS analysis with OVCAR-3 cells did detect moderate activity (~25% positive cells).

Related to Le^y is the KH-1 antigen, which contains both the Le^y tetrasaccharide and the Le^x trisaccharide epitopes. Overexpression of the KH-1 antigen has been seen in a variety of human adenocarcinomas. The KH-1 nonasaccharide antigen was synthesized [33] and fashioned into vaccines as shown in Fig. 14.6 by conjugation via the reductive amination procedure or through use of the MMCCH linker to KLH [34]. Immunological studies of the two constructs initiated in mice showed high titers of both IgM and IgG antibodies for the MMCCH-construct, while the other only generated IgM antibodies [35]. The antibodies elicited by both constructs recognized not only the KH-1 antigen, but also the Le^y antigen, which is similar to the four saccharides at the non-reducing end of the KH-1 nonasaccharide. On the basis of these results, the KH-1-KLH vaccine plus adjuvant is being prepared for clinical evaluation.

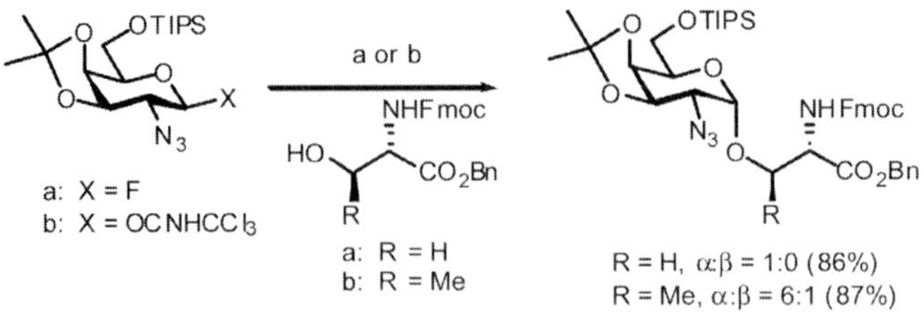

Scheme 14.1 Synthesis of cassette: (a) X=$OCNHCCl_3$, R=H, TMSOTf, THF, –78 °C, (b) X=F, R=Me, Cp_2ZrCl_2, AgOTf, CH_2Cl_2.

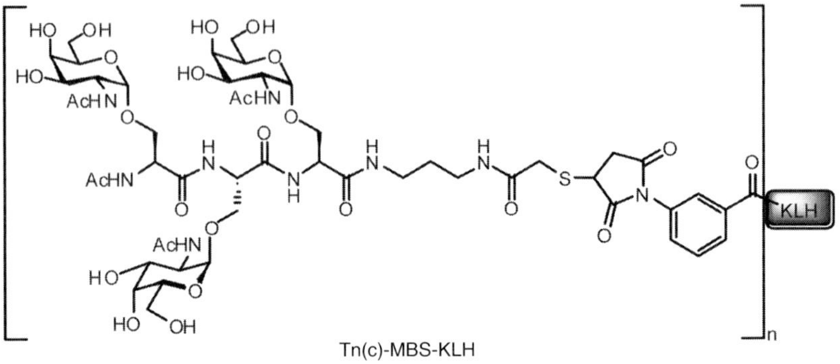

Tn(c)-MBS-KLH

Fig. 14.3 Structure of clustered Tn vaccine.

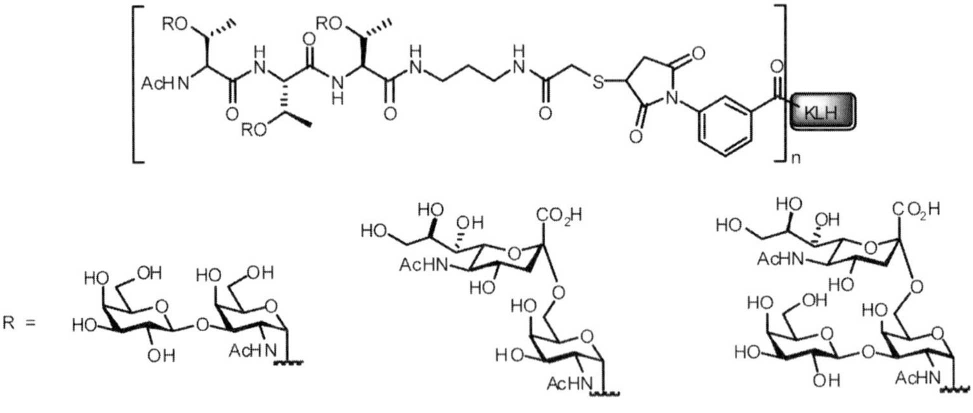

TF(c)-MBS-KLH   STn(c)-MBS-KLH   2,6-STF(c)-MBS-KLH

Fig. 14.4 Structures of other clustered glycophorin vaccines.

Globo-H, a hexasaccharide first isolated in sub-milligram quantities from a human breast cancer cell line, is expressed at the cancer cell surface as a glycolipid and possibly a glycoprotein and was subsequently immunocharacterized by means of the monoclonal antibody MBr1. Immunohistological analysis determined that

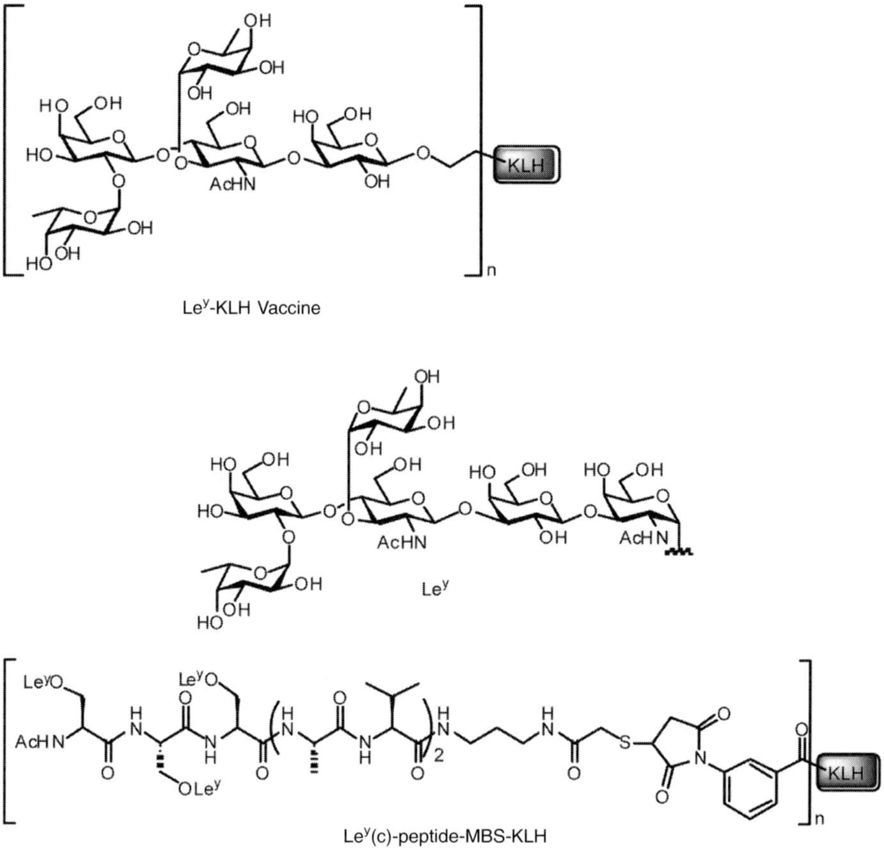

Fig. 14.5 Le^y KLH conjugate vaccines.

Globo-H is additionally expressed in colon, lung, ovary, prostate, and small cell lung cancers and led to its interest for vaccine development. Synthetic chemists were able to corroborate the structure of the complex hexasaccharide by total chemical synthesis [36]. Moreover, these syntheses have allowed for significant quantities of the antigen to be generated for use in development of a vaccine shown in Fig. 14.7. The syntheses have produced both allyl- and pentenyl-glycosides [37] for attaching the linker to the protein carrier. As in the case of other carbohydrate antigen-based vaccines, the KLH conjugate was superior to the BSA conjugate in terms of immunogenicity when administered with the adjuvant QS-21. Studies in mice with Globo-H-KLH showed high titer IgM and IgG responses against the Globo-H antigen. The antibodies reacted with MCF-7 cancer cells (Globo-H positive) but not with B78.2 cells (Globo-H negative) [38]. Additionally, the antibodies were highly effective at inducing complement-mediated cytotoxicity (48% lysis). The Globo-H-KLH vaccine is progressing through clinical trials in breast cancer settings [39].

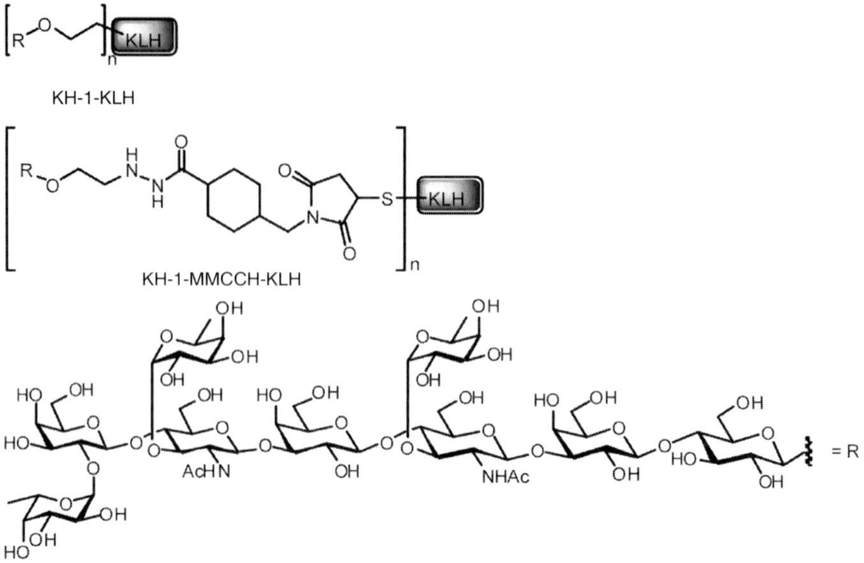

Fig. 14.6 KH-1 nonasaccharide vaccines with variable linkage to carrier protein KLH.

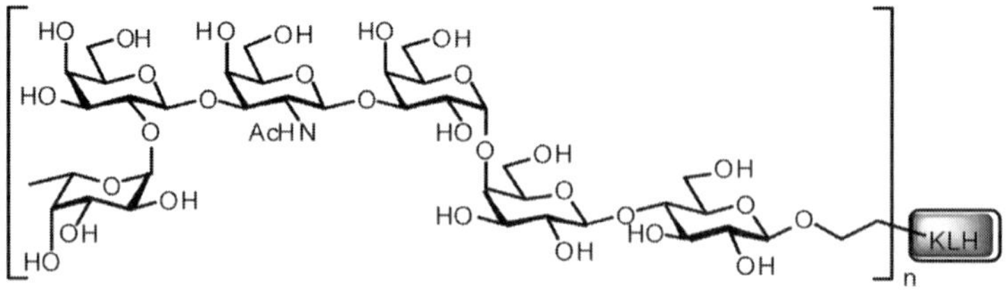

Fig. 14.7 Globo-H-KLH vaccine used in clinical trials for treatment of breast, ovarian, and prostate cancer.

In addition, a full Phase I trial completed by 18 patients with progressive and reoccurring prostate cancer gave promising results [40, 41]. All immunized patients exhibited good IgM responses against Globo-H, with the antibodies recognizing Globo-H-expressing cell lines, and in some cases induced complement-mediated lysis. In addition to the ELISA and FACS assays, the PSA levels of the prostate cancer patients were evaluated. The vaccine may have brought about a decline in the slope of the plot of log PSA concentration, although the conduct of the trial was not designed to document such a claim. Additionally, observation of patients six to nine months after vaccine treatment suggested the possibility of favorable changes in PSA slopes having occurred in patients initially presented in a non-metastatic state. While promising data is still being obtained from patients who continue to receive booster injections and are disease free, the relationship between PSA slopes and early biological efficacy is, at best, anecdotal at this stage.

14.2 Cancer Vaccines

Another complex carbohydrate antigen, fucosyl GM1, has been identified as a highly specific marker associated with small cell lung carcinoma (SCLC) cells. The first total synthesis of this sialyl-containing hexasaccharide was accomplished by methodology first developed in our group for synthesis of the Globo-H antigen [42]. The introduction of a pentenyl glycoside at the reducing end allowed for a more efficient synthesis of potential conjugation precursors. The antigen has undergone KLH conjugation to form the vaccine shown in Fig. 14.8, and mouse studies are currently underway to investigate its immunogenicity relative to previous non-synthetic fucosyl GM1 vaccines [43, 44].

All of the carbohydrate-based cancer vaccines described above were built to target one antigen per vaccine. It is now well established that several different carbohydrate antigens can be associated with any given cancer type. These antigens may possibly show differential levels of expression during the phases of cellular development, and it is believed that an approach that is monovalent in nature may not be sufficient for targeting a population of transformed cells. In contrast, a polyvalent approach involving more than one antigen could well provide a heightened and more varied immune response as well as an improved prognosis for survival. Two different approaches have been taken in the search for vaccines that will target more than one carbohydrate antigen. In one approach, a polyvalent vaccine is constructed from a mixture of existing monovalent conjugate vaccines. *In an alternative method, which we strongly prefer, a multivalent conjugate vaccine is constructed, in which a single molecule contains a subset or a full set of antigens.*

Preclinical results from a polymolecular method involving four KLH conjugate vaccines in mice showed a promising outcome at the immunological level. The conjugate vaccines used were GD3-KLH, Le^y-KLH, and two peptidic antigens MUC1-KLH and MUC2-KLH, along with the immunological adjuvant QS-21. The results demonstrated that the immunogenicities of the individual components were not decreased in the polyvalent vaccine. This conclusion was evidenced by high titer IgM and IgG antibody induction regardless of the administration meth-

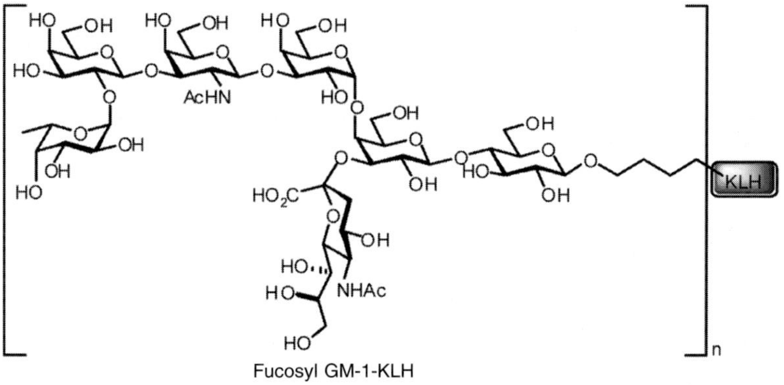

Fig. 14.8 Synthetic fucosyl GM-1 conjugate used to investigate development of a SCLC vaccine.

od (singly in separate mice, separate sites in same mouse, or the same site in mouse) [45]. The antibodies reacted specifically with the respective antigens in tumor cells lines expressing the antigens. On the basis of these results a series of Phase II clinical trials in breast, ovarian, and prostate cancer patients are being initiated. These feature the use of three to seven individual antigen-KLH conjugates (all of which have been previously investigated in Phase I clinical trials) mixed with an immunoadjuvant and administered at a single site. The antibody responses obtained will be evaluated relative to the antibody response of the same conjugates previously administered as monovalent vaccines.

The use of monomolecular multivalent conjugate vaccines for the treatment of cancer is now possible thanks to various advances in relevant synthetic chemistry. Two different glycopeptides (Fig. 14.9) have been elaborated, each one containing three different carbohydrate antigens. One construct utilized the natural mucin-type architecture that involves the linkage of Le^y, TF, and Tn carbohydrate domains to the peptide backbone through serine hydroxyl groups [46]. In this case the serine-GalNAc cassette (vide supra) was incorporated into each glycoamino acid. The other glycopeptide utilized a non-natural tetramethylene linker between the Globo-H, Le^y, and Tn carbohydrate antigens and the peptide backbone [47]. The non-natural amino acid comprising this linker can be called hydroxynorleucine or tris homoserine. The focus of investigations with these constructs was to determine whether the immune system would stimulate a multifaceted response against a single entity. Preliminary investigations of the two constructs showed the conjugate based on the hydroxynorleucine construct to be considerably more antigenic than the mucin-derived construct, and so the former was selected for additional investigation in murine hosts. Evaluation of the hydroxynorleucine-based conjugate vaccine confirmed that both IgM and IgG antibodies were generated, and FACS analysis revealed their ability to react selectively with cancer cells expressing those antigens [48]. In addition to the inclusion of QS-21 [12] as an adjuvant, GPI-0100 [49], which is structurally similar to QS-21 but considerably less toxic, was tested and found to be more effective in antibody generation. The results of these experiments have suggested that a single vaccine construct composed of several different carbohydrate-based antigens may have the potential to stimulate a multifaceted immune response. Further constructs incorporating additional antigens, as well as taking account of the cluster effect seen with the smaller glycophorin family of antigens, are actively being pursued and preclinical data will be forthcoming.

14.2.2
Lipid Carriers

Another approach used in the construction of vaccines has focused on lipids. Lipid carriers are attractive for a number of reasons. Most lipids used are B-cell stimulators, capable of amplifying the proliferation of B-lymphocytes in response to stimulation with antigens. Additionally, many lipids are completely synthetic and their use may result in vaccines that are wholly synthetic, thereby allowing for

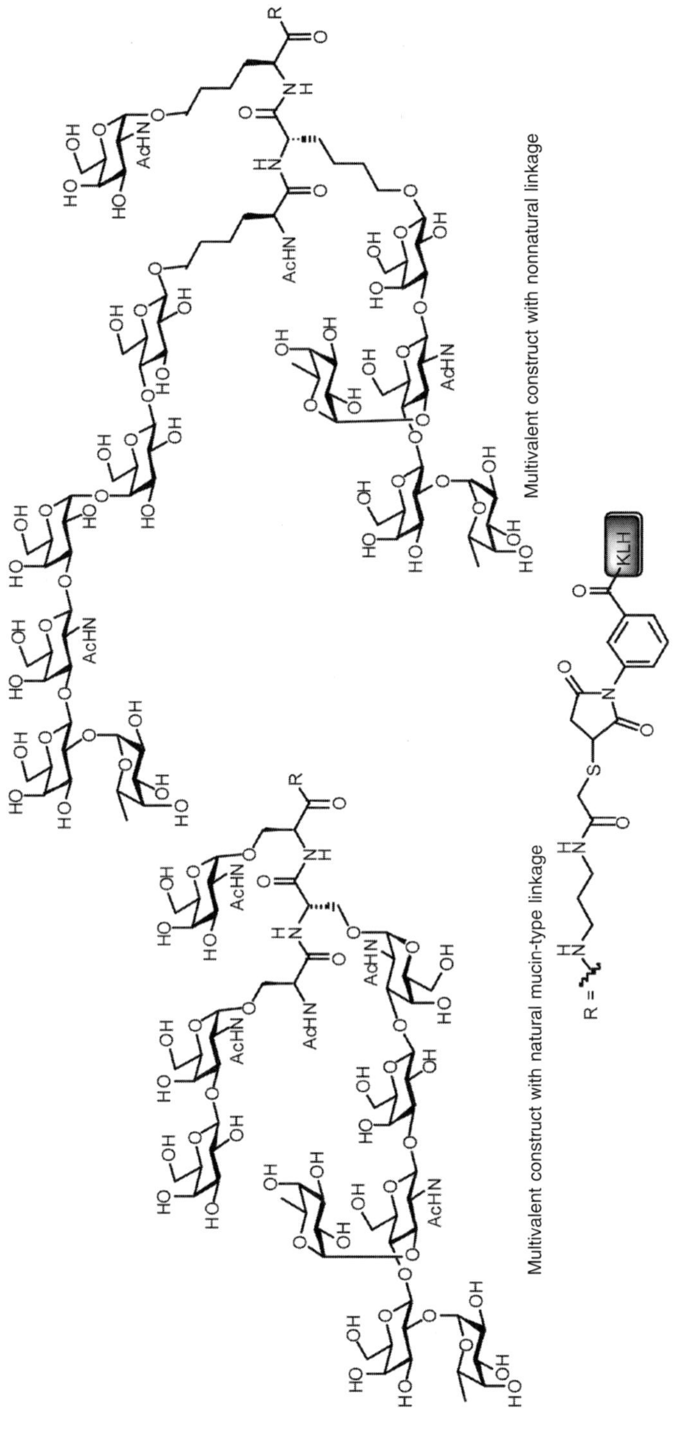

Fig. 14.9 Structures of two multiantigenic unimolecular vaccines utilizing natural and non-natural amino acids in the peptide backbone.

SAR analyses as to efficacy. Fig. 14.10 displays different carbohydrate-based vaccines developed by use of lipid carriers.

In 1994, the development of a totally synthetic carbohydrate vaccine was achieved through the use of the monomeric, dimeric, and trimeric Tn antigens on serine with a 4-aminobutyric acid spacer at the C-terminus [50]. The synthetic antigens were conjugated to ovine serum albumin (OSA). Mice immunized with dimeric or trimeric Tn antigen conjugated to OSA showed a stronger antibody (IgM and IgG) response to a Tn-glycoprotein than mice immunized with monomeric Tn antigen. In an attempt to generate a completely synthetic vaccine, the dimeric Tn antigen was conjugated to tripalmitoylglycerylcysteinylserine (Pam$_3$Cys), which is known as a highly potent B-cell and macrophage activator, derived from the immunologically active N-terminal sequence of the principal lipoprotein of E. coli [51, 52]. The Di-Tn-Pam$_3$Cys conjugate, which is a protein carrier-free immunogen, elicited an immune response against Tn-expressing glycoproteins [53]. The lipopeptide conjugate produced not only a high IgM response but also a significant IgG anti-Tn response without any carrier molecules or additional adjuvants.

The same lipopeptide, Pam$_3$Cys, has also been successfully used in the construction of a Le^y vaccine [54]. The synthetic Le^y(c)-peptide-Pam$_3$Cys was shown to be more immunogenic than the corresponding Le^y-Pam$_3$Cys and Le^y(c)-KLH vaccines when co-administered with adjuvant QS-21, as suggested by the profile of IgM and IgG antibodies produced, and their subsequent recognition of tumor cells bearing Le^y epitopes [32].

Additionally, a novel conjugate molecule in which the ganglioside GM2 antigen is linked through a spacer (ω-hydroxynonanoate) to an immunostimulant (B-cell stimulatory glycolipid BAYR1005) has been synthesized [14]. Vaccination of rabbits with this completely synthetic construct in complete Freund's adjuvant resulted in the induction of IgG antibodies against GM2.

14.2.3
T-Cell Epitopes

In an effort to induce a cytotoxic T-cell immune response, a number of groups have investigated the incorporation of known T-cell epitopes into vaccine constructs in place of the immunogenic protein or lipid carrier.

A synthetic conjugate of a tumor-associated MUC1 glycopeptide antigen and a tetanus toxin epitope connected by a spacer was synthesized in order to investigate the feasibility of this approach [55]. Tn-, TF-, and sialyl-Tn-antigen glycopeptides were incorporated into the tandem repeat region of MUC1 at a threonine residue. Only the STn-containing glycopeptide exhibited a proliferating effect on peripheric blood lymphocytes, and this was further used in the development of a novel vaccine. The strategy involved combination of the B-cell epitope (STn-glycododecapeptide tumor-associated antigen) with a T-cell epitope of tetanus toxin, tethered together through a flexible spacer to produce the construct shown in Fig. 14.11. Immunological evaluation showed that, for the synthetic conjugate, vaccine prolif-

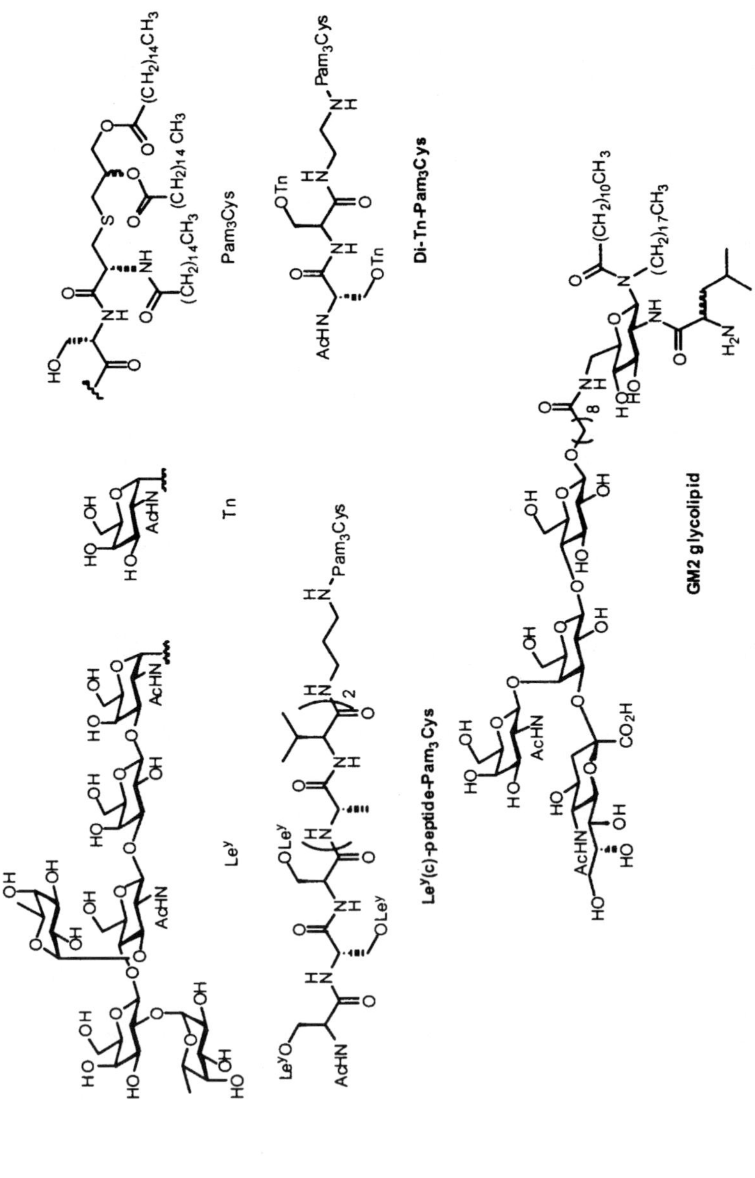

Fig. 14.10 Structures of vaccines utilizing lipid carriers.

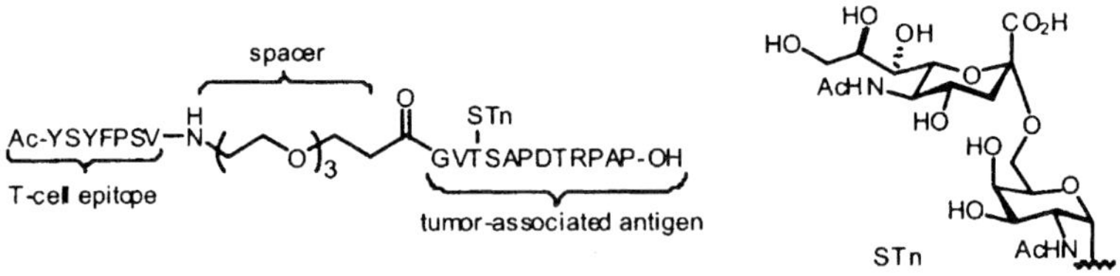

Fig. 14.11 Vaccine containing a B-cell domain as well as a T-cell epitope to induce a T-cell-dependent immune response.

eration only proceeded in the presence of antigen-producing cells and was not found for purified T-cells. This was considered to constitute evidence of an antigen-specific reactivity. The proliferating lymphocytes were characterized by use of monoclonal antibodies directed against surface antigens. FACS analysis showed that the conjugate induced proliferation of up to 100% for $CD3^+$ and 53% for $CD8^+$ T-cells. These preliminary results support the anticipation that conjugates of T-cell epitopes and tumor-associated MUC glycopeptide antigens might induce a cytotoxic T-cell response and that on this basis an efficient synthetic antitumor vaccine may be developed.

This approach has also been studied with use of the Tn, TF, STn, and 2,3-STF antigens. One team of investigators has incorporated a carbohydrate alkylated homocysteine into a peptide sequence known to bind class I MHC molecules on antigen-presenting cells [56, 57]. Another research group has synthesized four different glycopeptides containing variations in both the peptide amino acids and the distance between the T-antigen and the peptide scaffold [58]. Unfortunately, there are no immunological data yet available for these constructs.

14.2.4
Dendrimers

The use of a dendrimer as a scaffold for conjugation to carbohydrate domains is gaining interest as a method for vaccine construction. As in the cases with lipids and T-cell epitopes, the dendrimeric vaccines are completely synthetic in nature, and homogeneity can be strictly controlled. When dendrimers are used in place of proteogenic carriers the likelihood of carrier-induced immune suppression is decreased.

The synthesis of a dendrimeric glycopeptide containing multiple O-linked Tn carbohydrate antigens along with a $CD4^+$ T-cell epitope as exemplified in Fig. 14.12 has been described [59, 60]. The fully synthetic immunogen is highly defined in composition and carries a high saccharidic epitope ratio over the entire molecule. The construct was able to induce anti-Tn IgG antibodies that recognize human tumor cell lines [61, 62]. A therapeutic immunization procedure performed with this fully synthetic immunogen increased the survival of tumor-bearing mice. When used in active specific immunotherapy, the dendrimer carrying the tri-Tn glycotope was much more efficient than the mono-Tn analogue in pro-

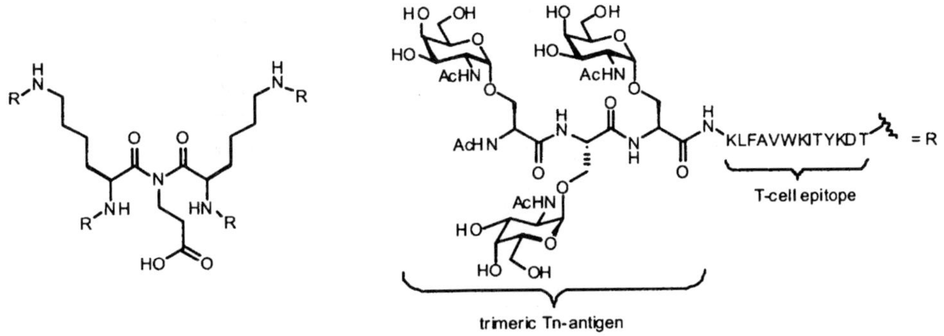

Fig. 14.12 Dendrimer containing a clustered Tn motif as well as a T-cell epitope used in vaccine development.

moting mouse survival. Furthermore, a linear glycopeptide carrying two copies of the tri-Tn glycotope was shown to be poorly efficient in relation to the dendrimeric construct. Other investigators have also used this approach, but their Starburst dendrimer did not elicit the desired response, presumably due to the presence of only dimeric-Tn antigens [50]. The accurately defined and versatile dendrimeric system represents a potentially efficient strategy for induction of carbohydrate-specific antitumor immune responses. It is becoming increasingly apparent that both the clustering of the carbohydrate antigens and the way in which they are displayed seem to be important parameters in stimulating efficient anti-saccharide immune responses.

14.3
Bacterial Polysaccharide Vaccines

Polysaccharide-encapsulated bacteria such as *Haemophilus influenzae* type b, *Streptococcus pneumoniae*, *Salmonella typhi*, and *Neisseria meningitidis* comprise a major class of human pathogens. The level of mortality associated with infections caused by these organisms is substantial, especially in populations with weak or diminished immune systems. Vaccine development has focused on the generation of antibodies specific for and reactive with capsular polysaccharide epitopes of these bacteria. The carbohydrate epitopes found on the surface of these prokaryotes are characteristically composed of repeating units of oligosaccharides that are extensively hydrated and polyanionic in nature. These oligosaccharides, used by the bacteria for numerous functions such as ion transport and adhesion, are also responsible for interfering with the early elimination of the pathogen from circulation by the body. Capsular polysaccharides alone are poor immunogens and the response generated is highly dependent on age. Children aged three months to two years, who are no longer provided with maternal antibodies derived from passive immunization and do not possess mature immune systems, are highly susceptible to bacterial infections. The same effect can be observed with elderly and otherwise immunocompromised patients.

Glycoconjugate vaccines would allow for the possibility of protecting at-risk populations against these dangerous pathogens. By combining the saccharide moiety with peptidic motifs in a single entity, the glycoconjugate could convert what was once a T-cell-independent immune response into a T-cell-dependent response. Much effort is currently being focused on the development of conjugate vaccines resulting from the chemical linking of immunologically dominant epitopes present in lipopolysaccharides to protein carriers. If successful, this approach may provide class switching, increased affinity of the specific antibodies, and B-cell memory. In certain populations, such as those with weak or immature immune systems, conjugate vaccines may be very desirable.

Conjugate vaccines against *Salmonella typhi* and *Streptococcus pneumoniae* are currently in clinical trials and others are being developed against *Escherichia coli*, *Klebsiella pneumoniae*, and *Pseudomonas aeruginosa*. In addition to conjugate vaccines comprised of polysaccharide carbohydrate domains, there is an increasing emergence of carbohydrate domains arrived at by total chemical synthesis. These are briefly illustrated in the following sections.

Haemophilus influenzae is a Gram-negative organism characterized by six different serotypes (a, b, c, d, e, f), classified according to their repeating acidic disaccharide structures, which are linked through glycosidic or phosphodiester bonds. Type b (Hib) is the only cause of common and serious diseases such as meningitis, septicemia, pneumonia, and infectious arthritis. A pure Hib polysaccharide vaccine is protective in children of 18 months of age and there are several Hib conjugate vaccines licensed in the United States for use in primary immunization series [63]. The success of these conjugate vaccines has resulted in the development of similar conjugate vaccines against other encapsulated pathogens [63, 64]. Although not a common occurrence, children occasionally do not show the desired anti-PRP (poly-3-β-D-ribose-(1,1)-D-ribitol-5-phosphate) antibody responses after two doses. In these instances the Hib vaccine is considered a failure. One group is investigating this problem and has seen positive results when synthetic PRP oligomers are chemically linked to peptides as shown in Fig. 14.13. Their main focus is on creating and evaluating synthetic PRP-peptide conjugates in which the peptidic moiety contains an autologous T-cell epitope to enhance cellular immune responses [65]. Preliminary studies in a rabbit model show a protective level of anti-PRP antibody response.

Neisseria meningitidis, a Gram-negative organism, is the leading cause of bacterial meningitis in persons of all ages. Without antibiotic treatment, the likelihood of survival is low, and severe neurological defects occur in those who survive. There are twelve known organism-associated serogroups, with five (A, B, C, Y and W-135) responsible for most invasive meningococcal disease. Currently there are two vaccines licensed in the United States (bivalent A/C and tetravalent A/C/Y/W-135), both of which are effective for persons of two years and older. Younger children, however, are usually unable to generate a sufficient antibody response even with a booster [64]. In order to improve immune responses in children, conjugate

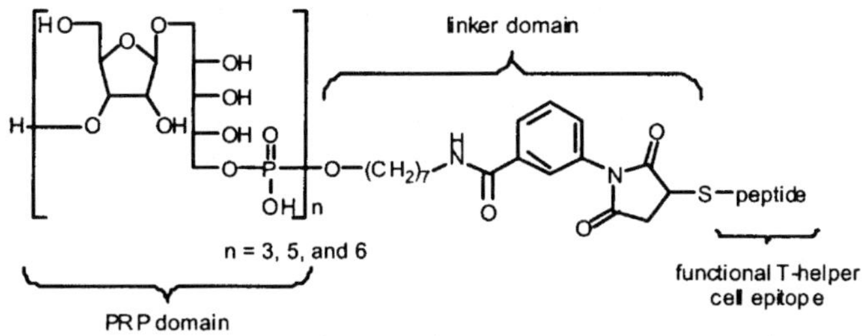

Fig. 14.13 A synthetic PRP glycopeptide conjugate vaccine targeted against *H. influenzae* type b.

vaccines synthesized by the successful Hib methodology have resulted in three different meningococcal serogroup C vaccines being licensed in the United Kingdom for use in children under two. Currently, bivalent meningococcal vaccines are also being developed [66, 67]. The development of a serotype B vaccine has been difficult because it is not very immunogenic in humans, quite possibly due to its similarity to certain human gangliosides [68]. An approach to improvement of immunogenicity currently under investigation has focused on vaccines that contain a chemically modified serogroup B antigen [69]. Jennings and co-workers first prepared a synthetic N-propionylated group B meningococcal polysaccharide (NPr-GBMP) in 1986 and have completed numerous experiments in attempts to overcome the antigen's inherent poor immunogenicity and to create a competent vaccine [70]. Their most recent analogues, depicted in Fig. 14.14, are the (NeuPr)$_4$-conjugates attached to TT or HSA at the reducing or non-reducing end, and these are being investigated for their ability to produce mAbs that recognize GBMP and are both bactericidal and protective [69].

A fully synthetic glycopeptidolipid construct as shown in Fig. 14.14 has also been synthesized for investigation as a vaccine against *N. meningitidis* [71]. The construct consists of a carbohydrate-containing B-epitope, a spacer, a T-cell epitope, and a lipopeptide. The carbohydrate moiety is an L-*glycero*-D-*manno*-heptose sugar, a saccharide representing a partial structure of the inner-core oligosaccharide of meningococcal lipopolysaccharides, shown to evoke an immune response when conjugated to tetanus toxoid [72]. The T-cell epitope is a peptide sequence identified as an MHC class II restricted site for human T-cells and derived from an outer-membrane protein of *N. meningitidis* [73, 74]. The lipopeptide, Pam$_3$Cys, in addition to its known B-cell and macrophage activating properties [51, 52], was chosen to serve as a membrane anchor in the lipid layer of cell membranes to ensure efficient and long-term delivery of the antigen.

Group B *Streptococcus*, a Gram-positive organism, is a leading cause of neonatal disease. There are five different serotypes, all of which contain a common trisaccharide core, with *Streptococcus* type III accounting for most of the infections. Babies can acquire passive immunity to these pathogens when the mother is vaccinated prior to pregnancy. A synthesis of the type III serotype carrying an artificial spacer, as shown in Fig. 14.15, has been accomplished, and this compound is cur-

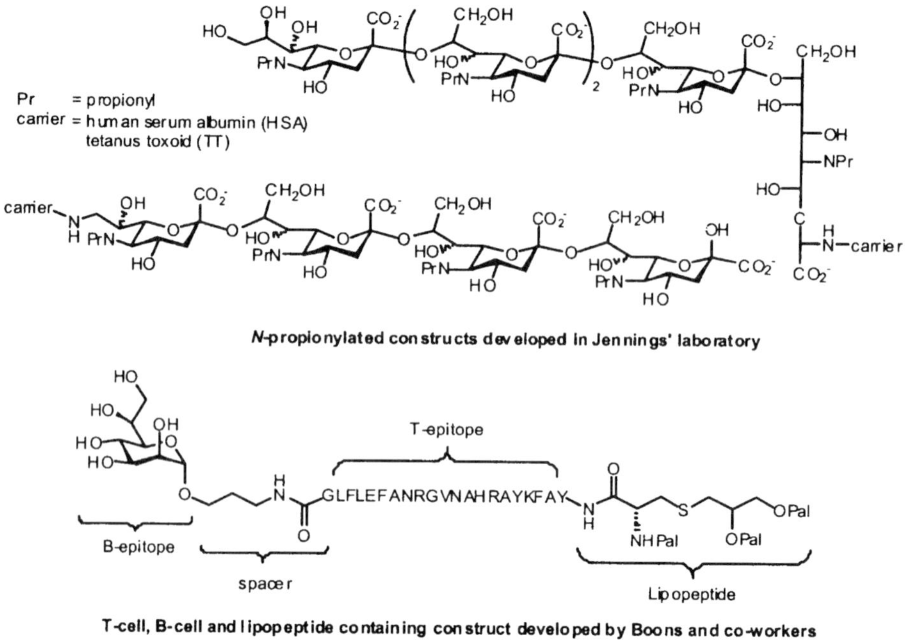

Fig. 14.14 Structures of synthetic vaccines against *N. meningitidis* currently being investigated.

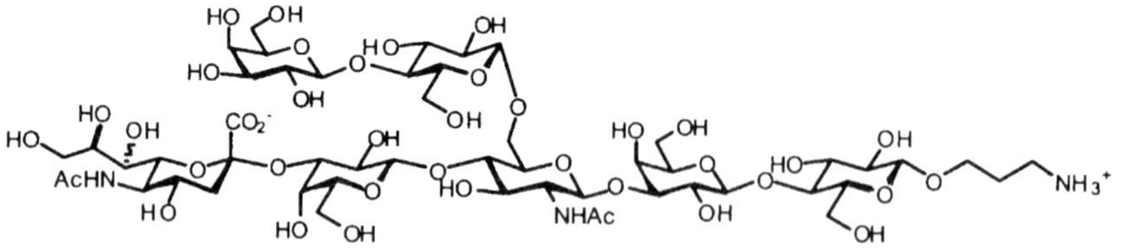

Fig. 14.15 Group B *Streptococcus* synthetic oligosaccharide created for diagnostic purposes.

rently being evaluated for use in an assay to detect antibodies against Group B type III Streptococcus in pregnant women [75]. It is likely that the synthetic heptasaccharide may find other uses in vaccine development.

Although *Shigella dysenteriae*, the cause of dysentery or shigellosis, has been known for more than a century, there is still no licensed vaccine for its prevention. A recent approach has involved the synthesis of the hexadecasaccharide fragment of the O-specific polysaccharide (O-SP) found on the surface of *S. dysenteriae* type 1 [76]. The saccharide moiety, bearing a flexible carbon linker, was conjugated to HSA by use of a heterobifunctional linker to form the vaccine construct shown in Fig. 14.16. Immunology studies in mice using saccharide-protein conjugates with varying levels of conjugation showed that the hexadecamer with a molar ratio of 9:1 (saccharides/protein) gave the highest IgG titers, while 19:1 and 4:1 gave significantly lower titers [77]. Higher levels of serum IgG lipopolysac-

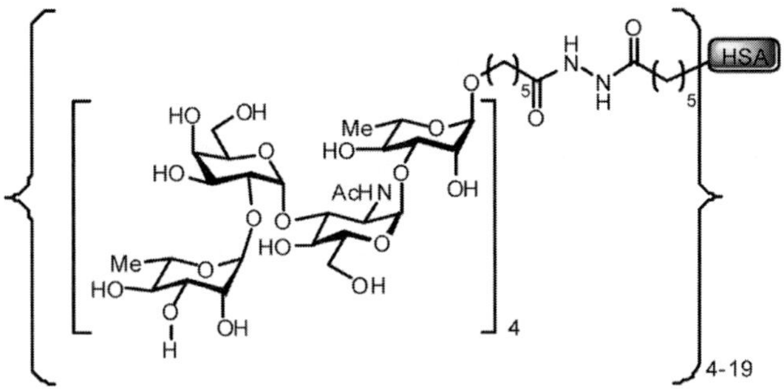

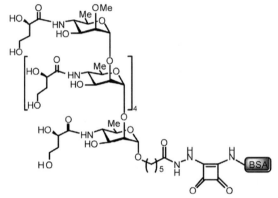

Fig. 14.16 Synthetic saccharide-protein conjugate for vaccine development against *Shigella dysenteriae*.

Fig. 14.17 Vaccine conjugate against *Vibrio cholerae* O1, serotype Ogawa.

charide (LPS) antibodies were generated in mice in response to the conjugate vaccine, relative to those induced by the O-SP-HSA or O-SP-tetanus toxoid conjugates which utilized O-SP isolated from LPS of the bacteria. Plans to evaluate the clinical efficacy of these vaccines with medically useful carriers are currently being investigated.

New progress in cholera vaccine development includes the synthesis of neoglycoconjugates from the hexasaccharide determinant of *Vibrio cholerae* O1, serotype Ogawa. The synthetic carbohydrate domain [78] was conjugated by attachment at a single-point or in a clustered motif to the carrier protein (CSA or BSA) by reductive amination or through a linker, by use of squaric acid diethyl ester conjugation [79]. Immunology studies indicated the BSA vaccine construct depicted in Fig. 14.17 to be immunogenic in mice, with higher and earlier antibody responses generated when given in conjunction with adjuvant [80].

14.4
Synthetic Parasitic Polysaccharide Conjugate Vaccine

The parasite has proved to be a very formidable adversary in the development of vaccines. It is clear that the complexity of the life cycle, the heterogeneity of the surface antigens expressed, and the obvious requirement for eliciting not only a protective antibody response, but a T-cell mediated one, is required to meet this challenge. It also reflects the still incomplete knowledge both of the host protective immune response and of the parasite's protective and evasive mechanisms.

The malaria parasite *Plasmodium falciparum*, infecting nearly 10% of the human population yearly, accounts for the deaths of more than two million people by inflammation initiated by the malarial toxin. In efforts to develop a vaccine against malaria, researchers have made a synthetic glycosylphosphatidylinositol (GPI), a domain known to function as a malarial toxin [81]. Conjugation of the synthetic hexasaccharide malarial toxin, depicted in Fig. 14.18, to KLH provided the vaccine construct, which was subsequently used for immunization of murine subjects [82]. In addition to the generation of antibodies against the GPI glycan, the mice were protected against many of the characteristics associated with the disease, including malarial acidosis, pulmonary edema, cerebral syndrome, and fatality.

Leishmaniasis, a tropical disease spread by the bite of infected sandflies, is quickly becoming a therapeutic target, due to the yearly increase in diagnoses. On its cell surface, the *Leishmania* parasite contains lipophosphoglycans (LPGs) composed of a GPI anchor, a repeating phosphorylated disaccharide, and a cap tetrasaccharide [83]. Seeberger and co-workers have developed a synthesis of the unique tetrasaccharide cap moiety, which contains an unusual galactose β-(1–4) mannosidic linkage, and conjugated it to Pam$_3$Cys and KLH to form two different potential vaccines as shown in Fig. 14.19 [84, 85]. Both constructs are currently undergoing immunological experiments in mice to evaluate their potential in the treatment of Leishmaniasis.

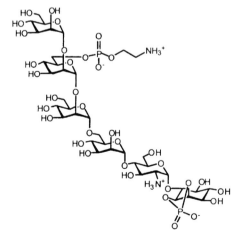

Fig. 14.18 Synthetic hexasaccharide malarial toxin used to develop a conjugate vaccine.

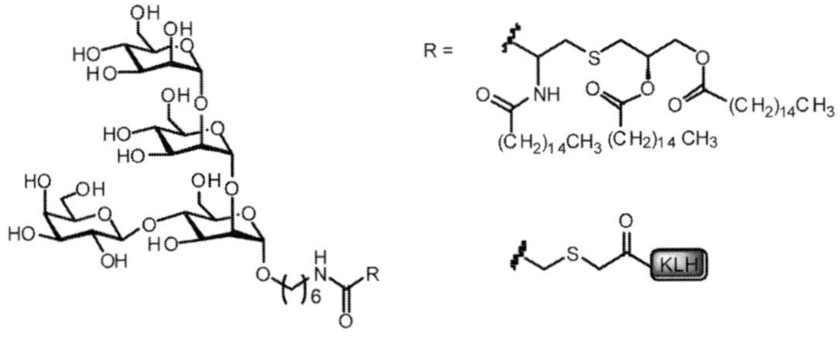

Fig. 14.19 Potential synthetic vaccines against the *Leishmania* parasite.

14.5
Conclusions

There are numerous vaccines being actively investigated for the treatment of cancer and bacterial and parasitic infections. Advances at the forefront of organic chemistry have allowed researchers to fashion completely synthetic carbohydrate-based antigens. The resulting antigens have been used extensively for investigating different strategies for competent vaccine construction. Although there have been many advancements in the field of immunotherapy, much remains to be explored.

14.6
References

1 SCHMIDT, R. R. in *Comprehensive organic synthesis: selectivity, strategy and efficiency in modern organic chemistry*, TROST, B. M., FLEMING, I., Eds.; Pergamon Press, Inc., **1991**, Vol. 6, pp. 33–64.
2 DANISHEFSKY, S. J.; BILODEAU, M. T. *Angew. Chem., Int. Ed. Engl.* **1996**, *35*, 1380.
3 SEEBERGER, P. H. *J. Carbohydr. Chem.* **2002**, *21*, 613.
4 HAKOMORI, S.-I. *Adv. Cancer Res.* **1989**, *52*, 257.
5 LIVINGSTON, P. O.; ZHANG, S. L.; LLOYD, K. O. *Cancer Immunol. Immunother.* **1997**, *45*, 1.
6 DANISHEFSKY, S. J.; ALLEN, J. R. *Angew. Chem., Int. Ed. Engl.* **2000**, *39*, 836.
7 MARKL, J.; LIEB, B.; GEBAUER, W.; ALTENHEIN, B.; MEISSNER, U.; HARRIS, J. R. *J. Cancer Res. Clin. Oncol.* **2001**, *127*, R3.
8 MUSSELLI, C.; LIVINGSTON, P. O.; RAGUPATHI, G. *J. Cancer Res. Clin. Oncol.* **2001**, *127*, R20.
9 RAGUPATHI, G. *Cancer Immunol. Immunother.* **1996**, *43*, 152.
10 LIVINGSTON, P. O.; RAGUPATHI, G. *Cancer Immunol. Immunother.* **1997**, *45*, 10.
11 ZHANG, S.; CORDON-CARDO, C.; ZHANG, H. S.; REUTER, V. E.; ADLURI, S.; HAMILTON, W. B.; LLOYD, K. O.; LIVINGSTON, P. O. *Int. J. Cancer* **1997**, *73*, 42.
12 KENSIL, C. R.; PATEL, U.; LENNICK, M.; MARCIANI, D. *J. Immunol.* **1991**, *146*, 431.
13 CASTRO-PALOMINO, J. C.; RITTER, G.; FORTUNATO, S. R.; REINHARDT, S.; OLD, L. J.; SCHMIDT, R. R. *Angew. Chem., Int. Ed. Engl.* **1997**, *36*, 1998.
14 DULLENKOPF, W.; RITTER, G.; FORTUNATO, S. R.; OLD, L. J.; SCHMIDT, R. R. *Chem. Eur. J.* **1999**, *5*, 2432.
15 SPRINGER, G. F. *Science* **1984**, *224*, 1198.
16 ORNTOFT, T. F.; HARVING, N.; LANGKILDE, N. C. *Int. J. Cancer* **1990**, *45*, 666.

17 ZHANG, S.; ZHANG, H. S.; CORDON-CARDO, C.; REUTER, V. E.; SINGHAL, A. K.; LLOYD, K. O.; LIVINGSTON, P. O. *Int. J. Cancer* **1997**, *73*, 50.

18 WERTHER, J. L.; RIVERA-MACMURRAY, S.; BRUCKNER, H.; TATEMATSU, M.; ITZKOWITZ, S. H. *Br. J. Cancer* **1994**, *69*, 613.

19 FUNG, P. Y. S.; MADEJ, M.; KOGANTY, R. R.; LONGENECKER, B. M. *Cancer Res.* **1990**, *50*, 4308.

20 HOLMBERG, L. A.; SANDMAIER, B. M. *Expert Opin. Biol. Th.* **2001**, *1*, 881.

21 ZHANG, S. L.; WALBERG, L. A.; OGATA, S.; ITZKOWITZ, S. H.; KOGANTY, R. R.; REDDISH, M.; GANDHI, S. S.; LONGENECKER, B. M.; LLOYD, K. O.; LIVINGSTON, P. O. *Cancer Res.* **1995**, *55*, 3364.

22 RAGUPATHI, G.; KOGANTY, R. R.; QIU, D. X.; LLOYD, K. O.; LIVINGSTON, P. O. *Glycoconjugate J.* **1998**, *15*, 217.

23 RAGUPATHI, G.; HOWARD, L.; CAPPELLO, S.; KOGANTY, R. R.; QIU, D. X.; LONGENECKER, B. M.; REDDISH, M. A.; LLOYD, K. O.; LIVINGSTON, P. O. *Cancer Immunol. Immunother.* **1999**, *48*, 1.

24 CHEN, X. T.; SAMES, D.; DANISHEFSKY, S. J. *J. Am. Chem. Soc.* **1998**, *120*, 7760.

25 KUDUK, S. D.; SCHWARZ, J. B.; CHEN, X. T.; GLUNZ, P. W.; SAMES, D.; RAGUPATHI, G.; LIVINGSTON, P. O.; DANISHEFSKY, S. J. *J. Am. Chem. Soc.* **1998**, *120*, 12474.

26 GLUNZ, P. W.; HINTERMANN, S.; SCHWARZ, J. B.; KUDUK, S. D.; CHEN, X. T.; WILLIAMS, L. J.; SAMES, D.; DANISHEFSKY, S. J.; KUDRYASHOV, V.; LLOYD, K. O. *J. Am. Chem. Soc.* **1999**, *121*, 10636.

27 SCHWARZ, J. B.; KUDUK, S. D.; CHEN, X. T.; SAMES, D.; GLUNZ, P. W.; DANISHEFSKY, S. J. *J. Am. Chem. Soc.* **1999**, *121*, 2662.

28 SLOVIN, S. F., personal communication.

29 SAMES, D.; CHEN, X. T.; DANISHEFSKY, S. J. *Nature* **1997**, *389*, 587.

30 KUDRYASHOV, V.; KIM, H. M.; RAGUPATHI, G.; DANISHEFSKY, S. J.; LIVINGSTON, P. O.; LLOYD, K. O. *Cancer Immunol. Immunother.* **1998**, *45*, 281.

31 SABBATINI, P.; KUDRYASHOV, V.; RAGUPATHI, G.; DANISHEFSKY, S.; LIVINGSTON, P.; BORNMANN, W.; SPASSOVA, M.; ZATORSKI, A.; SPRIGGS, D.; AGHAJANIAN, C.; SOLGNET, S.; PEYTON, M.; O'FLAHERTY, C.; CURTIN, J.; LLOYD, K. *Clin. Cancer Res.* **2000**, *6*, 4560.

32 KUDRYASHOV, V.; GLUNZ, P. W.; WILLIAMS, L. J.; HINTERMANN, S.; DANISHEFSKY, S. J.; LLOYD, K. O. *Proc. Natl. Acad. Sci. USA.* **2001**, *98*, 3264.

33 DESHPANDE, P. P.; DANISHEFSKY, S. J. *Nature* **1997**, *387*, 164.

34 DESHPANDE, P. P.; KIM, H. M.; ZATORSKI, A.; PARK, T. K.; RAGUPATHI, G.; LIVINGSTON, P. O.; LIVE, D.; DANISHEFSKY, S. J. *J. Am. Chem. Soc.* **1998**, *120*, 1600.

35 RAGUPATHI, G.; DESHPANDE, P. P.; COLTART, D. M.; KIM, H. M.; WILLIAMS, L. J.; DANISHEFSKY, S. J.; LIVINGSTON, P. O. *Int. J. Cancer* **2002**, *99*, 207.

36 PARK, T. K.; KIM, I. J.; HU, S. H.; BILODEAU, M. T.; RANDOLPH, J. T.; KWON, O.; DANISHEFSKY, S. J. *J. Am. Chem. Soc.* **1996**, *118*, 11488.

37 ALLEN, J. R.; ALLEN, J. G.; ZHANG, X. F.; WILLIAMS, L. J.; ZATORSKI, A.; RAGUPATHI, G.; LIVINGSTON, P. O.; DANISHEFSKY, S. J. *Chem. Eur. J.* **2000**, *6*, 1366.

38 RAGUPATHI, G.; PARK, T. K.; ZHANG, S. L.; KIM, I. J.; GRABER, L.; ADLURI, S.; LLOYD, K. O.; DANISHEFSKY, S. J.; LIVINGSTON, P. O. *Angew. Chem., Int. Ed. Engl.* **1997**, *36*, 125.

39 GILEWSKI, T.; RAGUPATHI, G.; BHUTA, S.; WILLIAMS, L. J.; MUSSELLI, C.; ZHANG, X. F.; BENCSATH, K. P.; PANAGEAS, K. S.; CHIN, J.; HUDIS, C. A.; NORTON, L.; HOUGHTON, A. N.; LIVINGSTON, P. O.; DANISHEFSKY, S. J. *Proc. Natl. Acad. Sci. USA* **2001**, *98*, 3270.

40 RAGUPATHI, G.; SLOVIN, S. F.; ADLURI, S.; SAMES, D.; KIM, I. J.; KIM, H. M.; SPASSOVA, M.; BORNMANN, W. G.; LLOYD, K. O.; SCHER, H. I.; LIVINGSTON, P. O.; DANISHEFSKY, S. J. *Angew. Chem., Int. Ed. Engl.* **1999**, *38*, 563.

41 SLOVIN, S. F.; RAGUPATHI, G.; ADLURI, S.; UNGERS, G.; TERRY, K.; KIM, S.; SPASSOVA, M.; BORNMANN, W. G.; FAZZARI, M.; DANTIS, L.; OLKIEWICZ, K.; LLOYD, K. O.; LIVINGSTON, P. O.; DANISHEFSKY, S. J.; SCHER, H. I. *Proc. Natl. Acad. Sci. USA* **1999**, *96*, 5710.

42 ALLEN, J. R.; DANISHEFSKY, S. J. *J. Am. Chem. Soc.* **1999**, *121*, 10875.

43 Cappello, S.; Liu, N.X.; Musselli, C.; Brezicka, F.T.; Livingston, P.O.; Ragupathi, G. *Cancer Immunol. Immunother.* **1999**, *48*, 483.
44 Dickler, M.N.; Ragupathi, G.; Liu, N.X.; Musselli, C.; Martino, D.J.; Miller, V.A.; Kris, M.G.; Brezicka, F.T.; Livingston, P.O.; Grant, S.C. *Clin. Cancer Res.* **1999**, *5*, 2773.
45 Ragupathi, G.; Cappello, S.; Yi, S.S.; Canter, D.; Spassova, M.; Bornmann, W.G.; Danishefsky, S.J.; Livingston, P.O. *Vaccine* **2002**, *20*, 1030.
46 Williams, L.J.; Harris, C.R.; Glunz, P.W.; Danishefsky, S.J. *Tetrahedron Lett.* **2000**, *41*, 9505.
47 Allen, J.R.; Harris, C.R.; Danishefsky, S.J. *J. Am. Chem. Soc.* **2001**, *123*, 1890.
48 Ragupathi, G.; Coltart, D.M.; Williams, L.J.; Koide, F.; Kagan, E.; Allen, J.; Harris, C.; Glunz, P.W.; Livingston, P.O.; Danishefsky, S.J. *Proc. Natl. Acad. Sci. USA* **2002**, *99*, 13699.
49 Marciani, D.J.; Press, J.B.; Reynolds, R.C.; Pathak, A.K.; Pathak, V.; Gundy, L.E.; Farmer, J.T.; Koratich, M.S.; May, R.D. *Vaccine* **2000**, *18*, 3141.
50 Toyokuni, T.; Hakomori, S.-I.; Singhal, A.K. *Bioorg. Med. Chem.* **1994**, *2*, 1119.
51 Bessler, W.G.; Cox, M.; Lex, A.; Suhr, B.; Wiesmuller, K.H.; Jung, G. *J. Immunol.* **1985**, *135*, 1900.
52 Hoffmann, P.; Wiesmuller, K.H.; Metzger, J.; Jung, G.; Bessler, W.G. *Biol. Chem.* **1989**, *370*, 575.
53 Toyokuni, T.; Dean, B.; Cai, S.P.; Boivin, D.; Hakomori, S.; Singhal, A.K. *J. Am. Chem. Soc.* **1994**, *116*, 395.
54 Glunz, P.W.; Hintermann, S.; Williams, L.J.; Schwarz, J.B.; Kuduk, S.D.; Kudryashov, V.; Lloyd, K.O.; Danishefsky, S.J. *J. Am. Chem. Soc.* **2000**, *122*, 7273.
55 Keil, S.; Claus, C.; Dippold, W.; Kunz, H. *Angew. Chem., Int. Ed. Engl.* **2001**, *40*, 366.
56 George, S.K.; Schwientek, T.; Holm, B.; Reis, C.A.; Clausen, H.; Kihlberg, J. *J. Am. Chem. Soc.* **2001**, *123*, 11117.
57 George, S.K.; Holm, B.; Reis, C.A.; Schwientek, T.; Clausen, H.; Kihlberg, J. *J. Chem. Soc., Perkin Trans. 1* **2001**, 880.
58 Hilaire, P.M.S.; Cipolla, L.; Franco, A.; Tedebark, U.; Tilly, D.A.; Meldal, M. *J. Chem. Soc., Perkin Trans. 1* **1999**, 3559.
59 Bay, S.; Lo-Man, R.; Osinaga, E.; Nakada, H.; Leclerc, C.; Cantacuzene, D. *J. Pept. Res.* **1997**, *49*, 620.
60 Vichier-Guerre, S.; Lo-Man, R.; Bay, S.; Deriaud, E.; Nakada, H.; Leclerc, C.; Cantacuzene, D. *J. Pept. Res.* **2000**, *55*, 173.
61 Lo-Man, R.; Bay, S.; Vichier-Guerre, S.; Deriaud, E.; Cantacuzene, D.; Leclerc, C. *Cancer Res.* **1999**, *59*, 1520.
62 Lo-Man, R.; Vichier-Guerre, S.; Bay, S.; Deriaud, E.; Cantacuzene, D.; Leclerc, C. *J. Immunol.* **2001**, *166*, 2849.
63 Lee, C.J.; Lee, L.H.; Koizumi, K. *Infect. Med.* **2002**, *19*, 179.
64 Lee, C.J.; Lee, L.H.; Koizumi, K. *Infect. Med.* **2002**, *19*, 127.
65 Kandil, A.A.; Chan, N.; Klein, M.; Chong, P. *Glycoconjugate J.* **1997**, *14*, 13.
66 Choo, S.; Zuckerman, J.; Goilav, C.; Hatzmann, E.; Everard, J.; Finn, A. *Vaccine* **2000**, *18*, 2686.
67 Lieberman, J.M.; Chiu, S.S.; Wong, V.K.; Partridge, S.; Chang, S.J.; Chiu, C.Y.; Gheesling, L.L.; Carlone, G.M.; Ward, J.I. *JAMA* **1996**, *275*, 1499.
68 Wyle, F.A.; Artenstein, M.S.; Brandt, B.L.; Tramont, E.C.; Kasper, D.L.; Altieri, P.L.; Berman, S.L.; Lowenthan, J.P. *J. Infect. Dis.* **1972**, *126*, 514.
69 Pon, R.A.; Lussier, M.; Yang, Q.L.; Jennings, H.J. *J. Exp. Med.* **1997**, *185*, 1929.
70 Jennings, H.J.; Roy, R.; Gamian, A. *J. Immunol.* **1986**, *137*, 1708.
71 Reichel, F.; Ashton, P.R.; Boons, G.J. *Chem. Comm.* **1997**, 2087.
72 Verheul, A.F.M.; Boons, G.J.P.H.; Vandermarel, G.A.; Vanboom, J.H.; Jennings, H.J.; Snippe, H.; Verhoef, J.; Hoogerhout, P.; Poolman, J.T. *Infect. Immun.* **1991**, *59*, 3566.
73 Wiertz, E.J.H.J.; Van Gaans-Van den Brink, J.A.M.; Schreuder, G.M.T.H.; Termijtelen, A.A.M.; Hoogerhout, P.; Poolman, J.T. *J. Immunol.* **1991**, *147*, 2012.
74 Wiertz, E.J.H.J.; Van Gaans-Van den Brink, J.A.M.; Gausepohl, H.; Proch-

nicka-Chalufour, A.; Hoogerhout, P.; Poolman, J.T. *J. Exp. Med.* **1992**, *176*, 79.
75 Demchenko, A.V.; Boons, G.J. *J. Org. Chem.* **2001**, *66*, 2547.
76 Pozsgay, V. *Angew. Chem., Int. Ed. Engl.* **1998**, *37*, 138.
77 Pozsgay, V.; Chu, C.Y.; Pannell, L.; Wolfe, J.; Robbins, J.B.; Schneerson, R. *Proc. Natl. Acad. Sci. USA* **1999**, *96*, 5194.
78 Zhang, J.A.; Kovac, P. *Carbohydr. Res.* **1999**, *321*, 157.
79 Zhang, J.A.; Kovac, P. *Bioorg. Med. Chem. Lett.* **1999**, *9*, 487.
80 Chernyak, A.; Kondo, S.; Wade, T.K.; Meeks, M.D.; Alzari, P.M.; Fournier, J.M.; Taylor, R.K.; Kovac, P.; Wade, W.F. *J. Infect. Dis.* **2002**, *185*, 950.
81 Hewitt, M.C.; Snyder, D.A.; Seeberger, P.H. *J. Am. Chem. Soc.* **2002**, *124*, 13434.
82 Schofield, L.; Hewitt, M.C.; Evans, K.; Siomos, M.A.; Seeberger, P.H. *Nature* **2002**, *418*, 785.
83 Turco, S.J.; Descoteaux, A. *Annual Review of Microbiology* **1992**, *46*, 65.
84 Hewitt, M.C.; Seeberger, P.H. *Org. Lett.* **2001**, *3*, 3699.
85 Hewitt, M.C.; Seeberger, P.H. *J. Org. Chem.* **2001**, *66*, 4233.

15
Chemistry, Biochemistry, and Pharmaceutical Potentials of Glycosaminoglycans and Related Saccharides

TASNEEM ISLAM and ROBERT J. LINHARDT

15.1
Introduction

The first description of glycosaminoglycans (GAGs) was by J. Müller in 1836, who isolated "chondrin", a sugar-related substance from cartilage that was later shown to contain a sulfo group by Mörner (1889) and renamed "chondroitsäure" [1]. It was not until 1935 that Karl Meyer discovered hyaluronic acid, initiating the exploration of GAG biochemistry and the identification of different types of GAGs. The past half-century has resulted in substantial progress in the elucidation of GAG fine structure, biosynthesis, and biological functions.

GAGs are unbranched, polydisperse, acidic polysaccharides, often covalently linked to a protein core to form proteoglycans (PGs). GAGs extend from a protein core in a brush-like structure. The core protein size ranges from 10 kDa to >500 kDa, and the number of attached GAG chains varies from 1 to >100 [2]. Except for hyaluronic acid, all GAGs are biosynthesized as PGs. The linkage region is the same in all PGs (except for keratan sulfate) and consists of the tetrasaccharide – glucuronic acid (GlcAp), galactose (Galp), Galp, and xylose (Xylp) – linked to the hydroxyl group of serine in the polypeptide core (Fig. 15.1) [3]. PGs occur in the membranes of all animal tissues, intracellularly in certain cells (usually in secretory granules) or extracellularly in the matrix, where they are exported to perform a variety of biological functions.

GAG biosynthesis is initiated with the synthesis of the core protein, rich in serine-glycine repeats [4], to which the linkage region and GAG are attached. GAGs are characterized (with the exception of keratan sulfate) by a repeating core disaccharide structure comprised of uronic acid and hexosamine residues. The amino group of the hexosamine residue is either N-acetylated or N-sulfonated, the uronic acid being either D-glucuronic acid or L-iduronic acid. Moreover, the repeating disaccharide units are O-sulfonated to varying degrees at the 3-, 4-, or 6-positions of the hexosamine residue and at the 2-position of the uronic acid residues. The most common GAGs are heparin, heparan sulfate (HS), hyaluronic acid (HA), chondroitin sulfate (CS), dermatan sulfate (DS), and keratan sulfate (KS) (Tab. 15.1).

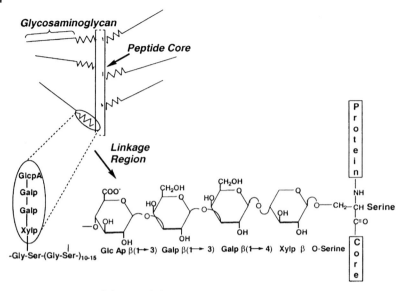

Fig. 15.1 Structure of the GAG linkage to protein in PGs.

Tab. 15.1 The structures of the sulfated GAGs.

GAG	Disaccharide Unit	Modifications
Heparan sulfate (HS)/ Heparin	[GlcAp/IdoApβ/α1-4GlcNpAcα1–4]	N-deacetylation, N-sulfation C5-epimerization of GlcAp C2-sulfonation on GlcAp/IdoAp C3,C6-sulfonation on GlcNpAc
Chondroitin sulfate (CS)/ Dermatan sulfate (DS)	[GlcAp/IdoApβ/α1–3GalNpAcα1–4]	C5-epimerization of GlcAp C2-sulfonation on GlcAp/IdoAp C4,C6-sulfonation on GlcNpAc
Keratan sulfate (KS)	[Galpβ1-4GlcNpAcβ1–3]	C6-sulfonation on Galp/GlcNpAc

15.1.1
Biological Activities

The PG family consists of 30 members, which display a wide variety of biological functions [5]. They play important roles in the extracellular matrix organization, influence cell growth and tissue maturation, and participate in the regulation of matrix turnover by binding, inactivating protease inhibitors.

The biological interactions and events mediated by PGs are believed to be due primarily to the presence of GAG chains. Since the PGs are often localized on cell surfaces and in the extracellular matrix, they function in the important role of cell-cell interaction, binding a variety of proteins and localizing these at the cell surface [6, 7]. The heparin/HS GAGs are known to bind over 100 different proteins, including enzymes, protease inhibitors, lipoproteins, growth factors, chemokines, selectins, extracellular matrix proteins, receptor proteins, viral coat proteins, and nuclear proteins [8]. Some of the biological roles of GAGs have been exploited for the design and preparation of therapeutic drugs.

15.1.2
Heparin and Heparan Sulfate

15.1.2.1 Structure and Properties

Heparin, an anticoagulant isolated from animal tissue, is an important and chemically unique polysaccharide of considerable biological significance. It was discovered in 1916 by Jay McLean, working under the directions of William Howell at John Hopkins University [9]. To ascertain the origin of a substance causing blood coagulation, McLean isolated fractions from mammalian tissues. These, however, instead of clotting blood, prevented its coagulation [10]. Howell recognized the importance of his student's discovery, suggesting heparin's therapeutic use to treat coagulation disorders.

Heparin is a polydisperse, highly sulfated, linear polysaccharide made up of repeating 1→4 linked uronic acid and glucosamine residues (Fig. 15.2) [11, 12]. Although heparin has been used clinically as an anticoagulant for the past 70 years, its precise structure remains unknown. The failure to understand heparin's structure completely is not the result of a lack of effort, but rather is due to its extremely complex nature. The structural complexity of heparin can be considered

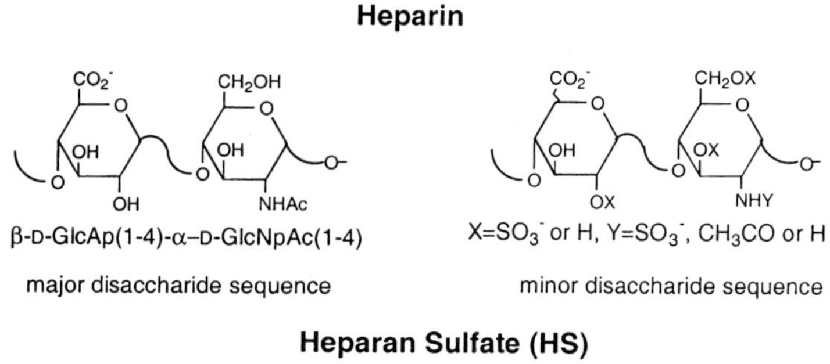

Heparin

α-L-IdoAp2S(1-4)-α-D-GlcNpS6S(1-4)
major disaccharide sequence

$X=SO_3^-$ or H, $Y=SO_3^-$, CH_3CO or H
minor disaccharide sequence

Heparan Sulfate (HS)

β-D-GlcAp(1-4)-α-D-GlcNpAc(1-4)
major disaccharide sequence

$X=SO_3^-$ or H, $Y=SO_3^-$, CH_3CO or H
minor disaccharide sequence

Fig. 15.2 Structures of heparin and heparan sulfate glycosaminoglycans.

at several levels. At the PG level, different numbers of polysaccharide chains (possibly with different saccharide sequences) can be attached to the various serine residues present in the protein core. Once freed from the protein core, through the action of tissue proteases, peptidoglycan heparin (a small peptide to which a single long polysaccharide chain (M_r 100,000) is attached) is formed [13]. This peptidoglycan is short-lived, as it is immediately processed by a β-endoglucuronidase to form a number of smaller polysaccharide chains (only one of which, corresponding to the original site of attachment to the core protein, should contain peptide) called GAG heparin. At the GAG level, some of heparin's structural complexity results from its polydispersity. GAG heparin has a molecular weight (MW) ranging from 5–40 kDa (degree of polymerization (dp) 10–80) with an MW (average) of 13 kDa. Even the heparin chain corresponding to the most prevalent dp represents a mere 5 mol% of a typical GAG heparin preparation [14]. GAG heparin has a second level of structural complexity associated with its primary structure or sequence [15]. The heparin/HS structure has been partially characterized by study both of its biosynthesis [16, 17] and of its chemical structure, by chemical, enzymatic, and spectroscopic techniques [10].

While GAG HS is structurally similar to GAG heparin (Fig. 15.2) [12], the core proteins of PG HS and heparin are different. HS is primarily found in the extracellular matrix and in cell membranes [18, 19], while heparin is only intracellular. Although structurally similar, HS and heparin have different ratios of N-acetyl to O-sulfo groups and can be often distinguished by differences in their sensitivity to heparin lyases [20].

15.1.2.2 Biosynthesis and Biological Functions

Heparin is synthesized in connective tissue-type mast cells, as part of the serglycin PG [2]. HS is produced by most animal cells and is bound to a variety of core proteins, corresponding to syndecan, glypican, perlecan, and agrin PGs [21]. After translation in the rough endoplasmic reticulum, the core proteins are transported to the Golgi apparatus, where the enzymes responsible for HS/heparin biosynthesis are located. Selected serine units are O-substituted with the GlcAp-Galp-Galp-Xylp-"linkage region" that connects the GAG chain to the core protein [22]. Next, the stepwise transfer of monosaccharide units from the appropriate UDP-sugars to the nonreducing termini of the nascent chains generates a precursor polysaccharide, [β1,4-GlcAp-α1,4GlcNpAc-]$_n$. The linear polysaccharide chain is extended by approximately 300 sugars before its synthesis is terminated [17]. Conversion of the (GlcAp-GlcNpAc)$_n$ precursor structure into the products recognized as heparin/HS occurs through a series of polymer-modification reactions initiated while the chain is still under elongation [23]. The first modification step is the N-deacetylation and N-sulfonation of GlcNAc residues. The resultant N-sulfo groups are prerequisite to all subsequent modifications, which include C-5 epimerization of GlcAp to L-iduronic acid (IdoAp) units, 2-O-sulfonation of GlcAp and IdoAp units, and 6-O- and 3-O-sulfonation of GlcNp residues. The process occurs in a stepwise fashion, with the products of a given reaction providing the substrate for subsequent reactions [21]. However, most of these reactions do not go to completion, and a fraction of the potential substrate residues in each step escapes modification. While it is unclear what control (if any) is exerted on the extent of modification, such partial polymer modification is a fundamental feature of the biosynthetic process, and results in the diversified domain structure of HS.

PG heparin is primarily found in the granules of mast cells. When mast cells degranulate, heparin is released as GAG heparin, the result of processing by proteases and endo-β-glucuronidases [13, 24]. Although the GAG heparin released on mast cell degranulation demonstrates anticoagulant activity, the role of this activity is unclear. There is no evidence that endogenous mast cell heparin plays a role in maintaining blood flow through the vasculature, even though this is the primary application for exogenously administered GAG heparin [25, 26]. The true biological function of heparin still remains contested [8].

The biological activities of HS PGs result primarily from the specific interaction of proteins with their GAG chains. Some of the functions of syndecan, a HS PG, are: (1) organization of extracellular matrix, through binding to collagens, fibronectin, thrombospondin, tenascin, etc., (2) organization of the epithelia, (3) affording non-thrombogenic vascular endothelial surfaces, through binding to antithrombin III (ATIII), protein C, and protease nexins, (4) as a co-receptor for fibroblast growth factor (FGF), and (5) regulation of development in early embryogenesis and in cancer [8]. Syndecans are expressed in tissue-specific patterns; their expression is highly regulated and this regulation may be affected by their cell surface residence time. Syndecans are shed from the cell surface as the result of their cleavage at a protease-susceptible site near the plasma membrane. The extracellular matrix at the basal cell surface might slow syndecan release.

15.1.2.3 Applications of Heparin and Heparan Sulfate

Although the biological roles of endogenous PG heparin and PG HS are not completely understood, this has not precluded the use of GAGs derived from these natural products – as well as GAG fractions, oligosaccharides, and synthetic analogues – for a variety of medical applications. Heparin is the most commonly used clinical anticoagulant. Over 33 metric tons of heparin, representing over 500 million doses, are manufactured worldwide each year [10]. Since orally administered heparin is inactive and heparin has a low bioavailability when administered *subcutaneously* [27, 28], it is usually injected *intravenously*. The success of low molecular weight (LMW) heparins is primarily a result of their high *subcutaneous* bioavailability [29]. LMW heparins are prepared by the controlled chemical or enzymatic depolymerization of heparin. The clinical use of LMW heparin has recently surpassed the use of heparin in the US. In addition to heparin's anticoagulant activity, it has a wide variety of other activities (Tab. 15.2).

Heparin and its Antithrombotic Activity The antithrombotic action of heparin is due mainly to its ATIII-mediated anticoagulant activity. ATIII, a serine proteinase

Tab. 15.2 Potential therapeutic applications of heparin and heparin analogues.

Application	Status	Reference
Anticoagulant/antithrombotic	Currently in use	30
Antiatherosclerotics	Clinical trials	31
Complement inhibitors	Clinical trials	32
Anti-inflammatory	Animal studies	33
Antiangiogenic agents	Animal studies	34
Anticancer agents	Animal studies	35
Antiviral agents	Animal studies	36
Anti-Alzheimer agents	Animal studies	37

inhibitor (SERPIN) [38], is an anionic [39] glycoprotein of molecular weight 58,000 [40]. This SERPIN forms tight, irreversible, equimolar complexes with its target enzymes (thrombin, factor Xa, etc.) through the formation of an ester between an arginine residue of its active site and the serine residue of the active site of the enzyme. This slow, time-dependent inhibition process is accelerated 2000-fold by heparin [39]. The reversible interaction of heparin with ATIII induces a conformational change in ATIII, which enhances its anticoagulant activity [39].

Apart from its ATIII-mediated activity, heparin (and dermatan sulfate) stimulates the inactivation of thrombin by heparin cofactor II (HCII). HCII has a molecular weight and pI similar to those of ATIII [41, 42]. The physiological role of HCII might be as a reserve of thrombin inhibitor when the plasma concentration of ATIII becomes abnormally low [43]. Unlike ATIII, HCII specifically inhibits thrombin and no other coagulation proteases [44, 45].

The antithrombotic activity of heparin is not limited to its anticoagulant effect; heparin also influences other factors significant in thrombogenesis. Heparin increases the electronegative potential and consequently the antithrombogenic character of the vessel wall [46]. It may increase the production and release of anticoagulant-active HS from the endothelium [47]. Heparin affects platelet function and the release of substances important to homeostasis from the platelets [48]. For instance, heparin releases the tissue factor pathway inhibitor (TFPI), which enhances antithrombogenesis [49].

Antiatherosclerotic Activity Atherosclerosis develops from a disturbed homeostasis between the blood and the vessel wall. Blood-borne constituents cause repeated injuries to the vessel wall in various target organs, provoking a chronic, inflammatory fibroproliferative response [50], which ultimately results in arterial obstruction and insufficient blood supply to the organs. A chronically elevated plasma low-density lipoprotein (LDL) cholesterol level plays a key role in damaging the arterial wall [51]. Inflammatory heparin-binding constituents, such as complement, C-reactive protein, fibrinogen, tumor necrosis factor, viruses, and lipoprotein a, are implicated, together with homocysteine and mechanical shear forces, in promoting the progression of atherogenesis [52, 53]. Repeated injuries to the endothelium weaken its resistance, resulting in transmigration of blood-borne constituents, such as LDLs, into the artery walls [52, 53].

When heparin is administered *intravenously*, lipoprotein lipase (LPL) is mobilized from the vascular endothelial surface into the blood. This may result in increased triglyceride lipolysis in the bloodstream, lowering the concentration of cholesterol-rich remnant particles in contact with the arterial wall [54–57]. The effect of heparin on the release and activation of LPL has been well studied [54, 55]. Heparin's application as an antiatherosclerotic agent is limited by its primary activity as an anticoagulant and its lack of oral bioavailability [58]. Heparin analogues might be useful in circumventing these problems.

The proliferation of smooth muscle cells (SMCs) after damage to endothelium is an important part of atherogenesis, resulting in further occlusion of vessels [59–61]. Both anticoagulant and non-anticoagulant heparin have demonstrated the ability to inhibit the proliferation of SMCs [62]. This activity results from heparin's interaction with growth factors including fibroblast growth factor (FGF) and endothelial cell growth factor (ECGF) [61].

Ability to Inhibit Complement Activation The complement system, important in the immune and inflammatory responses, involves proteins present in the bloodstream in an inactive form [32]. Activation of these complement proteins results in increased vascular permeability, activation of neutrophils, and alterations in cell membranes, ultimately causing cell lysis and death [63]. Activation of the complement system involves the sequential interaction of the serum complement proteins. Since complement activation can potentially produce profound effects, the system has inhibitors. C1 esterase inhibitor is a serum protein that inhibits the activation of the first component of complement. Heparin greatly potentiates this inhibitory activity [64]. Heparin is believed to bind simultaneously to C1 and to C1 esterase inhibitor, bringing the molecules into close proximity, kinetically favoring and stabilizing their interaction [65]. The C1 macromolecule may also be directly inactivated by heparin [66]. Purified C1q has two high-affinity binding sites for heparin, and heparin inhibits interaction of C1q with other C1 components to form hemolytically active C1 [67]. Complement activation ultimately involves amplification of the third component of complement (C3) through the formation of an amplification convertase. Heparin inhibits the generation of this amplification convertase [68] and this is independent of its antithrombin-binding activity [69].

Anti-inflammatory Activity Heparin may act as an anti-inflammatory agent through its interaction with selectins and chemokines [70]. Selectins are a family of transmembrane glycoproteins found on endothelium, platelets, and leukocytes [71]. When cells are activated, L-selectin on the cell surface increases, allowing their migration to the lymph nodes. In acute inflammation, neutrophils move through the vascular wall [8]. Initially, neutrophils anchor to the endothelium through a carbohydrate-protein interaction [71]. Although the putative ligand on the endothelium responsible for leukocyte interaction with selectins is sialyl Lewis X, HS has also been shown to play a role in this interaction [72].

Chemokines are soluble GAG-binding proteins involved in the recruitment and activation of leukocytes [73]. There are at least 15 known chemokines and in vitro they exhibit overlapping activities. Some selectivity is observed in vivo, and may correlate to differential localization of chemokines in tissues. The GAG chains of cell surface PGs have been postulated to play a role in this differential localization. The heparin/HS family of GAGs are most responsible for chemokine binding.

Role of Heparin/HS in Angiogenesis Angiogenesis (neovascularization) involves the growth of new capillary blood vessels. It plays an important role in normal development and in the physiology of reproduction [74]. Angiogenesis is also essential in wound repair, peptic ulcers, and myocardial infarction [75, 76]. In both physiologic and repair conditions, angiogenesis is regulated to switch on and off at predictable times. In a variety of disease processes, such as tumor growth and metastasis, however, angiogenesis is unregulated. After a tumor takes hold it grows slowly under oxygen limitations and remains quite small. When new capillaries come sufficiently close for oxygen to diffuse into the tumor, though, the vascularized tumor cells multiply rapidly. The process of angiogenesis requires the induction of proteases, degradation of the basement membrane, migration of the endothelial cells into the interstitial space, endothelial cell proliferation, lumen formation, generation of new basement membrane with the recruitment of pericytes, fusion of the newly formed vessels, and initiation of blood flow.

Heparin may play a variety of roles in angiogenesis. Immediately before capillary ingrowth, mast cells containing heparin accumulate at the site of the tumor [77]. The heparin from these mast cells can stimulate endothelial cell migration. Protamine and the chemokine platelet factor 4, both of which bind and inactivate heparin, can inhibit angiogenesis [78]. Heparin can localize, activate, stabilize, and stimulate angiogenic growth factors such as FGF and ECGF [79–81].

The FGFs are a multi-ligand, multi-receptor family in which one receptor can bind several ligands with high affinity. The assembly of 2FGF-2FGFR-2HS chains comprises a signal transduction complex that results in cell replication [82]. Interactions are coordinated through the HS PG, which can either promote or restrict growth factor binding to a particular receptor [83]. Heparin and heparin oligosaccharides (in the presence of angiostatic steroids) [84, 85] and even heparinase [86] can inhibit angiogenesis. In the presence of other factors such as steroids, the HS PG lining the endothelium restrains capillary growth [79]. This quiescent microvasculature can rapidly respond to heparin-modulated growth factors produced during ovulation, by wounds [87], or in inflammation (as occurs in stroke, where the damaged blood brain barrier requires repair [88]), and to the release of endogenous angiogenesis inhibitors (such as angiostatin and endostatin) [89].

Heparin and Cancer Heparin affects the progression of cancer in many ways. Because of its anticoagulant function, heparin can inhibit thrombin and fibrin formation induced by cancer cells. Heparins may therefore potentially inhibit intravascular arrest of cancer cells and thus promote metastasis. In addition to their anticoagulant function, heparins bind to growth factors and extracellular matrix proteins and consequently can affect proliferation and migration of cancer and angiogenesis in tumors [90]. Heparin has been found to inhibit expression of oncogenes and to affect the immune system [91]. Heparin has both stimulatory and inhibitory effects on proteolytic enzymes essential for invasion of cancer cells through the extracellular matrix [92]. Heparin also reduces tumor cell-platelet adhesion, reducing metastasis [93]. This wide variety of activities makes the ultimate effect of heparin on cancer still uncertain [34].

Antiinfective Activity   HS, ubiquitously found on the surfaces of animal cells, represents an ideal means for pathogens to localize on the membrane of the target cells that they infect [94, 95]. Moreover, heparin in mast cells [77] also binds pathogens and one of its roles might be to adhere to pathogens and to target them to dendritic or phagocytic cells.

Viruses gain entry into cells by using the HS PGs, which line the surface of most mammalian tissues, as receptors. Herpes simplex virus (HSV) anchors onto HS and localizes on the cell surface prior to its entry [96, 97]. Human immunodeficiency virus (HIV) binds HS PG on T-lymphocytes through an extended loop in its GP120 coat protein [96], and heparin can block this interaction [99]. Heparin has been clinically tested for the treatment of AIDS and is a potent and selective inhibitor of HIV-1 replication in cell culture [100]. Heparin, a highly sulfated heparin decasaccharide, and suramin, a small synthetic heparin analogue, can block dengue virus infection [36] by binding dengue envelope protein [101].

Malaria parasite circumsporozoites infect human liver by binding to the highly sulfated human liver HS PG [102]. Liver HSPG also appears to act as the receptor for the apolipoprotein E (apoE) [103].

The bacterial protein BGP from *Borrelia burgdorferi*, which causes Lyme disease, is a heparin-binding protein [104], suggesting that infection takes place through bacterial interaction with a HS acceptor.

Alzheimer's Disease   Alzheimer's disease (AD) is characterized by the deposition of amyloid plaque and neurofibrillary tangles in the brain [105]. The amyloid peptide contains a specific sequence capable of binding heparin at the low pH values present within these plaques [106]. Heparin also inhibits a protease nexin found predominantly in the brain and possessing an activity associated with AD [107]. Basic FGF is found bound to HS PGs in plaques as well as in inclusions in Parkinson disease, suggesting a possible role of HS PG in the formation of intraneuronal inclusions [108]. ApoE4, encoded by a genetic marker closely associated with late onset AD, binds tightly to heparin [103], and this interaction is associated with the neurotoxicity of apoE4 [109]

Interactions of Heparin with Proteins   The biological activity of heparin is usually attributed to its interaction with heparin-binding proteins [94, 110]. Heparin has been found to bind to a large number of proteins (Tab. 15.3). These proteins can be classified as: (1) enzymes, (2) protease inhibitors, (3) lipoproteins, (4) growth

Tab. 15.3 Selected heparin/HS binding enzymes and proteins.

1. Enzymes Lipolytic enzymes Kinase Phosphatases Enzymes acting on carbohydrates Proteases/esterases Nucleases, polymerases, and topoisomerases Other enzymes, oxidases, synthases	**6. Selectins** L-selectin P-selectin **7. Extracellular matrix proteins** Collagens I–VI Fibronectin Laminin Tenascin Vitronectin (S-protein)
2. Protease inhibitors (serpins) Antithrombin III (AT III) C1 Inhibitor proteins Heparin cofactor II (HCII) Protease nexin Thrombomodulin	**8. Receptor proteins** CD4 receptor FGF receptor (FGF1–4) Glycoprotein 330 (LDL receptor) **9. Viral coat protein** gp120 of HIV-1 gp140 and gp160 of HIV-2 Herpes simplex virus-1 (HSV-1) Dengue envelope protein
3. Lipoproteins Low and very low density lipoproteins Apolipoprotein B-100 Apolipoprotein E	
4. Growth factors Fibroblast growth factors Epidermal growth factors Hepatocyte growth factor Platelet-derived growth factor Smooth muscle cell growth factor Transforming growth factor Vascular endothelial growth factor	**10. Nuclear proteins** Histones Transcription factors **11. Other proteins** Fibrin Immunoglobulin G Protein C inhibitor Alzheimer β-amyloid precursor protein (APP) Platelet/endothelial cell adhesion molecule-1 (GMP-140)
5. Chemokines Interleukin 8 (IL-8) Neutrophil activating peptide 2 Platelet factor IV	

factors, (5) chemokines, (6) selectins, (7) extracellular matrix proteins, (8) receptor proteins, (9) viral coat proteins, (10) nuclear proteins, and (11) others [111]. The specificities and strengths of these interactions are probably the result of ionic binding between the sulfo groups of heparin and basic amino acid residues of the interacting protein [112]. In some cases binding can occur through nonionic interactions such as hydrogen bonding [94].

15.2
Dermatan and Chondroitin Sulfates

15.2.1
Structure and Biological Role

Dermatan sulfate (DS) and chondroitin sulfate (CS) make up a second GAG family, called galactosaminoglycans. CS is the most abundant GAG in the body, and occurs in both skeletal and soft tissue. Much of our knowledge of GAGs is derived from studies of chondroitin 4-sulfate (CS-A), the first member of this class to be isolated in a pure state, from cartilage. CS consists of repeating units of GlcAp and GalNpAc. The two most common isomers contain O-sulfo groups at positions 4 (CS-A) or 6 (CS-C) of the galactosyl residue (Fig. 15.3, structure I). The size of the CS chain varies greatly, with an average of about 40 repeating disaccharide units for the cartilage proteoglycan, corresponding to a molecular weight of about 20,000 [113]. The number of sulfo groups also varies, with some galactosyl residues containing none, and some containing multiple sulfo groups at positions 2 and 4 of GalNpAc and at position 2 of the uronic acid residue. The average number of O-sulfo groups in CS is ~ 0.8 per disaccharide [113].

DS (chondroitin sulfate B), found mainly in mucosa and skin, is a polydisperse, microheterogeneous sulfated copolymer of D-GalpNAc and primarily L-IdoAp acid, with O-sulfo groups most commonly found on the 4-position of D-GalpNAc residues and occasionally on the 6-position of D-GalpNAc and the 2-position of L-IdoAp residues [114] (Fig. 15.3, structure II). Several DS core proteins have been identified [115, 116]. These include large PGs with up to 25–30 DS polysaccharide chains and small PGs such as decorin and biglycan, with one and two DS polysaccharide chains, respectively.

Decorin DS PG binds collagen types I and II and plays an important role in the organization of collagen fibrils. The specific blend of PG and collagen determines the elasticity and transparency properties of the tissue [115]. Biglycan DS PG is found in cell surface and pericellular environments and its function is not well understood. DS PGs, biglycan, and decorin may play roles in regulating the extravascular activities of thrombin. In solution, biglycan, decorin, and their GAG chains accelerate HCII inhibition of thrombin. Both biglycan and decorin exhibit the same activity when bound to type V collagen. This observation suggests that

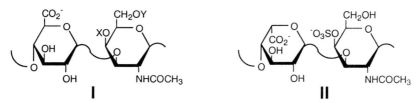

Fig. 15.3 The major disaccharide repeating units of chondroitin sulfates (structure I) and dermatan sulfate (structure II). In CS-A, $X=SO_3^-$ and $Y=H$, and in CS-C $X=H$ and $Y=SO_3^-$.

one function of these DS PGs is to provide a thromboresistant extravascular surface [117]. Thrombomodulin is a PG found on the luminal surface of the vascular endothelium and on underlying SMCs. It contains a single CS/DS GAG chain. Thrombomodulin binds thrombin, presumably through interaction both with its protein core and with its GAG chain [118]. Once bound, thrombin can act on protein C to form protein Ca, an activated serine protease, which inactivates factors Va and VIIIa, preventing the generation of factor Xa and thrombin, and inhibiting coagulation. Thrombin bound to thrombomodulin is also sensitive to inhibition by antithrombin III [114].

The GAG chains of decorin, biglycan, and thrombomodulin can act as anticoagulants by inhibiting thrombin, either directly through HCII or ATIII, or indirectly through protein C activation [117]. Since these PGs are found both on the luminal surface and on the subluminal surface, they provide a localized anticoagulant affect, affording thromboresistant surfaces at site in both intact and damaged vessels.

DS also acts as an antithrombotic agent by inhibiting the thrombin-induced aggregation of platelets and may activate the fibrinolytic pathway by causing the release of tissue plasminogen activator (tPA) [119]. The most thoroughly studied activity associated with DS is its acceleration of HC II-mediated inhibition of thrombin [120]. DS may also play a role in lipid metabolism by binding and releasing endothelial lipoprotein lipase into the circulation [119]. The compositions of the endothelial cell surface HS and DS PGs change during atherogenesis [121].

15.2.2
Therapeutic Applications

15.2.2.1 Dermatan Sulfate
Like HS, DS is a relatively weak anticoagulant in vitro (70 times less potent than heparin). Apart from its inhibition of thrombin-induced platelet aggregation [122], DS does not interact with platelets [48, 123]. Its anticoagulant effect is mainly based on enhancement of HCII activity [124]. To enhance HCII-mediated thrombin inhibition, DS requires at least seven to eight disaccharide units in addition to an IdoAp-GalNpAc4S disaccharide unit and IdoAp2S residues [125].

Recently, several DS preparations have been developed for prophylaxis of venous thromboembolism. In comparison with heparin, DS is a less active, but safer antithrombotic drug for *intravenous* administration, due to reduced hemorrhagic complications [126]. The high molecular weight of DS inhibits its absorption when administered *subcutaneously*. Low molecular weight DSs, such as Desmin 370 (OP370) [124], are less potent in vitro, but show improved pharmacokinetic properties, including increased bioavailability and duration of action [127]. Oversulfated DS derivatives, with two to three sulfo groups per disaccharide unit, have also been examined [128, 129]. Antithrombotic activity increases with increasing O-sulfo group content [128], but these derivatives show a concomitant increase in hemorrhagic complications [130].

Other applications for DS include the preparation of medical devices and artificial tissues. Stone [131] has patented a prosthetic meniscus for use as a knee im-plant, which acts as a scaffold for the regrowth of native meniscal tissue. The material is composed of collagen fibrils interspersed with DS. DS has also been useful in the development of artificial tissues [132]. Small arterial prostheses composed of a microporous polyurethane tube coated with a gel containing a mixture of type I collagen and DS have been designed. This gel promoted the adhesion and growth of endothelial cells, and reduced platelet adhesion in vitro. Grafts seeded with endothelial cells were highly antithrombotic when implanted into the carotid arteries of dogs.

15.2.2.2 Chondroitin Sulfates

Human plasma contains free CS at a concentration of ~ 0.1 mg/100 ml [133]. As a component of thrombomodulin, CS is essentially involved in the inhibition of thrombin clotting activity [134]. The moderate antithrombotic action of chondroitin 4-sulfate in vivo is partly due to its anticoagulant effect, which is only partially mediated by ATIII [135]. To improve its antithrombotic activity, CS was chemically sulfonated, resulting in a so-called semisynthetic analogue (SSHA) with a mean MW of 7 kDa [136]. In several clinical studies comparing the prophylactic effect of SSHA with that of standard low dose heparin, SSHA was shown to be as effective as heparin, but without heparin's increased bleeding risk [136]. Thus, despite its weak anticoagulant activity, SSHA has significant antithrombotic potential.

CS has been widely used as a neutriceutical for the treatment of osteoarthritis [137]. Clinical trials have suggested that CS may have some efficacy in treating osteoarthritis symptoms [138]. While it is unclear how CS works, particularly with regard to its low oral bioavailability, it may act as a weak anti-inflammatory [139]. CS from human milk has also been found to inhibit HIV glycoprotein gp120 binding to its host cell CD4 receptor in vitro [140]. CS-C has been used as a component of artificial skin [141].

15.3 Hyaluronan

15.3.1 Structure and Properties

Hyaluronan (HA), first isolated from the vitreous body of the eye by Meyer and Palmer in 1934 [142], consists of repeating disaccharide units of $[\rightarrow 4)$-β-D-GlcAp$(1 \rightarrow 3)$-β-D-GlcNpAc$(1 \rightarrow]_n$, where n can be up to 25,000 (Fig. 15.4). The contour length of an HA (M_r 4×10^6) is 10 µm [143]. HA is polymerized by plasma membrane-bound HA synthase and is not subjected to any type of covalent modification during its synthesis.

Nuclear magnetic resonance studies of the shape of HA performed by Scott [144] have shown the existence of internal hydrogen bonds that stabilize the chain in a stiffened helical conformation. HA chains have the capacity to self-aggregate

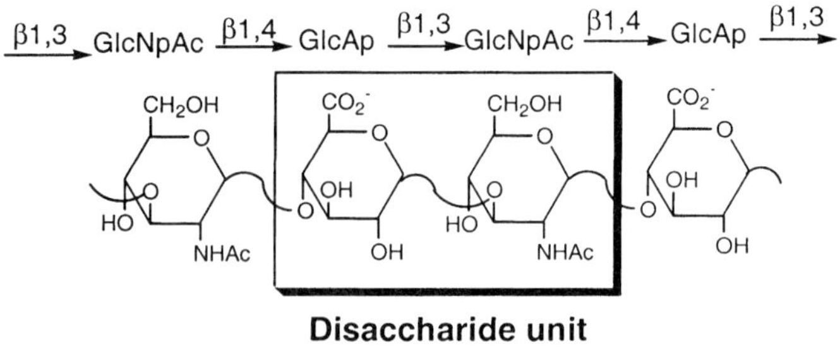

$\beta1,3$ GlcNpAc $\beta1,4$ GlcAp $\beta1,3$ GlcNpAc $\beta1,4$ GlcAp $\beta1,3$

Disaccharide unit

Fig. 15.4 The chemical structure of hyaluronan. The polymer is made up of alternating glucuronic acid (GlcAp) and N-acetylglucosamine (GlcNpAc).

Tab. 15.3 HA concentrations in various human organs and fluids [154].

Organ or fluid	Concentration ($\mu g\, g^{-1}$)
Aqueous humor	0.3–2.2
Brain	35–115
Dermis	200
Plasma (serum)	0.01–0.1
Synovial fluid	1400–3600
Thoracic lymph	8.5–18
Umbilical cord	4100
Urine	0.1–0.3
Vitreous body	140–340

in aqueous solutions and can be visualized by rotary shadow electron microscopy [145]. The aggregation of two anti-parallel HA molecules is promoted through hydrogen bonds between the acetamido group on one chain and the carboxyl group on the other [146]. High molecular weight HA is extremely viscous.

15.3.2
Tissue Distribution and Biosynthesis

HA is synthesized by almost all animals, certain bacteria, and viruses [147, 148]. HA is found mainly in the extracellular space, where it accumulates, but the polymer can also be bound to the cell surface or be located intracellularly around the nucleus and in the lysosomes [149–151]. The largest storage of HA in humans is in the skin, constituting about 50% of the body's HA [152] (Tab. 15.4).

The first cell-free studies of HA biosynthesis used Group A streptococcal bacteria. Markovitz and co-workers showed that the streptococcal HA synthase, located in the cell membrane, required Mg^{++} ions and used the two sugar nucleotide substrates UDP-GlcAp and UDP-GlcNpAc to polymerize a HA chain [154]. HA synthase has two different glycosyltransferase activities and exhibits at least six different functions (Fig. 15.5).

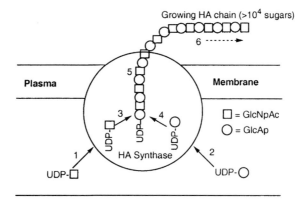

Fig. 15.5 Enzyme functions needed for hyaluronan biosynthesis. The diagram shows the membrane-bound hyaluronan synthase and the six independent activities required for the enzyme to make a disaccharide unit and extend the growing hyaluronan chain.

In 1983, Prehm [155] proposed a novel mechanism, distinctly different from that of other GAGs, for HA biosynthesis. He proposed that HA synthesis occurs at the reducing terminus of a growing HA chain by a two-site mechanism. In this mechanism, the reducing end sugar of the growing HA chain (either in the GlcNpAc or GlcAp site) would remain covalently bound to a terminal UDP, and the next sugar to be added from the second site would be transferred as the UDP-sugar onto the reducing end sugar with displacements of its terminal UDP [156]. The HA chain would then be in the second site. HA synthase assembles this high M_r HA, which is simultaneously extruded through the membrane into the extracellular space [156]. The HA chain has some hydrophobic character but is primarily hydrophilic, so it was a dilemma as to how this growing chain could be transferred across the hydrophobic lipid barrier of the plasma membrane. Weigel and co-workers proposed that cardiolipin, a common phospholipid, helps transfer HA by creating a pore-like passage within the enzyme through which the growing HA chain passes [157, 158].

15.3.3
Functions and Applications

Although the major biological function of HA is still unclear, many roles have been suggested. One important function is its ability to immobilize specific proteins (aggrecan, versican, neurocan, brevican, CD44) in desired locations within the body [159, 160]. The networks that HA forms are efficient insulators, since other macromolecules have trouble in finding room inside the HA network [145, 147]. With this property, HA (and other GAGs) can regulate (for example) the distribution and transport of major part of plasma proteins into the tissues. In the joints, HA probably has an important role as a lubricant between the joint surfaces [161].

HA is produced at high levels during cell proliferation, especially during mitosis. HA may help the cells to detach from the matrix, making it easier for them to divide, while some cell surface receptors (i.e., CD44 and RHAMM) bind HA, immobilizing them in the desired location [162].

HA has been reported to be involved in various events during morphogenesis and differentiation. Its concentrations increase in the areas where cell migration begins, suggesting that HA opens paths for cells to migrate through [163]. Cancer cells are often enriched with HA, and intense intracelluar staining for HA is a weak prognostic indicator for cancer therapy [164].

The production of HA is increased during inflammation, and generally, the viscous solutions seem to inhibit cell activities. HA increases phagocytosis in monocytes and granulocytes, but the importance of this phenomenon is unknown [165].

HA is often present in the pericellular space, probably protecting cells from lymphocytes and viruses [166]. The embryo is also covered by a thick HA coating during certain stages of development, which is probably important in differentiation [167]. How these coats are attached to the cell surface and what other molecules they contain is unknown.

Since the HA polymer does not itself exhibit any sequence diversity, its function is in part due to its chain length. The inductive role of HA in angiogenesis is attributed to oligosaccharides with four to 25 disaccharide residues [168], whereas high molecular weight HA exerted an inhibitory effect [169]. HA fragments have also been reported to evoke an inflammatory response in macrophages [170].

15.3.3.1 Medical Applications

The high water-binding capacity of HA and its high viscoelasticity give HA a unique profile among biological materials and make it suitable for various medical and pharmaceutical applications. One of the most successful medical applications of HA is the use of sodium hyaluronate (NaHA) for the treatment of osteoarthritis [144]. NaHA suppresses cartilage degeneration and release of proteoglycans from the extracellular matrix in cartilage tissues, protects the surface of articular cartilage [171], normalizes the properties of synovial fluids [172], and reduces pain perception [173, 174]. The mechanisms of these effects, however, have not yet been fully elucidated.

The application of HA in ophthalmology represents another medical application. In cataract surgery known as "viscosurgery," viscoelastic materials such as NaHA are used to maintain operative space and to protect the endothelial layer of the cornea or other tissues from physical damage [175]. In cataract surgery, a new technique known as phacoemulsification and aspiration (PEA) uses ultrasound to emulsify the nucleus of the opaque lens, which is then removed. Highly viscoelastic NaHA is used to protect the endothelium of the cornea from injury during PEA [176].

15.3.3.2 Hyaluronic Acid Biomaterials

HA has been blended with other materials to produce novel biomaterials with desirable physicochemical, mechanical, and biocompatible properties. These include blends with poly(vinylalcohol) for ophthalmic use, and with carboxymethylcellulose (carbodiimide crosslinked) to produce a bioabsorbable film (Seprafilm®) for prevention of postsurgical adhesions, for wound-healing applications (with collagen) and for preparing immunologically 'unrecognizable' liposomes [177].

Conversion of the carboxylic groups of HA to N-acylhydrazides affords derivatives useful for controlled drug release [177]. HA esters have been prepared and fabricated into hydrophobic gauzes and microspheres, for use in transmucosal drug delivery. HA or sulfonated HA have been conjugated to the surface polymers used in medical devices, by use of either chemical or photochemical activation, to furnish coated materials with novel cell adhesive (or non-adhesive) properties.

Several different crosslinking strategies have been applied to the preparation of HA-derived hydrogels [178]. Bis-epoxides, formaldehyde, and divinylsulfone have been used under alkaline conditions to crosslink hydroxy groups in the preparation of hydrogels. Milder crosslinking conditions using polyhydrazides afford biocompatible HA hydrogels that allow covalent attachment of therapeutic molecules [178].

In vivo, HA is degraded by the specific enzyme hyaluronidase and by hydroxy radicals, resulting in degradation of the interstitial HA network at sites of inflammation [177]. Hyaluronidase has a high affinity for polyanionic substrates; it binds and processes chondroitin sulfate and heparin more slowly than HA. Chemical modification of the carboxylic acid and hydroxyl groups of HA can significantly reduce its susceptibility towards hyaluronidase. For example, while a bis-epoxide crosslinked HA hydrogel is degraded in response to inflammation, it is barely affected in healthy tissues. Similarly, surface-immobilized HA and fully esterified HA are not degraded by hyaluronidase.

15.4 Keratan Sulfate

15.4.1 Structure and Distribution

The term keratan sulfate (KS), originally applied to the major GAG of corneal tissue, is now generally used to describe oligo- or polysaccharides containing repeating sulfated Gal$p\beta(1 \rightarrow 4)$GlcNpAc$\beta(1 \rightarrow 3)$ disaccharides (Fig. 15.6). Such KS sequences are found in several dissimilar PGs from cornea, cartilage, brain, and bone [179], as well as in carbohydrate moieties of less well characterized sulfated glycoproteins.

KS is typically heterogeneous in charge and size. It is usually of relatively low molecular weight, with nearly equal amounts of sulfo groups on the 6-positions of the D-Gal$p$ and the D-GlcNpAc [180]. It differs from other GAGs in the linkage region to the protein core, lacking the Xyl-serine linkage. In corneal tissue, the

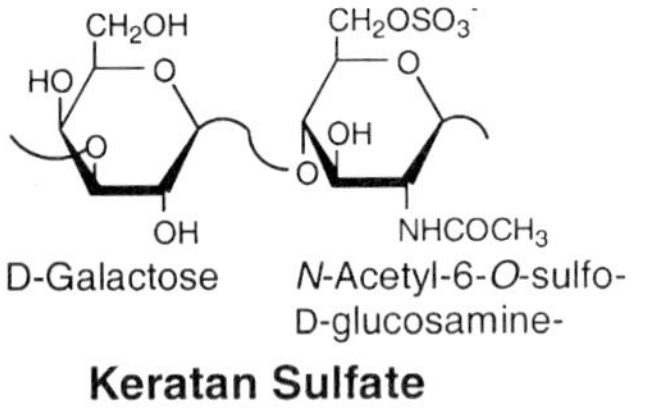

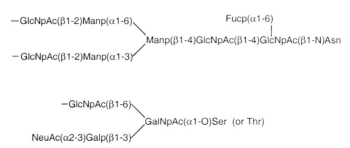

Keratan Sulfate

Fig. 15.6 The disaccharide repeating unit of keratan sulfate.

```
—GlcNpAc(β1-2)Manp(α1-6)                Fucp(α1-6)
                            \                |
                             Manp(β1-4)GlcNpAc(β1-4)GlcNpAc(β1-N)Asn
—GlcNpAc(β1-2)Manp(α1-3)/
```

```
        —GlcNpAc(β1-6)
                      \
                       GalNpAc(α1-O)Ser  (or Thr)
   NeuAc(α2-3)Galp(β1-3)/
```

Fig. 15.7 Linkage region of corneal and skeletal keratan sulfate. The upper structure shows the linkage region in KS type I (corneal), the lower structure the linkage region in KS type II (skeletal).

linkage is *N*-glycosidic between GlcNpAc and an asparagine [181] (Fig. 15.7), while in skeletal tissue KS is linked to the hydroxy groups of serine or threonine residues by an *O*-glycosidic bond to GalNpAc [182].

15.4.2
Chemistry and Biosynthesis of Linkage Regions

15.4.2.1 Keratan Sulfate on Cartilage Proteoglycans
In addition to CS chains and *N*-linked oligosaccharides, the large PG from cartilage contains more than 100 serine and threonine residues substituted with *O*-linked KS or oligosaccharide chains [183]. The structures of *O*-linked oligosaccharides indicate that they provide the linkage region for initiating elongation of KS chains. Synthesis of the *O*-linked oligosaccharides and KS occurs entirely in the Golgi complex [179]. The *O*-linked oligosaccharides and KS chains enhance the ability of the PGs to resist compressive deformation and protect the core protein against proteolysis.

15.4.2.2 Keratan Sulfate on Corneal Proteoglycans
The small KS PG from corneal tissue contains 1–2 attachment sites, each with the possibility of carrying two chains, and 1–2 *N*-linked oligosaccharides bound to a 45 kDa core protein [184]. The linkage region for the KS chains is a *N*-linked oli-

gosaccharide of the complex type [185], synthesized in the endoplasmic reticulum, while the KS chains are elaborated in the Golgi. This KS PG appears to interact at specific sites along the collagen fibrils in the stroma and has a function in maintaining the optical transparency of the cornea [186].

15.4.3
Biological Roles of Keratan Sulfate

The cartilage matrix is rich in KS PGs that interact with HA and link protein to form aggregates. The exact function of the KS GAG chains is not well understood. While the highly negatively charged KS contributes to the physicochemical properties of the PG, it does not appear to be essential for the ability to bear load effectively [187]. Although KS is found mostly in cartilage, small amounts of KS, possibly resulting from cartilage breakdown [188], can be detected in blood, providing important diagnostic information [189]. Patients with polyarticular osteoarthritis have abnormally high levels of serum KS, suggesting that a high rate of PG catabolism during turnover may predispose the development of arthritis [190].

15.4.3.1 Role of KS in Macular Corneal Dystrophy
The corneal stroma contains a KS PG and a DS PG as major non-collagenous components of the extracellular matrix [191]. These PGs are absent from opaque corneal wounds [192]. Macular corneal dystrophy (MCD) is an inherited disorder, clinically characterized by the accumulation of opaque deposits in the corneal stroma [193, 194]. A post-translational error in KS PG biosynthesis affords an unsulfated glycoconjugate [195, 196].

15.5
Other Acidic Polysaccharides

15.5.1
Acharan Sulfate

Acharan sulfate is a GAG isolated from the giant African snail *Achatina fulica*. This polysaccharide has a repeating →4)-a-D-GlcNpAc (1 → 4)-a-L-IdoAp2S (1 →) disaccharide structure (Fig. 15.8). The natural function of this molecule in the snail, while still unclear, may be as an anti-desiccant, a metal chelator, an anti-infective, or a locomotive (slime) agent. The unusual structure of acharan sulfate, similar but distinctive from both heparin and HS has resulted in studies of its biological activities. Chemically modified acharan sulfate, *N*-sulfoacharan sulfate, shows a heparin-like effect on basic FGF2 mitogenicity but at a greatly reduced level [197] and is a moderately active inhibitor of thrombin [198].

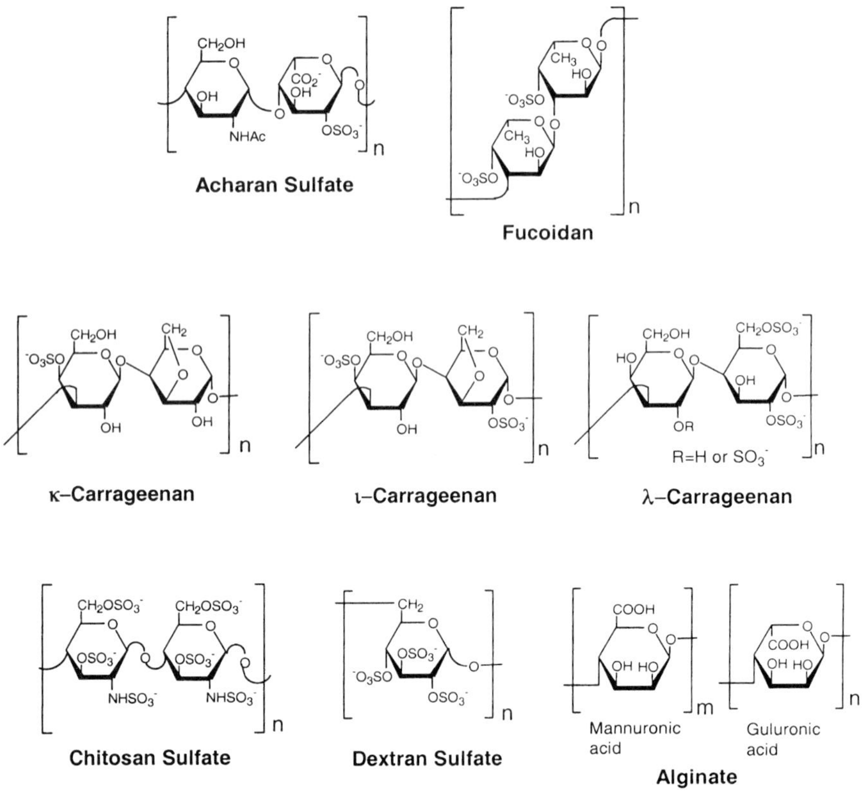

Fig. 15.8 Structures of other acidic polysaccharides.

15.5.2
Fucoidins

Fucoidin is a complex sulfated polysaccharide, derived from marine brown algae [199–201], the jelly coat from sea urchin eggs [202], and the sea cucumber body wall [203]. These glucuronofucoglycans and fucoglucuronans comprise a wide, continuous spectrum of low sulfate polymers. Fucoidins are primarily composed of a-$(1 \rightarrow 3)$-linked units of 4-sulfo-L-fucose with branching or a second sulfo group at position 3 (Fig. 15.8) [204]. Most investigations into their biological activity have focused on fucoidans from brown algae, *Fucus vesiculosus*, which exhibit a variety of biological effects on mammalian cells. *Fucus* fucoidan has anticoagulant activity [205–207], and is a potent activator of both ATIII and HCII [207]. Fucoidan inhibits both the initial binding of sperm and subsequent recognition [208]. It also prevents infection of human cell lines by several enveloped viruses [209]. Fucoidan blocks cell-cell binding mediated by P- or L-selectin but not by E-selectin [210]. Furthermore, it demonstrates differential binding to interleukins [211] and the hepatocyte growth factor [212]. Since this polysaccharide causes no toxicity or irritation, it may be useful as an anticoagulant, antiviral, anti-inflammatory, or contraceptive agent [213–215]. On oral administration fucoidans have been effective in healing and preventing gastric ulcers in animal models [201].

15.5.3
Carrageenans

Carrageenans are sulfated polysaccharides derived from various species of red algae. There are three main types: ι, κ, and λ. Both ι- and κ-carrageenan contain O-3-substituted O-4-sulfo-β-D-Galp units, while λ-carrageenan contains O-3-substituted O-2-sulfo-β-D-Galp units. In addition to the above moieties, ι-carrageenan also contains O-4-substituted O-2-sulfo-3,6-anhydro-α-D-Galp units, κ-carrageenan also contains O-4-substituted 3,6-anhydro-α-D-Galp units, and λ-carrageenan also contains O-4-substituted O-2,6-disulfo-α-D-Galp units (Fig. 15.8) [216]. An otherwise perfect alternating sequence in each of these polymers is complicated by occasional modifications in the placement and number of sulfo groups or – for ι- and κ-carrageenans – by the absence of the 3,6-anhydro-linkage [217].

Carrageenans (Irish moss) have been used medicinally for centuries. The moss has been used as a cough medicine [218], and a degraded ι-carrageenan is marketed in Europe as an anti-ulcer preparation [217]. High molecular weight carrageenans have a wide variety of applications in the food industry, serving as thickeners, stabilizers, and emulsifiers. Carrageenan activates Hageman factor, one of the blood coagulation factors, which has cardiotonic activity [219]. On oral administration in animals, both κ- and ι-carrageenan show anti-tumor activity [220]. Some concerns have recently been raised about the toxicity of low molecular weight carrageenans [221].

15.5.4
Sulfated Chitins

Chitin, poly(2-acetamido-2-deoxy-D-glucopyranose), is the main structural element of the cuticles of crab, shrimp, and insects, and is also widespread in the cell walls of fungi [222]. Chitin can be de-N-acetylated to prepare chitosan, which on chemical sulfation and/or carboxymethylation results in a polymer with certain structural similarities to heparin (Figs. 15.2 and 15.8) [223]. These chitosan derivatives show anticoagulant activity related to their degree of sulfation. Carboxymethylated sulfochitosan inhibits thrombin activity through ATIII to almost the same degree as heparin.

15.5.5
Dextran Sulfate

Dextran, a $(1 \rightarrow 4)$-β-D-,$(1 \rightarrow 3)$-α-D-branched Glcp polymer [222], can be chemically sulfonated to prepare dextran sulfate (Fig. 15.8) [224]. Both dextran (a plasma extender) and dextran sulfate have been used in pharmaceuticals. Dextran sulfate has low anticoagulant activity with high LPL-releasing activity [225]. This has permitted the exploitation of this agent as an anti-atherosclerotic in Japan [226]. Dextran sulfate has been used as a heparin replacement in anticoagulation and has been immobilized on plastic tubes to prepare non-thrombogenic surfaces [227]. Dextran sulfate is an inhibitor of HIV binding to T-lymphocytes, but its low oral bioavailability has precluded its use in the treatment of AIDS [228].

15.5.6
Alginates

Alginate is a commercially important component of brown seaweeds and is the most important mucilaginous polysaccharide, preventing desiccation of the seaweed when it is exposed to air [201]. It consists of D-mannuronic acid (ManAp) and L-guluronic acid (GulAp). The distribution of ManAp and GulAp in alginate chains give rise to three different block types, namely blocks of poly-ManAp, blocks of poly-GulAp and alternating blocks of the ManAp-GulAp (Fig. 15.8) [229]. Alginates are used as low-price viscosifiers or thickeners in a wide range of products. In the pharmaceutical field, alginates have been used for many years to treat wounds and gastric ulcers [201]. Alginate also depresses the plasma cholesterol level. Alginic acid strongly inhibits hyaluronidase and mast cell degranulation, involved in allergic reactions [230].

15.5.7
Fully Synthetic Sulfated Molecules

15.5.7.1 Polymers
Synthetic polymers such as poly(vinyl sulfate) and poly(anethole sulfonate) (Fig. 15.9) exhibit anticoagulant activity [231] and have been exploited in vitro to collect blood and plasma samples for assay. A biphenyl disulfonic acid urea copolymer shows potent anti-HIV activity [232]. These agents are highly toxic because of their long half-lives, the result of their failure to be cleared either through filtration or metabolism.

15.5.7.2 Small Sulfonated Molecules
Suramin (Fig. 15.9) was the first heparin analogue used clinically in a wide variety of applications [233]. These applications include its activity as an anthelmintic, an antiprotozoal, an antineoplastic, and an antiviral agent [36, 102]. Despite its potent activities, suramin has a very long half-life in the body and exhibits a wide range of toxic effects.

Naphthalene sulfonates (Fig. 15.9) show potent anti-HIV activity, but limited toxicological data are currently available [234]. A series of simple aliphatic disulfates and disulfonates have been tested for their ability to arrest amyloidosis in vivo as potential agents for the treatment of Alzheimer's disease [235]. Several alkyl malto-laminaro-α oligosaccharides, such as the highly sulfated dodecyl laminaropentaoside (Fig. 15.9), have also been investigated for anti-HIV activity [236].

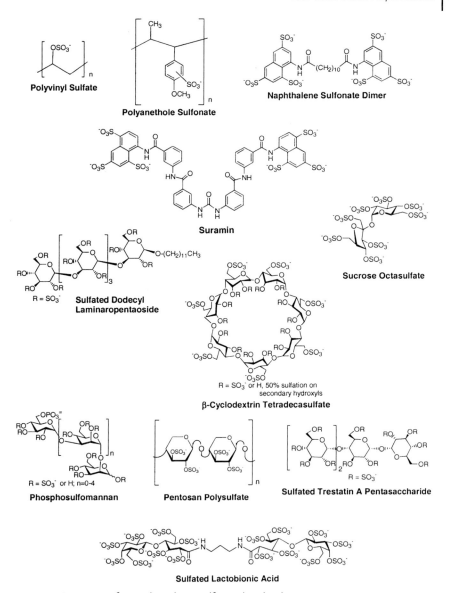

Fig. 15.9 Structures of several synthetic sulfonated molecules.

Sucralfate (Carafate®), insoluble aluminium sucrose octasulfate (SOS) (Fig. 15.9), is used in the treatment of ulcers [8]. The water-soluble sodium SOS binds in place of heparin/HS within the FGF-FGFR signal transduction complex. SOS derivatives lacking sulfo groups at specific positions prevent the assembly of active complex and are being investigated as potential anticancer agents [237].

Sulfated cyclodextrins such as β-cyclodextrin tetradecasulfate (Fig. 15.9) have demonstrated anticancer activity, presumably through an anti-angiogenic mechanism [238]. Sulfated cyclodextrins also inhibit complement activation [239].

Phosphosulfomannan, PI-88 (Fig. 15.9), is a chemically sulfonated phosphomannan oligosaccharide mixture derived from the yeast *Pichia holstii* [240]. PI-88 has recently been identified as a promising inhibitor of tumor cell growth and metastasis and is currently undergoing Phase I clinical trials. PI-88 is believed to block tumor growth by interfering in HS interaction with FGF and FGFR [241] and prevents metastasis by inhibiting heparanase [242], blocking the breakdown of extracellular matrix, preventing the spread of tumor cells. PI-88 also shows substantial antithrombotic activity through its catalysis of HCII-mediated inhibition of thrombin [243].

Pentosan, a linear Xylp polymer extracted from the bark of the birch tree *Fagus sylvantica*, when fully sulfonated and partially depolymerized (Fig. 15.9), is an anticoagulant with one tenth of heparin's activity on a weight basis [244]. Its primary anti-IIa activity has been postulated to be HCII-mediated [245]. Other xylans derived from corncobs, birchwood, and oatspelts provided heparinoids more active than pentosan polysulfate.

Trestatin A (Fig. 15.9), a pseudo-nonasaccharide obtained from strains of *Streptomyces dimorphogenes*, is a potent α-amylase inhibitor. A highly sulfated maltotriosyl trehalose pentasaccharide chemically modified from the Trestatin substructure has an antiproliferative activity comparable to that of heparin [246].

Sulfated lactobionic acid (Aprosulfate, LW10082) (Fig. 15.9), prepared through the chemical sulfonation of a lactose dimer, is an antithrombotic agent in animal models. It acts primarily through HCII but showed some toxicity precluding its clinical use [247, 48].

15.6
Pharmaceutical Potential and Challenges

This chapter describes a multiplicity of important biological functions that are associated with GAGs. Heparin and LMW heparin, for example, command a one billion dollar market worldwide, and are the preeminent clinical anticoagulants in use today. Furthermore, GAG-based (and potentially other carbohydrate-based) agents have unique properties – such as high specificity, delocalized binding sites, low antigenicity, and multivalency – difficult to replicate with other classes of molecules. The biological functions of GAGs might be enhanced by administration of exogenous GAGs, modified GAGs, GAG oligosaccharides, or GAG analogues. Such agonists might be useful, for example, as anticoagulants, as anti-infectives, in promoting cell growth, and for other important GAG-related activities. Alternatively, molecules might be designed as antagonists, to block normal GAG function, for use as procoagulants, or in inhibiting the assembly of signal transduction complexes and preventing cell replication in the treatment of cancers. Despite the great potential of new GAG-based drug therapies, the development of

LMW heparins represent the only significant success story in this class of molecules in the past 20 years of extensive research effort. Moreover, this success is limited to anticoagulant/antithrombotic agents, a role that was already being fulfilled by heparin itself. Thus, the agents could be viewed as merely "improved heparins" or a second-generation replacement for a relatively successful drug. The slow development of new GAG-based therapeutic agents is primarily associated with a number of difficult problems that still need to be addressed.

15.6.1
GAG-Based Agents Are Heterogenous

Heparin and LMW heparins are heterogeneous mixtures. This complicates their application as drugs in a number of ways. Firstly, the production of these agents must be controlled in order to obtain reproducible mixtures corresponding to those desired for the drug products. Secondly, analytical and quality control issues, while solvable, are more complicated than for homogenous products. Thirdly, biological evaluation of mixtures (half-life, pharmacokinetics, metabolism, and bio-activity) is more complicated than for a single entity. Fourthly, regulatory questions for drug mixtures can be formidable, making the drug approval process more difficult. Fifthly, a patent position is more difficult to establish for a mixture in which composition of matter is difficult to define.

15.6.2
GAG-Based Agents and Sulfonated Analogues Have Low Bioavailability

Heparin is only used by *intravenous* administration, while LMW heparins are effective when given either by *intravenous* or *subcutaneous* administration. This has resulted in a movement, particularly in the US, away from heparin and towards LMW heparins. Much of the biological potential of heparin, particularly its prophylactic uses (anti-atherosclerosis, anti-infective, etc.), cannot easily be exploited by a drug requiring injection. Heparin is not orally bioavailable because of its highly charged nature and its high molecular weight. Moreover, GAG-based analogues, while often having reduced molecular weights, are frequently still highly charged molecules. Three approaches might be used to solve these bioavailability problems. Firstly, excipients, drug carriers, or salt forms might be used to carry GAG-based drugs through membranes, facilitating their oral absorption. Approaches have ranged from ion-pairing agents (such as quaternary ammonium salts) [248], surfactants (such as saponins) [249], to small hydrogen-bonding peptoids (such as SNAC) [250]. Up to now, none of these approaches has afforded a clinically useful oral drug product. Secondly, the GAG-based drugs can be simplified by decreasing the number of required sulfo groups, reducing charge and molecular weight to make oral bioavailability possible. While this is a realistic and promising approach, no one has yet successfully designed an orally bioavailable GAG analogue for clinical use. Thirdly, it might be possible to mimic GAG binding without the use of charged sulfo groups and in the absence of a large extended linear template.

15.6.3
GAGs Have a Myriad of Biological Activities

The large number of biological activities of GAGs is a double-edged sword: while GAGs are interesting drug targets, it is difficult to obtain an agent with a single activity in the absence of complicating side effects. The multi-pharmacological profile of GAGs may be advantageous in treating complex disease processes in which all the properties align in the same direction. Furthermore, as in the case of the synthetic ATIII pentasaccharide, it may be possible to pair away other activities to obtain a homogeneous agent with a single prominent pharmacological activity.

15.6.4
Carbohydrate-Based Drugs Are Expensive and Difficult to Prepare

The synthetic heparin pentasaccharide, corresponding to the ATIII binding site, originally required over 30 synthetic steps [251]. This is over three times the number of synthetic steps required for prostaglandin synthesis, and the heparin pentasaccharide has to be administered in multi-milligram (not microgram) doses. Improvements in carbohydrate synthesis, including new enzymatic approaches [252, 253] and solid-phase synthesis [254], might one day make complex synthetic carbohydrate-based drugs economically viable. Indeed, the synthetic ATIII pentasaccharide is now clinically used as a highly specific anticoagulant agent in Europe.

The conformational flexibility of carbohydrates and their polyol structures give them many of their unique biological properties, such as low antigenicity, but limit the binding affinity to their protein targets. One approach is to design an alternative molecular scaffold, either flexible or inflexible, that can be used to recognize and occupy carbohydrate binding sites or protein-based receptors and act either as an agonist or as an antagonist. Our improved understanding of heparin-protein interactions [94] should aid in the design of such new agents.

15.7
Conclusion

In conclusion, while GAGs and their analogues have current therapeutic value, their further exploitation to treat a myriad of diseases depends on a number of new chemical, biological, and pharmaceutical advances. Future research on the fundamental physical, chemical, and biological properties of this important class of molecules should certainly provide the necessary knowledge to expand their therapeutic applications.

15.8 References

1. P.-D. KITTLICK, *Glycosaminoglycans: Recent Biochemical Results in the Fields of Growth and Inflammation*, VEB Gustav Fischer, Germany, **1985**.
2. L. KJELLÉN, U. LINDAHL, *Annu. Rev. Biochem.* **1991**, *60*, 443–475.
3. P. KRISTIAN, T. D. KNUT, *J. Cell Sci.* **2000**, *113*, 193–205.
4. J. H. KIMURA, L. S. LOHMANDER, V. C. HASCALL, *J. Cell Biochem.* **1984**, *26*, 261–278.
5. R. V. IOZZO, *Annu. Rev. Biochem.* **1998**, *67*, 609–652.
6. D. M. TEMPLETON, *Crit. Rev. Clin. Lab. Sci.* **1992**, *29*, 141–184.
7. T. E. HARDINGHAM, A. J. FOSANG, *FASEB J.* **1992**, *6*, 443–475.
8. R. J. LINHARDT, T. T. TOIDA, *Carbohydrates as Drugs*, Marcel Dekker, New York, **1996**.
9. A. B. FOSTER, A. J. HUGGARD, *Adv. Carbohydr. Chem. Biochem.* **1955**, *10*, 335–368.
10. R. J. LINHARDT, *Chem. & Indust.* **1991**, *2*, 45–50.
11. B. CASU, *Adv. Carbohydr. Chem. Biochem.* **1985**, *43*, 51–134.
12. W. D. COMPER, *Heparin (and related polysaccharides)* Vol. 7, Gordan and Breach Science Publ., Germany, **1981**.
13. A. A. HORNER, E. YOUNG, *J. Biol. Chem.* **1982**, *257*, 8749–8754.
14. R. J. LINHARDT, Z. M. MERCHANT, K. G. RICE, Y. S. KIM, G. L. FITZGERALD, A. C. GRANT, R. LANGER, *Biochemistry*, **1985**, *24*, 7805–7810.
15. R. J. LINHARDT, D. M. COHEN, K. G. RICE, *Biochemistr*, **1989**, *28*, 2888–2894.
16. L. KJELLÉN, I. PETTERSON, E. UNGER, U. LINDAHL, *Adv. Exp. Med. Biol.* **1992**, *313*, 107–111.
17. U. LINDAHL, D. S. FEINGOLD, L. RODEN, *TIBS* **1986**, *11*, 221–225.
18. J. T. GALLAGHER, J. E. TURNBULL, M. LYON, *Adv. Exp. Med. & Biol.* **1992**, *313*, 49–57.
19. M. BERNFIELD, R. KOBEYESI, M. KATO, M. T. HINKES, J. SPRING, R. L. GALLO, E. J. LOSE, *Ann. Rev. Cell Biol.* **1992**, *8*, 365–393.
20. R. J. LINHARDT, *Polysaccharide Lyases for Glycosaminoglycan Analysis*, Wiley Interscience, V 2, 17.13.17–17.13.32, **1994**.
21. B. CASU, U. LINDAHL, *Adv. Carbohydr. Chem. Biochem.* **2002**, *57*, 159–206.
22. R. N. KRISHNA, P. K. AGRAWAL, *Adv. Carbohydr. Chem. Biochem.* **2000**, *56*, 201–234.
23. K. LIDHOLT, L. KJELLÉN, U. LINDAHL, *Biochem. J.* **1989**, *261*, 999–1007.
24. R. L. STEVENS, C. C. FOX, L. M. LICHTENSTEIN, K. F. AUSTEN, *Proc. Natl. Acad. Sci. USA* **1988**, *85*, 2284–2287.
25. R. J. LINHARDT, D. LOGANATHAN, *Heparin, Heparinoids and Heparin Oligosaccharides: Structure and Biological Activities*, Plenum Press, New York, **1990**.
26. J. A. MARCUM, R. D. ROSENBERG, *Semin. Thromb. Haemostasis* **1987**, *13*, 464–474.
27. R. LANGER, R. J. LINHARDT, C. L. COONEY, M. KLEIN, D. TRAPPER, S. M. HOFFBERG, A. LARSEN, *Science* **1982**, *217*, 261–263.
28. M. D. FREEDMAN, *J. Clin. Pharmacol.* **1992**, *32*, 584–596.
29. N. S. GUNAY, R. J. LINHARDT, *Semin. Thromb. Hemostasis* **1999**, *25*, 5–16.
30. S. G. FRANGOS, A. H. CHEN, B. SUMPIO, *J. Am. Coll. Surg.* **2000**, *191*, 76–92.
31. B. R. JAEGER, *Ther. Apher.* **2001**, *5*, 207–211.
32. R. E. EDENS, R. J. LINHARDT, J. M. WEILER, *Compl. Prof.* **1993**, *1*, 96–120.
33. M. A. FATH, X. WU, R. E. HILEMAN, R. J. LINHARDT, M. A. KASHEM, R. M. NELSON, C. D. WRIGHT, W. M. ABRAHAM, *J. Biol. Chem.* **1998**, *273*, 13563–13569.
34. S. M. SMORENBURG, C. J. VAN NOORDEN, *Pharmacol. Rev.* **2001**, *53*, 93–105.
35. R. J. HETTIARACHCHI, S. M. SMORENBERG, J. GINSBERG, M. LEVINE, M. H. PRINS, H. R. BULLER, *Thromb. Haemost.* **1999**, *82*, 947–952.
36. Y. CHEN, T. MAGUIRE, R. E. HILEMAN, J. R. FROMM, J. D. ESKO, R. J. LINHARDT, R. M. MARKS, *Nat. Med.* **1997**, *3*, 866–871.
37. B. DUDAS, U. CORNELLI, J. M. LEE, M. J. HEJNA, M. WALZER, S. A. LORENS, R. F. MERVIS, J. FAREED, I. HANIN, *Neurobiol. Aging* **2002**, *23*, 97–104.
38. R. HUBER, R. W. CARRELL, *Biochemistry* **1989**, *28*, 8951–8966.
39. Y. S. KIM, K. B. LEE, R. J. LINHARDT, *Thromb. Res.* **1988**, *51*, 97–104.

40 K. Kurachi, G. Schmer, M. A. Hermodson, D. C. Teller, E. W. Davie, Biochem. J. **1976**, *15*, 368–372.
41 D. M. Tollefsen, M. K. Blank, J. Biol. Chem. **1982**, *257*, 2162–2169.
42 E. Ersdal-Badju, A. Lu, Y. Zuo, V. Picard, S. C. Bock, J. Biol. Chem. **1997**, *272*, 19393–19400.
43 T. H. Tran, B. Zbinden, B. Lammle, F. Duckert, Semin. Thromb. Haemost. **1985**, *11*, 342–346.
44 R. D. Rosenberg, P. S. Damus, J. Biol. Chem. **1973**, *248*, 6490–6505.
45 D. M. Tollefsen, Nouv. Rev. Fr. Hematol. **1984**, *26*, 233–237
46 L. B. Jaques, Hämostaseologie **1985**, *5*, 121–126.
47 H. B. Nader, L. Toma, M. S. Pinhal, V. Buonassisi, P. Colburn, C. P. Dietrich, Semin. Thromb. Hemost. **1991**, *17*, 47–56.
48 H. L. Messmore, B. Griffin, M. Koza, J. Seghatchian, J. Fareed, E. Coyne, Semin. Thromb. Hemost. **1991**, *17*, 57–64.
49 E. Gerlach, B. F. Becker, Zeitschrift für Kardiologie **1993**, *5*, 13–21.
50 R. Ross, Nature **1993**, *24*, 1447–1451.
51 B. R. Jaeger, Therapeutic Apheresis **2001**, *5*, 207–211.
52 V. Fuster, A. Lewis, Circulation **1994**, *90*, 2126–2145.
53 E. B. Smith, I. B. Massie, K. M. Alexander, Atherosclerosis **1976**, *25*, 71–84.
54 H. Engelberg, Pharmacol. Rev. **1984**, *36*, 91–110.
55 Z. M. Merchant, E. E. Erbe, W. P. Eddy, R. J. Patel, R. J. Linhardt, Atherosclerosis **1986**, *62*, 151–158.
56 S. Santamarina-Fiojo, H. B. Brewer Jr., Int. J. Clin. Lab. Res. **1994**, *24*, 143–147.
57 P. Sivaram, S. Y. Choi, L. K. Curtis, I. J. Goldberg, J. Biol. Chem. **1994**, *269*, 9409–9412.
58 T. K. Sue, L. B. Jaques, E. Yuen, Can. J. Physiol. Pharmacol. **1976**, *54*, 613–617.
59 J. J. Castellot Jr., D. L. Beeler, R. D. Rosenberg, M. J. Karnovsky, J. Cell Physiol. **1984**, *120*, 315–320.
60 J. J. Castellot Jr., J. Choay, J.-C. Lormeau, M. Petitou, E. Sache, M. J. Karnovsky, J. Cell Biol. **1986**, *102*, 1979–1984.
61 T. C. Wright, J. J. Castellot, J. R. Diamond, M. J. Karnovsky, Chemical and Biological Properties Clinical Applications, CRC Press, London, **1989**.
62 L. A. Pukac, G. M. Hirsch, J.-C. Lormeau, M. Petitou, J. Choay, M. J. Karnovsky, Am. J. Pathol. **1991**, *139*, 1501–1509.
63 S. Ruddy, I. Gigli, K. F. Austen, New Engl. J. Med. **1972**, *287*, 489–495.
64 H. Garg, B. T. Thompson, R. J. Linhardt, C. A. Hales, Proteoglycans and Lung Disease, Marcel Dekker, Inc., New York, **2002**.
65 G. B. Caughman, R. J. Boackle, J. Vesely, Mol. Immunol. **1982**, *19*, 287–295.
66 C. R. Minick, Immunity and Atherosclerosis, Academic Press, New York, **1980**.
67 S. Alameda, R. D. Rosenberg, D. H. Bing, J. Biol. Chem. **1983**, *258*, 785–791.
68 H. Engelberg, Adv. Lipid Res. **1983**, *20*, 219–255.
69 M. D. Kazatchkine, D. T. Fearon, D. D. Metcalfe, R. D. Rosenberg, K. F. Austen, J. Clin. Invest. **1981**, *67*, 223–228.
70 M. P. Bevilacqua, R. M. Nelson, A. Venot, R. J. Linhardt, I. Stamenkovic, Ann. Rev. Cell Biol. **1995**, 11601–11631.
71 R. P. McEver, Curr. Opin. Immunol. **1994**, *6*, 75–84.
72 R. M. Nelson, O. Cecconi, W. G. Roberts, A. Aruffo, R. J. Linhardt, M. P. Bevilacqua, Blood **1993**, *82*, 3253–3258.
73 D. P. Witt, A. D. Lander, Curr. Biol. **1994**, *4*, 394–400.
74 W. F. Long, F. B. Williamson, Med. Hypoth. **1984**, *13*, 385–394.
75 J. Folkman, Y. Shing, Heparin and Related Polysaccharides, Plenum Press, New York, **1992**.
76 T. K. Hunt, D. R. Knighton, K. K. Thakral, W. H. Goodson III, W. S. Andrews, Surgery, **1984**, *96*, 48–54.
77 D. Ribatti, A. Vacca, B. Nico, E. Crivellato, L. Roncali, F. Dammacco, Brit. J. Haematol. **2001**, *115*, 514–521.
78 S. Taylor, J. Folkman, Nature **1982**, *97*, 307–312.
79 J. Folkman, D. E. Ingber, Chemical and Biological Properties Clinical Applications, CRC Press, London, **1989**.
80 J. J. Mason, Cell **1994**, *78*, 547–552.
81 K. Shiokawa, M. Asano, C. Shiozaki, Jap. J. Clin. Med. **1992**, *50*, 1893–1901.
82 J. Schlessinger, A. N. Plotnikov, O. A. Ibrahimi, A. V. Eliseenkova, B. K. Yeh,

A. Yayon, R. J. Linhardt, M. Mohammadi, *Molecular Cell* **2000**, *6*, 743–750.
83 R. Reich-Slotky, D. Bonneh-Barkay, E. Shaoul, B. Bluma, C. M. Srahn, D. Ron, *J. Biol. Chem.* **1994**, *269*, 32279–32285.
84 J. Folkman, R. Langer, R. J. Linhardt, C. Haudenschild, S. Taylor, *Science* **1983**, *221*, 719–725.
85 R. Crum, S. Szabo, J. Folkman, *Science* **1985**, *230*, 1375–1378.
86 R. Sasisekharan, M. A. Moses, M. A. Nugent, C. L. Cooney, R. Langer, *Proc. Natl. Acad. Sci. USA* **1994**, *91*, 1524–1528.
87 J. Denekamp, *Prog. Appl. Microcirc.*, Karger, Basel, **1984**.
88 D. W. Beck, J. J. Olson, R. J. Linhardt, *J. Neuropath. Exper. Neurol.* **1986**, *45*, 503–512.
89 T. Sasaki, H. Larsson, J. Kreuger, M. Salmivirta, L. Claesson-Welsh, U. Lindahl, E. Hohenester, R. Timpl, *The EMBO Journal* **1999**, *18*, 6240–6248.
90 T. Miralem, A. Wang, *J. Biol. Chem.* **1996**, *271*, 17100–17106.
91 D. J. Tyrrell, A. P. Horne, K. R. Holme, J. M. Preuss, C. P. Page, *Adv. Pharmacol.* **1999**, *46*, 151–208.
92 M. Nakajima, T. Irimura, G. L. Nicolson, *J. Cell Biochem.* **1988**, *36*, 157–167.
93 L. Borsig, R. Wong, J. Feramisco, D. R. Nadeau, N. M. Varki, A. Varki, *Proc. Natl. Acad. Sci. USA* **2001**, *98*, 3352–3357.
94 I. Capila, R. J. Linhardt, *Angewandte Chemie* **2002**, *41*, 1000–1022.
95 D. Sawitzky, *Microbio. Immunol.* **1996**, *184*, 155–161.
96 B. C. Herold, R. J. Visallai, N. Susmarski, C. R. Brandt, P. G. Spear, *J. Gen. Virol.* **1994**, *75*, 1211–1222.
97 B. C. Herold, S. I. Gerber, T. Polonsky, B. J. Belval, P. N. Shaklee, K. Home, *Virology* **1995**, *206*, 1108–1116
98 C. C. Rider, C. R. Coombe, H. A. Harrop, E. F. Hounsell, C. Bauer, J. Feeny, B. Mulloy, N. Mahmood, A. Hay, C. R. Parish, *Biochemistry* **1994**, *33*, 6974–6980.
99 C. C. Rider, *Glycoconjugate J.* **1997**, *14*, 639–642.
100 M. Witvrouw, E. De Clercq, *Gen. Pharmac.* **1997**, *29*, 497–511.
101 R. M. Marks, H. Lu, R. Sundaresan, T. Toida, A. Suzuki, T. Imanari, M. J. Hernaiz, R. J. Linhardt, *J. Med. Chem.* **2001**, *44*, 2178–2187.
102 D. Rathore, T. F. McCutchan, M. J. Hernaiz, L. A. LeBrun, S. C. Lang, R. J. Linhardt, *Biochemistry* **2001**, *40*, 11518–11524.
103 J. Dong, C. A. Peters-Libeu, K. H. Weisgraber, B. W. Segelke, B. Rupp, I. Capila, M. J. Hernaiz, L. A. LeBrun, R. J. Linhardt, *Biochemistry* **2001**, *40*, 2826–2834.
104 N. Parveen, J. M. Leong, *Molecul. Microbiol.* **2000**, *35*, 1220–1234.
105 L. Buee, W. Ding, A. Delacourte, H. Fillit, *Brain Res.* **1993**, *601*, 154–163.
106 K. R. Brunden, N. J. Richter-Cook, N. Chaturvedi, R. C. Frederickson, *J. Neurochem.* **1993**, *61*, 2147–2154.
107 R. W. Scott, B. L. Bergmann, A. Bajpai, R. T. Hersh, H. Rodriguez, B. N. Jones, C. Barreda, S. Watts, J. B. Baker, *J. Biol. Chem.* **1985**, *260*, 7029–7034.
108 G. Perry, P. Richey, S. L. Siedlak, P. Galloway, M. Kawai, P. Cras, *Brain Res.* **1992**, *579*, 350–352.
109 H. G. Bazin, M. A. Marques, A. P. Owens, R. J. Linhardt, K. A. Krutcher, *Biochemistry* **2002**, *41*, 8203–8211.
110 B. Mulloy, R. J. Linhardt, *Curr. Opin. Struct. Biol.* **2001**, *11*, 623–628.
111 R. L. Jackson, S. J. Busch, A. D. Cardin, *Physiol. Rev.* **1991**, *71*, 481–537.
112 A. D. Cardin, H. J. R. Weintraub, *Arteriosclerosis* **1989**, *9*, 21–32.
113 R. V. Iozzo, *Lab. Invest.* **1985**, *53*, 373–396.
114 R. J. Linhardt, R. E. Hileman, *Gen. Pharmac.* **1995**, *26*, 443–451.
115 L, Cöster, *Biochem. Soc. Trans.* **1991**, *19*, 866–868.
116 H. Kresse, H. Hausses, E. Schönherr, *Experimentia* **1993**, *49*, 403–416.
117 H. C. Whinna, H. U. Choi, L. C. Rosenberg, F. C. Church, *J. Biol. Chem.* **1993**, *268*, 3920–3924.
118 M. C. Bourin, U. Lindahl, *Biochem. J.* **1993**, *289*, 313–330.
119 J. Pangrazzi, F. Gianese, *Haematologica* **1987**, *72*, 459–464.

120 D. M. Tollefsen, *Adv. Exp. Med. Biol.* **1992**, *313*, 167–176.
121 I. J. Edwards, W. D. Wagner, R. T. Owens, *Am. J. Path.* **1990**, *136*, 609–621.
122 E. Cofrancesco, M. Colombi, F. Gianese, M. Cortellaro, *Thromb. Res.* **1990**, *57*, 405–414.
123 D. Hoppensteadt, J. Walenga, J. Fareed, *Semin. Thromb. Hemost.* **1991**, *17*, 60–64.
124 P. Sié, D. Dupouy, C. Caranobe, M. Petitou, B. Boneu, *Blood* **1993**, *81*, 1771–1777.
125 G. Mascellani, L. Liverani, P. Bianchini, B. Parma, G. Torri, G. Bisio, M. Guerrini, B. Casu, *Biochem. J.* **1993**, *296*, 639–648.
126 S. Alban, *Carbohydrates as Drugs* Marcel Dekker: New York, **1996**.
127 M. Barbanti, F. Calanni, M. R. Milani, E. Marchi, N. Semeraro, M. Colucci, *Thromb. Haemost.* **1993**, *69*, 147–151.
128 F. Dol, M. Petitou, J. C. Lormeau, J. Choay, C. Caranobe, P. Sié, S. Saivin, G. Houin, B. Boneu, *J. Lab. Clin. Med.* **1990**, *115*, 43–51.
129 S. J. Brister, M. R. Buchanan, C. C. Griffin, C. L. van Gorp, R. J. Linhardt, US Patent 5922690, **1999**.
130 J. Van Ryn-McKenna, F. A. Ofosu, J. Hirsh, M. R. Buchanan, *Br. J. Haemat.* **1989**, *71*, 265–269.
131 K. R. Stone, US Patent 4880429, **1989**.
132 H. Miwa, T. Matsuda, *J. Vasc. Surg.* **1994**, *19*, 658–667.
133 N. Volpi, M. Cusmano, T. Venturelli, *Biochim. Biophys. Acta* **1995**, *1243*, 49–58.
134 K. Nawa, K. Sakano, H. Fujiwara, Y. Sato, N. Sugiyama, T. Teruuchi, T. Iwamoto, Y. Marumoto, *Biochim. Biophys. Res. Commun.* **1990**, *171*, 729–737.
135 T. D. Bjornsson, P. V. Nash, R. Schaten, *Thromb. Res.* **1982**, *27*, 15–21.
136 D. Bergqvist, B. Lindblad, T. Mätzsch, *Heparin and related Polysaccharides*, Plenum Press, New York, **1992**.
137 H. J. Hanselmann, *Best Pract. Res. Clin. Rheumatol.* **2001**, *15*, 595–607.
138 J. Y. Reginster, G. Gillot, O. Bruyere, Y. Henrotin, *Curr. Rheumatol. Rep.* **2000**, *2*, 472–477.
139 F. Ronca, L. Palmieri, P. Panicucci, *Osteoarthritis & Cartilage* **1998**, *6*, 14–21.
140 M. Konlee, *Positive Health News* **1998**, *17*, 4–7.
141 C. S. Osborne, W. H. Reid, M. H. Grant, *Biomaterials* **1999**, *20*, 283–290.
142 K. Meyer, J. W. Palmer, *J. Biol. Chem.* **1934**, *107*, 629–634.
143 J. H. Fessler, L. I. Fessler, *Proc. Natl. Acad. Sci. USA* **1966**, *56*, 141–147.
144 J. E. Scott, *The Biology of Hyaluronan*, Wiley, England, **1989**.
145 J. E. Scott, C. Cummings, A. Brass, Y. Chen, *Biochem. J.* **1991**, *274*, 699–705.
146 J. E. Scott, F. Heatley, *Proc. Natl. Acad. Sci. USA* **1999**, *96*, 4850–4855.
147 T. C. Laurent, J. R. Fraser, *Faseb J.* **1992**, *6*, 2397–2404.
148 P. L. DeAngelis, *Cell. Mol. Life Sci.* **1999**, *56*, 670–682.
149 J. A. Ripellinio, M. Bailo, R. U. Margolis, R. K. Margolis, *J. Cell Biol.* **1988**, *106*, 845–855.
150 P. S. Eggli, W. Graber, *J. Histochem. Cytochem.* **1995**, *43*, 689–697.
151 T. C. Laurent, J. R. E. Fraser, *Degradation of Bioactive Substance: Physiology and Pathology*, CRC Press, Boca Raton, **1991**.
152 R. K. Reed, K. Lilja, T. C. Laurent, *Acta Physiol. Scand.* **1988**, *134*, 405–411.
153 J. R. Fraser, T. C. Laurent, U. B. Laurent, *J. Intern. Med.* **1997**, *242*, 27–33.
154 M. Markovitz, J. A. Cifonelli, A. Dorfman, *J. Biol. Chem.* **1959**, *234*, 2343–2350.
155 P. Prehm, *Biochem. J.* **1983**, *211*, 191–198.
156 P. H. Weigel, V. C. Hascall, M. Tammi, *J. Biol. Chem.* **1997**, *272*, 13997–14000.
157 V. L. Tlapak-Simmons, E. S. Kempner, B. A Baggenstoss, P. H. Weigel, *J. Biol. Chem.* **1998**, *273*, 26100–26109.
158 V. L. Tlapak-Simmons, B. A. Baggenstoss, T. Clyne, P. Weigel, *J. Biol. Chem.* **1999**, *274*, 4239–4245.
159 A. S. Day, J. K. Sheehan, *Curr. Opin. Struct. Biol.* **2001**, *11*, 617–622.
160 M. I. Tammi, A. J. Day, E. A. Turley, *J. Biol. Chem.* **2002**, *277*, 4581–4584.
161 M. Hlavacek, *J. Biomechanics* **1993**, *26*, 1145–1150.
162 P. Herrlich, J. Sleeman, D. Wainwright, H. Konig, L. Sharman, F. Hilberg, H. Ponta, *Cell Adhes. Commun.* **1998**, *6*, 141–147.

163 B. P. Toole, *Cell Biology of Extracellular Matrix*, Plenum Press, New York, **1981**.
164 P. T. Bryan, T. M. Wright, M. I. Tammi, *J. Biol. Chem.* **2002**, *277*, 4593–4596.
165 M. R. Horton, C. M. McKee, C. Bao, F. Liao, J. M. Farber, J. Hodge-DuFour, E. Pure, B. L. Oliver, T. M. Wright, P. W. Noble, *J. Biol. Chem.* **1998**, *273*, 35088–35094.
166 B. J. Clarris, J. R. E. Fraser, *Exp. Cell Res.* **1968**, *49*, 181–193.
167 B. P. Toole, S. I. Munaim, S. Welles, C. B. Knudson, *The Biology of Hyaluronan*, Wiley, England, **1989**.
168 D. C. West, I. N. Hampson, F. Arnold, S. Kumar, *Science* **1985**, *225*, 1324–1326.
169 R. N. Feinberg, D. C. Beebe, *Science* **1983**, *220*, 1177–1179.
170 C. M. McKee, C. J. Lowenstein, M. R. Horton, J. Wu, C. Bao, B. Y. Chin, A. M. Choi, P. W. Noble, *J. Biol. Chem.* **1997**, *272*, 8013–8018.
171 K. Fukuda, H. Dan, M. Takayama, F. Kumano, M. Saitoh, S. Tanaka, *J. Pharmacol. Exp. Ther.* **1996**, *277*, 1672–1675.
172 A. Asari, S. Miyauchi, S. Matsuzaka, T. Ito, E. Kominami, Y. Uchiyama, *Arch. Histol. Cytol.* **1998**, *61*, 125–135.
173 S. Gotoh, J. Onaya, M. Abe, K. Miyazaki, A, Hamai, K. Horie, K. Tokuyasu, *Ann. Rheum. Dis.* **1993**, *52*, 817–822.
174 H. Iwata, *Clin. Orthop.* **1993**, *289*, 285–291.
175 E. A. Balazs, *Healon (Sodium Hyaluronate). A Guide to its use in Ophthalmic Surgery*, Wiley, New York, **1983**.
176 S. Miyauchi, K. Horie, M. Morita, M. Nagahara, K. Shimizu, *J. Ocul. Pharmacol. Ther.* **1996**, *12*, 27–34.
177 G. D. Prestwich, K. P. Vercruysse, *PSTT* **1998**, *1*, 42–43.
178 Y. Luo, K. R. Kirker, G. D. Prestwich, *J. Control. Release* **2000**, *69*, 169–184.
179 V. C. Hascall, *Progress in Clinical and Biological Research*, Alan R. Liss, New York, **1983**.
180 V. P. Bhavanandan, K. Meyer, *J. Biol. Hem.* **1968**, *243*, 1052.
181 R. Keller, T. Stein, H. W. Stuhlsatz, H. Greiling, E. Ohst, E. Muller, H.-D. Scharf, *Hoppe-Seyler's Z Physiol. Chem.* **1981**, *362*, 327–336.
182 N. Seno, N. Toda, *Biochim. Biophys. Acta* **1970**, *215*, 544.
183 L. S. Lohmander, V. C. Hascall, M. Yanagishita, K. E. Kuettner, J. H. Kimura, *Arch. Biochem. Biophys.* **1986**, *250*, 211–227.
184 K. Meyer, A. Linker, E. A. Davidson, *J. Biol. Chem.* **1953**, *205*, 611–616.
185 N. Seno, K. Meyer, B. Anderson, P. Hoffman, *J. Biol. Chem.* **1965**, *240*, 1005–1010.
186 J. E. Scott, *TIBS*, **1987**, *12*, 318–321.
187 E. J. Thonar, R. F. Meyer, R. F. Dennis, M. E. Lenz, D. Maldonado, J. R. Hassell, A. T. Hewitt, W. J. Stark, E. L. Stock, K. E. Kuettner, G. K. Klintworth, *Am. J. Ophthalmol.* **1986**, *102*, 561–569.
188 M. B. E. Sweet, A. Coelho, C. Schnitzler, T. J. Schnitzer, M. E. Lenz, I. Jakim, K. E. Kuettner, E. J. Thonar, *Arthritis Rheum.* **1988**, *31*, 648–652.
189 J. M. Williams, C. Downey, E. J. Thonar, *Arthritis Rheum.* **1988**, *31*, 557–560.
190 C. Balduini, G. De Luca, A. A. Castellani, *Keratan Sulfate Chemistry, Biology, Chemical Pathology*, The Biochemical Society, London, **1989**.
191 I. Axelsson, D. Heinegard, *Biochem. J.* **1975**, *145*, 491–500.
192 C. Cintron, H. Schneider, C. Kublin, *Exp. Eye Res.* **1973**, *17*, 251–259.
193 G. L. Klintworth, F. S. Vogel, *Am. J. Pathol.* **1964**, *45*, 565–576.
194 A. Gardner, *Invest. Ophthalmol.* **1969**, *8*, 475–483.
195 K. Nakazawa, J. R. Hassell, V. C. Hascall, L. S. Lohmander, D. A. Newsome, J. Krachmer, *J. Biol. Chem.* **1984**, *259*, 13751–13757.
196 T. O. Akama, J. Nakayama, K. Nishida, N. Hiraoka, M. Suzuki, J. McAuliffe, O. Hindsgaul, M. Fukuda, M. N. Fukuda, *J. Biol. Chem.* **2001**, *276*, 16271–16278.
197 H. Wang, T. Toida, Y. S. Kim, I. Capila, R. E. Hileman, M. Bernfield, R. J. Linhardt, *Biochem. Biophys. Res. Commun.* **1997**, *235*, 369–373.
198 S. J. Wu, M. W. Chun, K. H. Shin, T. Toida, Y. Park, R. J. Linhardt, Y. S. Kim, *Thrombosis Res.* **1998**, *92*, 273–281.

199 E. Percival, R. H. McDowell, *Chemistry and Enzymology of Marine Algal Polysaccharides*, Academic Press, New York, **1967**.

200 T. J. Painter, *The Polysaccharides*, Academic Press, New York, **1983**.

201 M. Nagaoka, H. Shibata, I. Kimura-Takagi, S. Hashimoto, R. Aiyama, S. Ueyama, T. Yokokura, *BioFactors* **2000**, *12*, 267–274.

202 G. K. SeGall, W. J. Lennarz, *Dev. Biol.* **1979**, *71*, 33–48.

203 P. S. A. Mourão, I. G. Bastos, *Eur. J. Biochem.* **1987**, *166*, 639–645.

204 M. S. Patankar, S. Oehninger, T. Barnett, R. L. Williams, G. F. Clark, *J. Biol. Chem.* **1993**, *268*, 21770–21776.

205 G. Bernardi, G. F. Springer, *J. Biol. Chem.* **1988**, *263*, 75–80.

206 S. Colliec, A. M. Fischer, J. Tapon-Bretaudiere, C. P. Boisson, P. Durand, J. Jozefonvicz, *Thromb. Res.* **1991**, *64*, 143–154.

207 T. Nishino, T. Nagumo, *Carbohydr. Res.* **1992**, *229*, 355–362.

208 M. C. Mahony, S. Oehninger, G. F. Clark, A. A. Acosta, G. D. Hodgen, *Contraception* **1991**, *44*, 657–665.

209 M. Baba, R. Snoeck, R. Pauwels, E. DeClerq, *Antimicrob. Agents Chemother.* **1988**, *32*, 1742–1745.

210 C. Foxall, S. R. Watson, D. Dowbenko, C. Fennie, L. A. Lasky, M. Kiso, D. Hasegawa, D. Asa, B. K. Brandley, *J. Cell. Biol.* **1992**, *117*, 895–902.

211 L. Ramsden, C. C. Rider, *Eur. J. Immun.* **1992**, *22*, 3027–3031.

212 T. Kobayashi, K. Honke, T. Miyazaki, K. Matsumoto, T. Nakamura, I. Ishizuka, A. Makita, *J. Biol. Chem.* **1994**, *269*, 9817–9821.

213 S. Oehninger, G. F. Clark, A. A. Acosta, G. D. Hodgen, *Fertil. Steril.* **1991**, *55*, 165–169.

214 M. C. Mahony, S. Oehninger, G. F. Clark, A. A. Acosta, G. D. Hodgen, *Contraception* **1991**, *44*, 657–665.

215 K. E. Arfors, K. Ley, *J. Lab. Clin. Med.* **1993**, *121*, 201–202.

216 D. J. Stancioff, N. F. Stanley, *Proc. Int. Seaweed Symp.* **1969**, *6*, 595.

217 K. A. Pittman, L. Golberg, F. Coulston, *Fd. Cosmet. Toxicol.* **1976**, *14*, 85–93.

218 E. Booth, *Chemical Oceanography*, Academic Press, London, **1975**.

219 H. J. Schwarz, R. W. Kellermeyer, *Proc. Soc. Exp. Biol. Med.* **1969**, *132*, 1021–1024.

220 T. H. Noda, H. Amano, K. Arashima, S. Hashimoto, K. Nisizawa, *Nippon, Suisan, Gakkaishi* **1989**, *55*, 1265–1271.

221 G. Yu, H. Guan, A. S. Ioanoviciu, S. A. Sikkander, C. Thanawiroon, J. K. Tobacman, T. Toida, R. J. Linhardt, *Carbohydr. Res.* **2002**, *337*, 433–440.

222 J. F. Kennedy, C. A White, *Bioactive Carbohydrates*, Ellis Horwood Limited, New York, **1983**.

223 S.-I. Nishimura, N. Nishi, S. Tokura, *Carbohydr. Res.* **1986**, *156*, 286–292.

224 C. R. Rickets, K. W. W. H. Walton, US Patent 2715091, **1955**.

225 A. Osol, *16th Remington's Pharmaceutical sciences*, Mack Publishing Co., Pennsylvania, **1980**.

226 M. Windholz, *The Merck Index, 9th Ed.*, Merck Publishing Co., Rahway, **1976**.

227 G. Oshima, *Thromb. Res.* **1988**, *49*, 353–361.

228 L. B. Jaques, L. M. Hiebert, S. M. Wice, *J. Lab. Clin. Med.* **1991**, *117*, 122–130.

229 G. Skjak-Braek, *Int. J. Biol. Macromol.* **1986**, *8*, 330–336.

230 M. Asada, M. Sugie, M. Inoue, K. Nakagomi, S. Hongo, K. Murata, S. Irie, T. Takeuchi, N. Tomizuki, S. Oka, *Biosci. Biotech. Biochem.* **1997**, *61*, 1030–1032.

231 R. Langer, R. J. Linhardt, M. Klein, P. M. Galliher, C. L. Cooney, M. M. Flanagan, *Biomaterials: Interfacial Phenomenon and Applications*, ACS Press, Washington D.C., **1982**.

232 D. L. Taylor, T. M. Brennan, C. G. Bridges, M. J. Mullins, A. S. Tyms, R. Jackson, A. D. Cardin, *Antiviral Res.* **1995**, *28*, 159–173.

233 F. Hawking, *Pharmacol. Chemother.* **1978**, *15*, 289–322.

234 G. T. Tan, A. Wickramasinghe, S. Verma, R. Singh, S. H. Hughes, J. M. Pezzuto, M. Baba, P. Mohan, *J. Med. Chem.* **1992**, *35*, 4846–4853.

235 R. Kisilevsky, L. J. Lemieux, P. E. Fraser, X. Kang, P. G. Hultin, W. A. Szarek, *Nature Med.* **1995**, *1*, 143–148.

236 K. Kasuraya, N. Ikushima, N. Takahashi, T. Shoji, H. Nakashima, N. Yamamoto, T. Yoshida, T. Uryu, *Carbodr. Res.* **1994**, *260*, 51–61.
237 B. K. Yeh, A. N. Plotnikov, A. V. Eliseenkova, D. Green, J. Pinnel, T. Polat, A. Gritli-Linde, R. J. Linhardt, M. Mohammadi, *Molec. Cell Biol.* **2002**, *22*, 7184–7192.
238 M. A. Mitchell, J. W. Wilks, *Ann. Reports in Medicinal Chemistry*, Academic Press Inc., San Diego, **1992**.
239 A. Gerloczy, T. Hoshino, J. Pitha, *J. Pharm. Sci.* **1994**, *83*, 193–196.
240 V. Ferro, C. Li, K. Fewings, M. C. Palemo, D. Podger, R. J. Linhardt, T. Toida, *Carbohydr. Res.* **2002**, *337*, 139–146.
241 C. R. Parish, C. Freeman, R. J. Brown, D. J. Francis, W. B. Cowlen, *Cancer Res.* **1999**, *59*, 3433–3441.
242 M. D. Hulett, C. Freeman, B. J. Hamdorf, R. T. Baker, M. J. Harris, C. R. Porish, *Nat. Med.* **1999**, *5*, 803–809.
243 G. Yu, N. S. Gunay, R. J. Linhardt, T. Toida, J. Fareed, D. A. Hoppensteadt, H. Shadid, V. Ferro, C. Li, K. Fewings, M. C. Polermo, D. Podger, *Eur. J. Med. Chem.* **2002**, *37*, 783–791.
244 M. F. Scully, M. Kumudini, K. M. Weerasinghe, V. Ellis, B. Djazaeri, V. V. Kakkar, *Thrombosis Res.* **1983**, *31*, 87–97.
245 F. Dol, P. Sie, D. Dupouy, B. Boneu, *Thromb. Haemostas.* **1986**, *56*, 295–301.
246 H. P. Wessel, *Tet. Lett.* **1990**, *31*, 6863–6866.
247 J. Giedrojc, K. Knipinski, H. K. Breddin, M. Bieawiec, *Polish J. Pharmacol.* **1996**, *48*, 317–322.
248 P. D. Ward, T. K. Tippin, D. R. Thakker, *Pharm. Sci. Tech. Today* **2000**, *3*, 346–358.
249 S. Y. Cho, J. S. Kim, H. Li, C. Shim, R. J. Linhardt, Y. S. Kim, *Arch. Pharmacol. Res.* **2002**, *25*, 86–98.
250 T. M. Rivera, A. Leone-Bay, D. R. Patow, H. R. Leipold, R. A. Baughman, *Pharm. Res.* **1997**, *14*, 1830–1834.
251 Y. Ichikawa, R. Monden, H. Kuzuhara, *Carbohydr. Res.* **1988**, *172*, 37–64.
252 K. M. Koeller, C. H. Wong, *Glycobiology* **2000**, *10*, 1157–1169.
253 M. D. Burkart, M. Izumi, E. Chapman, C. H. Lin, C. H. Wong, *Journal of Organic Chemistry* **2000**, *65*, 5565–5574.
254 P. H. Seeberger, W.-C. Haase, *Chem. Rev.* **2000**, *100*, 4349–4394.

16
A New Generation of Antithrombotics Based on Synthetic Oligosaccharides
MAURICE PETITOU and JEAN-MARC HERBERT

16.1
Introduction

Venous thrombosis, the formation of a blood clot, in the deep-veins of the leg is a common event in surgical patients. It is often painless (silent), and therefore undetected, but it may become a life-threatening condition when a piece of the clot (an embol) migrates to the heart, then to the lung, and triggers a pulmonary embolism. Venous thromboembolism (VTE) is thus a major cause of mortality and morbidity, with an incidence of 1.84/1000 (including deep vein thrombosis and pulmonary embolism), in western countries [1]. Epidemiological studies have demonstrated the essential role of prophylaxis in this domain [2]. They also indicate that aging of the population will increase the incidence of this illness.

Blood hypercoagulation plays a major role in VTE, and anticoagulants are prescribed to patients suffering from or susceptible to this disease. Two types of agents are used: vitamin K antagonists (VKAs) [3], which decrease the levels of some proteins involved in blood coagulation (blood coagulation factors, the carboxylation of which is under the influence of vitamin K), and heparin [4] and – particularly – low molecular weight heparins (LMWHs), which activate the anticoagulant protein antithrombin. None of these agents is perfect: VKAs have a delayed onset of action and the treatment is difficult to manage; heparin and low molecular weight heparin are complex polysaccharide mixtures of animal origin, must be injected once or twice daily, and the serious, but uncommon, side-effect Heparin Induced Thrombocytopenia (HIT) [5] necessitates careful monitoring of platelet count during the treatment. Clearly, there is room for improvement in antithrombotic therapy.

Current directions of research [6] in the field of antithrombotics include the search for agents that selectively inhibit blood coagulation factors (either directly or indirectly, through the intermediacy of plasma proteins such as antithrombin or heparin cofactor II), can be administered orally, and are of totally synthetic origin. In this article we focus on the approach that we have developed, based on the assumption that heparin and low molecular weight heparins being excellent drugs, if we could understand their precise mechanism of action we might then develop still better antithrombotic agents, particularly if the new active principles were entirely accessible through chemical synthesis. This approach turned out to be successful when the first agent produced in this way appeared to exhibit excellent antithrombotic properties in patients, superior to the reference treatment and with a better efficacy/safety ratio [7]. Here we successively discuss the heparin type of drugs, the discovery of these new antithrombotic agents that selectively inhibit blood coagulation factor Xa, and the discovery of another type of synthetic agents able to inhibit both thrombin and factor Xa.

16.2
Heparin and Its Mechanism of Action as an Antithrombotic Agent

16.2.1
Heparin, a Complex Polysaccharide with Blood Anticoagulant Properties

The use of heparin as an antithrombotic agent in man has long been impaired by the difficulties encountered in purifying it. The structure is now known [8], and it is understood that its long, negatively charged polysaccharide chains (Fig. 16.1) are prone to interact with a great variety of biological components, so the compound is very difficult to refine to the degree necessary for injection in humans. Following the discovery of efficient purification procedures, clinical trials in Sweden and in Canada in the late 1930s showed the antithrombotic properties of heparin, which has since become a widely used antithrombotic agent. However, because of the numerous potential interactions with the drug, side effects were observed (bleeding, platelet activation), resulting in rather poor tolerance of the treatment. In a first attempt to address this issue, clinicians tried reducing the doses, which, surprisingly, resulted in an improved tolerance and a sustained antithrombotic effect. As discussed in Section 16.2.2, this apparent paradox triggered much speculation.

The mechanism of the anticoagulant action of heparin remained essentially unknown until it was clearly shown in 1973 that the plasma protein antithrombin is involved as a cofactor [9]. Human antithrombin is a 432 amino acid glycoprotein present in plasma at fairly high concentrations (about 2.5 µM), and is regarded as a "suicide" inhibitor of several blood coagulation factors (see Section 16.2.3).

16.2.2
Which Coagulation Factor must be Inhibited?

After the identification of antithrombin as a heparin cofactor, many studies were initiated in order to identify which coagulation factors were inhibited by heparin and antithrombin [10]. Today, the list comprises factors IIa (thrombin), VIIa, IXa, Xa, and XIa (see Fig. 16.2) [11]. The rate of inhibition of these factors by antithrombin is considerably increased by heparin. This is particularly true for thrombin, as its rate of inhibition increases from $0.7–1.4\times10^4$ $M^{-1}s^{-1}$ to $1.5–4\times10^7$ $M^{-1}s^{-1}$ in the presence of optimal concentrations of the polysaccharide [12]. Thrombin has thus been regarded as the key enzyme in the antithrombotic effect of heparin.

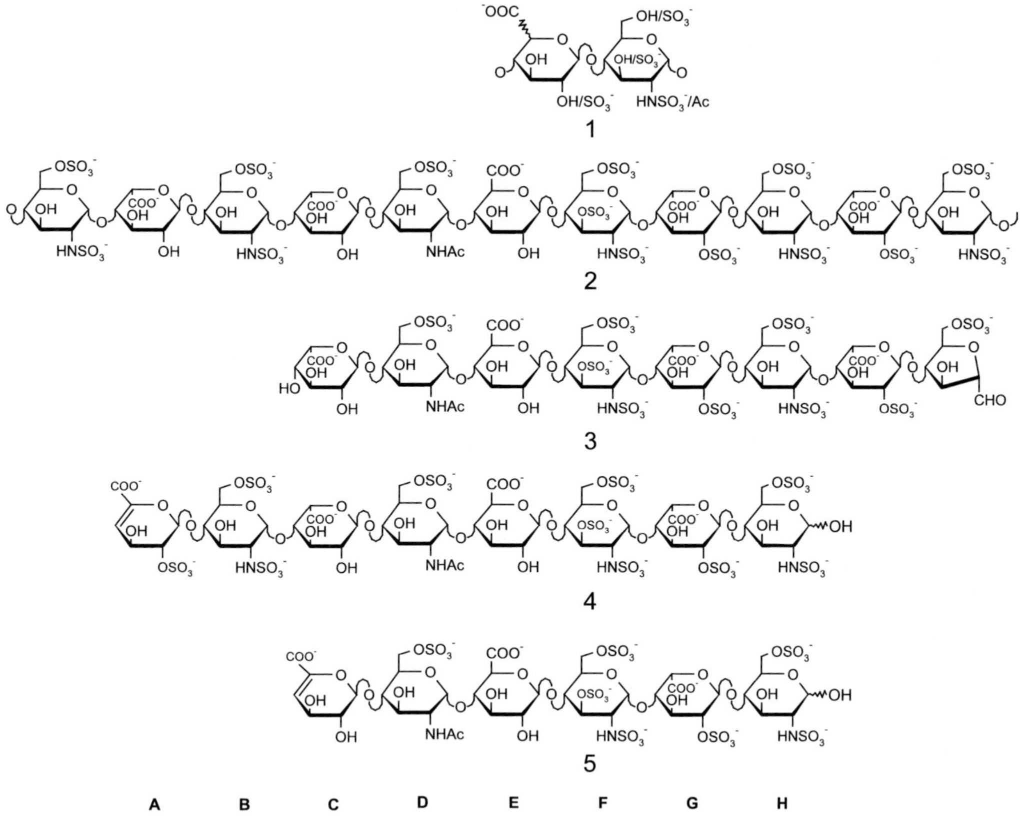

16.2 Heparin and Its Mechanism of Action as an Antithrombotic Agent | 443

Fig. 16.1 Structure of heparin. Heparin is made up of a repeating basic disaccharide unit (**1**) containing a uronic acid and a glucosamine. The structure **2** represents a hypothetical heparin molecule containing the antithrombin binding sequence DEFGH. This sequence was inferred from comparison of the octasaccharides **3** and **4** and the hexasaccharide **5**, which all bound to and activated antithrombin. However the pentasaccharide DEFGH could only be produced by chemical synthesis, which definitely established the structure of the antithrombin binding sequence.

However, when it was discovered that the low dose heparin regimen also resulted in excellent antithrombotic effects, inhibition of factor Xa was proposed as another key element in the antithrombotic action of heparin [13]. This proposal was further supported by the discovery that low molecular weight heparins, that preferentially inhibited factor Xa, were excellent antithrombotic agents. The pharmacology community was then split into two groups: on the one hand the advocates for the key role of factor Xa, and on the other hand, those who, sticking to the results of pharmacological studies in animal models of thrombosis, emphasized the role of thrombin inhibition and claimed that exclusive inhibition of factor Xa would probably result in very poor antithrombotic effects [14]. Hence the pending question concerning the rationale for developing anti-factor Xa agents as antithrombotic compounds.

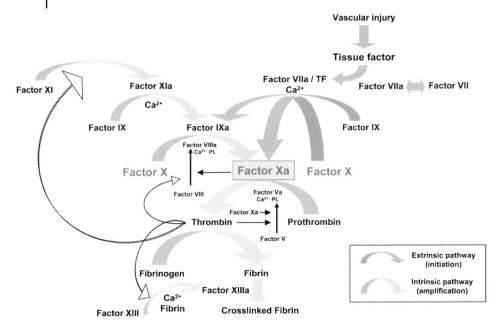

Fig. 16.2 Blood coagulation. Blood coagulation results from a cascade of enzymatic reactions. The system is regulated by serine proteinase inhibitors such as antithrombin III, which irreversibly inactivates factors IIa (thrombin), VIIa, IXa, and XIa.

16.2.3
The Structure of Heparin in Relation to Antithrombin Activation

While the debate about the nature of the coagulation factor that must be inhibited continued, structural studies on heparin revealed the presence of an antithrombin binding site responsible both for the binding to antithrombin and for the selective activation of this latter with respect to factor Xa inhibition [15]. More precisely, it was found that binding of this isolated pentasaccharide sequence (Fig. 16.1) induced a conformational change in the active site loop of antithrombin and that such an allosteric activation increased the bimolecular rate constant of the reaction of antithrombin with factor Xa by about 300 times, whereas it only multiplied the rate constant for thrombin inhibition by a factor of two [16]. It was also found that longer heparin sequences made up of 14–20 saccharide units were required for observation of a significant effect on thrombin inhibition by antithrombin according to a template mechanism [12] (see Scheme 16.1). Later, more precise studies demonstrated that at least a pentadecasaccharide was required (see Section 16.4). As a matter of fact, the respective role of the pentasaccharide sequence and the rest of the heparin chain with respect to inhibition of several coagulation enzymes is still under debate.

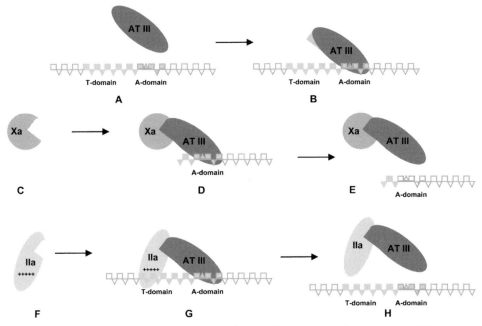

Scheme 16.1 Heparin activation of antithrombin III. Heparin (squares and triangles) binds to antithrombin through its antithrombin binding site (hatched units) and triggers a conformational change in antithrombin (B). Activated antithrombin can inhibit factor Xa (C), thrombin (F), and other coagulation factors. The inhibition proceeds through the formation of a ternary complex (D, G), in which, except for factor Xa (C), both the enzyme and antithrombin must simultaneously bind to heparin. This explains why short heparin fragments can only inhibit factor Xa. The enzyme-inhibitor complex then detaches (E, H) from the heparin molecule, which can enter another catalytic cycle.

16.2.4
The Limitations of Heparin

Although an essential drug, heparin has some limitations associated with the interactions that the long, negatively charged polysaccharic chains are susceptible to developing with a great number of biological components [17]. The shorter chains contained in low molecular weight heparin preparations give fewer undesirable interactions and are better tolerated. The interaction with platelet factor 4 (PF4), a protein released from the a-granules of platelets, is probably the cause of the poor antithrombotic activity of heparin in arterial thrombosis, in which platelet activation (and release) occurs. This interaction, or more precisely the resulting heparin-PF4 complexes recognized by some immunoglobulins, is at the origin of HIT [18]. It has also been shown that the release from endothelium of proteins such as PF4, lipoprotein lipase, and tissue-factor pathway inhibitor is directly related to the length and the charge of the heparin molecules. In the opposite direction, heparin molecules are trapped by endothelial cells with an affinity directly related to charge and length, causing complex and variable pharmacokinetics. The same factors, negative charge density and chain length, also explain the different susceptibility of different low molecular weight heparins with respect to neutralization by the heparin antidote protamine [19]. It is to be expected that heparin mimetics with well defined chemical structures should show simpler and more reproducible pharmacological profiles.

16.3
Synthetic Pentasaccharides, Selective Factor Xa Inhibitors, are Antithrombotic Agents

16.3.1
New Synthetic Oligosaccharides Required in Order to Validate a Pharmacological Hypothesis

While several pharmacologists were highly skeptical about selective inhibition of blood coagulation factor Xa resulting in a clinically relevant antithrombotic effect, our own observations in animal models of venous thrombosis with the octasaccharides described in Section 16.2.3 clearly suggested the validity of this approach. We were further encouraged to investigate it after our experience with low molecular weight heparins, for which similar concerns had initially been raised by some pharmacologists and clinicians, the same that later turned out to be among the best advocates of low molecular weight heparin therapy. In the present case, however, there was another obstacle: in order to test the hypothesis we first had to synthesize the active pentasaccharide required, possibly the most complex oligosaccharide ever synthesized. At the start of this work, seven publications dealing with synthesis in the field of heparin had appeared, between 1972 and 1976 [20]. However, in so far as the major structural features in heparin are the presence of N-sulfo-D-glucosamine, L-iduronic acid, and sulfate esters, the products prepared in these publications did not conform to criteria required for their designation as heparin fragments, and so the very first successful synthesis in the field may thus be considered to have reported by us in 1983 [21].

16.3.2
A Strategy for the Synthesis of an Active Pentasaccharide

The total synthesis of the pentasaccharide sequence represented a real challenge, with the presence of iduronic acid, glucuronic acid, and glucosamine, most of them N- and/or O-sulfonated on selected positions. This synthesis was primarily a problem of oligosaccharide synthesis, requiring the creation of interglycosidic bonds between uronic acid and glucosamine units, but it was complicated by the need to introduce sulfonate groups on the final oligosaccharide, and the protecting groups strategy had to allow the construction of the carbohydrate backbone and the selective introduction of these substituents. Actually, in the vast majority of syntheses so far reported, benzyl ethers have been utilized as permanent pro-

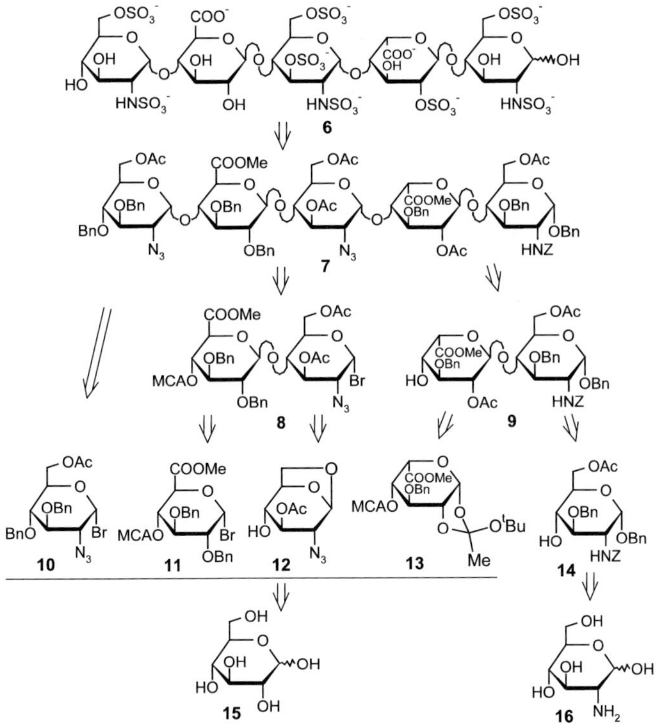

16.3 Synthetic Pentasaccharides, Selective Factor Xa Inhibitors, are Antithrombotic Agents

Fig. 16.3 The first synthesis of the pentasaccharide DEFGH was carried out from glucose and glucosamine.

tective groups, acetyl esters as semi-permanent and several others (chloroacetyl, levulinyl, allyl, methoxybenzyl, etc.) as temporary protective groups. A protected oligosaccharide was synthesized as a synthetic equivalent to the target oligosaccharide according to this selection (Fig. 16.3, in which acetyl groups substitute for O-sulfonato, benzyl for hydroxyl, azido and benzyloxycarbonyl for N-sulfonato, and methoxycarbonyl for carboxyl).

From the standpoint of oligosaccharide synthesis, the two classical types of glycosidic linkages are present in the target molecule: 1,2-*trans*, in the case of uronic acids, and 1,2-*cis*, in the case of glucosamine. We used a strategy by which disaccharide building blocks containing the 1,2-*trans* type of linkage were synthesized first, and were then linked together through the creation of the 1,2-*cis* bonds by a method developed by Paulsen and co-workers [22], which allows the synthesis of α-D-glucosaminides in high yields and with very good stereoselectivity. The applicability of this method in the field of heparin was, however, dependent on the reactivity, under these reaction conditions, of the hydroxy groups at position 4 in glucuronic and iduronic acids. The synthesis of model disaccharides indicated that this reactivity was high [23].

16.3.3
A Strategy for the Synthesis of the First Pentasaccharide

The synthesis of the first pentasaccharide [24] is depicted in Fig. 16.3. It having been shown that the coupling of disaccharide building blocks was potentially feasible, the problem was to elaborate building blocks bearing protective groups on appropriate positions. This was initially achieved by condensation of two suitably protected monosaccharides: a uronic acid, and a glucosamine derivative or a glucosamine precursor.

The D-glucuronic acid derivative **11** was obtained from a suitably protected D-glucose derivative, after Jones' oxidation at C-6 and bromination at C-1. Building block **8** was prepared from bromide **11** and alcohol **12** in the presence of silver carbonate in dichloromethane. The obtained disaccharide was then converted into the glycosyl donor **8** by acetolysis, followed by bromination at C-1.

The synthesis of the L-iduronic acid derivative **13** was achieved by epimerization of a suitably protected glucofurano 5-O-trifluoromethylsulfonate at C-5. The use of sodium trifluoroacetate as a nucleophile allowed easy conversion of the furano derivative into the corresponding pyrano form. The ortho-ester **13**, required in the glycosylation reaction in order to obtain the α-L configuration, also allowed selective protection by an acetyl group at C-2' in **9** while the hydroxyl group at C-4' was ready for glycosylation by **8**.

Coupling of **8** and **9** gave a tetrasaccharide that, after selective cleavage of the chloroacetyl group, was condensed with **10** to give the fully protected pentasaccharide **7**. The introduction or demasking of the functional groups present on the target molecule was achieved by the following sequence of deprotection and functionalization steps. Saponification was used to cleave the acetyl groups and the carboxylic acid blocking groups (methyl esters). O-Sulfonato groups were then introduced by use of mild sulfonating agents such as trimethylamine or pyridine/sulfur trioxide complexes [25]. After hydrogenolysis, to remove benzyl protecting groups and to generate the amino functions of glucosamine, N-sulfonation of glucosamine units was achieved in water at basic pH with the same sulfating agent as used for O-sulfonation. The final purification was easily performed by gel filtration and ion-exchange chromatography.

The synthesis just summarized represented the first example in the field. Variations of this process have been reported, by us and by others, aiming at a more efficient production of similar pentasaccharides (see Section 16.3.4). Among these variations, one might mention the introduction of the imidate method to couple the blocks, the use of neutral disaccharides as starting material for the preparation of building blocks, and the coupling of two partially functionalized monosaccharides followed by modification at the disaccharide level [26, 27].

16.3.4
Activation of Antithrombin: Structure/Activity Relationship

By the procedure described in Section 16.3.3 we were able to obtain the pentasaccharide **6**. This compound bound to antithrombin with the same affinity as heparin and selectively catalyzed inhibition of factor Xa [28]. Its antithrombotic activity could be demonstrated in animal models of venous thrombosis. To improve the efficiency of the synthesis, and also to stabilize the molecule with a view to pharmaceutical development, we also synthesized the methyl glycoside of this pentasaccharide **17**, which displayed identical biological properties. To study the structure/activity relationship of the interaction with antithrombin we also obtained many analogues with various sulfation patterns. We were thus able to establish that, of the various negatively charged groups, some played a prominent role while, in contrast, others were not required (Fig. 16.4). Notably, we were able to demonstrate the critical role played by the O-sulfonate group at position 3 of the glucosamine unit D, the most salient structural feature of the antithrombin binding sequence. From all these studies we were able to conclude that **17** played the same role as the antithrombin binding sequence of heparin regarding antithrombin activation [27]. Nevertheless, there is no evidence so far that the precise structure of **17** – that is, **17** with all its sulfate groups – is present in any heparin molecule [29].

16.3.5
Clinical Trials Results

Pentasaccharides such as **17** potentially represented a new class of pharmacological agents that selectively inhibit blood coagulation factor Xa. As always in such a situation, the validity of this new pharmacological approach had to be confirmed through clinical trials in humans; this was particularly necessary even though the heparin class of drugs was already in use, because as mentioned above (see Section 16.3.1) there were doubts about the antithrombotic properties of such compounds. We thus embarked on the chemical synthesis of the larger amounts of the pentasaccharide **17** (10–100 g) required to perform the mandatory preclinical and clinical studies. The innocuity of **17** having been demonstrated in animals,

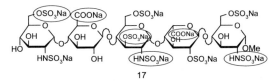

Fig. 16.4 Fondaparinux sodium. The methyl glycoside of the pentasaccharide **6** displayed similar biological properties, particularly with regard to factor Xa inhibition. Structure/activity relationship studies indicated that four sulfate groups (circled) played a critical role in the interaction with antithrombin. The two carboxyl groups and the configuration at C5 of the uronic acid units were also critical.

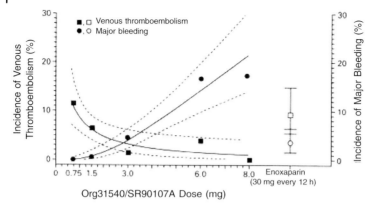

Fig. 16.5 Phase II clinical trial. This clinical trial definitively established that selective factor Xa inhibition results in an antithrombotic effect and that the pentasaccharide **17** had the potential to improve the risk/benefit ratio significantly for the prevention of venous thromboembolism. From these data the dose of 2.5 mg was chosen for phase III clinical trials.

clinical trials could start. Clinical phase I studies revealed that **17** possessed a half-life of 17 h in humans [30], a very interesting result, indicating that the precious material would remain in plasma for a relatively long time to carry out its beneficial work.

Then came the decisive phase II study, to demonstrate the clinical efficiency of a selective factor Xa inhibitor. This trial was conducted in North America, in 933 patients undergoing operations for hip replacement. It was carried out in comparison with the most widely used therapy: the low molecular weight heparin enoxaparin. As shown in Fig. 16.5, the pentasaccharide, injected in patients at doses from 0.75 mg to 3 mg, produced a better antithrombotic effect than the reference drug injected twice daily [7a]. From this study it was decided to use the dose of 2.5 mg once daily to perform the mandatory phase III trials requested to apply for a marketing authorization. These studies [7b–e], performed in patients undergoing operations for hip fracture, hip replacement, or major knee surgery, once again demonstrated the superiority of pentasaccharide treatment over low molecular weight heparin. Thus, meta-analysis indicated a 50% relative risk ratio improvement and equal safety (clinically relevant bleeding) in relation to the group of patients treated with enoxaparin. After these successful clinical trials, the pentasaccharide, the international chemical non-proprietary name of which is "fondaparinux sodium", was submitted for worldwide marketing and is available under the brandname Arixtra®. The future will tell if this pentasaccharide, a selective inhibitor of factor Xa devoid of non-specific interactions with blood and vessel components, will fulfil its promise as a drug.

16.3.6
The Second Generation of Antithrombotic Pentasaccharides

Chemical synthesis allows one to prepare any oligosaccharide required in order to investigate structure/activity relationships of natural structures, but it also offers medicinal chemists the possibility to create new structures endowed with agonistic or antagonistic properties. In the course of our research on antithrombotic oligosaccharides [27] we thus found that some sulfonate groups in the pentasaccharide **17** were critical for binding to antithrombin (Fig. 16.4), but we also found that others could be removed or, on the other hand, new ones introduced, sometimes resulting in improved affinity for antithrombin [31]. We also discovered that N-sulfonates could be replaced by O-sulfonates, and that free hydroxyl groups could be masked, in particular with alkyl groups, or simply removed (giving in this latter case the corresponding similarly active deoxy derivatives).

All these observations were put together to create a family of "non-glycosamino" glycan derivatives [32] (Fig. 16.6) displaying biological properties similar to those of the oligosaccharides described in Section 16.3.4, but the preparation of which was greatly simplified for two main reasons related to the pattern of protective groups used during the synthesis: (i) discrimination between O- and N-sulfonates was no longer required, and (ii) the free hydroxy groups could be permanently blocked as alkyl moieties, particularly methyl groups. By this approach, the pentasaccharide **18**, the counterpart of **17** in this series, was prepared and shown to possess the same pharmacological profile as **17** [33]. In this family, **19** was identified as possessing an affinity for antithrombin in the sub-nanomolar range, and a greatly increased elimination half-life [34, 35]. This compound (SanOrg34006, or idraparinux), is currently undergoing clinical trials in humans. Most remarkably, its use should allow a once a week administration for an efficient antithrombotic effect.

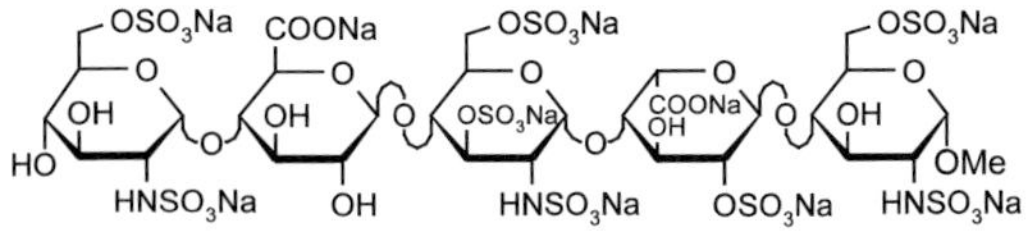

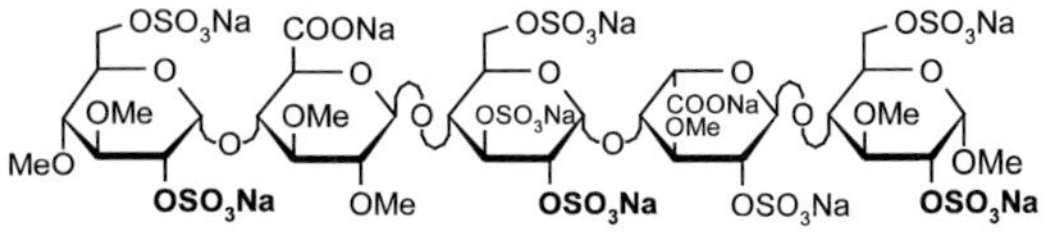

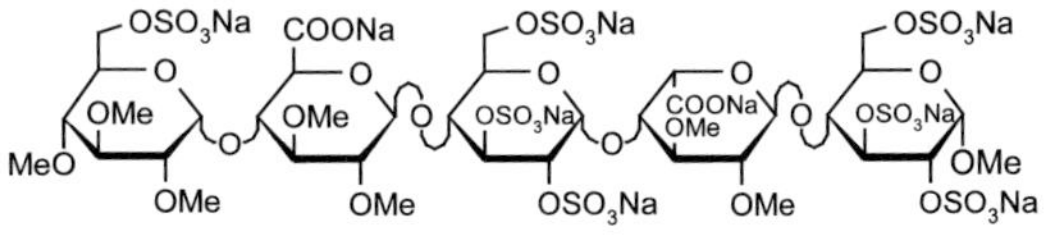

Fig. 16.6 "Non-glycosaminoglycan" anti-factor Xa pentasaccharides. These compounds display similar biological properties and are easier to synthesize than their hydroxylated and N-sulfonated counterparts. Chemical manipulations also allowed the modification of the pharmacological profile. Thus, SanOrg34006 (or idraparinux sodium) **19** has a elimination half-life of about 120 hours in humans, compared to 17 hours for **17**.

16.4
Synthetic Thrombin-Inhibiting Oligosaccharides: The Next Generation?

The pentasaccharides described above represent a new family of antithrombotic drugs, devoid of thrombin inhibitory properties, and free of interactions with blood and vessel components. However, having largely simplified the chemistry of heparin mimetics, we believed we were in a position to go further and to synthesize the longer oligosaccharides required to also inhibit thrombin according to the template mechanism of thrombin inhibition (Scheme 16.1). To produce a drug out of such a compound, we had to keep in mind that undesired interactions are, like thrombin inhibition, directly related to the charge and the size of the molecules [8]. Consequently, we had to design structures capable of discriminating between thrombin and other proteins, particularly PF4, which is known to interact with heparin in a non-specific manner. Chemical synthesis, through fine-tuning of the lengths and the charges of the synthetic oligosaccharides, should allow us to identify such structures. Our approach was also influenced by our desire to obtain a drug substance, which meant that we tried to keep the chemistry as simple as possible to facilitate scaling up of the preparation.

16.4.1
First Approach: Oligomerization of a Disaccharide

As shown in Scheme 16.1, the structure of a molecule capable of inhibiting thrombin comprises an antithrombin binding domain (A-domain) prolonged by a thrombin binding domain (T-domain) [36], which represents, according to literature data [37], between fourteen and twenty saccharide units. Looking at Scheme 16.1, one can imagine that one of the two possible ways of elongating the A-domain with a T-domain (either at the reducing or at the non-reducing end) should result in thrombin inhibition. In a first approach to the synthesis of such molecules, our desire to use a "short" synthetic route, caused us to think that an A-domain, since it was negatively charged, might also serve as a T-domain, and that a continuum of A-domains would necessarily feature the proper relative positions of the two domains, since antithrombin could bind at either end of the molecule, and thrombin be attracted by the remaining part of the chain. Thus, with such a family of oligosaccharides, thrombin inhibition should be observed as soon as the

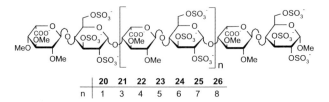

Fig. 16.7 Oligosaccharides synthesized to assess the size of a thrombin inhibitory oligosaccharide.

Tab. 16.1 Biological activities of compounds **20–32**.

	n (see Fig. 16.7)	Coagulation factor inhibition		Antithrombotic effect	
		Factor Xa (units mg^{-1})	Factor IIa (IC$_{50}$ ng mL^{-1})	Venous (ED$_{50}$ µg kg^{-1})	Arterial (ED$_{50}$ µg kg^{-1})
20	1	325±16	>10000	300±70	>1000
21	3	405±32	>10000	280±49	>1000
22	4	360±29	>10000	360±62	>1000
23	5	310±16	>10000	390±45	>1000
24	6	359±29	130±10	150±21	520±140
25	7	270±23	23±4	250±20	610±230
26	8	236±19	6.7±3	110±17	620±130
27	na	370±9	41±3	65.5±3	380±30
28	na	270±8	5.3±0.2	40±4	570±130
29	na	290±29	1.7±0.5	38±9	770±200
30	na	230±16	164±6.5	423±27	>1000
31	na	270±8	5.3±0.3	15±2	70±16
32	na	297±13	4.0±1.0	18±0.1	225±10

molecule is long enough to accommodate antithrombin and thrombin simultaneously.

Having identified the hexasaccharide **20** obtained by trimerization of a basic disaccharide unit and able to bind to and activate antithrombin [38], we synthesized larger homologous oligosaccharides (**21–26**, Fig. 16.7), and tested their anticoagulant properties [39]. All compounds, since they each contained an A-domain, bound to antithrombin and inhibited factor Xa (Tab. 16.1). Compounds **21–23**, shorter than a tetradecasaccharide, were inactive in the thrombin inhibition assay, whereas **24–26** displayed size-dependent activity in this assay, **26** being half as potent as standard heparin. These compounds displayed interesting biological properties, but were neutralized by PF4 and also activated platelets in the presence of plasma from HIT-patients, a test predictive of ability to induce HIT in patients. Thus, in spite of their interesting antithrombotic properties (**26** exhibited an antithrombotic activity similar to that obtained for standard heparin), these compounds were not suitable for drug development.

16.4.2
Second Approach: Molecules Containing Two Identified Domains

The results described in Section 16.4.1 prompted us to explore another family of related molecules possessing a specific A-domain prolonged by a T-domain not recognized by antithrombin. They also clearly showed that the oligosaccharides we were looking for must contain at least 15 or 16 saccharide units. Unlike in the approach described in Section 16.4.1, in which the A-domain and the T-domain were interchangeable, a key issue in the design of these structures was to attach the T-domain

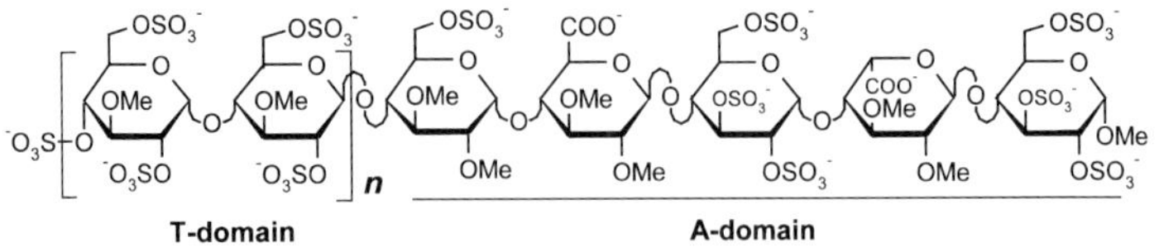

Fig. 16.8 Oligosaccharides containing well identified antithrombin binding and thrombin binding domains.

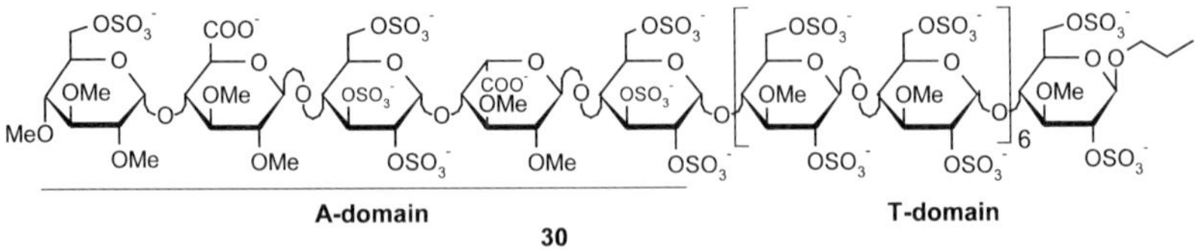

at the correct end of the A-domain to obtain efficient thrombin inhibition (see Scheme 16.1). Modeling studies of the ternary heparin/antithrombin/thrombin complex [36] and crystallography studies [40] suggested attachment of the T-domain at the non-reducing end of the A-domain. Aware that we had to synthesize at least a pentadecasaccharide, we selected as our targets compounds 27–29 (Fig. 16.8) in which the A-domain was a high-affinity analogue of the antithrombin binding sequence [34]. With regard to the T-domain, to mimic the "regular region" of the polysaccharide that binds to the anion-binding exosite II of thrombin [41], and to keep the chemistry simple, we used alternating α- and β-linked 3-O-methyl-2,6-di-O-sulpho-D-glucose units. Biological tests performed on 27–29 (Tab. 16.1) demonstrated that, as expected, they were inhibitors of both factor Xa and thrombin.

With these results to hand, we next wished to establish that the opposite arrangement of A- and T-domains would result in an inactive compound. We therefore synthesized the octadecasaccharide 30 (Fig. 16.8). The affinity for antithrombin and the anti-factor Xa activity of compound 30 were in the range observed for compounds 27–29 (Tab. 16.1) while, in contrast with 27–29, 30 hardly inhibited thrombin in the presence of antithrombin. However, as already noted with compounds 24–26, the anticoagulant activity of 27–29 was neutralized by PF4, indicating that the use of well identified A- and T-domains and reduction of the size down to the minimum still allowing thrombin inhibition was not sufficient to abolish the undesired interaction with PF4.

16.4.3
Introduction of a Neutral Domain

The above biological results were in full agreement with conclusions reached by others [18], who studied the interaction of heparin-like polyanionic compounds with PF4, and concluded that the density of negative charges along the saccharide backbone played a major role, and that the interaction was optimal when the size of the molecule was around that of a hexadecasaccharide. This conclusion was not favorable since it was approximately the size of the thrombin-inhibiting molecules we were hoping to synthesize.

There was, however, a hope of solving the problem. It was based primarily on molecular modeling experiments suggesting that, in heparin, the A-domain and the T-domain were separated by saccharide units not involved in interactions either with thrombin or with antithrombin [42, 43]. In addition, we knew that a tetrasaccharide sequence could constitute a T-domain, and the literature suggested that molecules shorter than an octasaccharide were practically devoid of interaction with PF444. Altogether, this indicated that an oligosaccharide around the size of a hexadecasaccharide, made up of a charged tetrasaccharide or hexasaccharide (T-domain) and a pentasaccharide (A-domain), the two separated by a neutral domain, should display the requested anticoagulant properties while avoiding interaction with PF4.

We synthesized such oligosaccharides [45] (Fig. 16.9) and discovered that they were potent thrombin and factor Xa inhibitors. Most remarkably, the thrombin inhibitory potency was not affected by PF4, even when this was added at a very high concentration (100 µg mL^{-1}). This is the reason why compounds **31** and **32** did not give a positive response in a HIT test, suggesting that they will not induce thrombocytopenia in patients. It may similarly explain the remarkable antithrombotic effects of such compounds in animal models of thrombosis (see Tab. 16.1). Compound **31** was found to be five and ten times more potent than standard heparin in venous and arterial thrombosis models, respectively. It is also worth mentioning that **31** and **32** displayed much less hemorrhagic activity than standard heparin and low molecular weight heparin [46], an observation possibly related to a lack of interaction with von Willebrand factor and fibrinogen, two interactions suspected to participate in the hemorrhagic activity of heparin. In conclusion, substitutes for heparin, endowed with potent antithrombotic activity but devoid of its major side effects, can be obtained by chemical synthesis. Clinical trials will tell whether these synthetic "heparin mimetics" will do better than the synthetic pentasaccharide **17**, currently available as a drug.

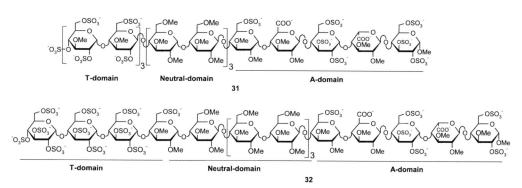

Fig. 16.9 Oligosaccharides comprising an antithrombin binding domain and a thrombin binding domain, separated by a neutral domain.

16.5
The Mechanism of Antithrombin Activation by Synthetic Oligosaccharides

In contrast to the situation with heparin, where the drug has been used for a long time whilst its mechanism of action was ignored, the new antithrombotics described in this article have a well defined mechanism of action. We would like briefly to emphasize here how the study of such well defined chemically synthesized oligosaccharides allowed the mechanism of antithrombin activation to be pinpointed.

It was tempting, looking at the structure of the antithrombin binding sequence of heparin (see Fig. 16.1), to divide it into a trisaccharide representing a very rare sequence in heparin (DEF) and a disaccharide (GH) that accounts for by far the largest part of the polysaccharide chains. We reasoned that antithrombin can bind to this rare sequence along a long heparin chain because the primary binding site is the trisaccharide DEF. Having synthesized this compound and also the other trisaccharide FGH, we were able to show that only DEF could induce the critical conformational change of antithrombin that results in activation [47]. We thus proposed the following precise mechanism of activation: DEF first binds to the protein and induces the conformational change, and in a second step antithrombin is locked in the active conformation by GH, after an appropriate change in conformation of the iduronate ring. This mechanism has been supported by enzyme kinetics experiments [48].

We were also able to demonstrate that the conformation of L-iduronic plays a critical role in the interaction of the above oligosaccharides and antithrombin. Using pentasaccharides containing conformationally locked L-iduronic acid units (through introduction of an intramolecular bridge) we found that only the $^2S_0$ conformation is compatible with antithrombin binding [49].

Finally, thanks to the high affinity of some pentasaccharides for antithrombin, it proved possible to obtain monocrystals of the corresponding complexes. X-ray diffraction studies on these crystals provided the three-dimensional structure of the pentasaccharide antithrombin complex [50]. The results nicely confirmed the data obtained by classical biochemical techniques on the structural elements involved in the interaction. The crystallographic data also confirm that the L-iduronic acid unit G adopts a $^2S_0$ conformation when the oligosaccharide is bound to the protein. They also reveal structural changes such as helix D elongation and formation of a new (P) helix occurring upon antithrombin activation.

16.6
Conclusion and Perspectives

We have described how, starting from a detailed analysis of the interaction of heparin and antithrombin, we were able to imagine and synthesize new antithrombotic oligosaccharides with a mechanism of action that, though mediated by antithrombin, was substantially different from that of heparin and the low molecular weight heparin class of drugs. The most advanced of these new antithrombotics is now a marketed drug, demonstrating that chemical synthesis of complex oligosaccharides is no longer an obstacle to their use as drug substances.

With regard to the heparin class of drugs, it is interesting to note that after half a century of clinical use, the complex mechanism of their antithrombotic activity is still under debate, while the new antithrombotic agents studied here have a precise and well known mechanism of action. In contrast with a widespread view, well defined molecules that do not enter into undesired interactions seem to show a better pharmacological profile than heparin. This represent a real change in perspective in a field in which clinical efficacy used to be associated with high molecular weight, high sulfation level, and multifactorial effect.

Heparin is a member of the glycosaminoglycan family, and has been widely studied because of its clinical interest. Other glycosaminoglycans, heparan sulfate, dermatan sulfate, chondroitin sulfate, and hyaluronic acid, have been neglected for a long time but seem to be fashionable today, as judged by the number of articles devoted to them in scientific journals. One of the most challenging questions is the biological significance of their fascinating structural complexity. One may imagine that, when new active oligosaccharide sequences are identified, oligosaccharide analogues will be synthesized with the potential to produce new drugs.

16.7 References

1 E. Oger, *Thromb. Haemost.* **2000**, *83*, 657–660.
2 W. H. Geerts, J. A. Heit, G. P. Clagett, G. F. Pineo, C. W. Colwell, F. A. Anderson Jr., H. B. Wheeler, *Chest* **2001**, *119* (Suppl.), 132S–175S.
3 J. Hirsh, *New Eng. J. Med.* **1991**, *324*, 1865–1875.
4 J. Hirsh, *New Eng. J. Med.* **1991**, *324*, 1565–1574.
5 T. E. Warkentin, B. H. Chong, A. Greinacher, *Thromb. Haemost.* **1998**, *79*, 1–7.
6 J. I. Weitz, J. Hirsh, *Chest* **2001**, *119* (Suppl.), 95S–107S
7 a) A. G. G. Turpie, A. S. Gallus, J. A. Hoek, *New Engl. J. Med.* **2001**, *344*, 619–625. b) B. I. Eriksson, K. A. Bauer, M. R. Lassen, A. G. G. Turpie, *New Engl. J. Med.* **2001**, *345*, 1298–1304. c) K. A. Bauer, B. I. Eriksson, M. R. Lassen, A. G. G. Turpie, *New Engl. J. Med.* **2001**, *345*, 1305–1310. d) M. R. Lassen, K. A. Bauer, B. I. Eriksson, A. G. G. Turpie, *Lancet* **2002**, *359*, 1715–1720. e) A. G. G. Turpie, K. A. Bauer, B. I. Eriksson, M. R. Lassen, *Lancet* **2002**, *359*, 1721–1726.
8 B. Casu, *Adv. Carbohydr. Chem. Biochem.* **1985**, *43*, 51–134.
9 R. D. Rosenberg, P. S. Damus, *J. Biol. Chem.* **1973**, *248*, 6490–6505.
10 P. S. Damus, M. Hicks, R. D. Rosenberg, *Nature* **1973**, *246*, 355–357.
11 I. Björk, S. T. Olson, J. D. Shore, in *Heparin*, D. A. Lane, U. Lindahl Eds, Edward Arnold, London, **1989**, pp. 229–255.
12 S. T. Olson, I. Björk, *Semin. Thromb. Haemost.* **1994**, *20*, 373–409.
13 S. Wessler, E. T. Yin, *Thromb. Diathes. Haemorrh.* **1974**, *32*, 71–78.
14 D. P. Thomas, R. E. Merton, E. Gray, T. W. Barrowcliffe, *Thromb. Haemost.* **1989**, *204*, 204–207.
15 a) J. Choay, J.-C. Lormeau, M. Petitou, P. Sinaÿ, J. Fareed, *Ann. N.Y. Acad. Sci.* **1981**, *370*, 644–649. b) L. Thunberg, G.

Bäckström, U. Lindahl, *Carbohydr. Res.* **1982**, *100*, 393–410.
16 S.T. Olson, I. Björk, R. Sheffer, P.A. Craig, J.D. Shore, J. Choay, *J. Biol. Chem.* **1992**, *267*, 12528–12538.
17 L.B. Jaques, *Pharmacol. Rev.* **1979**, *31*, 99–167.
18 A. Greinacher, S. Alban, V. Dummel, G. Franz, C. Mueller-Eckardt, *Thromb. Haemost.* **1995**, *74*, 886–892.
19 M.A. Crowther, L.R. Berry, P.T. Monagle, A.K.C. Chan, *Brit. J. Haematol.*, **2002**, *116*, 178–186.
20 For a more detailed account on the very first part of our work see: M. Petitou in *Heparin*, D.A. Lane, U. Lindahl Eds, Edward Arnold, London, **1989**, pp. 65–79.
21 a) M. Petitou, J.-C. Jacquinet, P. Duchaussoy, I. Lederman, J. Choay, P. Sinaÿ, in *Glycoconjugates. Proceedings of the 7th International Symposium on Glycoconjugates*, A. Chester, D. Heinegard, A. Lundblad, S. Svensson Eds, Lund-Ronneby, Sweden, July 17–23, **1983**, p. 379. b) J.-C. Jacquinet, M. Petitou, P. Duchaussoy, I. Lederman, J. Choay, P. Sinaÿ, *ibid.* p. 380.
22 H. Paulsen, *Angew. Chem. Int. Ed. Engl.* **1982**, *21*, 155–224.
23 M. Petitou, *Thèse de Doctorat dès Sciences*, Orléans, France, **1984**.
24 a) P. Sinaÿ, J.-C. Jacquinet, M. Petitou, P. Duchaussoy, I. Lederman, J. Choay, G. Torri, *Carbohydr. Res.* **1984**, *132*, C5–C9. b) M. Petitou, P. Duchaussoy, I. Lederman, J. Choay, P. Sinaÿ, J.-C. Jacquinet, G. Torri, *Carbohydr. Res.* **1986**, *147*, 221–236.
25 E.E. Gilbert, *Chem. Rev.* **1962**, *62*, 549–589.
26 Several synthesis of L-iduronic acid, more or less useful for the purpose of the synthesis of heparin fragments, have been proposed, but an exhaustive review is out of the scope of the present article.
27 C.A.A. van Boeckel, M. Petitou, *Angew. Chem. Int. Ed. Engl.* **1993**, *32*, 1671–1690.
28 J. Choay, M. Petitou, J.-C. Lormeau, P. Sinaÿ, B. Casu, G. Gatti, *Biochem. Biophys. Res. Commun.* **1983**, *116*, 492–499.
29 It has been shown that glucosamine unit D is preferentially N-acetylated in the antithrombin binding sequences of pig mucosa heparin.
30 B. Boneu, J. Necciari, R. Cariou, A.M. Gabaig, G. Kieffer, J. Dickinson, G. Lamond, H. Moelker, T. Mant, H. Magnani, *Thromb. Haemost.* **1995**, *74*, 1468–1473.
31 C.A.A. van Boeckel, T. Beetz, S.F. van Aelst, *Tetrahedron Lett.* **1988**, *29*, 803–806.
32 a) G. Jaurand, J. Basten, I. Lederman, C.A.A. van Boeckel, M. Petitou, *Bioorg. Med. Chem. Lett.* **1992**, *2*, 897–900. b) J. Basten, G. Jaurand, B. Olde-Hanter, M. Petitou, C.A.A. van Boeckel, *ibid.* 901–904. c) J. Basten, G. Jaurand, B. Olde-Hanter, P. Duchaussoy, M. Petitou, C.A.A. van Boeckel, *ibid.* 905–910.
33 M. Petitou, P. Duchaussoy, G. Jaurand, F. Gourvenec, I. Lederman, J.-M. Strassel, T. Barzû, B. Crépon, J.-P. Hérault, J.-C. Lormeau, A. Bernat, J.-M. Herbert, *J. Med. Chem.* **1997**, *40*, 1600–1607.
34 P. Westerduin, C.A.A. van Boeckel, J.E.M. Basten, M.A. Broekhoven, H. Lucas, A. Rood, H. van der Heijden, R.G.M. van Amsterdam, T.G. van Dinther, D.G. Meuleman, A. Visser, G.M.T. Vogel, J.B.L. Damm, G.T. Overklift, *Bioorg. Med. Chem.* **1994**, *2*, 1267–1280.
35 J.-M. Herbert, J.-P. Hérault, A. Bernat, R.G.M. van Amsterdam, J.-C. Lormeau, M. Petitou, C.A.A. van Boeckel, P. Hoffmann, D.G. Meuleman, *Blood* **1998**, *91*, 4197–4205.
36 P.D.J. Grootenhuis, P. Westerduin, D. Meuleman, M. Petitou, C.A.A. van Boeckel, *Nature Struct. Biol.* **1995**, *2*, 736–739.
37 a) T.C. Laurent, A. Tengblad, L. Thunberg, M. Höök, U. Lindahl, *Biochem. J.* **1978**, *175*, 691–701. b) G.M. Oosta, W.T. Gardner, D.L. Beeler, R.D. Rosenberg, *Proc. Natl. Acad. Sci. USA* **1981**, *78*, 829–833. c) D.A. Lane, J. Denton, A.M. Flynn, L. Thunberg, U. Lindahl, *Biochem. J.* **1984**, *218*, 725–732. d) A. Danielsson, E. Raub, U. Lindahl, I. Björk, *J. Biol. Chem.* **1986**, *261*, 15467–15473.

38 P. Duchaussoy, G. Jaurand, P.-A. Driguez, I. Lederman, F. Gourvenec, J.-M. Strassel, P. Sizun, M. Petitou, J.-M. Herbert, *Carbohydr. Res.* **1999**, *317*, 63–84.

39 M. Petitou, P. Duchaussoy, P.-A. Driguez, G. Jaurand, J.-P. Hérault, J.-C. Lormeau, C. A. A. van Boeckel, J.-M. Herbert, *Angew. Chem. Int. Ed. Engl.* **1998**, *37*, 3009–3014.

40 L. Jin, J.-P. Abrahams, R. Skinner, M. Petitou, R. N. Pike, R. W. Carrell, *Proc. Natl. Acad. Sci. USA* **1997**, *94*, 14683–14688.

41 M. T. Stubbs, W. Bode, *Trends Biochem. Sci.* **1995**, *20*, 23–28.

42 This was experimentally confirmed by the biological properties of conjugates comprising an A-domain and a T-domain separated by a non-carbohydrate, flexible, spacer (see ref. 36) (such flexible conjugates surprisingly being found to be able to interact with PF4; see ref. 43).

43 We have shown that the nature of the neutral domain may have a strong influence on the biological activity: C. M. Dreef-Tromp, J. E. M. Basten, M. A. Broekhoven, T. G. van Dinther, M. Petitou, C. A. A. van Boeckel. *Bioorg. Med. Chem. Lett.*, **1998**, *8*, 2081–2086.

44 M. Maccarana, U. Lindahl, *Glycobiology* **1993**, *3*, 271–277.

45 a) M. Petitou, J.-P. Hérault, A. Bernat, P.-A. Driguez, P. Duchaussoy, J.-C. Lormeau, J.-M. Herbert, *Nature* **1999**, *398*, 417–422, b) P.-A. Driguez, I. Lederman, J.-M. Strassel, J.-M. Herbert, M. Petitou, *J. Org. Chem.* **1999**, *26*, 9512–9520.

46 J.-M. Herbert, J.-P. Hérault, A. Bernat, P. Savi, P. Schaeffer, P.-A. Driguez, P. Duchaussoy, M. Petitou, *Thromb. Haemost.* **2001**, *85*, 852–860.

47 M. Petitou, T. Barzu, J.-P. Hérault, J.-M. Herbert, *Glycobiology* **1997**, *7*, 323–327.

48 a) U. R. Desai, M. Petitou, I. Björk, S. T. Olson, *Biochemistry* **1998**, *37*, 13033–13041. b) U. R. Desai, M. Petitou, I. Björk, S. T. Olson, *J. Biol. Chem.* **1998**, *273*, 7478–7487.

49 S. K. Das, J.-M. Mallet, J. Esnault, P.-A. Driguez, P. Duchaussoy, P. Sizun, J.-P. Herault, J.-M. Herbert, M. Petitou, P. Sinaÿ, *Angew. Chem. Int. Ed.* **2001**, *40*, 1670–1673.

50 Later, crystals of the pentasaccharide 17 in complexation with antithrombin were obtained and showed that this compound and higher affinity compounds bound in identical ways (J. Huntington and R. Carrell, unpublished).